Springer Geophysics

The Springer Geophysics series seeks to publish a broad portfolio of scientific books, aiming at researchers, students, and everyone interested in geophysics. The series includes peer-reviewed monographs, edited volumes, textbooks, and conference proceedings. It covers the entire research area including, but not limited to, applied geophysics, computational geophysics, electrical and electromagnetic geophysics, geodesy, geodynamics, geomagnetism, gravity, lithosphere research, paleomagnetism, planetology, tectonophysics, thermal geophysics, and seismology.

Giorgio Spada · Daniele Melini

Glacial Isostatic Adjustment

Theory for a Spherically Symmetric Earth and Numerical Results

Springer

Giorgio Spada
Department of Physics and Astronomy
"Augusto Righi"
University of Bologna
Bologna, Italy

Daniele Melini
Istituto Nazionale di Geofisica e
Vulcanologia
Rome, Italy

ISSN 2364-9119 ISSN 2364-9127 (electronic)
Springer Geophysics
ISBN 978-3-031-90537-7 ISBN 978-3-031-90538-4 (eBook)
https://doi.org/10.1007/978-3-031-90538-4

This Springer imprint is published by the registered company Springer Nature Switzerland AG
The registered company address is: Gewerbestrasse 11, 6330 Cham, Switzerland

I remember him in the hotel corridor, with a mathematics book in his hand, sometimes looking at the irrecoverable colors of the sky. One afternoon, we spoke of the duodecimal system of numbering (in which 12 is written as 10). Ashe said that he was converting some kind of tables from the duodecimal to the sexagesimal system (in which 60 is written as 10).

Jorge Luis Borges, Tlön, Uqbar, Orbis Tertius

Preface

The main purpose of this book is to describe the physical process known as "Glacial Isostatic Adjustment" (GIA). The phenomenon is due to the imbalance caused by large glacial masses at the Earth's surface, responsible for crustal deformations and variations in the gravity field on various scales of space and time. The basic process underlying GIA is therefore, in itself, simple. However, at the same time, GIA is a phenomenon of some complexity, having important implications on our understanding of the physics of the "Earth system".

Discovered during the 19th century, the phenomenon of isostasy mainly provides useful tools for understanding the structure of solid Earth. In particular, it allows us to investigate the long-term rheological characteristics of the lithosphere and mantle, about which seismology provides limited information. However, since ice masses evolve over a broad spectrum of space-time scales, the GIA phenomenon also affects sea-level variations and consequently also has implications for our understanding of past and future climate variations. For example, a correct interpretation of instrumental sea-level observations over the last century would be impossible without "decontaminating" the tide gauge data from the isostatic contribution in response to millennial variations in the ice volume. Furthermore, the changes of the Earth's gravity field observed today by means of satellite techniques are directly affected by GIA, which can be modelled and corrected for by applying the methodologies illustrated in our study.

The GIA phenomenon is characterised by a rich variability of spatial and temporal scales and, being non-linear in nature, it is sometimes not immediately comprehensible. The cryosphere, i.e., the portion of the Earth's surface in which water is found in the solid state, is continuously evolving and this implies an isostatic imbalance on a global scale. However, there are also various regional and local impacts due to the incessant interaction between the solid and fluid parts of the planet. Therefore, although the physics of GIA is in itself governed by simple equations, which we will study in detail in this book, the phenomenon manifests itself in a complex way, and concerns a large number of variables useful for the description of the "Earth system". The evolution of the global sea level documented since the Last Glacial Maximum ~ 21,000 years ago has been substantially due to the GIA, and has manifested itself in different ways in different places on the planet, driven by the melting of the large glacial masses of the late Pleistocene. At the same time, the deformations currently underway in polar regions affected by ice melting are largely governed by isostatic phenomena, strictly dependent on the rheological characteristics of the most superficial layers of the Earth. Furthermore, glacio-isostatic phenomena will not cease as long as the cryosphere will change its state, therefore they will also have a role in the centuries to come, in which we expect to witness major changes in the climate of our planet.

The book is aimed primarily at master's and doctoral students in geophysics or physics who intend to undertake the study of these disciplines, or who already have some experience and are aware of some basic concepts. Naturally it is also aimed at more expert colleagues who already work in the field of geophysics, geodynamics, geodesy or geology. It is a book of physics, so the topics are treated in a quantitative manner, sometimes using advanced mathematical concepts. However, the authors have tried to make available to readers all the basic material, which is organised in various appendices and in a rich bibliography. For this reason, the text can also be used by students and colleagues from other fields of science, who intend to approach this discipline for the first time. The details in the mathematical developments, necessary to substantiate the text, are presented to facilitate the reproducibility of the main results presented in the various chapters, and to make the notation used as homogeneous as possible. Although it is a predominantly theoretical text, the reader has the possibility of accessing calculation tools developed by the authors, which allow the practical implementation of numerical experiments to support the theoretical parts. These tools, in the form of open source computer codes (`ALMA`, `REAR`, `SELEN` and `TABOO`), have been designed in such a way as to facilitate student understanding and the creation of case studies of interest.

The contents of the book are partly original, the result of our research work in this field and our experience, and partly inspired by results from the vast literature existing. The central topic, which concerns the "Sea Level Equation", is presented in the form of an expanded and improved version of the work of Spada and Melini (2019b). The content of the various previous chapters is directly aimed at the development of the Sea Level Equation, avoiding as much as possible the introduction of material that is not strictly necessary. To make reading easier, many materials and various insights are collected in rather technical appendices, which always maintain a style consistent with the main text. The bibliography is broad, but perhaps not exhaustive, rather focused on articles relating to physical and physical-mathematical aspects, which constitute the main theme of this book, and less oriented towards geological aspects, given our training. The authors have strived to maintain a consistent formalism throughout the chapters of the book. This inevitably means that our conventions and the notation employed may somehow differ from works in the literature dealing with the same topics.

The book consists of six chapters. Chapter 1 contains introductory material relating to the Earth's gravitational field, the rheology of geophysical materials, and surface mass distributions. Chapter 2 is concerned with the "Love numbers" and the Green's functions, i.e., the quantities useful for characterising the response of a spherically symmetric Earth to surface masses localised in space and time. Chapter 3 addresses the problem of distributed surface loads and the Earth's response to these loads, in terms of 3-D displacements and variations in the gravitational field. Chapter 4 concerns the rotation of the Earth and the way in which the rotation is perturbed by the redistribution of mass at the surface, in the case of small geophysical perturbations. Chapter 5 introduces the "Sea Level Equation" and discusses its meaning, describing various forms in an increasing complexity taking advantage of the concepts developed in the previous chapters. Finally, in Chap. 6 the "Sea Level Equation" is solved numerically using a computer program developed by the authors (`SELEN`[4]), taking into consideration a vast set of geophysical quantities associated with the GIA. The six chapters are closely interconnected and make constant reference to five appendices which contain material that can be omitted on the first reading.

That of GIA is an exciting but an extremely complex geophysical topic, in its various aspects. The complexity resides in the fact that GIA involves many space-time scales and results from the interaction between some of the components of the "Earth System", i.e., the Solid Earth, the Ices and the Oceans. In this area, much has been done, much is being done but certainly much still remains to be done. We hope, with our contribution, to stimulate the interest of novice students, eager to learn something more about the physical processes that govern this phenomenon. At the same time, we hope that this book shall constitute a reference for more expert researchers looking for a text that brings together the most essential topics, without neglecting application aspects that can be addressed with the help of the software that the authors have developed over the years.

Bologna, Italy
Rome, Italy
February 2025

Giorgio Spada
Daniele Melini

Acknowledgements We have greatly benefited from discussions with a number of colleagues who participated to the Glacial Isostatic Adjustment and Sea Level Equation benchmark activities (see Spada et al., 2011a; Martinec et al., 2018), namely: Valentina R. Barletta, Paolo Gasperini, Thomas S. James, Samuel B. Kachuck, Matt A. King, Volker Klemann, Bjorn Lund, Zdeneck Martinec, Riccardo E. M. Riva, Karen Simon, Yu Sun, Wouter van der Wal, Bert Vermeersen, and Detlef Wolf.

A special acknowledgement goes Florence Colleoni for help and insightful discussions about the problem of the shoreline migration in GIA modeling.

The authors are also indebted to Micheal Bevis and Erik Ivins for encouragement and advice and to Glenn Milne for an insightful discussion about the meaning of the "GIA fingerprints".

We are grateful to Volker Klemann, Sam Kachuck and Geruo A for their constructive review and advice, and to Meike Bagge and Ewan Gowan for their very helpful comments and advice.

We thank Max Tegmark for making available his pixelization routines,[1] which had an essential role in the development of program `SELEN`. We also thank Mark Wieczorek for distributing the SHTOOLS (the Spherical Harmonics Tools) to the community.[2]

GS has greatly benefited from discussion, advice and encouragement from the participants to the 2023 Glacial Isostatic Adjustment (GIA) Training School (3–7 July 2023, Lantmäteriet, Gävle, Sweden) and of all editions of the school, hosted by Rebekka Steffen and Holger Steffen.

GS is indebted to all his collaborators and to his former students Anastasia Consorzi, Gabriele Bentivogli, Federico Disconzi, Luca Francioni, Gaia Galassi, Laura Girometti, Veronica Guerra, Claudia Imbrescia, Fernando Linsalata, Carla Morgana, Marco Olivieri, Claudia Piromallo, Irene Puliti, Beniamino Rocchi, Giada Salciccia, Marco Vanni and many others for stimulating and constructive discussions.

GS owes a lot to his friend and colleague Francesco Mainardi for advice, discussion and help on the theory of linear viscoelasticity and other subjects but most of all for his patience during the years.

GS says a special *grazie* to Paola for her patience during the long preparation of the original manuscript.

GS is profoundly indebted to Prof. Francesco Dalla Valle (*un maestro*), for continuous encouragement and advice since the late '70s, when GS was a student at the Liceo Scientifico St. G. Ricci Curbastro in Lugo di Romagna.

DM is deeply indebted to Gabriele Ferrara from the "Antonino Lo Surdo" INGV library in Roma for his patience and precious help in retrieving digital copies of bibliographic references.

The authors acknowledge Anastasia Consorzi for carefully revising a late version of the manuscript.

GS and DM acknowledge Valentina R. Barletta for insightful discussions about the surface boundary conditions and the GIA Liouville equations.

[1] See https://space.mit.edu/home/tegmark/icosahedron.html [last accessed on 28 August, 2024].

[2] See https://shtools.github.io/SHTOOLS/ [last accessed on September, 2024].

Last, the authors also thank Raffaello Mascetti for having patiently revised the manuscript, providing invaluable inspiration.

References

Douglas B (1997) Global sea level rise: a redetermination. Surveys in Geophysics 18:279–292
Farrell W, Clark J (1976) On postglacial sea level. Geophysical Journal of the Royal Astronomical Society 46(3):647–667
Martinec Z, Klemann V, van der Wal W, Riva R, Spada G, Sun Y, Melini D, Kachuck S, Barletta V, and Simon K (2018) A benchmark study of numerical implementations of the sea level equation in GIA modelling. Geophysical J Inter 215(1):389–414
Spada G, Barletta VR, Klemann V, Riva R, Martinec Z, Gasperini P, Lund B, Wolf D, Vermeersen L, and King M (2011a) A benchmark study for glacial isostatic adjustment codes. Geophysical J Inter 185(1):106–132
Spada G, Melini D (2019b) SELEN^4 (SELEN version 4.0): a Fortran program for solving the gravitationally and topographically self-consistent sea-level equation in glacial isostatic adjustment modeling. Geoscientific Model Development 12(12):5055–5075

Competing Interests This work is partly funded by a FFABR (Finanziamento delle Attività Base di Ricerca) grant of MIUR (Ministero dell'Istruzione, dell'Università e della Ricerca) by a research grant of DiSPeA (Dipartimento di Scienze Pure e Applicate) of the Urbino University "Carlo Bo", and by a RFO grant of the Dipartimento di Fisica e Astronomia "Augusto Righi" (DIFA) of the Alma Mater Studiorum University of Bologna.

The authors have no competing interests to declare that are relevant to the content of this manuscript.

Contents

Acronyms

ALP	Associated Legendre Polynomial
BC	Boundary Condition
CF	Continent Function
CM	Centre of Mass
CMB	Core-Mantle Boundary
CSH	Complex Spherical Harmonic
D97	Douglas (1997)
EF	Excitation function
ESL	Equivalent Sea Level
FC76	Farrell and Clark (1976)
FNALP	Fully Normalised ALP
GF	Green's Function
GIA	Glacial Isostatic Adjustment
GMSLR	Global Mean Sea Level Rise
GNSS	Global Navigation Satellite System
IPR	Instantaneous Pole of Rotation
LGM	Last Glacial Maximum
LLN	Loading Love Number
LN	Love Number
LOD	Length of Day
LP	Legendre Polynomial
LT	Laplace Transform
OF	Ocean Function
PC	Piecewise Constant
PGR	Post Glacial Rebound
PMTF	Polar Motion Transfer Function
PT	Paleo-topography
RRF	Rotational Response Function
RSH	Real Spherical Harmonic
RSL	Relative Sea Level
SNREI	Solid Non-Rotating Elastic Isotropic

SRF	Surface Response Function
TG	Tide Gauge
TLN	Tidal Love Number
VLM	Vertical Land Motion

Physical Constants

Symbol	Constant or parameter	Value	SI units
G	Universal gravitational constant	6.674×10^{-11}	$\mathrm{N\,m^2\,kg^{-2}}$
a	Earth's average radius	6.371×10^{6}	m
M, m^{e}	Earth's mass	5.9722×10^{24}	kg
g_0	Earth's surface gravity acceleration	9.816	$\mathrm{m\,s^{-2}}$
A	Earth's polar moment of inertia	8.0131×10^{37}	$\mathrm{kg\,m^2}$
C	Earth's equatorial moment of inertia	8.0394×10^{37}	$\mathrm{kg\,m^2}$
Ω	Earth's mean angular velocity	7.292115×10^{-5}	$\mathrm{s^{-1}}$
k^{ee}	Earth's excess ellipticity correction	0.0082	–
Hu	Haskell viscosity unit	10^{21}	Pa s
ρ^{i}	Ice density	931	$\mathrm{kg\,m^{-3}}$
ρ^{w}	Water density	1000	$\mathrm{kg\,m^{-3}}$
ρ^{e}	Earth's average density	5513	$\mathrm{kg\,m^{-3}}$

Chapter 1
Gravity, Rheology and Mass Distributions

What do you read, my lord?
Words, words, words...

(Hamlet: Act II)

1.1 Introduction

This chapter has the purpose of introducing a few basic principles concerning the gravity field of the Earth, the mechanical behaviour of geophysical materials, the mass distributions over the Earth's surface and their evolution in space and time. The three sections contain essential material finalised at understanding the response of the Earth to surface and tidal loads, and to the development of Sea Level Equation in later chapters, which constitutes the central topic of this book. Here we are only providing those definitions and concepts that are strictly required, avoiding unnecessary formalism.

In Sect. 1.2 we focus on the expansion of the gravity potential in multipoles, introducing the Stokes coefficients and the relationships of the long-wavelength harmonics with the basic physical properties of the Earth's mass distribution, and particularly with the moment of inertia. We are only providing the essential ideas and materials, referring the readers to more comprehensive presentations in general treaties of physical geodesy.

In describing the mechanical behavior of materials (Sect. 1.3), we limit ourselves to a phenomenological approach providing the stress-strain relationships for a few basic linear bodies. These are characterised by their response to mechanical tests, providing expressions for the creep functions and relaxation moduli in the time and in the Laplace domains. The correspondence principle of linear viscoelasticity is also

G. Spada and D. Melini, *Glacial Isostatic Adjustment*, Springer Geophysics,
https://doi.org/10.1007/978-3-031-90538-4_1

introduced. The readers are referred to fundamental treaties, quoted in the following, for a broader presentation of the Earth's rheology.

Last, in Sect. 1.4, we consider the surface mass distributions and their properties. In this framework, we introduce some fundamental geophysical quantities as topography and sea level and describe their interconnections, also defining the ocean function. By imposing the mass conservation principle, a physically self-consistent expression is obtained for the surface load in response to mass redistributions at the Earth's surface, which is necessary for the development of the Sea Level Equation.

1.2 The Gravitational Field

Here we consider the classical problem of describing the gravitational field of a quasi-spherical mass distribution by means of harmonic functions. The purpose is to provide only the basic elements for the understanding of the material presented in the central section of the book. The subject has been addressed in a number of fundamental treaties of physical geodesy (see Heiskanen and Moritz 1981; Lambeck 1988; Hofmann-Wellenhof and Moritz 2006; Torge and Müller 2012), to which the readers are referred for a more complete presentation.

1.2.1 Mass Distributions

It is useful to recall that the following relations exist between the spherical coordinates (r, θ, λ) and the rectangular coordinates (x, y, z) of a point P in space

$$\begin{aligned} x &= r \sin\theta \cos\lambda \\ y &= r \sin\theta \sin\lambda \\ z &= r \cos\theta \,, \end{aligned} \tag{1.1}$$

where r is the modulus of the position vector[1] $\vec{r}$ of P, θ is *co-latitude* (sometimes referred to as *polar distance* in physical geodesy), and λ the *geocentric longitude*. The geometry is sketched in Fig. 1.1. The *geocentric latitude* is defined as

$$\phi = \frac{\pi}{2} - \theta \,. \tag{1.2}$$

[1] In this book we will use the arrow notation $\vec{v}$ for *geometrical vectors* (such as the position vector) and the boldface notation $\mathbf{v}$ for *algebraic vectors* (an ordered list of scalar quantities). Of course, the distinction is not always sharp, since to every geometrical vector $\vec{v}$ corresponds an algebraic vector $\mathbf{v}$ once a reference frame is defined, and *vice versa*.

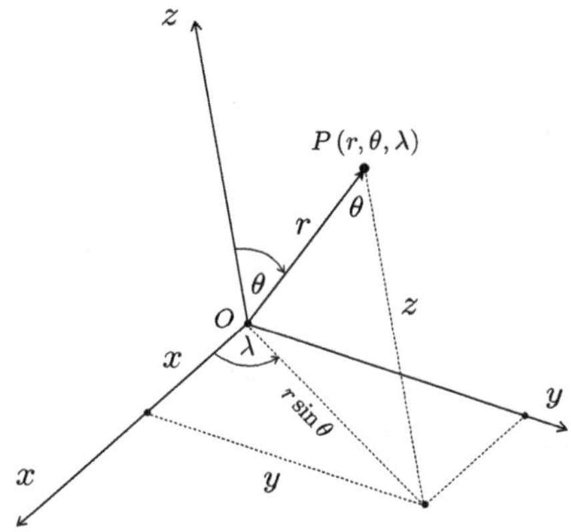

Fig. 1.1 Spherical (r, θ, λ) and rectangular coordinates (x, y, z) of a point P in space; note that $r \geq 0$, while $0 \leq \theta \leq \pi$ and $0 \leq \lambda \leq 2\pi$

We note that Eq. (1.1) can be inverted to obtain

$$\begin{aligned} r &= \sqrt{x^2 + y^2 + z^2} \\ \theta &= \arctan\left(\frac{\sqrt{x^2 + y^2}}{z}\right) \\ \lambda &= \arctan\left(\frac{y}{x}\right) . \end{aligned} \tag{1.3}$$

1.2.1.1 Mass

The mass (or *zeroth order moment*) of the arbitrary mass distribution shown in Fig. 1.2a is

$$M \equiv \int_V \mathrm{d}m , \tag{1.4}$$

where the integral is over the volume V of the mass distribution. The mass element is

$$\mathrm{d}m = \rho(\vec{r}')\,\mathrm{d}V , \tag{1.5}$$

where $\rho(\vec{r}')$ is the density at radius $\vec{r}' = (x', y', z')$, which denotes the position of $\mathrm{d}m$ in a cartesian right-handed coordinate system $Oxyz$, of arbitrary origin O.

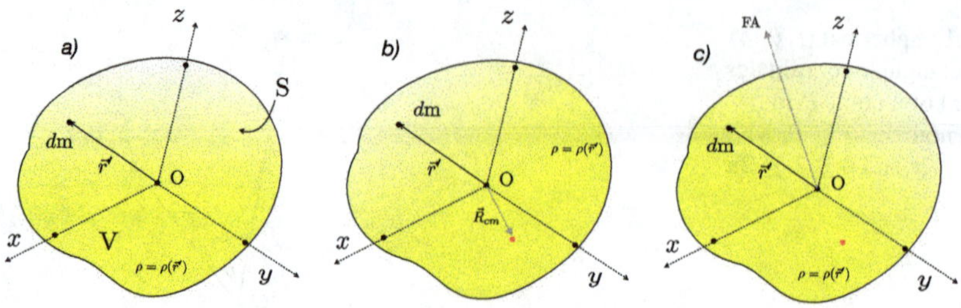

Fig. 1.2 **a** Geometry for the evaluation of the mass M, **b** of the centre of mass $\vec{R}_{cm}$ and **c** of the inertia tensor I_{ij} of a continuous and quasi-spherical mass distribution of volume V, bounded by a closed surface S

1.2.1.2 Centre of Mass

The position vector $\vec{R}_{cm}$ of the centre of mass (CM, or *first order moment*) of a continuous mass distribution with density $\rho(\vec{r}')$ is defined as the weighted average of the position vectors $\vec{r}'$ of each mass element dm (see Fig. 1.2b). In formulae:

$$\vec{R}_{cm} \equiv \frac{\int_V \vec{r}' \mathrm{d}m}{\int_V \mathrm{d}m} = \frac{1}{M} \int_V \vec{r}' \mathrm{d}m \,, \tag{1.6}$$

where M is the mass of the distribution. Hence the cartesian components of $\vec{R}_{cm}$ are:

$$(x_{cm}, y_{cm}, z_{cm}) \equiv \frac{1}{M} \left(\int_V x' \mathrm{d}m \,, \int_V y' \mathrm{d}m \,, \int_V z' \mathrm{d}m \right) . \tag{1.7}$$

1.2.1.3 Inertia Tensor

The inertia tensor (or *second order moment*) I_{ij} of a continuous mass distribution with density $\rho(\vec{r}')$ is the rank 2 cartesian tensor defined as

$$I_{ij} \equiv \int_V \left(x'_k x'_k \, \delta_{ij} - x'_i x'_j \right) \mathrm{d}m \,, \tag{1.8}$$

where for convenience we have set $(x_{1,2}, x_3) \equiv (x, y, z)$, we use the Einstein convention of implicit summation on the repeated indices, and δ_{ij} is the Kronecker delta[2] (e.g., Landau et al. 1986).

By its own definition, the inertia tensor is symmetric (i.e., $I_{ij} = I_{ji}$), hence it has effectively only six independent elements. The *diagonal elements* are

[2] The Kronecker delta is $\delta_{ij} = 0$ for $i \neq j$ and $\delta_{ij} = 1$ for $i = j$.

$$I_{11} = \int_V \left(y'^2 + z'^2\right) \, dm \tag{1.9}$$

$$I_{22} = \int_V \left(x'^2 + z'^2\right) \, dm \tag{1.10}$$

$$I_{33} = \int_V \left(x'^2 + y'^2\right) \, dm \,, \tag{1.11}$$

while the off-diagonal elements (or *inertia products*) are:

$$I_{12} = I_{21} = -\int_V x' \, y' \, dm \tag{1.12}$$

$$I_{13} = I_{31} = -\int_V x' \, z' \, dm \tag{1.13}$$

$$I_{23} = I_{32} = -\int_V y' \, z' \, dm \,. \tag{1.14}$$

Since I_{ij} is real and symmetric, its eigenvalues $A \leq B \leq C$ are real, and the corresponding eigenvectors are mutually orthogonal. The directions of the eigenvectors of I_{ij} define the *principal axes of inertia*, with the eigenvector relative to the largest eigenvalue (C) defining the *figure axis* of the body, labelled by FA in Fig. 1.2c. In the reference frame of the eigenvectors, we have $I_{ij} = \mathrm{Diag}\,(A,\, B,\, C)$.

The trace of the inertia tensor:

$$\mathrm{Tr}\,(I_{ij}) \equiv I_{kk} = 2 \int_V r'^2 \, dm \,, \tag{1.15}$$

is invariant under a rigid rotation of the reference frame. As shown by Rochester and Smylie (1974), the trace remains unchanged under a wide class of Earth's deformations (see comment to Eq. 1.46).

1.2.2 Gravity Potential

Following Hofmann-Wellenhof and Moritz (2006) and Torge and Müller (2012), the *gravity potential* is

$$W\,(\vec{r}) = W^{\mathrm{g}} + W^{\mathrm{c}} \,, \tag{1.16}$$

where $W^{\mathrm{g}}\,(\vec{r})$ is the *gravitational potential* of the mass distribution and $W^{\mathrm{c}}(\vec{r})$ is the *centrifugal potential* due to Earth's rotation. By definition, the *gravity vector* – also referred to as *gravity* or *gravity field* – is the gradient of the gravity potential

$$\vec{g} = \nabla W \tag{1.17}$$

and represents the total force per unit mass, including the gravitational and the centrifugal contribution.[3] Here we are concerned with the gravitational potential W^g; the centrifugal potential W^c is defined by Eq. (4.55) and its time variations are discussed in Appendix D.1.

1.2.2.1 Equipotential Surfaces

The surfaces over which the gravity potential $W(\vec{r})$ assumes a constant value are defined as the equipotential surfaces (or *level surfaces*) of the gravity field. By its own definition, gravity $\vec{g}$ is orthogonal to the equipotential surfaces, and defines the local direction of plumb lines. Note that neither the modulus of $\vec{g}$ nor its components are necessarily constant on the equipotential surfaces.

Ideally, the free equilibrium surface of the oceans represents a portion of an equipotential surface (see Hofmann-Wellenhof and Moritz 2006), which Carl Friedrich Gauss (1777–1855) defined as the *mathematical surface of the Earth* and that we now call *geoid*,[4] a smooth closed surface defined by

$$W(\vec{r}) = \text{constant}\,. \tag{1.18}$$

Following Torge and Müller (2012), we note that the one above is a physical definition of geoid, which relies upon the idea of freely adjustable, homogeneous, invariable ocean masses only subject to gravity.

1.2.2.2 The Gravitational Potential

In the treatment of the gravitational potential W^g, we shall refer to the geometry displayed in Fig. 1.3. It shows a quasi-spherical mass distribution of volume V, bounded by a closed surface S. The mass distribution rotates with an angular velocity $\vec{\omega}$, with the intersection of $\vec{\omega}$ with surface S corresponding to the instantaneous pole of rotation (IPR). The arbitrary internal mass distribution is described by the density field $\rho(\vec{r}')$. To describe the system, we choose a non-inertial rotating right-handed cartesian coordinate system[5] $Ox_1x_2x_3$ whose origin O is not necessarily coincident with the position of the centre of mass (CM) of the body. Furthermore, since the x_3

---

[3] Two different conventions about the choice of the sign of the gravity potential exist in the geophysical literature. The one that we are using in (1.17) is commonly employed in geodesy, while in physics the gravity potential ϕ is usually defined such that $\vec{g} = -\nabla\phi$. The two definitions are of course related by $W = -\phi$. In this book we will employ both conventions, choosing for each topic the one that is commonly used in the reference works in literature. To avoid ambiguities, we will always use the different symbols W and ϕ (or φ) for the potentials defined according to the geodetic and physical conventions, respectively.

[4] The term *geoid* was coined by Johann Benedict Listing (1808–1882).

[5] We shall also use, in some instances, the notation $Oxyz$.

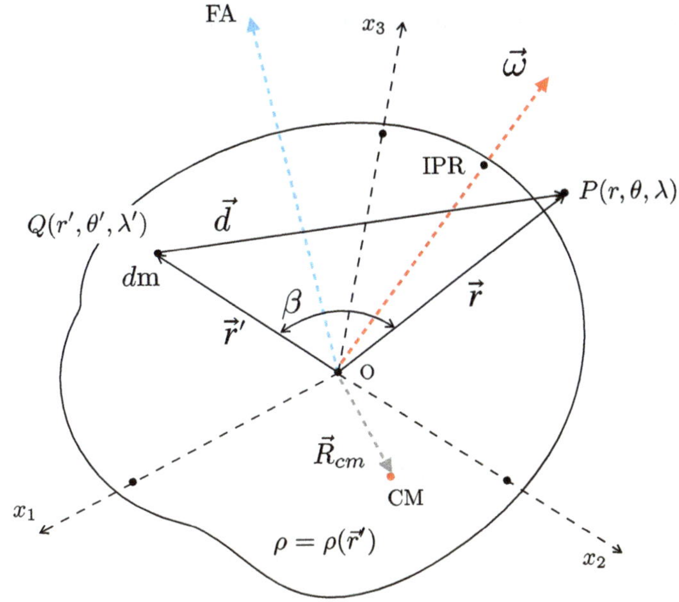

Fig. 1.3 Geometry for the description of the gravitational potential W^g external to a quasi-spherical mass distribution with density $\rho = \rho(\vec{r}')$. Points P and Q denote the position of an *observer* outside the mass distribution and of a mass element dm, respectively. As explained in the text, at point P the gravitational potential W^g is obtained by integration of Newton's law (1.19) over all the mass elements dm

axis is not in general coincident with the figure axis (FA), the inertia tensor is not necessarily diagonal when expressed in the $Ox_1x_2x_3$ frame.

We now consider a small (point-like) mass element dm located at an arbitrary point $Q \equiv (r', \theta', \lambda')$. By Newton's law of gravitation, at a point $P \equiv (r, \theta, \lambda)$ the contribution to gravitational potential due to dm is

$$\mathrm{d}W^g = \frac{G\,\mathrm{d}m}{d}, \tag{1.19}$$

where

$$G = 6.674 \times 10^{-11}\ \mathrm{N\,m^2\,kg^{-2}} \tag{1.20}$$

is Newton's constant and $d = \sqrt{d_i d_i}$, where $\vec{d} = \overrightarrow{QP}$ is the position vector of P relative to Q. By applying the cosines theorem in plane geometry to triangle $\widehat{OPQ}$, the *reciprocal distance* is

$$\frac{1}{d} = \frac{1}{r} \cdot \frac{1}{\sqrt{1 - 2\left(\frac{r'}{r}\right)\cos\beta + \left(\frac{r'}{r}\right)^2}}, \tag{1.21}$$

which, for $r'/r < 1$, can be expanded in a infinite series

$$\frac{1}{d} = \frac{1}{r} \cdot \sum_{n=0}^{\infty} \left(\frac{r'}{r}\right)^n P_n(\cos\beta), \tag{1.22}$$

where $P_n(\cos\beta)$ is the Legendre polynomial of harmonic degree n (see Sect. A.1).

1.2.2.3 Multipole Expansion of Gravitational Potential

Integrating Eq. (1.19) over the whole mass distribution and taking (1.22) into consideration, we obtain the *multipole expansion* of gravitational potential[6]

$$W^g(r, \theta, \lambda) = \frac{G}{r} \sum_{n=0}^{\infty} \int_V \left(\frac{r'}{r}\right)^n P_n(\cos\beta)\, dm = W_0^g + W_1^g + W_2^g \ldots, \tag{1.23}$$

where we note that since $P_0(\cos\beta) = 1$, the zeroth order multipole is

$$W_0^g(r) = \frac{GM}{r}, \tag{1.24}$$

which represents the spherically symmetric Newtonian term, describing the gravitational potential outside a localised mass having the same mass M of the general mass distribution considered here. Contrary to W_0^g, the multipoles W_n^g of degree $n \geq 1$ also depend upon co-latitude θ and longitude λ of the observer through angle β. They account for the effects of a-sphericity and of the irregular mass distribution on the gravitational field.

Since according to (1.23) the multipoles decay with distance as $1/r^{n+1}$, their detectability by means of satellite gravity methods becomes increasingly difficult with increasing harmonic degree n (see Lambeck 1988; Hofmann-Wellenhof and Moritz 2006; Torge and Müller 2012; Kaula 2013). As we shall see in the following, multipoles W_1^g and W_2^g have a particular physical meaning, being related with the coordinates of the CM (see Eq. 1.6) and with the elements of the inertia tensor (1.8), respectively.

---

[6] It can be shown that the condition $r > r'$ is *sufficient* for the convergence of the series (1.22), but it is not *necessary*. In other words, for some configurations of the mass distribution, the infinite series (1.23) converges to the potential also if there are regions where $r \leq r'$; that would be the case, for instance, near the polar regions, if the ellipsoidal shape of the Earth is taken into account. It is also important to remark that (1.23) is a valid representation of the potential only *outside* of the mass distribution, where the potential is an harmonic function. For a detailed discussion, see Chaps. 6 and 7 of Moritz (1980).

1.2.3 Spherical Harmonic Expansion

The multipole expansion (1.23) is expressed in terms of angle β and radius r. However, using the addition theorem for spherical harmonics (A.28) it is possible to write the gravitational potential in terms of the spherical coordinates θ (co-latitude) and λ (longitude) of point P, which leads to the expansion

$$W^{g}(r,\gamma) = \frac{GM}{r}\sum_{nm}\left(\frac{a}{r}\right)^{n} W^{g}_{nm}\, Y_{nm}(\gamma)\,, \tag{1.25}$$

where $\gamma \equiv (\theta, \lambda)$, $Y_{nm}(\gamma)$ is the 4π-normalised spherical harmonic function of degree n and order m defined by (A.22) and where we have introduced the non-dimensional complex expansion coefficients

$$W^{g}_{nm} \equiv \frac{1}{M}\cdot\frac{1}{2n+1}\int_{V}\left(\frac{r'}{a}\right)^{n} Y^{*}_{nm}(\gamma')\,\mathrm{d}m'\,, \tag{1.26}$$

where $\gamma' = (\theta', \lambda')$ and a is a conventional reference length which is normally assumed to be the equatorial radius of the (quasi-spherical) mass distribution or its average radius (see e.g., Lambeck 1980; Hofmann-Wellenhof and Moritz 2006).

Using Eqs. (A.44–A.47), the complex expansion (1.25) can be transformed into a real expansion involving the associated Legendre polynomials (ALPs, see Sect. A.2) $P_{nm}(\cos\theta)$, as

$$W^{g}(r,\gamma) = \frac{GM}{r}\sum_{\substack{nm\\ m\geq 0}}\left(\frac{a}{r}\right)^{n}\left(c_{nm}\cos m\lambda + s_{nm}\sin m\lambda\right) P_{nm}(\cos\theta)\,, \tag{1.27}$$

where the non-dimensional constants

$$\begin{Bmatrix} c_{nm} \\ s_{nm} \end{Bmatrix} = (2-\delta_{0m})\frac{(n-m)!}{(n+m)!}\frac{1}{M}\int_{V}\left(\frac{r'}{a}\right)^{n} P_{nm}(\cos\theta')\begin{Bmatrix} \cos m\lambda' \\ \sin m\lambda' \end{Bmatrix}\mathrm{d}m \tag{1.28}$$

are often referred to, after George Gabriel Stokes (1819–1903), as *Stokes coefficients*[7] of harmonic degree n ($n = 0, 1, \ldots$) and order m ($m = 0, 1, \ldots, n$) of the gravitational field (see e.g., Lambeck 1980). We note that sometimes they are also called *Stokes constants* or *harmonic coefficients of the gravitational field* (Torge and Müller 2012).

Based upon the symmetries displayed by the corresponding spherical harmonic function, coefficients with $m = 0$ are referred to as *zonal coefficients*, those with $0 < m < n$ are the *sectoral coefficients* whereas *tesseral coefficients* are those with $m = n$

[7] Since in the expansion (1.27) we have used the un-normalised form of the ALPs $P_{nm}(\cos\theta)$, the Stokes coefficients given in (1.28) should be referred to as *un-normalised Stokes coefficients*.

(see e.g., Hofmann-Wellenhof and Moritz 2006). For real harmonics, these ranges of m correspond to well defined spatial patterns, illustrated and shortly discussed in Sect. A.4.

1.2.4 Low Degree Stokes Coefficients

The Stokes coefficients of harmonic degree $n = 1$ and 2 have a particular importance, since they are related, in a non obvious way, with the moments of order 0, 1 and 2 of the mass distribution. In the following, we compute these Stokes coefficients explicitly; the results are summarised in Table 1.1.

1.2.4.1 Degree $n = 0$ Stokes Coefficients

By setting $n = m = 0$, Eq. (1.28) shows that

$$c_{00} = 1 \tag{1.29}$$

and

$$s_{00} = 0\,, \tag{1.30}$$

which hold for any mass distribution. We note that, because of the sine term in Eq. (1.28), all the zonal sine Stokes coefficients vanish:

Table 1.1 Low-degree ($0 \le n \le 2$) cosine (c_{nm}) and sine (s_{nm}) Stokes coefficients, expressed in terms of the components of the CM position vector (x_{cm}, y_{cm}, z_{cm}) and of the components of the inertia tensor I_{ij} of a mass distribution

degree n	order m	cosine c_{nm}	sine s_{nm}
0	0	1	0
1	0	$+\frac{z_{cm}}{a}$	0
1	1	$-\frac{x_{cm}}{a}$	$-\frac{y_{cm}}{a}$
2	0	$-\frac{I_{33} - (I_{11} + I_{22})/2}{Ma^2}$	0
2	1	$\frac{I_{13}}{Ma^2}$	$\frac{I_{23}}{Ma^2}$
2	2	$\frac{I_{22} - I_{11}}{4Ma^2}$	$-\frac{I_{12}}{2Ma^2}$

$$s_{n0} = 0\,, \quad n = 0,\ 1,\ 2,\ \ldots \tag{1.31}$$

1.2.4.2 Degree $n = 1$ Stokes Coefficients

There are three non-vanishing degree $n = 1$ Stokes coefficients. Equation (1.28) gives

$$\begin{aligned} Ma\,c_{11} &= \int_V r' P_{11}(\cos\theta') \cos\lambda' dm \\ &= -\int_V r' \sin\theta' \cos\lambda' dm = -\int_V x' dm \end{aligned} \tag{1.32}$$

$$\begin{aligned} Ma\,s_{11} &= \int_V r' P_{11}(\cos\theta') \sin\lambda' dm \\ &= -\int_V r' \sin\theta' \sin\lambda' dm = -\int_V y' dm \end{aligned} \tag{1.33}$$

and

$$Ma\,c_{10} = \int_V r' P_{10}(\cos\theta') dm = \int_V r' \cos\theta' dm = \int_V z' dm\,, \tag{1.34}$$

where we have used Table A.1. Hence, from (1.7), we have:

$$c_{11} = -\frac{x_{cm}}{a} \tag{1.35}$$

$$s_{11} = -\frac{y_{cm}}{a}\,, \tag{1.36}$$

and

$$c_{10} = \frac{z_{cm}}{a}\,, \tag{1.37}$$

where (x_{cm}, y_{cm}, z_{cm}) are the coordinates of the CM. Thus, we see that choosing the CM as the origin of the coordinate system O, no terms of harmonic degree $n = 1$ will appear in the expansion of the gravitational potential (e.g., Hofmann-Wellenhof and Moritz 2006).

1.2.4.3 Degree $n = 2$ Stokes Coefficients

Coefficient c_{20}. Setting $n = 2$ and $m = 0$ in (1.28) and using table (A.1), we have

$$
\begin{aligned}
c_{20} &= \frac{1}{M}\int_V \left(\frac{r'}{a}\right)^2 P_{20}(\cos\theta')\,dm \\
&= \frac{1}{2Ma^2}\int_V r'^2\left(3\cos^2\theta' - 1\right)\,dm \\
&= \frac{1}{2Ma^2}\int_V \left(3z'^2 - r'^2\right)\,dm\,.
\end{aligned} \tag{1.38}
$$

But from (1.9–1.11) and (1.15) we note that

$$
\int_V z'^2\,dm = \frac{I_{11} + I_{22} - I_{33}}{2}\,, \tag{1.39}
$$

and

$$
\int_V r'^2\,dm \equiv \frac{I_{11} + I_{22} + I_{33}}{2}\,, \tag{1.40}
$$

hence:

$$
c_{20} \equiv \frac{I_{11} + I_{22} - 2I_{33}}{2Ma^2}\,. \tag{1.41}
$$

Coefficients c_{21} and s_{21}. Setting $n = 2$ and $m = 1$ in (1.28) and using the definition of $P_{21}(\cos\theta)$, for the cosine tesseral coefficient c_{21} we have:

$$
\begin{aligned}
c_{21} &= \frac{1}{3Ma^2}\int_V r'^2 P_{21}(\cos\theta')\,\cos\lambda'\,dm \\
&= \frac{1}{3Ma^2}\int_V r'^2\cdot(-3\sin\theta'\cos\theta')\cdot\cos\lambda'\,dm \\
&= -\frac{1}{Ma^2}\int_V \underbrace{r'\sin\theta'\cos\lambda'}_{x'}\cdot\underbrace{r'\cos\theta'}_{z'}\,dm \\
&= -\frac{1}{Ma^2}\int_V x'z'\,dm = \frac{I_{13}}{Ma^2}\,,
\end{aligned} \tag{1.42}
$$

where we have used definition (1.13). Similarly, for the sine tesseral coefficient s_{21} we have

$$
\begin{aligned}
s_{21} &= \frac{1}{3Ma^2}\int_V r'^2 P_{21}(\cos\theta')\,\sin\lambda'\,dm \\
&= \frac{1}{3Ma^2}\int_V r'^2\cdot(-3\sin\theta'\cos\theta')\cdot\sin\lambda'\,dm \\
&= -\frac{1}{Ma^2}\int_V \underbrace{r'\sin\theta'\sin\lambda'}_{y'}\cdot\underbrace{r'\cos\theta'}_{z'}\,dm \\
&= -\frac{1}{Ma^2}\int_V y'z'\,dm = \frac{I_{23}}{Ma^2}\,,
\end{aligned} \tag{1.43}
$$

where we have recalled the definition of I_{23} (1.14).

Coefficients c_{22} and s_{22}. Imposing $n = m = 2$ in Eq. (1.28) and using Table (A.1), for the sectoral cosine Stokes coefficient of degree $n = 2$ we have

$$\begin{aligned} c_{22} &= \frac{1}{12Ma^2}\int_V r'^2 P_{22}(\cos\theta')\,\cos 2\lambda'\,\mathrm{d}m \\ &= \frac{1}{12Ma^2}\int_V r'^2 \cdot 3\sin^2\theta' \cdot \left(\cos^2\lambda' - \sin^2\lambda'\right)\,\mathrm{d}m \\ &= \frac{1}{4Ma^2}\int_V \big(\underbrace{r'^2\sin^2\theta'\cos^2\lambda'}_{x'^2} - \underbrace{r'^2\sin^2\theta'\sin^2\lambda'}_{y'^2}\big)\mathrm{d}m \\ &= \frac{1}{4Ma^2}\int_V \left(x'^2 - y'^2\right)\,\mathrm{d}m = \frac{I_{22} - I_{11}}{4Ma^2}, \end{aligned} \tag{1.44}$$

where we have used (1.9) and (1.10). Along the same lines, the sine sectoral coefficient is

$$\begin{aligned} s_{22} &= \frac{1}{12Ma^2}\int_V r'^2 P_{22}(\cos\theta')\,\sin 2\lambda'\,\mathrm{d}m \\ &= \frac{1}{12Ma^2}\int_V r'^2 \cdot 3\sin^2\theta' \cdot 2\sin\lambda'\cos\lambda'\,\mathrm{d}m \\ &= \frac{1}{2Ma^2}\int_V \big(\underbrace{r'\sin\theta'\cos\lambda'}_{x'} \cdot \underbrace{r'\sin\theta'\sin\lambda'}_{y'}\big)\mathrm{d}m \\ &= \frac{1}{2Ma^2}\int_V x'\,y'\mathrm{d}m = -\frac{I_{12}}{2Ma^2}. \end{aligned} \tag{1.45}$$

where we have used definition (1.12).

Explicit forms of the Stokes coefficients of the gravitational field for $n = 0, 1, 2$ and their relation with coordinates of the CM and elements of the inertia tensor are summarised in Table 1.1. Note that knowledge of all the five degree $n = 2$ Stokes coefficients does not suffice *per se* to determine the six elements of the inertia tensor.

1.2.4.4 A Note on the Trace of the Inertia Tensor

The zonal Stokes coefficient c_{20} can be written in terms of the trace of the inertia tensor, i.e., the sum of its diagonal elements. This is straightforward since

$$c_{20} = \frac{I_{11} + I_{22} - 2I_{33}}{2Ma^2} = \frac{\mathrm{Tr}(I)}{2\,Ma^2} - \frac{3\,I_{33}}{2\,Ma^2}. \tag{1.46}$$

It is of some interest to consider the changes of $\mathrm{Tr}(I)$ in response to some geophysical perturbation that induces a displacement field $\vec{u}$. In 1910, George Howard Darwin (1845–1912) showed that for a solenoidal displacement field (i.e., a field for which $\nabla \cdot \vec{u} = 0$; see Appendix B), the trace does not vary (Darwin 1908). By extending

Darwin's theorem, Rochester and Smylie (1974) have shown that the trace remains unchanged under a wider class of deformations (see also Lambeck 1980). Since in this work we generally assume incompressibility, in which case the displacement field is solenoidal (see Sect. C.1.3), we have

$$\Delta c_{20} = -\frac{3\,\Delta I_{33}}{2\,Ma^2}\,, \tag{1.47}$$

where Δc_{20} is the variation of the zonal Stokes coefficient of degree $n = 2$, associated to whatsoever geophysical process causing a variation ΔI_{33} of the axial component of the inertia tensor. This note will have some consequences in Sect. 4.2.5, where we consider the inertia variations induced by surface loading.

1.3 Basic Rheological Models

Rheology, or the "study of flow", plays a very important role in the context of glacial isostasic adjustment. Since the response of the Earth to the stress exerted by the large continental ice sheets reflects its physical properties, observing gravity variations, sea-level changes and crustal deformations provides invaluable constraints on the rheological properties of the planet (see e.g., Spada 2017; Whitehouse 2018).

In the following, we limit ourselves to a very basic phenomenological description of the laws of rheology, only giving the essentials for the understanding of the GIA theory. Since this is sufficient for our purposes, we shall only present a one-dimensional description of the rheological models. For a full account of rheology and of its relevance in geophysics and geodynamics, the readers are referred to the treatises of Ranalli (1995) and Karato (2008). An historical perspective of rheology is given by Tanner and Walters (1998) and Walters (2010); some other important references shall be given later.

1.3.1 Rheological Equation of State

A *rheological equation of state* (or *constitutive equation*) is a relation between kinematic and dynamic quantities characterizing the state of a certain material. Following Reiner (1949) and Ranalli (1995), a general rheological equation of state reads

$$\mathscr{R}\,(\epsilon, \dot{\epsilon}, \ddot{\epsilon} \ldots ; \sigma, \dot{\sigma}, \ddot{\sigma} \ldots ; m_1, m_2, \ldots) = 0\,, \tag{1.48}$$

where $\mathscr{R}$ is referred to as *material function*, $\epsilon = \epsilon(t)$ is the strain (a kinematic quantity), $\sigma = \sigma(t)$ is the stress (a dynamic quantity), the dot indicates the time derivative, cartesian indices are omitted, and $m_1, m_2, \ldots$ are *material parameters*. As pointed out by Ranalli (1995), the term *material constants* would not be appropriate for the m_i's, since these may vary with varying extrinsic conditions (temperature, pressure and *time*), even considering constant intrinsic conditions (i.e., the same materials). As we shall discuss in the following, for a *linear visco-elastic material* the constitutive equation (1.48) is a linear relation.

1.3.2 Idealised Mechanical Tests

In the study of the mechanical behaviour of materials, it is convenient to refer to some idealised experiments (or *tests*) which are schematically represented in Fig. 1.4.

In a *creep test*, the stress is suddenly applied at time $t = 0^+$ and subsequently left unchanged[8] (see Fig. 1.4a). Hence:

$$\sigma(t) \equiv \begin{cases} 0 & \text{if } t \leq 0 \\ \sigma_0 & \text{if } t > 0\,, \end{cases} \tag{1.49}$$

where $\sigma_0 > 0$ is a constant. Thus, in this type of mechanical test, the time-varying strain $\epsilon = \epsilon(t)$ is the unknown variable. It can be determined experimentally or, in the case the rheological equation of state (1.48) of the material is known a priori, it can be obtained solving a corresponding (and possibly non-linear) differential equation. By the definition of (left-continuous) Heaviside step function (E.1), we note that in a creep test the stress can be simply expressed as

$$\sigma(t) = \sigma_0\, H(t)\,. \tag{1.50}$$

In a *creep-recovery test*, schematically displayed in Fig. 1.4b, the stress is imposed at time $t = 0^+$ and subsequently released at the time of unloading $t = t_u^+$, so that:

$$\sigma(t) \equiv \begin{cases} 0 & \text{if } t \leq 0 \\ \sigma_0 & \text{if } 0 < t \leq t_u \\ 0 & \text{if } t > t_u\,. \end{cases} \tag{1.51}$$

Using the Heaviside step function, the imposed stress can be equivalently written as

$$\sigma(t) = \sigma_0\,(H(t) - H(t - t_u))\,. \tag{1.52}$$

In a *stress relaxation* test (see Fig. 1.4c), the *strain* is imposed at time $t = 0^+$, and kept constant afterwards:

$$\epsilon(t) \equiv \begin{cases} 0 & \text{if } t \leq 0 \\ \epsilon_0 & \text{if } t > 0\,; \end{cases} \tag{1.53}$$

---

[8] With the notation $t = t_0^+$ we denote a time $t = t_0 + t_\epsilon$, with $t_\epsilon > 0$ being infinitesimally small.

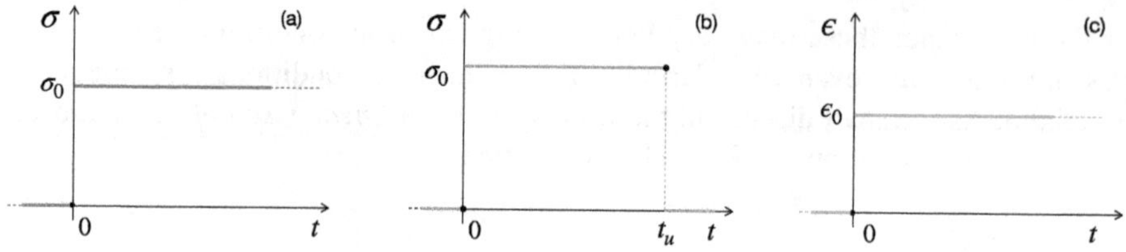

Fig. 1.4 Idealised mechanical tests: **a** *creep* test, **b** *creep-recovery* test and **c** *stress relaxation* test. Note that in **a** and **b** the stress is given, while in **c** the strain is given

therefore, in a stress relaxation test, the problem consists in the determination of the stress $\sigma(t)$. Again, with the aid of the Heaviside step function (E.1), in a stress relaxation test the strain is equivalently expressed as

$$\epsilon(t) = \epsilon_0 \, H(t) \, . \tag{1.54}$$

Finally, in a *cyclic test*, the stress varies in time as

$$\sigma(t) = \sigma_0 \sin(\omega t + \varphi) \, , \qquad t \geq 0 \, , \tag{1.55}$$

where $\omega = 2\pi/T$ is the angular frequency, T is the oscillation period and φ is an arbitrary phase. Here we shall only consider the cyclic test marginally, although in Sect. 2.3.4 we shall compute the Love numbers for an oscillating source of excitation. The readers are referred to Mainardi (2010) and Kelly (2019) for the important implications of the cyclic test upon the fundamental problem of energy dissipation in solids (Knopoff 1964).

1.3.3 Classical Mechanical Models

In the following, we shall describe the properties of some fundamental *linear rheological bodies*, for which the rheological equation of state (1.48) can be written as an operator equation

$$F(D)\sigma = G(D)\epsilon \, , \tag{1.56}$$

where σ is stress, ϵ is strain, $D \equiv \frac{\mathrm{d}}{\mathrm{d}t}$ and $F(D)$ and $G(D)$ are polynomials in D, whose coefficients are material parameters that do not depend upon σ or ϵ (Mainardi 2010; Mainardi and Spada 2011).

The classical mechanical models that we shall consider (the Hooke, Newton, Maxwell, Kelvin-Voigt and Burgers models) can be either described by a linear rheological equation of state like (1.56) or by *mechanical analogues* that involve proper arrangements of *springs* and *dashpots* (Mainardi 2010). As pointed out by Reiner (1949) and later by Ranalli (1995), these elements have no counterpart in the real materials and are only useful to describe their phenomenological behaviour. For a microphysical description of the rheological laws, the readers are referred to the treaties of Ranalli (1995) and Karato (2008).

Since Eq. (1.56) is a linear equation, it can be solved by well established methods, once suitable initial conditions are given. The integral transform method is often employed, either based upon Laplace or Fourier transforms, since in this way differential equations are translated into algebraic equations that often (but not *always*) can be anti-transformed analytically. The Laplace transform shall be used throughout this book; its definition is given in Appendix B.3 along with some of its useful properties.

1.3.4 Hooke

Among the classical mechanical models, the Hooke model has a fundamental role. It is named after Sir Robert Hooke (1635–1703), who first stated the law of elasticity by his famous anagram *ceiiinosssttuv* (“*Ut tensio, sic vis*”) meaning “As the extension, so the force”.

The Hooke (H) body is a *solid*, with constitutive equation

$$\epsilon(t) = \frac{\sigma}{E}, \tag{1.57}$$

where E is a generic *elastic constant* having physical dimensions of a stress. Figure 1.5 shows the stress-strain relationship (1.57), valid for stress values below the elastic limit, and the mechanical analogue for the Hookean solid i.e., a spring of elastic constant E.

Since for the Hooke body the constitutive equation (1.57) is expressed by a law of linear proportionality, the outcomes of the idealised experiments shown in Fig. 1.4 are straightforward. When subject to a creep-recovery test, the Hooke body undergoes an instantaneous strain at the time of loading $t = 0$ and to an instantaneous un-straining upon un-loading at time $t = t_u$, with a *full recovery* of the initial strain (see Fig. 1.7a). In a stress relaxation test, the stress shall not *relax*, i.e., it shall remain constant to the initial value $\sigma = E\epsilon_0$. In a cyclic test, the stress and the strain shall be in phase, since they are proportional one to another: $\sigma(t) = E\epsilon(t)$ (see e.g., Kelly 2019).

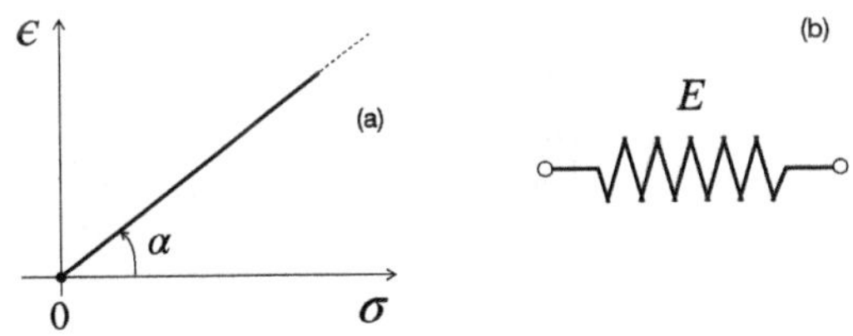

Fig. 1.5 Stress-strain diagram of the Hookean solid (**a**) and the spring, its mechanical analogue (**b**). The angle α is such that $\tan\alpha = 1/E$

1.3.5 Newton

Isaac Newton's law for fluids has been formulated in the *Principia* (§IX) as follows:

> *Resistentiam, quae oritur ex defectu lubricitatis partium fluidi, caeteris paribus, proportionalem esse velocitati, qua partes fluidi separantur ab invicem.*

... to mean that "*The resistance arising from the want of lubricity in the parts of a fluid, is, others things being equal, proportional to the velocity with which the parts of the fluid are separated from one another*" (see Franco and Partal 2010, for notes about translation).

With respect to the Hooke body, the Newton (N) body, also referred to as *linear viscous* body, lies at the opposite side of the spectrum of rheological behaviors. In fact, the Newtonian body is a *fluid*, with constitutive equation

$$\dot{\epsilon}(t) = \frac{\sigma}{\eta}, \tag{1.58}$$

where the intrinsic parameter η is the *newtonian viscosity*, having physical dimensions of stress $\times$ time (see Fig. 1.6). In the creep and creep-recovery tests, the behaviour of the Newtonian fluid can be conveniently studied by integrating the constitutive equation:

$$\epsilon(t) = \frac{1}{\eta} \int_{-\infty}^{t} \sigma(t')\, dt', \tag{1.59}$$

where we have assumed that at $t = -\infty$ the material was in its natural configuration i.e., $\epsilon(-\infty) = 0$.

1.3.5.1 Creep Test (Newton)

In a creep test, since for $t \leq 0$ we have $\sigma(t) = 0$, from (1.59) we obtain $\epsilon(t) = 0$. For $t > 0$, we have $\int_{-\infty}^{t} \sigma(t')\, dt' = \sigma_0 \int_0^t dt'$, hence:

$$\epsilon(t) = \frac{\sigma_0}{\eta} t, \quad t > 0, \tag{1.60}$$

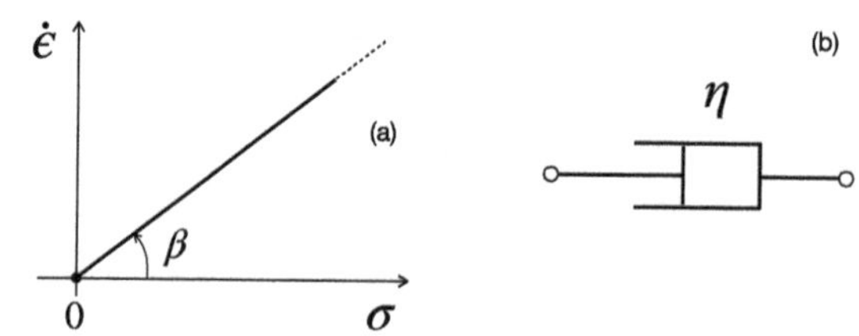

Fig. 1.6 Stress-strain diagram for the Newtonian fluid (**a**) and the dashpot, its mechanical analogue (**b**). The angle β is such that $\tan\beta = 1/\eta$

showing that, in a creep test, a Newtonian body flows at a constant strain rate $\dot{\epsilon} = \frac{\sigma_0}{\eta}$, which characterises a *steady state flow*.

1.3.5.2 Creep-Recovery Test (Newton)

In a creep-recovery test, before un-loading, the response shall be the same as for the creep test, hence $\epsilon(t) = 0$ for $t \leq 0$, and

$$\epsilon(t) = \frac{\sigma_0}{\eta} t \,, \quad 0 < t \leq t_u \,, \tag{1.61}$$

where t_u is the time span of the load. After un-loading, Eq. (1.59) gives

$$\epsilon(t) = \frac{\sigma_0}{\eta} t_u \,, \quad t > t_u \,, \tag{1.62}$$

meaning that for a Newtonian fluid the strain built until time $t = t_u$ is permanent and cannot be recovered. This behavior, shown in Fig. 1.7b, is diametrically opposed to that of a Hookean body, for which in a creep-recovery test there is *full recovery* of the initial strain (see Fig. 1.7a).

1.3.6 Maxwell

The Maxwell rheological model (Maxwell 1867) describes a material in which elastic and viscous behaviours are combined (Boyaval 2021). The Maxwell rheology has an important role in this book, since most of the modern GIA models assume this type of rheology for the mantle and for the lithosphere (e.g., Whitehouse 2018). After loading, the Maxwell material accounts for *steady state creep*, i.e., for a flow characterised by a constant strain rate (Ranalli 1995).

The Maxwell (M) mechanical analogue, shown in Fig. 1.8, consists of a spring (an Hookean element) of elastic modulus E connected in series to a dashpot or damper

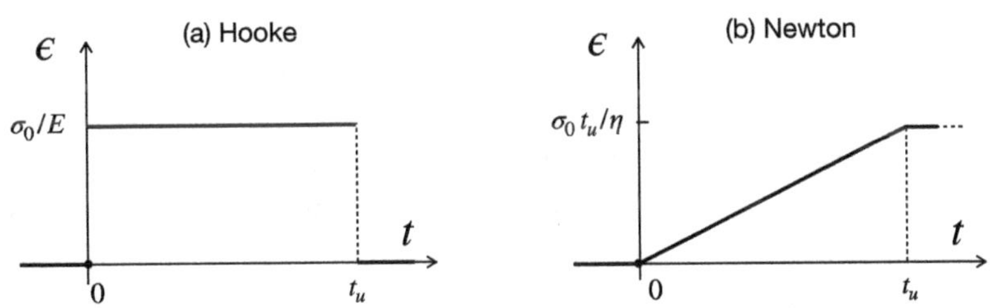

Fig. 1.7 Strain response of a Hookean body (**a**) and of a Newtonian body (**b**) in a creep-recovery test

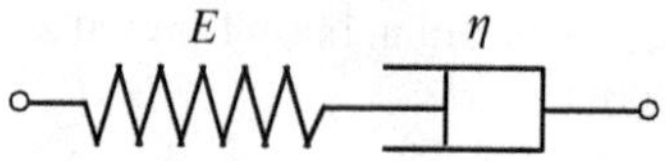

Fig. 1.8 Mechanical analogue of the Maxwell rheological model. Using Reiner's notation (Reiner 1945), M = H − N, where M, H and N stand for Maxwell, Hooke and Newton, respectively, and "−" symbolises a connection in series

(a Newtonian element) of viscosity η. The Maxwell body is sometimes referred to as *elastoviscous* body and the corresponding mechanical analogue to as *relaxation element* (Mainardi 2010).

The two mechanical elements are subject to the same stress, and the total strain ϵ is the sum of the individual strains; hence the total strain rate is

$$\dot{\epsilon}(t) = \dot{\epsilon}_{\rm H} + \dot{\epsilon}_{\rm N}\,, \tag{1.63}$$

where $\dot{\epsilon}_{\rm N}$ and $\dot{\epsilon}_{\rm H}$ are the Newtonian and the Hookean contributions, respectively. Using (1.57) and (1.58), the constitutive equation for the Maxwell body therefore reads:

$$\dot{\epsilon}(t) = \frac{\sigma}{\eta} + \frac{\dot{\sigma}}{E}\,, \tag{1.64}$$

which by time integration gives

$$\epsilon(t) = \frac{1}{\eta}\int_{-\infty}^{t} \sigma(t')\mathrm{d}t' + \frac{1}{E}\sigma(t)\,, \tag{1.65}$$

where we have assumed the conditions $\epsilon(-\infty) = \sigma(-\infty) = 0$.

1.3.6.1 Creep Test (Maxwell)

In a creep test, (1.65) gives $\epsilon(t) = 0$ for $t \le 0$. Otherwise, we obtain

$$\epsilon(t) = \sigma_0\left(\frac{t}{\eta} + \frac{1}{E}\right), \quad t > 0\,, \tag{1.66}$$

showing that after an instantaneous elastic reaction, the Maxwell body flows at a constant strain rate, exactly as a Newtonian fluid (*steady-state creep*); see Fig. 1.9a.

It is now convenient to introduce the *creep compliance* function, which for a general viscoelastic material is defined as the strain per unit stress applied in a creep test:

$$J(t) = \frac{\epsilon(t)}{\sigma_0}\,. \tag{1.67}$$

Then, from (1.66), for the Maxwell body the creep compliance is:

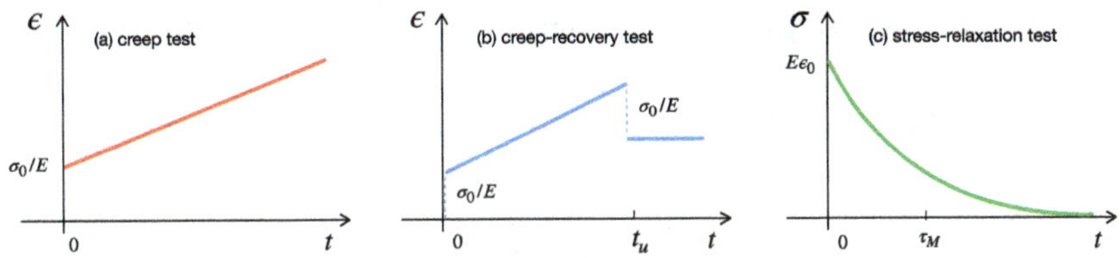

Fig. 1.9 Qualitative response of a Maxwell body to the basic mechanical tests

$$J_{\mathrm{M}}(t) = \frac{t}{\eta} + \frac{1}{E}, \quad t \geq 0. \tag{1.68}$$

We note that, according to (1.57) and (1.60), for the H and N bodies the creep compliances are

$$J_{\mathrm{H}}(t) = \frac{1}{E}, \quad t \geq 0, \tag{1.69}$$

and

$$J_{\mathrm{N}}(t) = \frac{t}{\eta}, \quad t \geq 0, \tag{1.70}$$

respectively. We also note that since the creep compliance represents a (normalised) strain, creep compliances add in series.

1.3.6.2 Creep-Recovery Test (Maxwell)

For a creep-recovery test, (1.65) gives $\epsilon(t) = 0$ for $t \leq 0$. Before un-loading, the M body response is the same as in the creep test:

$$\epsilon(t) = \sigma_0 \left(\frac{t}{\eta} + \frac{1}{E} \right), \quad 0 < t \leq t_u, \tag{1.71}$$

whereas, after unloading

$$\epsilon(t) = \frac{\sigma_0}{\eta} t_u, \quad t > t_u, \tag{1.72}$$

showing that only the initial instantaneous elastic deformation σ_0/E is recovered, leaving a permanent viscous strain (see Fig. 1.9b).

1.3.6.3 Stress Relaxation Test (Maxwell)

Setting $\epsilon = \epsilon_0$, for $t > 0$ the Maxwell constitutive equation (1.64) gives

$$0 = \frac{\sigma}{\eta} + \frac{\dot{\sigma}}{E}, \tag{1.73}$$

which upon integration yields

$$\sigma(t) = \sigma_0 \, \mathrm{e}^{-t/\tau_\mathrm{M}}, \quad t > 0, \tag{1.74}$$

where $\sigma_0 = \sigma(0^+) = E\epsilon_0$ is the initial stress[9] and

$$\tau_\mathrm{M} = \frac{\eta}{E} \tag{1.76}$$

is the *Maxwell relaxation time*. Hence, under a constant strain, in a Maxwell body the stress decays with time (it *relaxes*) following an exponentially law with a time constant τ_M (see Fig. 1.9c). Note that relaxation is peculiar of the Maxwell model; it is not observed for the Hookean and Newtonian bodies. According to (1.74), τ_M is the time required to reach a stress level $\sigma \approx {}^1\!/_3\,\sigma_0$.

In analogy with the creep compliance in a creep test, in a stress relaxation test it is convenient to introduce a function known as *relaxation modulus*, defined as the stress per unit strain in a relaxation test:

$$G(t) = \frac{\sigma(t)}{\epsilon_0}, \tag{1.77}$$

so that, according to (1.74), for the M body the relaxation modulus is:

$$G_\mathrm{M}(t) = E \, \mathrm{e}^{-t/\tau_\mathrm{M}}. \tag{1.78}$$

We note that since the relaxation moduli represent (normalised) stresses, they add in parallel.

---

[9] The relationship $\sigma_0 = E\epsilon_0$ is rather intuitive and can be obtained from the constitutive equation (1.64). The idea is to discretise the equation in the vicinity of $t = 0$, obtaining

$$\epsilon(\Delta t) - \epsilon(-\Delta t) = \frac{1}{\eta}\left[\sigma(\Delta t) - \sigma(-\Delta t)\right]\Delta t + \frac{1}{E}\left[\sigma(\Delta t) - \sigma(-\Delta t)\right], \tag{1.75}$$

where Δt is a small time interval. For $\Delta t \to 0$, $\epsilon(\Delta t)$ goes to $\epsilon(0^+) \equiv \epsilon_0$ and $\epsilon(-\Delta t)$ tends to zero, according to the definition of stress-relaxation test (see 1.53). Furthermore, for $\Delta t \to 0$ the first term on the *r.h.s.* vanishes, in the hypothesis that the stress remains finite. Assuming that, for $\Delta t \to 0$, $\sigma(\Delta t)$ tends to the finite limit $\sigma(0^+) \equiv \sigma_0$, Eq. (1.75) gives $\sigma_0 = E\epsilon_0$. A similar argument is presented by Kelly (2019).

1.3.7 *Kelvin-Voigt*

The Kelvin-Voigt (K) model is a two-element mechanical model in which a Hookean spring of elastic modulus E and a Newtonian dashpot of viscosity η are arranged in parallel (see Fig. 1.10). The model was introduced by Sir William Thomson, Baron of Kelvin (Thomson 1865) and refined by Woldemar Voigt in 1892. The Kelvin-Voigt material is also referred to as *firmoviscous* material and the corresponding mechanical analogue to as *retardation element* (Mainardi 2010).

As we shall see afterwards, in response to loading the K element shows *elastic afterworking*, i.e., time-dependent strain approaching asymptotically a limiting value corresponding to the purely elastic response (Ranalli 1995). As we have discussed in Sect. 1.3.6.1, this behaviour is not displayed by the Maxwell model, which is characterised by a steady state (constant strain rate) response to creep instead. The K material is thus the simplest example of rheological *transient* behaviour (see e.g., Ranalli 1995).

In the K material, the two mechanical elements are subject to the same strain ϵ and the total stress σ results from the sum of the individual stresses. Hence,

$$\sigma(t) = \sigma_{\mathrm{H}} + \sigma_{\mathrm{N}}\,, \tag{1.79}$$

where σ_{H} and σ_{N} are the Hookean and the Newtonian contributions, respectively. Thus, the constitutive equation for the K material is:

$$\sigma(t) = E\epsilon + \eta\dot{\epsilon}\,, \tag{1.80}$$

where we have used the constitutive equations for the H body (1.57) and for the N body (1.58), respectively.

1.3.7.1 Creep Test (Kelvin-Voigt)

In a creep test, $\sigma = \sigma_0$ for $t > 0$, so that (1.80) takes the form of a non-homogeneous ODE, with $\sigma_0 = E\epsilon + \eta\dot{\epsilon}$. The homogeneous equation has general solution $\epsilon(t) = C\mathrm{e}^{-t/\tau_{\mathrm{K}}}$ where C is a constant and

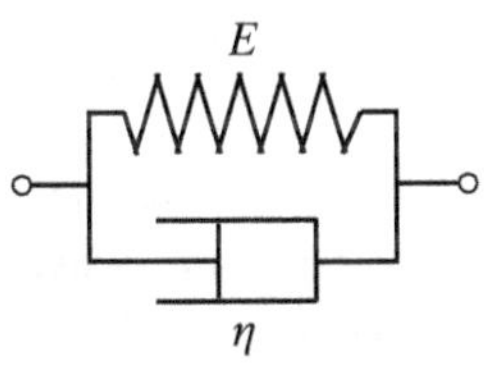

Fig. 1.10 Mechanical analogue of the Kelvin-Voigt rheological model. Using Reiner's notation, K = H|N, where symbol | denotes the connection in parallel

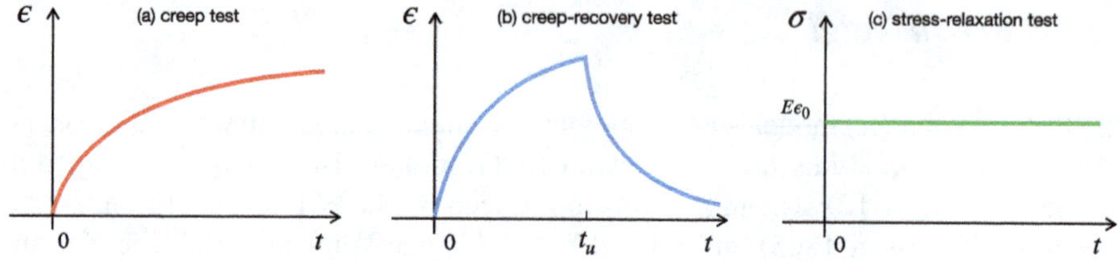

Fig. 1.11 Qualitative response of the Kelvin-Voigt body to the three elementary mechanical tests considered in Fig. 1.4

$$\tau_K = \frac{\eta}{E} \tag{1.81}$$

is known as *retardation time*. Since the non-homogeneous equation has particular solution $\epsilon(t) = \sigma_0/E$, the general solution of (1.80) for a creep test is $\epsilon(t)=\sigma_0/E + Ce^{-t/\tau_K}$. Assuming that both elements are unstrained before the stress is applied i.e., $\epsilon(0) = 0$, the response of a K material to a creep test is

$$\epsilon(t) = \frac{\sigma_0}{E}\left(1 - e^{-t/\tau_K}\right), \quad t \geq 0, \tag{1.82}$$

showing that there is no instantaneous elastic strain and that the elastic response of the spring is delayed by the damping action of the Newtonian element. The response of a K material to a creep test is shown in the diagram of Fig. 1.11a. According to (1.82), the strain grows following a *delayed exponential law* (see Ranalli 1995) with the asymptotic value σ_0/E. Following definition (1.67), for the K material the creep compliance is

$$J_K(t) = \frac{1}{E}\left(1 - e^{-t/\tau_K}\right), \quad t \geq 0. \tag{1.83}$$

1.3.7.2 Creep-Recovery Test (Kelvin-Voigt)

In a creep-recovery test (Fig. 1.11b), before un-loading occurs, the response of the Kelvin-Voigt material is the same as in the creep test. Hence

$$\epsilon(t) = \frac{\sigma_0}{E}\left(1 - e^{-t/\tau_K}\right), \quad 0 \leq t \leq t_u\,. \tag{1.84}$$

After un-loading, i.e., for $t > t_u$, the stress is $\sigma(t) = 0$, so that the Kelvin-Voigt equation (1.80) takes the form $0 = E\epsilon + \eta\dot{\epsilon}$, which has solution $\epsilon(t) = C'e^{-t/\tau_K}$ where C' is a constant. Imposing continuity with (1.84) at $t = t_u$ provides $C' = (\sigma_0/E)\left(e^{t_u/\tau_K} - 1\right)$, thus giving

$$\epsilon(t) = \frac{\sigma_0}{E}\left(\mathrm{e}^{t_u/\tau_K} - 1\right)\mathrm{e}^{-t/\tau_K}\,, \quad t \geq t_u\,, \tag{1.85}$$

which is portrayed in the diagram of Fig. 1.11b, where we note that there is a delayed recovery of the transient creep cumulated during the loading phase, but no instantaneous nor permanent strain (see e.g., Kelly 2019).

1.3.7.3 Stress-Relaxation Test (Kelvin-Voigt)

In such experiment, $\epsilon(t) = \epsilon_0$ for $t \geq 0$. Thus, from the constitutive equation (1.80) we have $\sigma(t) = E\,\epsilon_0$, i.e., the stress is a constant. Hence, the Kelvin-Voigt element does not relax (see Fig. 1.11c).

1.3.8 General Viscoelastic Models

Although only seldom utilised in GIA studies, there are important rheological models in addition to the two-element bodies discussed so far. In this section we briefly summarise their properties; the readers are referred to the treaties of Ranalli (1995) and Karato (2008) for a more thorough discussion and for an illustration of the geophysical implications.

1.3.8.1 Three-Element Models

Following Mainardi (2010), two three-element rheological models can be introduced. The first is known as Zener model after Zener (1948). It consists of a spring in series with a K element, and since it has the properties of a solid, it is also refereed to as Standard Linear Solid or sometimes *three-element elastic model*. The second, the anti-Zener model, is composed by a dashpot in series with a K element, and it is also known as *three-element viscous model*. Both the Zener and the anti-Zener models can take mechanically equivalent forms involving a M analogue, as shown in Fig. 2.2 of Mainardi (2010). Their constitutive equations, given for instance by Kelly (2019), are somewhat cumbersome to obtain, being of the second order in the time-derivatives. Expressions for the creep compliances and relaxation moduli are explicitly presented by Mainardi (2010).

1.3.8.2 The Burgers Model

A quite general rheological model can be obtained combining a Maxwell analogue with a Kelvin-Voigt analogue in series. This combination, known as Burgers (or

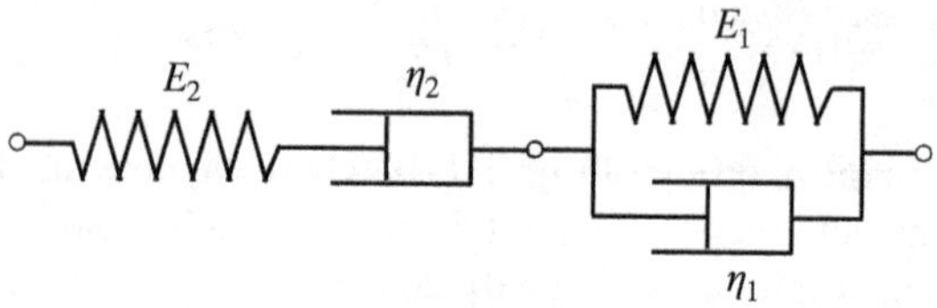

Fig. 1.12 Mechanical analogue of the Burgers (B) rheological model. The one shown is known as the *creep representation* of the B body. A mechanically equivalent representation exists, the *relaxation representation* (Müller 1986), in which $\text{B} = M_1|M_2$, where M_1 and M_2 are two Maxwell elements with appropriate material constants (see also Mainardi 2010)

general linear) body after Johannes Martinus Burgers (1895–1981), is a mechanical system characterised by two elastic (E_1, E_2) and two viscous elements (η_1, η_2). Hence, using Reiner's notation, B = M − K. The B mechanical analogue is displayed in Fig. 1.12. According to Ranalli (1995), the corresponding constitutive equation is:

$$\eta_1\ddot{\epsilon} + E_1\dot{\epsilon} = \frac{\eta_1}{E_2}\ddot{\sigma} + \left(\frac{\eta_1}{\eta_2} + \frac{E_1}{E_2} + 1\right)\dot{\sigma} + \frac{E_1}{\eta_2}\sigma\,. \tag{1.86}$$

The Burgers model exhibits *(i)* an instantaneous elastic response, *(ii)* an intermediate transient response and *(iii)* a long-term viscous response, which represent the general features of typical experimental creep curves (e.g., Ranalli 1995). For this reason, it has been widely employed in the geosciences (see the review of Karato 2010). In particular, it has been useful to model the post-seismic deformations (see e.g., Pollitz 2003; Hetland and Hager 2005; Pollitz 2005; Pollitz et al. 2006; Hetland and Hager 2006; Melini et al. 2008; Cannelli et al. 2010; Sun et al. 2014) and post-glacial rebound (see e.g., Peltier et al. 1980, 1981; Sabadini et al. 1985; Peltier 1985a; Yuen et al. 1986; Peltier et al. 1986; Sabadini et al. 1987; Körnig and Müller 1989; Spada et al. 2011b; Caron et al. 2017; Boughanemi and Mémin 2024).

Solving the constitutive equation under constant stress (creep test, see Fig. 1.13a) gives, for the Burgers body, $\epsilon(t) = \sigma_0 J_{\text{B}}(t)$, where the creep compliance is:

$$J_{\text{B}}(t) = \frac{1}{E_2} + \frac{1}{\eta_2}t + \frac{1}{E_1}\left(1 - \mathrm{e}^{-t/\tau_{\text{K}}}\right), \quad t \geq 0\,, \tag{1.87}$$

with $\tau_{\text{K}} = \eta_1/E_1$ being the retardation time of the K element (see Fig. 1.12). In a creep-recovery experiment (1.13b), there is an instantaneous recovery of the elastic strain, followed by a recovery of the transient creep; the viscous deformation is, of course, permanent. In a relaxation test (see Fig. 1.13c), the B material relaxes following a double exponential decay law. Indeed, after some algebra, one obtains $\sigma(t) = \epsilon_0 G_{\text{B}}(t)$ where the Burgers relaxation modulus is

$$G_{\text{B}}(t) = G_1\mathrm{e}^{-t/\tau_1} + G_2\mathrm{e}^{-t/\tau_2}\,, \quad t \geq 0\,, \tag{1.88}$$

where constants G_i and τ_i are functions of material parameters E_i and η_i ($i = 1, 2$).

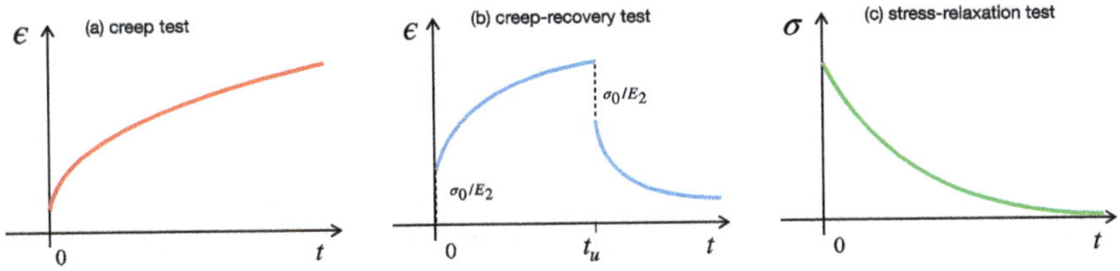

Fig. 1.13 Qualitative response of a Burgers material to the three fundamental mechanical tests shown in Fig. 1.4

1.3.9 Correspondence Principle

The *correspondence principle of linear viscoelasticity* is a powerful tool in the analysis of viscoelastic problems. In practical terms, by the correspondence principle, the Laplace transform of any viscoelastic constitutive equation takes the form of the Hooke's law, provided that the elastic modulus E is substituted by an appropriate frequency-dependent complex modulus $E(s)$, whose specific form depends upon the viscoelastic rheological law assumed for the material (Fung 1965; Tschoegl 2012). In this way, as discussed by Christensen (1982), the vast catalogue of static elastic solutions can be immediately converted to quasi-static viscoelastic solutions by replacing elastic moduli with appropriate frequency-dependent moduli, and reinterpreting elastic field variables as transformed viscoelastic field variables.

To illustrate the essence of the correspondence principle, we consider the example of the Maxwell rheological model. From Sect. 1.3.6, we recall that the Maxwell constitutive equation in the time domain reads

$$\dot{\epsilon}(t) = \frac{\sigma}{\eta} + \frac{\dot{\sigma}}{E}, \tag{1.89}$$

where $\epsilon(t)$ is the strain, $\sigma(t)$ is the stress, η is viscosity and E is the elastic modulus. Taking the Laplace transform (see Appendix B.3) of both sides, assuming vanishing pre-initial conditions (i.e., $\epsilon(0^-) = 0$ and $\sigma(0^-) = 0$), and re-arranging gives[10]

$$\tilde{\sigma}(s) = E_{\rm M}(s)\,\tilde{\epsilon}(s)\,, \tag{1.90}$$

where $\tilde{\sigma}(s)$ and $\tilde{\epsilon}(s)$ are the Laplace transforms of $\sigma(t)$ and $\epsilon(t)$, respectively,

[10] In this book we will generally use the tilde to denote a function of the Laplace variable s, except for some cases where the s-dependence is immediately clear without any possible ambiguity. The tilde will also be used to distinguish a function $f(t)$ from its Laplace transform $\tilde{f}(s)$.

$$E_M(s) = \frac{Es}{s + 1/\tau_M} \tag{1.91}$$

is the *complex modulus* for the Maxwell material and we recall that $\tau_M = \eta/E$ is the Maxwell relaxation time. It is apparent that the Laplace domain equation (1.90) has the same form that Hooke's law takes in the time domain (i.e., $\sigma = E\,\epsilon$), provided that E is substituted with $E_M(s)$. This "proves" the validity of the correspondence principle.

Some fundamental relations exist among $E(s)$, $\tilde{J}(s)$ and $\tilde{G}(s)$, where these latter are the Laplace-transformed creep compliance and relaxation modulus of a viscoelastic material, respectively. Again, we consider the case of the maxwellian material as an example. We first note that the Laplace transform of (1.68) is

$$\tilde{J}_M(s) = \frac{1}{\eta s^2} + \frac{1}{Es} = \frac{1}{s E_M(s)}, \tag{1.92}$$

where we have used Table B.1 and (1.91). Similarly, the Laplace transform of (1.78) is

$$\tilde{G}_M(s) = \frac{E}{s + 1/\tau_M} = \frac{E_M(s)}{s}, \tag{1.93}$$

hence the remarkable identity:

$$\tilde{J}_M(s)\,\tilde{G}_M(s) = \frac{1}{s^2}, \tag{1.94}$$

which in the time domain reads

$$J_M(t) * G_M(t) = t, \tag{1.95}$$

where $*$ denotes time-convolution, defined in Table B.1. It is well known (see e.g., Mainardi 2010), that the relations $\tilde{J}(s) = 1/E(s)$, $\tilde{G}(s) = E(s)/s$ and $\tilde{J}(s)\tilde{G}(s) = 1/s^2$ are general and hold for all materials obeying a linear constitutive law.

1.4 Surface Mass Distributions

Here we define various quantities useful to describe the distribution of the ice and water masses over the surface of the Earth and their time evolution. We first introduce the *ocean function*, which in a given location is determined by the ice thickness and by the Earth's *topography*, and its complementary function known as *continent function*. The surface mass distribution is described in terms of the *surface load* function. With the aid of the ocean function, the time-variations of the surface load with respect to a reference state are expressed in terms of changes in the *ice thickness*, variations in *sea level*, and changes in the *oceans geometry*.

1.4.1 *Topography and Sea Level*

1.4.1.1 The Ocean Function

The *ocean function* (OF) characterises the water and the ice mass distributions across the Earth's surface. At a given location of co-latitude θ and longitude λ, the OF can only take two possible values, being defined as follows:

$$O(\gamma, t) \equiv \begin{cases} 1, \text{ if } T + \dfrac{\rho^{\mathrm{i}}}{\rho^{\mathrm{w}}} I < 0 \\ 0, \text{ if } T + \dfrac{\rho^{\mathrm{i}}}{\rho^{\mathrm{w}}} I \geq 0 , \end{cases} \tag{1.96}$$

where $\gamma = (\theta, \lambda)$, t is time, ρ^{i} and ρ^{w} are the density of ice and water, respectively,[11] $I(\gamma, t) \geq 0$ denotes the *ice thickness*, $T(\gamma, t)$ is the topographic relief (henceforth abbreviated by *topography*), with

$$T(\gamma, t) = -B , \tag{1.97}$$

where B is *sea level*, defined as

$$B(\gamma, t) = r^{\mathrm{ss}} - r^{\mathrm{se}} , \tag{1.98}$$

where $r^{\mathrm{ss}}(\gamma, t)$ and $r^{\mathrm{se}}(\gamma, t)$ are the radii of the *sea surface* and of the *solid Earth surface* in a geocentric reference frame, respectively. It is important to note that both r^{ss} and and r^{se} are defined at all locations on the surface of the Earth, since r^{ss} is the radius of the equipotential surface which on the oceans coincides with the mean sea level. Hence, by (1.98), also sea level B is defined on both ocean and land regions.

As far as we know, the OF has been first introduced by Prey (1922), who has also decomposed it in series of spherical harmonics to a maximum degree $n_{max} = 16$. The readers are referred to Balmino et al. (1973), Munk and MacDonald (1975), Lambeck (1980) and to Okuno and Nakada (2001) for more recent determinations of the OF and of its harmonic spectrum.

1.4.1.2 The Continent Function

The *continent function* (CF) is the complement to one of the OF:

$$C(\gamma, t) = 1 - O , \tag{1.99}$$

where the ocean function $O(\gamma, t)$ is given by (1.96).

---

[11] Henceforth ρ^{i} and ρ^{w} are assumed to be constants.

1.4.2 *Ice and Water Distributions*

According to the local values of T and I, five different mutually exclusive conditions can be met, which are discussed in detail below and illustrated in Table 1.2 and Fig. 1.14.

1.4.2.1 Condition 1: Ice-Free Ocean

In this situation, the topography is $T < 0$ and since there is no ice ($I = 0$), from Eqs. (1.96) and (1.99) we have $O = 1$ and $C = 0$.

Table 1.2 Values of the ocean function OF (Eq. 1.96) and of its complement CF (1.99) according to the values of topography T and of ice thickness I, with a description of the five possible conditions discussed in the text. Here "bsl" and "asl" stand for *below sea level* and *above sea level*, respectively

Condition	Sea level	Topography	Ice thickness	$T + \frac{\rho^i}{\rho^w} I$	OF	CF
1 - Ice-free ocean	$B > 0$	$T < 0$	$I = 0$	< 0	1	0
2 - Floating ice	$B > 0$	$T < 0$	$I > 0$	< 0	1	0
3 - Ice grounded bsl	$B > 0$	$T < 0$	$I > 0$	≥ 0	0	1
4 - Ice grounded asl	$B < 0$	$T > 0$	$I > 0$	≥ 0	0	1
5 - Ice-free land	$B < 0$	$T > 0$	$I = 0$	≥ 0	0	1

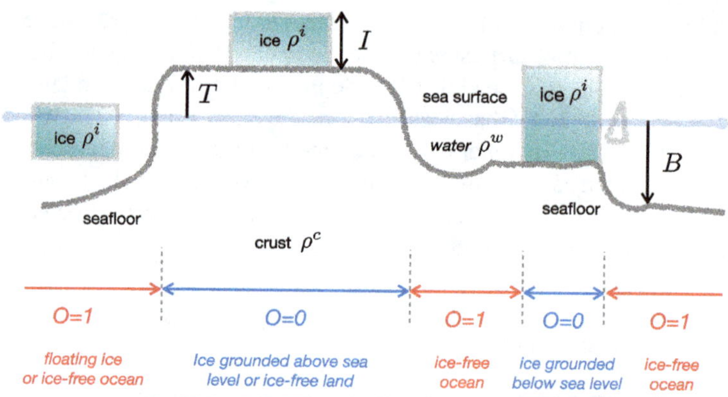

Fig. 1.14 Possible values of the OF (O), corresponding to the five different conditions listed in Table 1.2, according to the values of T (topography) and I (ice thickness). Note that the continent function is $C = 1 - O$. With $B = -T$ we denote sea level

1.4.2.2 Condition 2: Floating Ice

Where ice is floating, $T < 0$ and $I > 0$. By the condition of isostatic equilibrium, all the columns have the same mass:

$$\rho^{\mathrm{w}} B(\gamma, t) = \rho^{\mathrm{i}} I + \rho^{\mathrm{w}} y\,, \tag{1.100}$$

where y is the thickness of the water layer beneath the ice column (see Fig. 1.15a). Since $y > 0$, using Eq. (1.97), the floating ice condition reads

$$T(\gamma, t) + \frac{\rho^{\mathrm{i}}}{\rho^{\mathrm{w}}} I(\gamma, t) < 0\,, \tag{1.101}$$

stating that the ice is lighter than the water column ($\rho^{\mathrm{i}} I < \rho^{\mathrm{w}} B$). Hence, Eqs. (1.96) and (1.99) give $O = 1$ and $C = 0$, respectively. Since from Fig. (1.15a) we have $y = B - (I - x)$, where x is the thickness of the ice above sea level, Eq. (1.100) gives $x = \left(1 - \rho^{\mathrm{i}}/\rho^{\mathrm{w}}\right) I$.

1.4.2.3 Condition 3: Ice Grounded Below Sea Level

In this condition, illustrated in Fig. 1.15b, the topography is $T < 0$ and the ice thickness is $I > 0$, like in the case of floating ice (Fig. 1.15a). However, the system is not in isostatic equilibrium, and the ice column turns out to be heavier than the water column, i.e., $\rho^{\mathrm{i}} I \geq \rho^{\mathrm{w}} B$. This implies

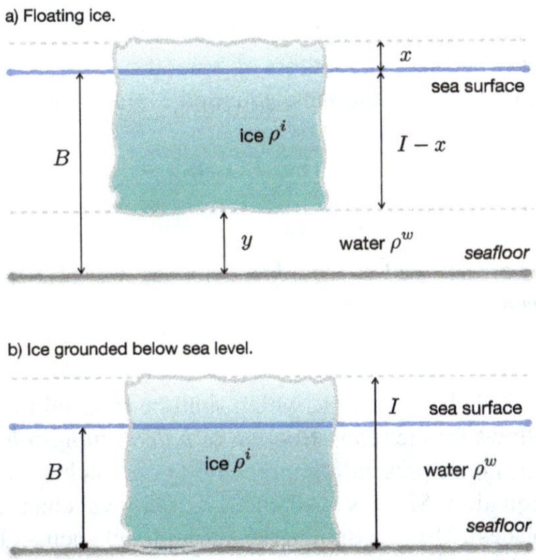

Fig. 1.15 Diagrams showing floating ice (**a**) and ice grounded below sea level (**b**), where I and B are the ice thickness and sea level, respectively

$$T(\gamma, t) + \frac{\rho^{\mathrm{i}}}{\rho^{\mathrm{w}}} I(\gamma, t) \geq 0\,, \tag{1.102}$$

hence according to definitions (1.96) and (1.99), we have $O = 0$ and $C = 1$.

1.4.2.4 Condition 4: Ice Grounded Above Sea Level

When ice is grounded above sea level, the topography and the ice thickness are $T > 0$ and $I > 0$, respectively, so that from Eqs. (1.96) and (1.99), we have $O = 0$ and $C = 1$.

1.4.2.5 Condition 5: Ice-Free Land

In ice-free elevated lands, where $T > 0$ and $I = 0$ (see Fig. 1.14), Eqs. (1.96) and (1.99) give $O = 0$ and $C = 1$, respectively.

1.4.2.6 The PT Equation

From the definition of topography given by Eq. (1.97), it is possible to draw a few important consequences.

We first define *sea-level change* by the difference

$$\mathcal{S}(\gamma, t) = B - B_0\,, \tag{1.103}$$

where $B_0 = B(\gamma, t_0)$ is sea level in a reference equilibrium state attained for $t \leq t_0$ (we shall set, without loss of generality, $t_0 = 0$). Then, taking the difference between Eq. (1.97) and the same equation evaluated at present time (i.e., $t = t_N$), we obtain

$$\begin{aligned} T(\gamma, t) - T(\gamma, t_N) &= -(B - B_N) \\ &= -((B - B_0) - (B_N - B_0))\,, \end{aligned} \tag{1.104}$$

where $B_N = B(\gamma, t_N)$. Hence, using Eq. (1.103) and defining $T_N = T(\gamma, t_N)$, we have

$$T(\gamma, t) = T_N - (\mathcal{S} - \mathcal{S}_N)\,. \tag{1.105}$$

This last equation, which shall be referred to as "PT equation" in the following, allows for the reconstruction of *paleo-topography* (i.e., the time history of topography) once present topography $T(\gamma, t_N)$ is known and the solution of the Sea Level Equation (SLE) is available (i.e., sea-level change $\mathcal{S}(\gamma, t)$ is known at all times and places). Using a digital elevation model such as ETOPO1 (see Amante and Eakins 2009; Eakins and Sharman 2012) as a representation for the present-day relief T_N, the PT equation can be used to refine the value of $T(\gamma, t)$ using successive iterations of the SLE (see Chap. 6 and Appendix E).

The method of reconstruction based on Eq. (1.105) has been first introduced by Peltier (1994) and since then applied to paleogeographic investigations made in the framework of GIA studies (see e.g., Spada and Galassi 2017, and references therein). It is worth to note that Peltier (1994) has also defined another quantity associated with topography, namely *true paleo-topography*, as follows

$$PT(\gamma, t) = T + I\,, \tag{1.106}$$

where T is topography and I is ice thickness. It is clear that $PT(\gamma, t)$ can be easily reconstructed once T has been determined by solving the SLE, since in GIA studies $I(\gamma, t)$ is usually given a priori.

1.4.3 Surface Loads

This section is devoted to the description of the *surface load function* $L(\gamma, t)$ associated to the ice and water distributions at the Earth's surface. We shall also consider its variation $\mathcal{L}(\gamma, t)$ with respect to a reference state since it has a central role in the following chapters.

1.4.3.1 The Surface Load Function

We define the *surface load function* $L(\gamma, t)$ by means of

$$M(t) = \int_e L\, \mathrm{d}A\,, \tag{1.107}$$

where $M(t)$ is the mass distributed over the whole Earth's surface in the form of ice or water, and

$$\mathrm{d}A = a^2 \mathrm{d}\gamma \tag{1.108}$$

is the area element, being a the average Earth's radius and

$$\mathrm{d}\gamma = \sin\theta\, \mathrm{d}\theta\, \mathrm{d}\lambda. \tag{1.109}$$

Hence, according to Eq. (1.107), the surface load represents the mass acting on the Earth's surface per unit area, with:

$$L(\gamma, t) = \frac{\mathrm{d}M}{\mathrm{d}A}\,. \tag{1.110}$$

1.4.3.2 Mass Distribution

At a given time t, the mass $M(t)$ distributed over the Earth's surface can be expressed as

$$M(t) = M^{\mathrm{i}} + M^{\mathrm{w}}\,, \tag{1.111}$$

where M^{i} and M^{w} are the mass of the ice and of the water at time t, respectively. These two terms are now analysed separately.

The ice mass term stems from two contributions, with

$$M^{\mathrm{i}}(t) = M^{\mathrm{i,gr}} + M^{\mathrm{i,fl}}\,, \tag{1.112}$$

where superscript gr denotes grounded ice—either above or below sea level—while fl indicates floating ice. By the definition of CF (see Eq. 1.99), we can write

$$M^{\mathrm{i}}(t) = \rho^{\mathrm{i}} \int_{\mathrm{e}} IC\mathrm{d}A + \rho^{\mathrm{i}} \int_{\mathrm{fl}} I\mathrm{d}A\,, \tag{1.113}$$

where the first integral is over the whole Earth's surface e and it accounts for all the grounded ice $M^{\mathrm{i,gr}}$, while the second is over marine regions with floating ice and represents the total mass of the floating ice $M^{\mathrm{i,fl}}$.

In a similar way, M^{w} stems from two terms:

$$M^{\mathrm{w}}(t) = M^{\mathrm{w,if}} + M^{\mathrm{w,fl}}, \tag{1.114}$$

where superscripts if and fl indicate contributions from the ice-free oceans and from ocean regions where there is floating ice, respectively. Using the condition of isostasy for floating ice (1.100), we obtain

$$M^{\mathrm{w}}(t) = \rho^{\mathrm{w}} \int_{\mathrm{if}} B\mathrm{d}A + \int_{\mathrm{fl}} \left(\rho^{\mathrm{w}} B - \rho^{\mathrm{i}} I\right) \mathrm{d}A\,. \tag{1.115}$$

Hence, using Eqs. (1.113) and (1.115) in (1.111), the mass of the (ice+water) system is

$$M(t) = \rho^{\mathrm{i}} \int_{\mathrm{e}} IC\mathrm{d}A + \cancel{\rho^{\mathrm{i}} \int_{\mathrm{fl}} I\mathrm{d}A} + \rho^{\mathrm{w}} \int_{\mathrm{if}} B\mathrm{d}A + \rho^{\mathrm{w}} \int_{\mathrm{fl}} B\mathrm{d}A - \cancel{\rho^{\mathrm{i}} \int_{\mathrm{fl}} I\mathrm{d}A}\,, \tag{1.116}$$

where we note that the integrals of I over region fl cancel out and those involving B can be merged since $\int_{\mathrm{if}} B\mathrm{d}A + \int_{\mathrm{fl}} B\mathrm{d}A = \int_{\mathrm{o}} B\mathrm{d}A$, where the subscript o denotes integration over the oceans. Taking advantage of the OF definition given by Eq. (1.96), we thus obtain

$$M(t) = \int_{\mathrm{e}} \left(\rho^{\mathrm{i}} IC + \rho^{\mathrm{w}} BO\right) \mathrm{d}A\,. \tag{1.117}$$

When compared to Eq. (1.107), this equation provides a compact and symmetrical expression for the surface load, which reads

$$L(\gamma, t) = \rho^{\mathrm{i}} IC + \rho^{\mathrm{w}} BO\,, \tag{1.118}$$

where the two terms on the *r.h.s.* describe the ice and the water contributions, respectively. This is an important result.

1.4.3.3 Surface Load Variation

We define the *surface load variation* as the difference between the surface load at a given time t and its value in a previous reference state:

$$\mathcal{L}(\gamma, t) = L - L_0\,, \tag{1.119}$$

where $L_0 = L(\gamma, t_0)$. Using Eq. (1.118) gives

$$\mathcal{L}(\gamma, t) = \rho^{\mathrm{i}}(IC - I_0 C_0) + \rho^{\mathrm{w}}(BO - B_0 O_0)\,. \tag{1.120}$$

We now define the *ice thickness variation* as

$$\mathcal{I}(\gamma, t) = I - I_0\,, \tag{1.121}$$

and the *ocean function variation*

$$\mathcal{O}(\gamma, t) = O - O_0\,, \tag{1.122}$$

where subscript 0 indicates reference values, i.e., $I_0 \equiv I(\gamma, t_0)$ and $O_0 \equiv O(\gamma, t_0)$. Also using the definition of topography (1.97) and of sea-level change (1.103), we obtain

$$\begin{aligned} \mathcal{L}(\gamma, t) &= \rho^{\mathrm{i}}\left((I_0 + \mathcal{I})C - I_0 C_0\right) + \rho^{\mathrm{w}}\left((B_0 + \mathcal{S})O - B_0 O_0\right) \\ &= \rho^{\mathrm{i}}\mathcal{I}C + \rho^{\mathrm{i}} I_0 (C - C_0) + \rho^{\mathrm{w}}\mathcal{S}O + \rho^{\mathrm{w}} B_0 (O - O_0) \\ &= \rho^{\mathrm{i}}\mathcal{I}C - \rho^{\mathrm{i}} I_0 (O - O_0) + \rho^{\mathrm{w}}\mathcal{S}O - \rho^{\mathrm{w}} T_0 (O - O_0) \\ &= \rho^{\mathrm{i}} C\mathcal{I} + \rho^{\mathrm{w}} O\mathcal{S} - (\rho^{\mathrm{i}} I_0 + \rho^{\mathrm{w}} T_0)\mathcal{O}\,. \end{aligned} \tag{1.123}$$

Thus, the load variation can be conveniently written as the sum of three terms:

$$\mathcal{L}(\gamma, t) = \mathcal{L}^{\mathrm{i}} + \mathcal{L}^{\mathrm{s}} + \mathcal{L}^{\mathrm{o}}, \tag{1.124}$$

with

$$\mathcal{L}^{\mathrm{i}}(\gamma, t) \equiv \rho^{\mathrm{i}} \mathcal{W} \tag{1.125}$$
$$\mathcal{L}^{\mathrm{s}}(\gamma, t) \equiv \rho^{\mathrm{w}} \mathcal{Z} \tag{1.126}$$
$$\mathcal{L}^{\mathrm{o}}(\gamma, t) \equiv \rho^{\mathrm{r}} \mathcal{X}, \tag{1.127}$$

where

$$\mathcal{W}(\gamma, t) \equiv \mathcal{C}\mathcal{I} \tag{1.128}$$
$$\mathcal{Z}(\gamma, t) \equiv \mathcal{O}\mathcal{S} \tag{1.129}$$
$$\mathcal{X}(\gamma, t) \equiv \mathcal{Q}\mathcal{O}, \tag{1.130}$$

and we have defined the auxiliary variable

$$\mathcal{Q}(\gamma) \equiv -\left(\frac{\rho^{\mathrm{i}}}{\rho^{\mathrm{r}}} I_0 + \frac{\rho^{\mathrm{w}}}{\rho^{\mathrm{r}}} T_0\right), \tag{1.131}$$

being $\rho^{\mathrm{r}} > 0$ an arbitrary reference density. We note that adopting the decomposition (1.124) for the surface load variation $\mathcal{L}$, variables $\mathcal{W}$, $\mathcal{Z}$ and $\mathcal{X}$ (hence $\mathcal{L}^{\mathrm{i}}$, $\mathcal{L}^{\mathrm{s}}$ and $\mathcal{L}^{\mathrm{o}}$) are uniquely associated to variations of the ice thickness, of sea level and of the OF, respectively.

Variable $\mathcal{Z}$, defined by (1.129), can be interpreted as the "projection" of sea-level change $\mathcal{S}$ over the OF. It will play an important role in the development of the pseudo-spectral approach to the SLE in Chap. 5. We note that, symmetrically, variable $\mathcal{W}$ can be viewed as the projection of the ice thickness variation $\mathcal{I}$ over the CF.

1.4.4 Mass Conservation

The mass conservation principle is paramount in the framework of the Sea Level Equation. It takes several forms, which are discussed in this section.

Essentially, the mass conservation principle states that

$$M(t) = M_0\,, \tag{1.132}$$

where $M(t)$ is the mass of the (ice+water) system at a given time t and M_0 is the mass at the reference time $t_0 \le t$. Note that we assume that the mass of the solid Earth is unvaried.

Defining the *mass variation* as

$$\mathcal{M}(t) = M - M_0\,, \tag{1.133}$$

the mass conservation principle can be stated in an integral form as

$$\mathcal{M}(t) \equiv \int_{e} \mathcal{L} \, dA = 0, \tag{1.134}$$

where $\mathcal{L}$ is the surface load variation defined by Eq. (1.119). An equivalent expression of the mass conservation principle is

$$\langle \mathcal{L}(\gamma, t) \rangle^{e} = 0, \tag{1.135}$$

where the average over the *whole Earth's surface* of a scalar function $F = F(\gamma, t)$, evaluated at time t, is defined as

$$\langle F \rangle^{e}(t) \equiv \frac{1}{A^{e}} \int_{e} F(\gamma, t) \, dA. \tag{1.136}$$

Using Eq. (A.38), the mass conservation principle (1.135) can be also expressed as

$$\mathcal{L}_{00}(t) = 0, \tag{1.137}$$

where $\mathcal{L}_{00}$ is the coefficient of degree $n = 0$ and order $m = 0$ of the expansion of $\mathcal{L}(\gamma, t)$ in series of complex spherical harmonics (CSHs). In this work we shall only consider *physically plausible* loads, which obey the principle of mass conservation either given by (1.134) or by the equivalent forms (1.135) or (1.137).

Using the expression for the surface load variation given by Eq. (1.120), the mass conservation principle can be also stated as

$$\mathcal{M}(t) = \rho^{i} \int_{e} (IC - I_0 C_0) \, dA + \rho^{w} \int_{e} (BO - B_0 O_0) \, dA = 0, \tag{1.138}$$

which can be rearranged as follows

$$\mu(t) + \rho^{w} \int_{e} (BO - B_0 O_0) \, dA = 0, \tag{1.139}$$

where

$$\mu(t) = \rho^{i} \int_{e} (IC - I_0 C_0) \, dA \tag{1.140}$$

represents the variation of the *grounded ice mass* $M^{i,gr}$ (see Eq. 1.113).

Now it is useful to transform the mass conservation condition (1.139) in another form that involves the sea-level variation $\mathcal{S}$ explicitly. Indeed, using the definition of sea-level change (1.103) and of topography (1.97), we have

$$\begin{aligned} BO - B_0 O_0 &= (B - B_0 + B_0)O - B_0 O_0 \\ &= (B - B_0)O + B_0(O - O_0) \\ &= \mathcal{S}O + B_0 O \\ &= \mathcal{S}O - T_0 O, \end{aligned} \tag{1.141}$$

so that[12] (1.139) becomes

$$\mu(t) + \rho^{\mathrm{w}} \int_{\mathrm{e}} \mathcal{S}O \, \mathrm{d}A - \rho^{\mathrm{w}} \int_{\mathrm{e}} T_0 O \, \mathrm{d}A = 0. \tag{1.142}$$

We now introduce the operation of *ocean average* of a scalar function $F(\gamma, t)$, defined as

$$\langle F \rangle^{\mathrm{o}}(t) \equiv \frac{1}{A^{\mathrm{o}}} \int_{\mathrm{o}} F \, \mathrm{d}A, \tag{1.143}$$

where label o stands for ocean, the integration is over the region where $O = 1$ at time t, and

$$A^{\mathrm{o}}(t) = \int_{\mathrm{o}} \mathrm{d}A \tag{1.144}$$

is the *area of the oceans* at the same time. Using the definition of OF, the average $\langle F \rangle^{\mathrm{o}}(t)$ can be also expressed as an integral over the whole Earth's surface, with

$$\langle F \rangle^{\mathrm{o}}(t) \equiv \frac{1}{A^{\mathrm{o}}} \int_{\mathrm{e}} F \, O \, \mathrm{d}A, \tag{1.145}$$

or, equivalently, as

$$\langle F \rangle^{\mathrm{o}}(t) \equiv \frac{A^{e}}{A^{o}} \langle \, F \, O \, \rangle^{\mathrm{e}}, \tag{1.146}$$

where we have used the definition of whole-Earth surface average given by Eq. (1.136). By (1.146), the mass conservation condition (1.142) can be cast in the compact form

$$\mu(t) + \rho^{\mathrm{w}} A^{\mathrm{o}} \langle \mathcal{S} \rangle^{o} - \rho^{\mathrm{w}} A^{\mathrm{e}} \langle T_0 \, O \rangle^{\mathrm{e}} = 0, \tag{1.147}$$

which shall be useful in ensuing chapters to impose the mass conservation constraint.

---

[12] It is easily verified that $BO - B_0 O_0 = \mathcal{S} O_0 - TO$ is also true, in addition to (1.141). Hence the following interesting, more symmetrical identity exists, which would demand some attention: $BO - B_0 O_0 = {}^1\!/_2 \left[\mathcal{S} (O + O_0) + (T + T_0) O \right]$.

References

Amante C, Eakins B (2009) ETOPO1 arc-minute global relief model: procedures, data source and analysis. Technical Memorandum NESDIS NGDC; 24, National Oceanic and Atmospheric Administration

Balmino G, Lambeck K, Kaula WM (1973) A spherical harmonic analysis of the Earth's topography. J Geophys Res 78(2):478–481

Boughanemi A, Mémin A (2024) Effects of Andrade and Burgers rheologies on glacial isostatic adjustment modeling in Antarctica. Geodesy Geodyn 15(5):429–440

Boyaval S (2021) Viscoelastic flows of Maxwell fluids with conservation laws. ESAIM: Math Model Numer Anal 55(3):807–831

Cannelli V, Melini D, Piersanti A (2010) Post-seismic stress relaxation with a linear transient rheology. Ann Geophys 53(2):89–99

Caron L, Métivier L, Greff-Lefftz M, Fleitout L, Rouby H (2017) Inverting Glacial Isostatic Adjustment signal using Bayesian framework and two linearly relaxing rheologies. Geophys J Int 209(2):1126–1147

Christensen R (1982) Theory of viscoelasticity. Dover, Mineola, New York

Darwin GH (1908) Scientific papers, vol II. University Press

Eakins B, Sharman G (2012) Hypsographic curve of Earth's surface from ETOPO1. NOAA National Geophysical Data Center, Boulder, CO

Franco JM, Partal P (2010) The Newtonian fluid. Rheology 1:74–95

Fung Y (1965) Foundation of solid mechanics. Prentice-Hall, Englewood Cliffs, NJ

Heiskanen WA, Moritz H (1981) Physical geodesy. Institute of Physical Geodesy, Technical University

Hetland EA, Hager BH (2005) Postseismic and interseismic displacements near a strike-slip fault: a two-dimensional theory for general linear viscoelastic rheologies. J Geophys Res Solid Earth 110(B10)

Hetland E, Hager BH (2006) The effects of rheological layering on post-seismic deformation. Geophys J Int 166(1):277–292

Hofmann-Wellenhof B, Moritz H (2006) Physical geodesy. Springer

Karato S-I (2008) Deformation of Earth materials: an introduction to the rheology of solid Earth. Cambridge University Press

Karato S-I (2010) Rheology of the Earth's mantle: a historical review. Gondwana Res 18(1):17–45

Kaula W (2013) Theory of satellite geodesy: applications of satellites to geodesy, Dover Earth Science. Dover Publications

Kelly P (2019) Mechanics lecture notes – an introduction to solid mechanics. The University of Auckland

Knopoff L (1964) Q Rev Geophys 2:625–660

Körnig M, Müller G (1989) Rheological models and interpretation of postglacial uplift. Geophys J Int 98(2):243–253

Lambeck K (1980) The Earth's variable rotation: geophysical causes and consequences. Cambridge University Press

Lambeck K (1988) Geophysical geodesy. Clarendon Oxford

Landau LD, Lifsic EM, Pitaevskii L, Kosevich A (1986) Course of theoretical physics: volume 7, theory of elasticity. Pergamon Press

Mainardi F (2010) Fractional calculus and waves in linear viscoelasticity: an introduction to mathematical models. World Scientific

Mainardi F, Spada G (2011) Creep, relaxation and viscosity properties for basic fractional models in rheology. Eur Phys J Spec Top 193(1):133–160

Maxwell JC (1867) IV. On the dynamical theory of gases. Philos Trans Royal Soc Lond 157:49–88

Melini D, Cannelli V, Piersanti A, Spada G (2008) Post-seismic rebound of a spherical Earth: new insights from the application of the Post-Widder inversion formula. Geophys J Int 174(2):672–695

Moritz H (1980) Advanced physical geodesy. Wichmann, Karlsruhe

Müller G (1986) Generalized Maxwell bodies and estimates of mantle viscosity. Geophys J Int 87(3):1113–1141

Munk WH, MacDonald GJ (1975) The rotation of the Earth: a geophysical discussion. Cambridge University Press

Okuno J, Nakada M (2001) Effects of water load on geophysical signals due to glacial rebound and implications for mantle viscosity. Earth Planets Space 53(12):1121–1135

Peltier WR, Wu P, Yuen DA (1981) The viscosities of the Earth's mantle. In: Anelasticity in the Earth, pp 59–77. American Geophysical Union (AGU)

Peltier W (1985) New constraints on transient lower mantle rheology and internal mantle buoyancy from glacial rebound data. Nature 318(6047):614–617

Peltier WR (1994) Ice age paleotopography. Science 265(5169):195–201

Peltier W, Yuen D, Wu P (1980) Postglacial rebound and transient rheology. Geophys Res Lett 7(10):733–736

Peltier W, Drummond R, Tushingham A (1986) Post-glacial rebound and transient lower mantle rheology. Geophys J Int 87(1):79–116

Pollitz FF (2003) Transient rheology of the uppermost mantle beneath the Mojave Desert, California. Earth Planet Sci Lett 215(1–2):89–104

Pollitz FF (2005) Transient rheology of the upper mantle beneath central Alaska inferred from the crustal velocity field following the 2002 Denali earthquake. J Geophys Res Solid Earth 110(B8)

Pollitz FF, Bürgmann R, Banerjee P (2006) Post-seismic relaxation following the great 2004 Sumatra-Andaman earthquake on a compressible self-gravitating Earth. Geophys J Int 167(1):397–420

Prey A (1922) Darstellung der Höhen-und Tiefenverhältnisse der Erde durch eine Entwicklung nach Kugelfunktionen bis zur 16. Weidmannsche Buchhandlung, Ordnung, Abhandlungen der königliche Gesellschaft der Wissenschaften zu Göttingen, Mathematisch-Physikalische Klasse

Ranalli G (1995) Rheology of the Earth. Chapman and Hall

Reiner M (1949) Twelve lectures on theoretical rheology. North-Holland Publishing Company

Reiner M (1945) A classification of rheological properties. J Sci Instrum 22(7):127

Rochester M, Smylie D (1974) On changes in the trace of the Earth's inertia tensor. J Geophys Res 79(32):4948–4951

Sabadini R, Yuen D, Gasperini P (1985) The effects of transient rheology on the interpretation of lower mantle viscosity. Geophys Res Lett 12(6):361–364

Sabadini R, Smith BK, Yuen DA (1987) Consequences of experimental transient rheology. Geophys Res Lett 14(8):816–819

Spada G (2017) Glacial isostatic adjustment and contemporary sea level rise: an overview. Surv Geophys 38(1):1–33

Spada G, Galassi G (2017) Extent and dynamic evolution of the lost land aquaterra since the last glacial maximum. Comptes Rendus Geosci 349(4):151–158

Spada G, Colleoni F, Ruggieri G (2011) Shallow upper mantle rheology and secular ice sheet fluctuations. Tectonophysics 511(3–4):89–98

Sun T, Wang K, Iinuma T, Hino R, He J, Fujimoto H, Kido M, Osada Y, Miura S, Ohta Y et al (2014) Prevalence of viscoelastic relaxation after the 2011 Tohoku-oki earthquake. Nature 514(7520):84–87

Tanner RI, Walters K (1998) Rheology: an historical perspective. Elsevier

Thomson W (1865) On the elasticity and viscosity of metals. Proc R Soc Lond 14:289–297

Torge W, Müller J (2012) Geodesy. Walter de Gruyter

Tschoegl NW (2012) The phenomenological theory of linear viscoelastic behavior: an introduction. Springer

Voigt W (1892) Ueber innere Reibung fester Körper, insbesondere der Metalle. Annalen der Physik 283(12):671–693

Walters K (2010) History of rheology. Hist Philos Sci Technol III:91

Whitehouse PL (2018) Glacial isostatic adjustment modelling: historical perspectives, recent advances, and future directions. Earth Surface Dyn 6(2):401–429

Yuen DA, Sabadini RCA, Gasperini P, Boschi E (1986) On transient rheology and glacial isostasy. J Geophys Res 91:11,420–11,438

Zener C (1948) Elasticity and anelasticity of metals. University of Chicago Press

Chapter 2
Love Numbers and Green's Functions

There is a tide in the affairs of men,
which, taken at the flood, leads on to fortune.

(Julius Caesar: Act IV)

2.1 Introduction

Love numbers, named after the English polymath Augustus Edward Hough Love (1863–1940), are non-dimensional quantities useful to describe the response of a spherically symmetric planet to various sources of excitation (Love 1909, 1911, 1944).

According to their classical definition, the Love numbers represent the response to the tidal attraction exerted by an external body or to a localised and impulsive surface load. In the first case, they are referred to as *tidal Love numbers* (TLNs), in the second as *loading Love numbers* (LLNs). However, in the literature the LLNs are often referred to as *load-deformation coefficients* (see e.g., Longman 1962, 1963) while the term *Love number* should be used to indicate the tidal numbers. The TLNs for horizontal displacements are also sometimes referred to as *Shida numbers*, after the Japanese geodesist Toshi Shida (1912a, b).

Some recent contributions demonstrate that the study of Love numbers is still a matter of active research (see Spada 2008; Na and Baek 2011; Spada et al. 2011a; Bevis et al. 2016; Chen et al. 2017; Kachuck and Cathles 2018, 2019; Fowler et al. 2019; Caron et al. 2025, and references therein). Furthermore, Love numbers have widespread application in the field of planetology (see Zhang 1992; Van Hoolst et al. 2003; Wahr et al. 2006; Sotin et al. 2009; Melini et al. 2022, just to mention a few

G. Spada and D. Melini, *Glacial Isostatic Adjustment*, Springer Geophysics,
https://doi.org/10.1007/978-3-031-90538-4_2

examples), since they provide an indirect probe of the large-scale structure of the interior of celestial bodies.

Love numbers and the associated Green's functions, sometimes referred to as *space-time Green's functions* (see e.g., Peltier 1974), have a paramount importance in context of GIA. However, the Love number formalism is not immediately intuitive. In this chapter we only give the essentials on this subject, necessary to describe the Earth's response to distributed surface loads and to external potentials, which are fundamental ingredients of the Sea Level Equation. The readers are referred to Love (1909, 1911, 1944), Longman (1962, 1963) and Farrell (1972) for the fundamental definitions, and to Peltier (1974), Lambeck (1980), Wu and Peltier (1982, 1984) for a detailed discussion about Love numbers in the context of GIA. For their role in the study of the rotational dynamics of the Earth, the readers are referred to the treaties of Munk and MacDonald (1975) and Lambeck (1980), while for a condensed review also describing the various analytical and numerical methods available for the computation of the Love numbers, see Spada et al. (2011a) and references therein. This same work also provides community-agreed, high-precision numerical values of Love numbers for a realistic 1-D Earth model with Maxwell rheology.

The chapter is organised as follows. In Sect. 2.2 we describe the geometrical and physical features of the Earth models that will be considered throughout this book, which are all characterized by a spherical symmetry before deformation occurs. Then, we introduce the loading Green's functions for an elastic Earth, which are subsequently extended to a general viscoelastic rheology in Sect. 2.3. In the final section Sect. 2.4, some methods for the computation of the Love numbers are discussed, also providing a few numerical results. All these topics are essential ingredients for the construction of the surface and of the rotational *response functions*, which shall be discussed in the next two chapters.

2.2 Spherically Symmetric Models

All the Earth models that we shall consider in this book are characterised by the spherically symmetric structure illustrated in Fig. 2.1. In the literature, these models are often referred to as "1-D Earth models" or sometimes "1-D GIA models", since by virtue of spherical symmetry, the density field, the elastic constants and the rheological parameters only depend upon radius r. However, since the surface loads or the gravitational pull from external bodies are unevenly distributed over the Earth's surface, the response of 1-D Earth models shall be 3-D and time-dependent, in general. We note that the 1-D Earth models are more general than the spherically symmetric non-rotating elastic and isotropic (SNREI) models, since these only account for the Earth's elastic response.

The 1-D models are still having a fundamental role in the development of the Post Glacial Rebound (PGR) and of the Glacial Isostatic Adjustment (GIA) theories for a spherical Earth (for a review, the readers are referred to Whitehouse 2009; Spada 2017; Whitehouse 2018). The reason is that for 1-D GIA models well established

Fig. 2.1 An example of 1-D spherically symmetric GIA model. The mantle is composed of various viscoelastic layers of arbitrary linear rheology and thickness, shear modulus and density. In the case shown, the lithosphere (in blue) is assumed to be homogeneous while the core (green) is uniform

analytical results exist, and in their numerical implementations the computational burden is generally modest (Spada et al. 2011a). At the same time, they are characterised by a sufficient degree of realism and they can be adapted satisfactorily to existing geophysical and geodetic datasets (Whitehouse 2018). Publicly available computer programs for the simulation of PGR (see Spada et al. 2011a, and references therein) and of GIA (Spada and Melini 2019b) are based on the assumption of spherical symmetry. References to fundamental papers on 1-D GIA models shall be given throughout the chapter.

In 1-D GIA modelling, it is *normally* assumed that: *(i)* the Earth is isotropic, self-gravitating, and incompressible (i.e., its volume is constant), *(ii)* the core is fluid, inviscid (i.e., of vanishing viscosity) and uniform,[1] *(iii)* the Earth's mantle is spherically layered and characterised by a Maxwell linear viscoelastic rheology, *(iv)* the lithosphere is purely elastic. However, it is important to remark that the Love numbers formalism has been extended to a compressible Earth (see e.g., Vermeersen et al. 1996b; Sabadini et al. 2016), to a realistically layered core (e.g., Melini et al. 2022), to "generalised Maxwell rheologies" (e.g., Spada and Boschi 2006) and to non-elastic layered lithospheres (see e.g., Barletta et al. 2006). Furthermore, *non-self-gravitating* and *non-gravitating*[2] 1-D GIA models can also be considered within the same theoretical framework (see Wu and Ni 1996). Some details about 1-D compressible models shall be discussed in Sect. C.3.

In the case of 1-D GIA models, the response of the Earth is conveniently described by Love numbers (Love 1911), for which closed-form solutions are available in some particular cases (Wu and Peltier 1982; Wu and Ni 1996; Vermeersen et al. 1996a; Vermeersen and Sabadini 1997). For the simplest 1-D GIA model, constituted by a homogeneous incompressible sphere (the so-called *Kelvin Earth model*), an elegant

[1] For a discussion about the effect of a stratified core on GIA, see e.g. Klemann and Martinec (2011) or Huang et al. (2023).

[2] A model is *non-self-gravitating* when gravity does not change according to the deformation state of the Earth. It is *non-gravitating* when the gravity force is simply neglected.

analytical expression of the Love numbers can be established (see e.g., Munk and MacDonald 1975), which shall be derived in Sect. C.6 of the Appendix and discussed in Sect. 2.4.1 below. However, the complexity of the analytical solutions increases considerably with increasing number of layers, even excluding the effects of self-gravitation (see Wu and Ni 1996). Some details about the semi-anaytical schemes for obtaining Love numbers of multi-layered models are given in Sect. C.7.

PGR or GIA simulators like TABOO (Spada 2003; Spada et al. 2011a) or SELEN[4] (Spada and Melini 2019b) are designed to assimilate Love numbers of general multi-layered 1-D models. These are often computed making use of the well established *viscoelastic normal modes* (VNM) theory, introduced by Peltier (1974) and discussed in Appendix C. The VNM approach can be extended to more advanced models, like those characterised by compressible (e.g., Vermeersen et al. 1996b) or transient rheologies (e.g., Yuen et al. 1986; Caron et al. 2017). However, valid alternatives to the VNM approach have been proposed where the traditional Laplace domain technique is complemented by time-domain methods (Spada 2008; Melini et al. 2022).

In 1-D spherically symmetric models, all physical parameters only depend upon radius r. This holds true for the Lamé constants

$$\lambda = \lambda(r) \tag{2.1}$$

$$\mu = \mu(r), \tag{2.2}$$

and for density

$$\rho = \rho(r), \tag{2.3}$$

but also for the rheological parameters, such as for example newtonian viscosity:

$$\eta = \eta(r). \tag{2.4}$$

For the developments of the viscoelastic field equations (see Sect. C.1), the background gravitational acceleration $g_0(r)$ is of some importance. For a 1-D spherically symmetric Earth, it is well known that the gravitational acceleration is

$$g_0(r) = \frac{G M_{\text{in}}}{r^2}, \tag{2.5}$$

where $M_{\text{in}} = M_{\text{in}}(r)$ is the mass inside the sphere of radius r. Hence

$$g_0(r) = \frac{4\pi G}{r^2} \int_0^r z^2 \rho(z)\,\mathrm{d}z, \tag{2.6}$$

which implies:

$$g_0'(r) + \frac{2g_0(r)}{r} = 4\pi G\rho(r), \tag{2.7}$$

where the prime denotes the radial derivative.[3]

2.2.1 *Elastic Green's Functions*

The Green's functions (GFs) for surface loading aim at solving the following problem:

> *Given a 1-D Earth model, determine its response to a unit, impulsive localised (point-like) load acting at the surface, as a function of angular distance and time.*

The solution of this problem is *not* straightforward. As we shall see, GFs are tightly interconnected with Love numbers. In this section, following Spada and Stocchi (2006), we first discuss the case of a rigid Earth, introducing the so-called *direct GF* for the potential of the gravity field. Then, we move to the case of an elastic Earth, where *incremental GFs* are introduced to describe the additional variation in the gravitational potential and the components of the displacement vector in response to surface loading. At the end of the section, we provide an example concerning the elastic GFs for a modern spherically symmetric non-rotating elastic and isotropic (SNREI) seismological model.

2.2.2 *Direct GF*

Here we consider a mass acting through its own weight on the surface of a rigid, spherically symmetric Earth. Due to the assumption of rigidity, the surface stress exerted does not cause deformation. However, the load induces a *direct variation* of the Earth's gravitational potential, which is worth to study before considering the effects of deformation further on.

[3] We note that (2.7) can be equivalently obtained from Poisson's equation $\nabla^2\phi_0 = 4\pi G\rho$, where ϕ_0 is the gravitational potential. Indeed, in a spherically symmetric model, ϕ_0 is a function of r only, hence by using the expression (B.32) for the laplacian operator in spherical coordinates we obtain $\phi_0'' + \frac{2}{r}\phi_0 = 4\pi G\rho$ which, taking into account that the magnitude of the downward-directed gravity vector $\vec{g}_0$ is $g_0 = \phi_0'$, is coincident with (2.7).

2.2.2.1 Unit Surface Load

The point mass is characterised by an impulsive time history

$$\mu(t) = \delta(t)\, \Delta m, \tag{2.8}$$

where δ is Dirac's delta and Δm is the intrinsic mass, which is assumed to be negligible with respect to the whole Earth's mass ($\Delta m/m^{\mathrm{e}} \ll 1$). The actual value of Δm is immaterial to our discussion; we can assume $\Delta m = 1$ kg without loss of generality. We note that, with this definition, $\mu(t)$ has the physical dimensions of $[\mathrm{MT}^{-1}]$. The mass is placed at an arbitrary point $\gamma' = (\theta', \lambda')$ on the Earth's surface. The corresponding localised surface load function $\mathcal{L}^{\Delta}$ can be expressed in terms of mass per unit area and unit time, again with the aid of Dirac's delta, as follows

$$\mathcal{L}^{\Delta}(\alpha, t) = \frac{\mu(t)}{a^2}\, \frac{\delta(\theta - \theta')\, \delta(\lambda - \lambda')}{\sin\theta}, \tag{2.9}$$

where the properties of Dirac's delta ensure that

$$\int_{\mathrm{e}} \mathcal{L}^{\Delta}(\alpha, t)\, \mathrm{d}A = \mu(t), \tag{2.10}$$

where the integral is over the whole surface of the Earth.

2.2.2.2 Direct Contribution to Gravitational Potential

It is now of interest to evaluate the variation of the gravitational potential at point O of spherical coordinates (θ, λ), where a hypothetical observer is located. The geometry of the problem is sketched in Fig. 2.2. By Newton's law of gravitation, the gravitational potential (per unit time) at O is

$$\varphi^{\mathrm{r}}(\alpha, t) = -\frac{G\, \mu(t)}{d(\alpha)}, \tag{2.11}$$

where superscript r reminds that we are dealing—at least for the moment—with a *rigid* Earth, α is the great circle angle between the load and the observation point (i.e., the co-latitude of the observer relative to the point mass), d is the distance separating the mass from the observer, and

$$G = 6.674 \times 10^{-11}\ \mathrm{N\, m^2\, kg^{-2}} \tag{2.12}$$

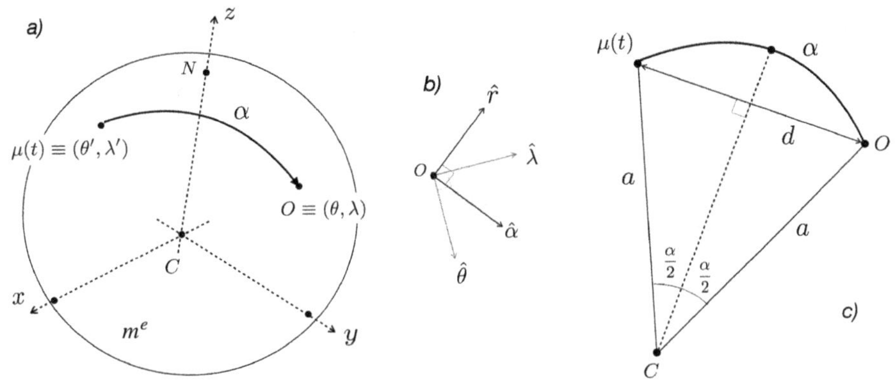

Fig. 2.2 Geometry of the source-observer system, useful for constructing the GFs. Here $\mu(t)$ is a localised impulsive mass and O is an observer at the Earth's surface. The two points are separated by a great circle angle α ($0 \leq \alpha \leq \pi$) and by a linear distance d ($0 \leq d \leq 2a$). With $\hat{\theta}$ and $\hat{\lambda}$ we denote the unit vectors in the direction of increasing geographical co-latitude θ and longitude λ, respectively. The tangent unit vector $\hat{\alpha}$ points to increasing values of α. The Earth's mass and radius are m^e and a, respectively. Points C and N denote the CM and the north pole, respectively

is Newton's constant. It is worth to note that α is implicitly a function of γ' and γ via the *cosine rule for sides* in spherical trigonometry, namely:

$$\alpha = \alpha(\gamma, \gamma') = \cos^{-1}\left[\cos\theta\cos\theta' + \sin\theta\sin\theta'\cos(\lambda - \lambda')\right] \tag{2.13}$$

(see e.g., Donnay 2011).

The potential $\varphi^r(\alpha, t)$ adds to the background gravitational potential, i.e., to the gravitational potential of the unperturbed 1-D Earth. In the literature, $\varphi^r(\alpha, t)$ is referred to as *direct* (or *rigid*) contribution to gravitational potential. By elementary plane trigonometry (see Fig. 2.2c), we have

$$d(\alpha) = 2a \sin\left(\frac{\alpha}{2}\right), \tag{2.14}$$

so that Eq. (2.11) gives

$$\varphi^r(\alpha, t) = -\delta(t)\,\frac{a\,g_0}{m^e}\,\frac{\Delta m}{2\,\sin\left(\frac{\alpha}{2}\right)}, \tag{2.15}$$

where

$$g_0 = \frac{Gm^e}{a^2} \tag{2.16}$$

is surface gravitational acceleration in the unperturbed state.

2.2.2.3 Green's Function for the Direct Gravitational Potential

The Green's function (GF) associated with the direct gravitational potential is simply defined as the potential variation per unit applied mass:

$$\Gamma^{\varphi,\mathrm{r}}(\alpha,t) \equiv \frac{\varphi^{\mathrm{r}}}{\Delta m}, \tag{2.17}$$

so that from (2.15) we have

$$\Gamma^{\varphi,\mathrm{r}}(\alpha,t) = -\delta(t)\,\frac{a\,g_0}{m^{\mathrm{e}}}\,\frac{1}{2\sin\left(\frac{\alpha}{2}\right)}, \tag{2.18}$$

having physical dimensions of a potential per unit time and unit mass.

A fully equivalent but more insightful expression for $\Gamma^{\varphi,\mathrm{r}}(\alpha,t)$ is obtained by recalling an elegant analytical result involving Legendre polynomials, namely:

$$\frac{1}{2\sin\left(\frac{\alpha}{2}\right)} \equiv \sum_{n=0}^{\infty} P_n(\cos\alpha), \tag{2.19}$$

which represents one of the so-called *Legendre sums*[4] that have been recently reconsidered by Fowler et al. (2019) to facilitate the numerical evaluation of the GFs. From (2.18), this result allows one to establish "in a natural way" the following *spectral representation* of the rigid GF

$$\Gamma^{\varphi,\mathrm{r}}(\alpha,t) = \delta(t)\sum_{n=0}^{\infty} \varphi_n^{r} P_n(\cos\alpha)\,, \tag{2.20}$$

where we observe that the coefficients of the expansion

$$\varphi_n^{\mathrm{r}} = -\frac{a\,g_0}{m^{\mathrm{e}}} \equiv -\frac{G}{a} \tag{2.21}$$

happen to be not actually dependent upon the harmonic degree n (i.e., their spectrum is *white*).

From Eq. (2.18) we note that $\Gamma^{\varphi,\mathrm{r}}(\alpha,t)$ diverges as α^{-1} for $\alpha \to 0$ in consequence of the localised, pointlike nature of the loading mass; furthermore, it is worth to observe that $\Gamma^{\varphi,\mathrm{r}}(\alpha,t)$ does not change sign over the interval $0 \le \alpha \le \pi$. As we shall see in the ensuing sections, the singularity of $\Gamma^{\varphi,\mathrm{r}}(\alpha,t)$ for $\alpha \to 0$ shall disappear as soon as finite-sized surface mass distributions are considered.

---

[4] More Legendre sums are reported in Sect. A.1.1, see in particular Eqs. (A.8)–(A.12) but see also Farrell (1972).

2.2.3 *Incremental GFs*

For a deformable Earth, the change in the mass distribution contributes an *indirect* (or *incremental*) gravitational potential variation, which adds up to the direct effect. Physically, this is a manifestation of the body *self-gravitation*, meaning that gravitational potential varies according to the mass redistribution forced by the surface load (see e.g., Wu and Peltier 1982). Henceforth, in the particular case of an elastic body, we construct a GF for the incremental gravitational potential and for the three components of the surface displacement vector. It is important to note that, due to the assumption of spherical symmetry, the displacement vector at the observer location O shall have components only along the $\hat{r}$ direction (increasing radius), and along $\hat{\alpha}$ (increasing source-observer angular separation α, see Fig. 2.2b). For the surface forcing considered here, a transverse component of displacement (normal to the plane defined by $\hat{r}$ and $\hat{\alpha}$) would only be observed in the case of a laterally heterogeneous, 3-D Earth.

In the following of this section, we shall limit ourselves to the basic elements necessary to construct the GFs for an elastic Earth. Results concerning the loading Love numbers and GFs for a suite of SNREI models currently employed in seismology, have been presented in the work of Wang et al. (2012), where a procedure for the computation of the elastic LLNs is also outlined. We remark that those presented below constitute the most basic set of GFs, necessary to undertake the study of the Sea Level Equation further on. Indeed, using suitable combinations of LLNs, it is possible to construct GFs corresponding to other geophysical variables of interest, like e.g., gravity anomalies and tilts induced by surface loading (Longman 1962, 1963). In the seminal work of Peltier (1974), these have been generalised to the case of a Maxwell Earth.

2.2.3.1 Incremental Potential

Due to the hypothesis of elasticity, we can certainly assert that the GF associated to the indirect gravitational potential varies in phase with the source. Hence, in analogy with (2.20), the contribution of elasticity to the GF can be expanded in series of LPs as

$$\Gamma^{\varphi,\mathrm{e}}(\alpha, t) = \delta(t) \sum_{n=0}^{\infty} \varphi_n^{\mathrm{e}} \, P_n(\cos\alpha)\,, \tag{2.22}$$

where now the harmonic coefficients φ_n^{e} are, in general, dependent upon the degree n. Their effective value is ultimately determined by solving the equilibrium equations for a SNREI Earth (Dahlen 1968), once the elastic constants and the density profile have been assigned and boundary conditions have been imposed (see Sun and Sjöberg 1998, and references therein). For a detailed description on how the equilibrium equations are solved for a general 1-D viscoelastic Earth model, and of the appropriate boundary conditions, the readers are referred to Appendix C. However, in order to obtain a suitable form for the incremental potential, and to define the elastic Love numbers, an explicit solution of the equations is not required.

Based upon the fundamental works of Love (1911, 1944), we can assume that without regard of the Earth's layering, at a given harmonic degree, coefficients φ_n^{e} are proportional to the ones of the direct GF, i.e.,

$$\varphi_n^{\mathrm{e}} = k_n^{\mathrm{L,e}} \, \varphi_n^{\mathrm{r}} \,, \tag{2.23}$$

where $k_n^{\mathrm{L,e}}$ is a dimensionless constant defined as the *elastic loading Love number* (or the *elastic load-deformation coefficient*) associated with the incremental potential (superscript L stands for loading). Hence

$$k_n^{\mathrm{L,e}} \equiv \frac{\varphi_n^{\mathrm{e}}}{\varphi_n^{\mathrm{r}}} = \frac{\text{Incremental potential at degree } n}{\text{Direct potential at degree } n} \,, \tag{2.24}$$

in agreement with Eq. (32) of Farrell (1972). With this definition, the harmonic coefficients of the total potential variation are

$$\varphi_n = \varphi_n^{\mathrm{e}} + \varphi_n^{\mathrm{r}} = \left(1 + k_n^{\mathrm{L,e}}\right) \varphi_n^{\mathrm{r}} = -\left(1 + k_n^{\mathrm{L,e}}\right) \frac{a g_0}{m^{\mathrm{e}}} \,, \tag{2.25}$$

where we have used (2.21).

The total GF for the potential is obtained combining (2.22) with (2.20), which gives:

$$\Gamma^{\varphi}(\alpha, t) \equiv \Gamma^{\varphi,\mathrm{r}} + \Gamma^{\varphi,\mathrm{e}} = -\delta(t) \, \frac{a g_0}{m^{\mathrm{e}}} \sum_{n=0}^{\infty} \left(1 + k_n^{\mathrm{L,e}}\right) P_n(\cos\alpha) \,, \tag{2.26}$$

where the terms "1" and "$k_n^{\mathrm{L,e}}$" account for the direct and for the indirect effects, respectively.[5] Equation (2.25) has been referred to as "Love's principle" by Chen et al. (2010).

According to Bruns formula (see e.g., Hofmann-Wellenhof and Moritz 2006), the variation of the geoid height corresponding to a small variation of potential $\varphi\,(\alpha, t)$ is given by[6] $-\frac{\varphi\,(\alpha,t)}{g_0}$. Hence, from (2.26), it is straightforward to obtain the expression for the GF relative to the geoid height variation:

[5] It should be remarked that lassically, symbols (h, k, ℓ) were used to indicate what we call TLNs, while for the LLNs the notation (h', k', ℓ') is often used. Symbols (h'', k'', ℓ'') are used for the shear Love numbers (see e.g., Lambeck 1980).

[6] Note that, in terms of the potential W defined according to the geodetic convention (see Sect. 1.2, and in particular the footnote 3), the variation of the geoid height would be $+\Delta W^g/g_0$. It is easily seen, however, that both conventions would lead to the same expression for (2.27).

$$\Gamma^g(\alpha, t) \equiv -\frac{\Gamma^\varphi}{g_0} = \delta(t)\,\frac{a}{m^{\mathrm{e}}} \sum_{n=0}^{\infty} \left(1 + k_n^{\mathrm{L,e}}\right) P_n(\cos\alpha)\,, \tag{2.27}$$

having physical dimensions [$\mathrm{LT^{-1}M^{-1}}$].

2.2.3.2 Vertical Displacement

To define the GF describing the vertical component of elastic displacement (along $\hat{r}$ in Fig. 2.2b), we assume the same form as for $\Gamma^{\varphi,\mathrm{e}}$, so that

$$\Gamma^{u,\mathrm{e}}(\alpha, t) = -\delta(t) \sum_{n=0}^{\infty} u_n^{\mathrm{e}}\, P_n(\cos\alpha)\,, \tag{2.28}$$

where u_n^{e} denotes a suitable harmonic coefficient, and the minus sign has been chosen for convenience. Of course, contrary to the case of the gravitational potential, there is no rigid contribution to vertical displacement

$$\Gamma^{u,\mathrm{r}}(\alpha, t) = 0, \tag{2.29}$$

so that we have

$$\Gamma^u(\alpha, t) = \Gamma^{u,\mathrm{r}} + \Gamma^{u,\mathrm{e}} = \Gamma^{u,\mathrm{e}}. \tag{2.30}$$

In analogy with Eq. (2.23), it is convenient to define another dimensionless elastic loading Love number, named $h_n^{\mathrm{L,e}}$, associated with vertical displacement (and *a priori* independent from $k_n^{\mathrm{L,e}}$), such that

$$u_n^{\mathrm{e}} = h_n^{\mathrm{L,e}}\,\frac{\varphi_n^{\mathrm{r}}}{g_0} = -h_n^{\mathrm{L,e}}\,\frac{a}{m^{\mathrm{e}}}, \tag{2.31}$$

where we have used (2.21) and dividing by the unperturbed surface gravitational acceleration g_0 ensures dimensional consistency (see Eq. 32 of Farrell 1972). Hence, the GF for vertical displacement turns out to have the spectral form

$$\Gamma^{\mathrm{u}}(\alpha, t) = \delta(t)\,\frac{a}{m^{\mathrm{e}}} \sum_{l=0}^{\infty} h_n^{\mathrm{L,e}}\, P_n(\cos\alpha)\,. \tag{2.32}$$

and has dimensions of [$\mathrm{LT^{-1}M^{-1}}$].

2.2.3.3 Horizontal Displacement

The GF for the horizontal components of the elastic displacement vector needs to be handled in a somewhat special way. Contrary to the GF for the gravitational potential $\Gamma^{\varphi}(\alpha, t)$ (2.22) and to the GF for vertical displacement $\Gamma^{u}(\alpha, t)$ (2.28), the horizontal GF is a vector parallel to the direction of increasing source-observer angular separation $\hat{\alpha}$ (see Fig. 2.2). As discussed in Sect. A.5, a vector field with axial symmetry can be expanded in series of

$$\partial_\alpha P_n(\cos\alpha) \equiv \frac{\partial P_n(\cos\alpha)}{\partial\alpha} \tag{2.33}$$

and not simply of $P_n(\cos\alpha)$. Hence, the horizontal GF takes the form

$$\vec{\Gamma}^{v,\mathrm{e}}(\alpha, t) \equiv -\delta(t) \sum_{n=0}^{\infty} v_n^{\mathrm{e}}\, \partial_\alpha P_n(\cos\alpha)\, \hat{\alpha}\,, \tag{2.34}$$

where v_n^{e} are appropriate harmonic coefficients.

Since, by definition, in the case of a rigid Earth there is no horizontal displacement, the direct GF is:

$$\vec{\Gamma}^{v,\mathrm{r}}(\alpha, t) = 0\,, \tag{2.35}$$

hence the horizontal displacement GF for an elastic Earth is

$$\vec{\Gamma}^{v}(\alpha, t) = \vec{\Gamma}^{v,\mathrm{r}} + \vec{\Gamma}^{v,\mathrm{e}} = \vec{\Gamma}^{v,\mathrm{e}}\,. \tag{2.36}$$

As done above for the incremental potential and for vertical displacement, we define a dimensionless loading Love number, $\ell_n^{\mathrm{L,e}}$ associated with horizontal displacement as follows

$$v_n^{\mathrm{e}} = \ell_n^{\mathrm{L,e}} \frac{\varphi_n^{\mathrm{r}}}{g_0} = -\ell_n^{\mathrm{L,e}} \frac{a}{m^{\mathrm{e}}}\,, \tag{2.37}$$

where we have used (2.21), in such a way that the horizontal displacement GF takes the form

$$\vec{\Gamma}^{v}(\alpha, t) \equiv \delta(t) \frac{a}{m^{\mathrm{e}}} \sum_{n=0}^{\infty} \ell_n^{\mathrm{L,e}}\, \partial_\alpha P_n(\cos\alpha)\, \hat{\alpha}\,, \tag{2.38}$$

with dimensions of $[\mathrm{LT}^{-1}\mathrm{M}^{-1}]$.

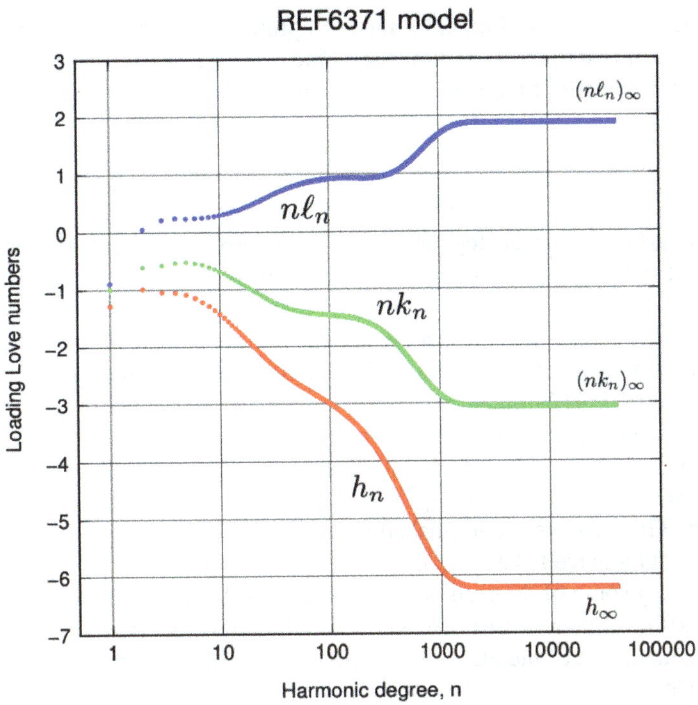

Fig. 2.3 Elastic loading Love numbers for the seismological model REF6371 (Kustowski et al. 2008), similar to the PREM model of Dziewonski and Anderson (1981). The Love numbers have been kindly provided by Pascal Gegout and are now included in the REAR package, available from: https://github.com/danielemelini/rear (see Melini et al. 2015a). Sets of elastic Love numbers for various SNREI models, taken from the literature, can be found at https://github.com/strawpants/snrei, a repository maintained by Roelof Rietbroek [last accessed on September 7, 2024]. Additional sets, including the PREM Love numbers to degree $n_{max} = 46,341$, can be found in Wang et al. (2012) as supplementary material

2.2.3.4 Elastic Green's Functions: An Example

Figure 2.3 shows a plot of the elastic loading Love numbers[7] (h_n, k_n, ℓ_n) for the seismological REF6371 model of Kustowski et al. (2008), as a function of harmonic degree n. The Love numbers, computed for degrees up to $n_{max} = 40,000$, are expressed in a reference frame with origin in the Earth's centre of mass (CM). Following the recipe of Wang et al. (2012), it is however possible to adopt other reference frames, as pointed out in detail by Blewitt (2003).

[7] In this section we drop the superscripts L and e of the elastic LLNs to simplify the notation.

A remarkable feature of Fig. 2.3 is the asymptotic behavior shown by $h_n, n\ell_n$ and nk_n for sufficiently large values of the harmonic degree n. According to Jeans' rule,[8] the spatial wavelength of a Legendre polynomial of degree n is

$$\lambda \approx \frac{2\pi a}{n + {}^1\!/_2}, \tag{2.39}$$

where a is Earth's radius. Hence the large n limit corresponds to displacements and gravitational potential perturbations of short wavelength ($n_{max} = 40,000$ implies $\lambda \approx 1$ km) that "do not see" the sphericity of the Earth and are only sensitive to the elastic properties of the outmost layers. Thus, the existence of asymptotes in the Love numbers plots is common to all SNREI models, and not limited to REF6371. As pointed by Farrell (1972), the asymptotic Love numbers reproduce the displacements that constitute the solution of the *Boussinesq's problem*, the archetypal of all surface loading problems, which shall be briefly discussed in Sect. 2.4.2.2.

The asymptotic properties of (h_n, k_n, ℓ_n) can be exploited in the numerical calculation of the GFs. This is accomplished by the so-called Kummer's transformation, first employed by Farrell (1972) in this context. In doing that, we follow Wang et al. (2012), proving some of the results reported in their Appendix B. Note that Fowler et al. (2019) have made a step beyond, since they have devised a method for the analytical calculation of the elastic response of the Earth to finite (i.e., disc-shaped) surface loads.

We first focus on the vertical displacement GF (2.32) that coherently with the simplified notation adopted here, reads

$$\Gamma^u(\alpha) = \frac{a}{m^e} \sum_{n=1}^{\infty} h_n \, P_n(\cos\alpha) \,, \tag{2.40}$$

where we recall that α denotes the co-latitude of the observer with respect to the applied point load (see Fig. 2.2), and we have temporarily dropped the factor $\delta(t)$ to simplify notation. Furthermore, the sum over the harmonic degrees only includes terms with $n \geq 1$ since degree $n = 0$ Love numbers would not contribute for a mass conserving surface load.[9]

Denoting by $h_\infty \approx -6.2160$ the asymptotic value of h_n evidenced in Fig. 2.3 (red curve), the above expression can be conveniently handled as follows:

---

[8] A nice demonstration of Jeans' rule, based on heuristics, can be found in the lecture notes by Adrian Borsa and Cathy Constable available at https://igppweb.ucsd.edu/~cathy/Classes/SIO229/appendix_jeans.p.pdf [last accessed on September 7, 2024].

[9] The GF that we are discussing here violates the mass conservation principle, since we are introducing an additional test mass Δm and therefore we are changing the total Earth mass from m^e to $m^e + \Delta m$. However, when considering convolutions of the GFs with *physically plausible* surface loads (see Chap. 3), the $n = 0$ harmonic component vanishes because of the mass conservation condition on the load model (see Sect. 1.4.4). This motivates dropping the $n = 0$ term also at the present stage.

$$
\begin{aligned}
\Gamma^{u}(\alpha) &= \frac{a}{m^{\mathrm{e}}} \sum_{n=1}^{\infty} (h_n - h_\infty + h_\infty)\, P_n(\cos\alpha) \\
&= \frac{a}{m^{\mathrm{e}}} \sum_{n=1}^{\infty} (h_n - h_\infty)\, P_n(\cos\alpha) + \frac{a}{m^{\mathrm{e}}} h_\infty \sum_{n=1}^{\infty} P_n(\cos\alpha) \\
&= \frac{a}{m^{\mathrm{e}}} \sum_{n=1}^{\infty} (h_n - h_\infty)\, P_n(\cos\alpha) + \frac{a}{m^{\mathrm{e}}} h_\infty \left(\sum_{n=0}^{\infty} P_n(\cos\alpha) - 1 \right) \\
&\simeq \frac{a}{m^{\mathrm{e}}} \sum_{n=1}^{n_\infty} (h_n - h_\infty)\, P_n(\cos\alpha) + \frac{a}{m^{\mathrm{e}}} h_\infty \left(\frac{1}{2\sin\frac{\alpha}{2}} - 1 \right), \qquad (2.41)
\end{aligned}
$$

where we have used the *Legendre sum* (A.8) explicitly. This result has important implications. In fact, the sum in the first term is practically limited to values $n \leq n_\infty$ where n_∞ is the degree such that for $n > n_\infty$ the condition $h_n \approx h_\infty$ is attained to a high precision (from Fig. 2.3, this condition is met for $n_\infty \approx 3,000$). Further, the use of the Legendre sum in the second term provides a closed-form solution that allows to avoid further numerical evaluation. Equation (2.41) matches Eq. (B1) of Wang et al. (2012).

Recalling Eq. (2.27) and the *Legendre sum* (A.8), the geoid height variation GF can be cast in the form:

$$
\Gamma^{g}(\alpha) = \frac{a}{m^{\mathrm{e}}} \left(\frac{1}{2\sin\frac{\alpha}{2}} - 1 \right) + \underbrace{\frac{a}{m^{\mathrm{e}}} \sum_{n=1}^{\infty} k_n\, P_n(\cos\alpha)}_{\text{2nd term}}, \qquad (2.42)
$$

with:

$$
\begin{aligned}
\text{2nd term} &= \sum_{n=1}^{\infty} \frac{n k_n}{n} P_n(\cos\alpha) \\
&= \sum_{n=1}^{\infty} \frac{1}{n} (n k_n - (n k_n)_\infty + (n k_n)_\infty)\, P_n(\cos\alpha) \\
&= \sum_{n=1}^{\infty} \left(\frac{n k_n - (n k_n)_\infty}{n} \right) P_n(\cos\alpha) + (n k_n)_\infty \sum_{n=1}^{\infty} \frac{1}{n} P_n(\cos\alpha) \\
&\simeq \sum_{n=1}^{n_\infty} \left(\frac{n k_n - (n k_n)_\infty}{n} \right) P_n(\cos\alpha) - (n k_n)_\infty \ln\left(\sin\frac{\alpha}{2} \sin\left(1 + \frac{\alpha}{2}\right)\right), \qquad (2.43)
\end{aligned}
$$

where $(n k_n)_\infty \approx -3.0562$ is the asymptotic value attained by $n k_n$ (see green curve in Fig. 2.3) and we have used the *Legendre sum* (A.10) explicitly. Hence, combining (2.43) with (2.42) we end up with

$$\Gamma^{g}(\alpha) \simeq \frac{a}{m^{e}}\left(\frac{1}{2\sin\frac{\alpha}{2}}-1\right)-\frac{a}{m^{e}}(nk_n)_{\infty}\ln\left(\sin\frac{\alpha}{2}\sin\left(1+\frac{\alpha}{2}\right)\right)+ \\ +\frac{a}{m^{e}}\sum_{n=1}^{n_{\infty}}\left(\frac{nk_n-(nk_n)_{\infty}}{n}\right)P_n(\cos\alpha)\,, \tag{2.44}$$

that matches Eq. (B3) of Wang et al. (2012), taking into consideration that they use symbol k_{∞} to denote $(nk_n)_{\infty}$.

Last, along the same lines, we can obtain a suitable expression for the horizontal displacement GF (2.38), which according to our simplified notation reads

$$\Gamma^{v}(\alpha) \equiv \sum_{n=1}^{\infty} \ell_n \frac{\partial P_n(\cos\alpha)}{\partial\alpha}\,. \tag{2.45}$$

Using the Kummer's transformation, this expression can be handled as follows:

$$\begin{aligned}\Gamma^{v}(\alpha) &= \sum_{n=1}^{\infty}\frac{n\ell_n}{n}\frac{\partial P_n(\cos\alpha)}{\partial\alpha} \\ &= \sum_{n=1}^{\infty}\frac{1}{n}\left(n\ell_n-(n\ell_n)_{\infty}+(n\ell_n)_{\infty}\right)\frac{\partial P_n(\cos\alpha)}{\partial\alpha} \\ &= \sum_{n=1}^{\infty}\frac{1}{n}\left(n\ell_n-(n\ell_n)_{\infty}\right)\frac{\partial P_n(\cos\alpha)}{\partial\alpha}+(n\ell_n)_{\infty}\sum_{n=1}^{\infty}\frac{1}{n}\frac{\partial P_n(\cos\alpha)}{\partial\alpha} \\ &\simeq \sum_{n=1}^{n_{\infty}}\frac{1}{n}\left(n\ell_n-(n\ell_n)_{\infty}\right)\frac{\partial P_n(\cos\alpha)}{\partial\alpha}-(n\ell_n)_{\infty}\frac{\cos\frac{\alpha}{2}(1+2\sin\frac{\alpha}{2})}{2\sin\frac{\alpha}{2}(1+\sin\frac{\alpha}{2})}\,,\end{aligned} \tag{2.46}$$

where we have used the Legendre sum (A.12) and $(n\ell_n)_{\infty} \approx 1.8922$ is the asymptotic value attained by $n\ell_n$ (see blue curve in Fig. 2.3). Equation (2.46) matches Eq. (B2) of Wang et al. (2012), where they use the symbol $l_{\infty} = (n\ell_n)_{\infty}$.

In Fig. 2.4, we show the GFs for model REF6371, evaluated according to Eqs. (2.41), (2.43), and (2.46), expressed in units of (m/kg).[10] Since the GFs are singular for $\alpha \to 0$ and attain small numerical values, a suitable normalization is usually applied to ease their visualization (Farrell 1972; Wang et al. 2012). Hence, the GFs have been scaled by a factor $(a\alpha) \times 10^{12}$, where a is Earth's radius and α is the angular distance expressed in radians. In all the computations we have conservatively set $n_{\infty} = 5,000$. The plots in Fig. 2.4 match quite well those shown by Wang et al. (2012) for model PREM of Dziewonski and Anderson (1981). Differences in

[10] Note that if we had not dropped the $\delta(t)$ factor, the units of the GFs would be m/kg/s.

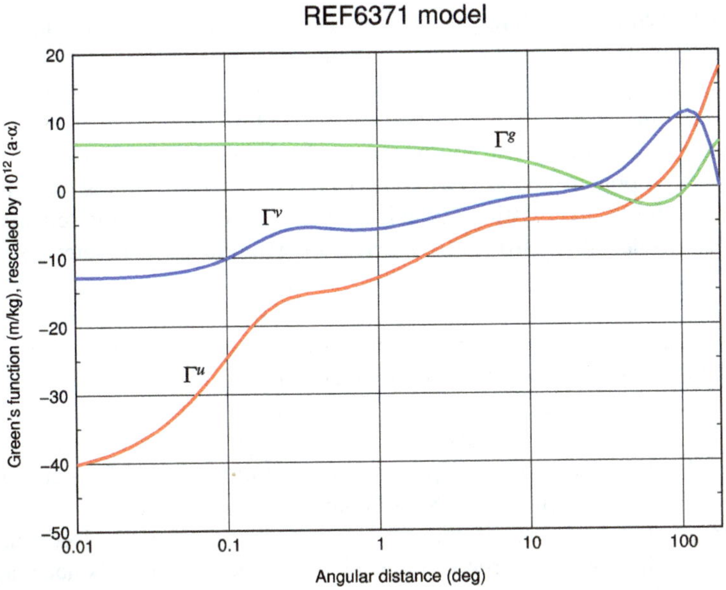

Fig. 2.4 GF for vertical displacement (Γ^u), horizontal displacement (Γ^v) and geoid height (Γ^g), pertaining to the REF6371 seismological model of Kustowski et al. (2008) and computed with the aid of the Kummer's transformation, as explained in the text

the details of the curves are to be attributed to differences in the internal structures of the two SNREI models PREM and REF6371 and, consequently, in the values of Love numbers.

2.3 Viscoelastic Green's Functions

This section is devoted to the study of the LNs and of the GFs for a 1-D GIA model characterised by a general linear viscoelastic rheology (either *steady state* or *transient*), whose structure is shown in Fig. 2.1. This shall be accomplished with the aid of the existing literature and by suitable generalisations of the results achieved for the elastic 1-D GIA models that we have considered in previous section. The LNs are described both in the time and in the Laplace domains, since the normal mode method commonly employed to compute them is based on the Laplace transform technique. The tidal LNs are also shortly discussed. The readers are referred to Sect. C.7 for an illustration of the method of solution for a layered Earth model based upon the *viscoelastic normal modes* (VNM) theory of Peltier (1974). In the framework of the VNM method, it is shown that for a 1-D GIA model the LLNs are causal functions, (i.e., one-sided functions that vanish for $t < 0$), characterised by two physically distinct components (see Wu and Peltier 1982; James 1991; Spada et al. 1992b; Wu and Ni 1996; Vermeersen et al. 1996a; Vermeersen and Sabadini 1997; Spada et al. 2011a, and references therein).

The first component of the LLNs (the *elastic component*, already described in Sect. 2.2.3) is in-phase with the (δ-like) impulsive load and only depends upon the value of the elastic constants and density within each layer of the Earth model. The second component (the *viscous component*) is discussed in this section, taking advantage of the results existing in the literature. This component is sensitive to the rheological profile of the Earth, but it depends upon the elastic parameters and density as well.[11] Due to the delayed readjustment of the Earth, the viscous component is lagged with respect to the impulsive surface load. The viscous components of all the viscoelastic LLNs have a multi-exponential form, analogous to a "Prony series" (e.g., Tschoegl 2012). By the viscoelastic normal mode method, it has been firmly verified that this is valid for the classical (steady-state) Maxwell rheology (see e.g., Spada et al. 2011a, and Sect. C.7) but also for some rheological models characterised by a transient behaviour and a discrete spectrum of relaxation like the Burgers' (see e.g., Sabadini et al. 1985; Yuen et al. 1986; Mainardi and Spada 2011).

We remark that the values of the LLNs of harmonic degree $n = 1$ depend on the choice of the origin of the reference frame adopted to describe the displacements and the change in the gravitational potential (e.g., Spada et al. 2011a). In the following, we assume that the Love numbers are expressed in the centre of mass (CM) reference frame of the (Earth + load) system. For a discussion, the readers are referred to Greff-Lefftz and Legros (1997), Greff-Lefftz (2000) and Blewitt (2003).

2.3.1 Geoid Height Variation

The time-dependent "k" LLN for a viscoelastic Earth has the general form

$$k_n^{\mathrm{L}}(t) = \underbrace{k_n^{\mathrm{L,e}}\,\delta(t)}_{\text{elastic component}} + \underbrace{H(t)\sum_{i=1}^{M} k_{ni}^{\mathrm{L}}\,\mathrm{e}^{s_{ni}t}}_{\text{viscous component}}, \tag{2.47}$$

where the first term represents the elastic component of the LLN and $k_n^{\mathrm{L,e}}$ is the elastic LLN. In the viscous component, $H(t)$ is the unit step (Heaviside) function defined by Eq. (E.1), the constants k_{ni}^{L} $(i = 1, \ldots M)$ describe the viscous parts of the LLN, each associated to the particular frequency

$$s_{ni} = -\frac{1}{\tau_{ni}}, \tag{2.48}$$

[11] For this reason, it would be better to call it *viscoelastic component*.

where the τ_{ni}'s are the *characteristic relaxation times* of the Earth's model, and M is the number of VNMs, determined by the nature of rheological layering (number and rheology of layers, *et cetera*; see Sect. C.7 for a detailed discussion). The characteristic relaxation times form the *spectrum of relaxation* of the 1-D model.

In the special case of a rigid (i.e., un-deformable) Earth, we have $k_n^{\mathrm{L,e}} = k_{ni}^{\mathrm{L}} = 0$. The non-rigid response is, in general, determined by the (model-dependent) values of $k_n^{\mathrm{L,e}}$, k_{ni}^{L}, and s_{ni}. In Eq. (2.47) we note that $k_n^{\mathrm{L}}(t)$ has dimensions of $[\mathrm{T}^{-1}]$; the same holds for viscous terms k_{ni}^{L}, while the elastic contribution $k_n^{\mathrm{L,e}}$ is non-dimensional.

2.3.1.1 The Geoid Height Green's Function

By a straightforward generalization of the elastic result (2.27), the geoid GF for a viscoelastic Earth reads

$$\Gamma^g(\alpha, t) = \frac{a}{m^{\mathrm{e}}} \sum_{n=0}^{\infty} \left(\delta(t) + k_n^{\mathrm{L}}(t)\right) P_n(\cos\alpha)\,, \tag{2.49}$$

where the gravity potential LLN $k_n^{\mathrm{L}}(t)$ is given by Eq. (2.47) and where a and m^{e} are the Earth's radius and its mass, respectively, and $P_n(\cos\alpha)$ is the degree n Legendre polynomial (see e.g., Peltier 1974; Wu and Peltier 1982). The impulsive term $\delta(t)$ accounts for the direct effect of the load on changes of the geoid height, while $k_n^{\mathrm{L}}(t)$ describes the total indirect effect associated with the elastic and the viscous deformations. It is useful to remark that, because of the direct effect, $\Gamma^g(\alpha, t)$ does not vanish in the case of a rigid Earth (i.e., for $k_n^{\mathrm{L}}(t) = 0$).

We observe that in the reference frame with origin in the CM of the whole Earth system including the solid and the fluid portions, which is conventionally adopted here since this seems to us to be the most natural choice, there will be no first-degree terms in the spherical harmonic expansion of the gravitational potential (Hofmann-Wellenhof and Moritz 2006), nor of its variation in response to surface loads (Greff-Lefftz and Legros 1997; Greff-Lefftz 2000). Hence, there are no degree $n = 1$ terms in the expansion of $\Gamma^g(\alpha, t)$ in Eq. (2.49). This means that $\delta(t) + k_1^{\mathrm{L}}(t)$ must vanish. Recalling the expression (2.47) for $k_n^{\mathrm{L}}(t)$, this condition is met if $k_1^{\mathrm{Le}} = -1$ and $k_{1i}^{\mathrm{Le}} = 0$ $(i = 1, 2, \ldots M)$. The numerical results that shall be presented in the following (see e.g., Sect. 2.4.3) are consistent with these conditions. Note that other choices are possible, as discussed in Greff-Lefftz and Legros (1997); for instance, in Farrell (1972), the LLNs are expressed in the reference frame with origin at the centre of the undeformed Earth (often this point is indicated as CE). The readers can find more on this topic in the works of Blewitt (2003) and Wang et al. (2012).

Taking advantage of the addition theorem for spherical harmonics (see A.28), it is possible to express $\Gamma^g(\alpha, t)$ explicitly as a function of the coordinates of the observer $\gamma = (\theta, \lambda)$ and of those of the point load $\gamma' = (\theta', \lambda')$. Such transformation is useful in view of the study of the response of the Earth to surface loads in next chapter, which in this way is more conveniently expressed in the geographical reference frame. In fact, using (A.28) into (2.49) we obtain

$$\Gamma^g(\gamma, \gamma', t) = \frac{a}{m^e} \sum_{nm} \Gamma^g_{nm}(\gamma', t)\, Y_{nm}(\gamma)\,, \tag{2.50}$$

where the a sum is over all harmonic degrees n and orders m (see A.35). Equation (2.50) represents an expansion in series of CSH for Γ^g, with complex coefficients

$$\Gamma^g_{nm}(\gamma', t) = \frac{\delta(t) + k^L_n(t)}{2n+1}\, Y^*_{nm}(\gamma')\,. \tag{2.51}$$

2.3.2 *Displacement Field*

2.3.2.1 The Vertical Displacement Green's Function

In a way similar to "k", it is possible to define a time-dependent viscoelastic loading Love number "h" that determines, for each harmonic degree n, the vertical displacement of the Earth's surface in response to a localised impulsive load. Such LLN, which along with "k" is an essential component of sea-level GF that enters into the Sea Level Equation, has also the multi-exponential form

$$h^L_n(t) = h^{L,e}_n\, \delta(t) + H(t) \sum_{i=1}^{M} h^L_{ni}\, e^{s_{ni} t}\,, \tag{2.52}$$

where $h^{L,e}_n$ and h^L_{ni} describe the elastic and the viscous parts of $h^L_n(t)$, respectively, and the s_{ni} are the same as for the 'k' LLN (see Eq. 2.47). For a rigid Earth, the surface load does not induce vertical displacements, hence in this case we would have $h^{L,e}_n = h^L_{ni} = 0$.

In analogy with (2.32), the general viscoelastic GF for vertical displacement is

$$\Gamma^u(\alpha, t) = \frac{a}{m^e} \sum_{n=0}^{\infty} h^L_n(t)\, P_n(\cos\alpha)\,, \tag{2.53}$$

where $h^L_n(t)$, given by (2.52), is the LLN for vertical displacement (see e.g., Peltier 1974; Wu and Peltier 1982). We observe that *i)* $h^L_n(t)$ and $k^L_n(t)$ are independent of one another, *ii)* contrary to $k^L_n(t)$, there is no "direct effect" accounted for in $h^L_n(t)$, hence the vertical displacement vanishes for a rigid Earth, i.e., for $h^L_n(t) = 0$, and that *iii)* unlike $k^L_1(t)$, the degree $n = 1$ LLN $h^L_1(t)$ is not *a priori* constrained by the choice of the origin of the reference frame in the CM of the whole Earth; its value being numerically determined by the solution of the equilibrium equations (see Spada et al. 2011a).

By the aid of the addition theorem for spherical harmonics (A.28), Eq. (2.53) can be suitably transformed into a harmonic series in the geographical reference frame:

$$\Gamma^u(\gamma, \gamma', t) = \frac{a}{m^{\mathrm{e}}} \sum_{nm} \Gamma^u_{nm}(\gamma', t)\, Y_{nm}(\gamma)\,, \tag{2.54}$$

with complex coefficients dependent upon the source coordinates:

$$\Gamma^u_{nm}(\gamma', t) = \frac{h^{\mathrm{L}}_n(t)}{2n+1}\, Y^*_{nm}(\gamma')\,. \tag{2.55}$$

2.3.2.2 The Horizontal Displacement Green's Function

Although horizontal displacements are not directly involved into the Sea Level Equation, it is useful to obtain an expression of the corresponding viscoelastic LLN "ℓ" since GIA is causing, in addition to gravity changes and vertical displacements, significant horizontal movements.

The detectability of GIA-induced horizontal crustal motions by means of Very Long Baseline Interferometry (VLBI) techniques was first addressed by James and Morgan (1990) (see also James 1991) and later investigated in other studies (see e.g., Spada et al. 1992b; James and Lambert 1993; Peltier 1995; Giunchi et al. 1997), especially in view of the recognised sensitivity of horizontal motions to the rheology of the upper mantle (Peltier 1998). Since the work of Sella et al. (2007), the importance of horizontal movements induced by GIA is fully recognised by the GNSS community (see e.g., King et al. 2010; Kreemer et al. 2018) while some theoretical aspects are still under investigation (Hermans et al. 2018).

In analogy with the expressions for "h" and "k" introduced in previous sections, the time-dependent "ℓ" viscoelastic LLN reads

$$\ell^{\mathrm{L}}_n(t) = \ell^{\mathrm{L,e}}_n\, \delta(t) + H(t) \sum_{i=1}^{M} \ell^{\mathrm{L}}_{ni}\, \mathrm{e}^{s_{ni} t}\,, \tag{2.56}$$

where $\ell^{\mathrm{L,e}}_n$ and ℓ^{L}_{ni} define the elastic and viscous parts of $\ell^{\mathrm{L}}_n(t)$, respectively. The generalization of the elastic horizontal GFs (see Eq. 2.38) to the viscoelastic case is straightforward, with

$$\vec{\Gamma}^v(\alpha, t) = \frac{a}{m^{\mathrm{e}}} \sum_{n=0}^{\infty} \ell^{\mathrm{L}}_n(t)\, \partial_\alpha P_n(\cos\alpha)\, \hat{\alpha}\,, \tag{2.57}$$

with $\ell^{\mathrm{L}}_n(t)$ given by (2.56).

Now, using the angular derivatives of the addition theorem for spherical harmonics, it is possible to project the horizontal displacements GFs along the usual unit vectors $\hat{\theta}$ (south-ward) and $\hat{\lambda}$ (east-ward) of the spherical polar coordinates system (see Fig. 2.2). In particular, using Eq. (A.30), for the $\hat{\theta}$ component we have

$$\Gamma^{v_\theta}(\gamma, \gamma', t) \equiv \vec{\Gamma}^v \cdot \hat{\theta} = \sum_{n=0}^{\infty} \ell_n^{\mathrm{L}}(t)\, \partial_\alpha P_n(\cos\alpha)\, \hat{\alpha} \cdot \hat{\theta} \tag{2.58}$$

$$= \frac{a}{m^{\mathrm{e}}} \sum_{nm} \Gamma_{nm}^{v}(\gamma', t)\, \nabla_\theta Y_{nm}(\gamma)\,, \tag{2.59}$$

while using Eq. (A.31), the $\hat{\lambda}$ component reads:

$$\Gamma^{v_\lambda}(\gamma, \gamma', t) \equiv \vec{\Gamma}^v \cdot \hat{\lambda} = \sum_{n=0}^{\infty} \ell_n^{\mathrm{L}}(t)\, \partial_\alpha P_n(\cos\alpha)\, \hat{\alpha} \cdot \hat{\lambda} \tag{2.60}$$

$$= \frac{a}{m^{\mathrm{e}}} \sum_{nm} \Gamma_{nm}^{v}(\gamma', t)\, \nabla_\lambda Y_{nm}(\gamma)\,, \tag{2.61}$$

with harmonic coefficients:

$$\Gamma_{nm}^{v}(\gamma', t) = \frac{\ell_n^{\mathrm{L}}(t)}{2n+1}\, Y_{nm}^{*}(\gamma')\,. \tag{2.62}$$

In closing, we note that the expansion (2.34) in series of $\partial_\alpha P_n(\cos\alpha)$ for the horizontal components (2.59) and (2.61) is consistent with the general expansions of a poloidal horizontal field in series of CSH, expressed by (B.63) and (B.64), respectively.

2.3.3 *Sea Level Change*

The sea-level GF has a particularly important role throughout this work, since it an essential ingredient of the Sea Level Equation that shall be introduced in Chap. 5. Such GF is defined by the difference

$$\Gamma^{s}(\alpha, t) \equiv \Gamma^{g} - \Gamma^{u}, \tag{2.63}$$

where Γ^g and Γ^u are the geoid and the vertical displacement GFs, given by Eqs. (2.49) and (2.53), respectively. Hence

$$\Gamma^{s}(\alpha, t) \equiv \frac{a}{m^{\mathrm{e}}} \sum_{n=0}^{\infty} \eta_n^{\mathrm{L}}(t)\, P_n(\cos\alpha)\,, \tag{2.64}$$

where we have defined the particular combination of LLNs

$$\eta_n^{\mathrm{L}}(t) \equiv \delta(t) + k_n^{\mathrm{L}}(t) - h_n^{\mathrm{L}}(t)\,, \tag{2.65}$$

with $k_n^{\mathrm{L}}(t)$ and $h_n^{\mathrm{L}}(t)$ given by (2.47) and (2.52), respectively.[12]

We observe that exploiting the addition theorem for spherical harmonics (A.28), similarly to the other GFs, the sea-level GF Γ^s can be suitably expanded in series of CSH in the geographical reference frame:

$$\Gamma^s(\gamma, \gamma', t) = \frac{a}{m^{\mathrm{e}}} \sum_{nm} \Gamma^s_{nm}(\gamma', t)\, Y_{nm}(\gamma)\,, \tag{2.66}$$

with coefficients

$$\Gamma^s_{nm}(\gamma', t) = \frac{\eta_n^{\mathrm{L}}(t)}{2n+1}\, Y^*_{nm}(\gamma')\,. \tag{2.67}$$

2.3.4 *Laplace Domain Love Numbers*

Since numerical methods for computing the LLNs often rely upon the correspondence principle of linear viscoelasticity (see, e.g., Tschoegl 2012, and Sect. 1.3.9), it is useful to consider the Laplace transform (LT) of the TLNs and of the LLNs (a general introduction to Laplace transforms is given by Widder 2015). Indeed, the multi-exponential form of the LLNs was first recognised after they were determined in the Laplace domain, in the framework of the viscoelastic normal modes method (Peltier 1974; Yuen and Peltier 1982). However, it is useful to remark that the use of LTs is not mandatory, since the correspondence principle in linear viscoelasticity can be equivalently expressed using the Fourier transforms formalism. For the theoretical aspects, the readers are referred to Tschoegl (2012); interesting applications to planets of the Fourier-transformed Love numbers can be found in Sotin et al. (2009) and, more recently, in Michel and Boy (2021) and Melini et al. (2022). The readers are referred to Sect. C.7.1 for a description of the Laplace domain method of solution for the Love numbers in the case of a spherically layered Earth.

2.3.4.1 Loading Love Numbers

Applying the basic transformation rules listed in Table B.1 to the time-domain expressions for the Love numbers (2.47), (2.52) and (2.56), one easily obtains the LLNs as finite series of functions having isolated poles along the real negative axis:

---

[12] It is interesting to note that, by virtue of the known relationship between LLNs and TLNs given by Eq. (2.75) below, we can equivalently write $\eta_n^{\mathrm{L}}(t) = \delta(t) + k_n^{\mathrm{T}}(t)$ where $k_n^{\mathrm{T}}(t)$ is the TLN for gravitational potential. Hence, $\Gamma^s(\alpha, t)$ can be expressed as a purely "tidal quantity", although in (2.63) it is defined as a purely "loading quantity".

$$\begin{Bmatrix} \tilde{k}_n^{\mathrm{L}} \\ \tilde{h}_n^{\mathrm{L}} \\ \tilde{\ell}_n^{\mathrm{L}} \end{Bmatrix} (s) \equiv \begin{Bmatrix} k_n^{\mathrm{L,e}} \\ h_n^{\mathrm{L,e}} \\ \ell_n^{\mathrm{L,e}} \end{Bmatrix} + \sum_{i=1}^{M} \frac{1}{s - s_{ni}} \begin{Bmatrix} k_{ni}^{\mathrm{L}} \\ h_{ni}^{\mathrm{L}} \\ \ell_{ni}^{\mathrm{L}} \end{Bmatrix}, \tag{2.68}$$

where s is the complex Laplace variable (note that s has the physical dimensions of a frequency), $(k_n^{\mathrm{L,e}}, h_n^{\mathrm{L,e}}, \ell_n^{\mathrm{L,e}})$ are the elastic LLNs, $(k_{ni}^{\mathrm{L}}, h_{ni}^{\mathrm{L}}, \ell_{ni}^{\mathrm{L}})$ describe the viscous component, s_{ni} are the inverse decay times (see 2.48) that form the *spectrum of relaxation*, M is the number of viscoelastic normal modes, and $n = 0, 1, \ldots$ is the harmonic degree. In the framework of Viscoelastic Normal Modes (VNMs), the $(k_{ni}^{\mathrm{L}}, h_{ni}^{\mathrm{L}}, \ell_{ni}^{\mathrm{L}})$ are also dubbed *viscoelastic residues* (see, e.g., Spada et al. 2011a).

For incompressible and stably layered 1-D GIA models (i.e., those in which density $\rho(r)$ is not decreasing with depth), the inverse decay times s_{ni} are numerically found to be real-valued and negative (e.g., Sabadini et al. 1982; Vermeersen et al. 1996a; Vermeersen and Sabadini 1997; Spada et al. 2011a), corresponding to an exponentially decaying response to the impulsive load. Indeed, density inversions would cause mantle overturns and in that case some of the s_{ni}'s may attain positive values, which would correspond to an indefinitely growing response to the load (see Eqs. 2.47, 2.52 and 2.56). We also note that compressible models are asymptotically affected by *Rayleigh-Taylor instability* even in the case the number of layers is limited and even for a stably stratified background density field (Hanyk et al. 1999, see also Sect. C.6). The numerical values of s_{ni} and the number of modes M in Eqs. (2.68) only depend upon the density and rheological profile of the Earth model and not on the type of forcing, being either loading or tidal, hence their numerical values are the same for the LLNs and the TLNs (see Sect. C.7).

At a given harmonic degree n, the s_{ni}'s that define the spectrum of relaxation of the Earth model are roots of an algebraic equation of degree M in the complex variable s, often referred to as *secular equation*, which reads

$$N_n(s) = 0, \tag{2.69}$$

where

$$N_n(s) = \prod_{k=1}^{M} (s - s_{nk}) = \sum_{k=0}^{M} n_{kn} s^k, \tag{2.70}$$

with constants n_{kn} being the (real-valued) coefficients of the *secular polynomial* $N_n(s)$. For the 1-D GIA models considered in this work (see Fig. 2.1), $M = 4N_V$ where N_V is the number of Maxwell viscoelastic layers that compose the mantle (see e.g., Vermeersen et al. 1996a; Spada et al. 2011a, with additional details given in Sect. C.7). For Earth models characterised by more complex rheologies, also possibly including transient effects, the secular equation is still an algebraic equation but its complexity is increased, and M increases with N_V following different laws (see e.g., Sabadini et al. 1985; Yuen et al. 1986). An example of relaxation spectrum for a realistic, multilayered 1-D model is provided in Fig. 2.6 below.

The *fluid* LLNs are defined as the values attained by the LLNs at long times, when a new state of isostatic equilibrium is reached after a unit step surface load has been applied (i.e., after the Earth has *creeped* under the load). The fluid LLNs and TLNs have therefore an important role in the GIA theory. The time needed to effectively reach the state of isostatic equilibrium shall depend on the values of the rheological parameters for the Earth model. Qualitatively, we expect that this time shall be increasing with bulk viscosity of the Earth. To evaluate the fluid LLNs, we first obtain the LLNs corresponding to a step-wise load history by convolving (2.47), (2.52) and (2.56) with the unit step function $H(t)$, *i.e.*

$$\begin{Bmatrix} k_n^{\mathrm{L,H}} \\ h_n^{\mathrm{L,H}} \\ \ell_n^{\mathrm{L,H}} \end{Bmatrix}(t) = \begin{Bmatrix} k_n^{\mathrm{L}} \\ h_n^{\mathrm{L}} \\ \ell_n^{\mathrm{L}} \end{Bmatrix}(t) * H(t)\,, \tag{2.71}$$

which in the Laplace domain, by virtue of (B.69), reads

$$\begin{Bmatrix} \tilde{k}_n^{\mathrm{L,H}} \\ \tilde{h}_n^{\mathrm{L,H}} \\ \tilde{\ell}_n^{\mathrm{L,H}} \end{Bmatrix}(s) = \frac{1}{s}\begin{Bmatrix} \tilde{k}_n^{\mathrm{L}} \\ \tilde{h}_n^{\mathrm{L}} \\ \tilde{\ell}_n^{\mathrm{L}} \end{Bmatrix}(s). \tag{2.72}$$

The fluid LLNs correspond to the limit for $t \to \infty$ of (2.71), which, by the *final value theorem* (B.75), can be written as

$$\begin{Bmatrix} k_n^{\mathrm{L,f}} \\ h_n^{\mathrm{L,f}} \\ \ell_n^{\mathrm{L,f}} \end{Bmatrix} \equiv \lim_{s\to 0}\begin{Bmatrix} \tilde{k}_n^{\mathrm{L}} \\ \tilde{h}_n^{\mathrm{L}} \\ \tilde{\ell}_n^{\mathrm{L}} \end{Bmatrix}(s). \tag{2.73}$$

Hence, the fluid LLNs correspond to the low frequency limit ($s \to 0$) of (2.68), which reads

$$\begin{Bmatrix} k_n^{\mathrm{L,f}} \\ h_n^{\mathrm{L,f}} \\ \ell_n^{\mathrm{L,f}} \end{Bmatrix} \equiv \begin{Bmatrix} k_n^{\mathrm{L,e}} \\ h_n^{\mathrm{L,e}} \\ \ell_n^{\mathrm{L,e}} \end{Bmatrix} - \sum_{i=1}^{M}\frac{1}{s_{ni}}\begin{Bmatrix} k_{ni}^{\mathrm{L}} \\ h_{ni}^{\mathrm{L}} \\ \ell_{ni}^{\mathrm{L}} \end{Bmatrix}, \tag{2.74}$$

for the three fluid LLNs, respectively.[13] We note that $(k_n^{\mathrm{L,f}}, h_n^{\mathrm{L,f}}, \ell_n^{\mathrm{L,f}})$ are dimensionless and their numerical value does not depend upon the viscosity profile adopted, although they are sensitive to the density layering of the mantle (see e.g., Spada et al. 2011a). We observe that, conversely, the elastic values $(k_n^{\mathrm{L,e}}, h_n^{\mathrm{L,e}}, \ell_n^{\mathrm{L,e}})$ represent the short time limit ($t \to 0^+$) of the time-domain LLNs convolved by a Heaviside

[13] The fluid limit of the LLNs can also be determined directly, by assuming an inviscid fluid rheology in the viscoelastic mantle and using the solution of the field equations in a fluid layer obtained in Sect. C.4. This method, as discussed by Wu and Peltier (1982), can be useful to verify that the all the components of the spectrum of relaxation have been correctly identified numerically.

step function (Eq. 2.71), which, by virtue of the *initial value theorem* (B.76), are equivalent to the high-frequency limit ($s \to \infty$) of the Laplace-domain LLNs given by (2.68).

2.3.4.2 Tidal Love Numbers

According to the VNMs theory, the TLNs have the same multi-exponential form of the LLNs, but they correspond to different surface boundary conditions, since in this case the forcing is an external potential that does not load the Earth (Peltier 1974; Munk and MacDonald 1975; Wu and Peltier 1982, 1984). A detailed discussion of the boundary conditions for the tidal and the loading problems is given in Sect. C.5.

The TLNs are, in general, numerically distinct (and independent) from the LLNs (see e.g., Spada et al. 2011a). Note, however, that

$$k_n^{\mathrm{L}}(t) = k_n^{\mathrm{T}} - h_n^{\mathrm{T}}, \tag{2.75}$$

and that more complex relationships exist between *loading*, *tidal*, and *shear* Love numbers (e.g., Saito 1974; Molodensky 1977; Saito 1978; Lambeck 1980). Of course, Eq. (2.75) also holds in the Laplace domain. Since changes in the centrifugal potential of the Earth are essentially expressed by harmonic functions of degree $n = 2$ (see Lambeck 1980), TLNs of higher harmonic degrees have no direct role in GIA studies. Along with the LLNs, the degree $n = 2$ TLNs are essential for the evaluation of the various forms of the Polar Motion Transfer Function in Sect. D.2, necessary to solve the Liouville equations for polar motion (see Sect. 4.3.3).

In close analogy with the expressions found for the LLNs, for the time domain TLNs we thus write

$$\left\{\begin{matrix} k_n^{\mathrm{T}} \\ h_n^{\mathrm{T}} \\ \ell_n^{\mathrm{T}} \end{matrix}\right\}(t) \equiv \left\{\begin{matrix} k_n^{\mathrm{T,e}} \\ h_n^{\mathrm{T,e}} \\ \ell_n^{\mathrm{T,e}} \end{matrix}\right\}\delta(t) + H(t)\sum_{i=1}^{M}\left\{\begin{matrix} k_{ni}^{\mathrm{T}} \\ h_{ni}^{\mathrm{T}} \\ \ell_{ni}^{\mathrm{T}} \end{matrix}\right\}\mathrm{e}^{s_{ni}t}, \tag{2.76}$$

with Laplace transforms

$$\left\{\begin{matrix} \tilde{k}_n^{\mathrm{T}} \\ \tilde{h}_n^{\mathrm{T}} \\ \tilde{\ell}_n^{\mathrm{T}} \end{matrix}\right\}(s) \equiv \left\{\begin{matrix} k_n^{\mathrm{T,e}} \\ h_n^{\mathrm{T,e}} \\ \ell_n^{\mathrm{T,e}} \end{matrix}\right\} + \sum_{i=1}^{M}\frac{1}{s - s_{ni}}\left\{\begin{matrix} k_{ni}^{\mathrm{T}} \\ h_{ni}^{\mathrm{T}} \\ \ell_{ni}^{\mathrm{T}} \end{matrix}\right\} \tag{2.77}$$

and fluid limits

$$\left\{\begin{matrix} k_n^{\mathrm{T,f}} \\ h_n^{\mathrm{T,f}} \\ \ell_n^{\mathrm{T,f}} \end{matrix}\right\} \equiv \left\{\begin{matrix} k_n^{\mathrm{T,e}} \\ h_n^{\mathrm{T,e}} \\ \ell_n^{\mathrm{T,e}} \end{matrix}\right\} - \sum_{i=1}^{M}\frac{1}{s_{ni}}\left\{\begin{matrix} k_{ni}^{\mathrm{T}} \\ h_{ni}^{\mathrm{T}} \\ \ell_{ni}^{\mathrm{T}} \end{matrix}\right\}. \tag{2.78}$$

Due to the harmonic nature of tides (see e.g., Melchior 1983), it is certainly of some interest to find an expression for the TLNs when the Earth is subject to a periodic forcing. With this aim, it is convenient to start from the time domain TLNs, although the same results can be obtained directly from (2.77) and working in the Laplace domain. To simplify the notation, we write the time-domain TLNs as follows

$$x(t) = x^{\mathrm{e}}\,\delta(t) + H(t)\sum_k x_k\,\mathrm{e}^{-\gamma_k t}, \tag{2.79}$$

where $x(t)$ represents any of the three TLNs in Eq. (2.76), $\sum_k$ implies a sum from $k = 1$ to $k = M$, x^{e} and x_k are the corresponding elastic and viscous components, respectively, γ_k is a frequency defined as $-s_{nk}$, and implicitly all these quantities are evaluated for harmonic degree $n = 2$, although the results that we shall obtain in the following of this section are valid for any value of n.

Since (2.79) holds for a δ-like tidal impulse, to obtain the solution for a harmonic tidal forcing of angular frequency $\omega = \frac{2\pi}{T}$, where T is the period of the oscillation, we evaluate

$$x^{\omega}(t) = x(t) \,*\, \mathrm{e}^{\mathrm{i}\omega t}, \tag{2.80}$$

where $*$ denotes, as usual, the time convolution.[14] Thus, from (2.80) we obtain

$$\begin{aligned}
x^{\omega}(t) =& x^{\mathrm{e}}\,\delta(t) \,*\, \mathrm{e}^{\mathrm{i}\omega t} + \sum_k x_k \left(H(t)\,\mathrm{e}^{-\gamma_k t}\right) * \mathrm{e}^{\mathrm{i}\omega t} \\
=& x^{\mathrm{e}}\,\mathrm{e}^{\mathrm{i}\omega t} + \sum_k x_k \int_{-\infty}^{+\infty} H(t')\,\mathrm{e}^{-\gamma_k t'}\,\mathrm{e}^{\mathrm{i}\omega(t-t')}\,\mathrm{d}t' \\
=& \left(x^{\mathrm{e}} + \sum_k x_k \int_0^{\infty} \mathrm{e}^{-(\mathrm{i}\omega+\gamma_k)t'}\,\mathrm{d}t'\right) \mathrm{e}^{\mathrm{i}\omega t} \\
=& \left(x^{\mathrm{e}} + \sum_k \frac{x_k}{\mathrm{i}\omega + \gamma_k}\right) \mathrm{e}^{\mathrm{i}\omega t},
\end{aligned} \tag{2.81}$$

where we have used the definition of time convolution (B.68) and the sifting property of Dirac's delta, i.e., $\int_{-\infty}^{+\infty} f(t)\delta(t - t_0)\mathrm{d}t = f(t_0)$.

It is now convenient to highlight the real and the imaginary parts of the expression in parentheses in Eq. (2.81). Some straightforward algebra gives

$$x^{\omega}(t) = [a(\omega) - ib(\omega)]\,\mathrm{e}^{\mathrm{i}\omega t}, \tag{2.82}$$

where functions

$$a(\omega) = x^{\mathrm{e}} + \sum_k \frac{x_k \gamma_k}{\omega^2 + \gamma_k^2} \tag{2.83}$$

[14] We note that by choosing a causal forcing like $H(t)\,\mathrm{e}^{\mathrm{i}\omega t}$ would introduce an exponentially decaying transient response in $x^{\omega}(t)$, which is not of interest here.

$$b(\omega) = \sum_k \frac{x_k \omega}{\omega^2 + \gamma_k^2} \tag{2.84}$$

depend on frequency and on the planet's properties (see Spada and Alfonsi 1998). From these expressions, we note that a sufficient condition for $a(\omega) > 0$ and $b(\omega) > 0$ is $x^e > 0$ and $x_k > 0$ for all values of k. Since elastic and viscous components of the TLNs are evaluated numerically, except in the very special case of the homogeneous Kelvin model discussed in Sect. 2.4.1, the occurrence of these conditions is to be verified case by case.

Using a polar form for the complex quantity $a(\omega) - ib(\omega)$, Eq. (2.82) can be equivalently written as

$$x^{\omega}(t) = c(\omega)\, \mathrm{e}^{\mathrm{i}(\omega t - \delta(\omega))}, \tag{2.85}$$

where

$$c(\omega) = \sqrt{a^2(\omega) + b^2(\omega)} \tag{2.86}$$

is the modulus of the Love number and

$$\delta(\omega) = \tan^{-1}\left[\frac{b(\omega)}{a(\omega)}\right], \tag{2.87}$$

is interpreted as the phase lag between the peak of the tide and the Earth's response, with $\delta(\omega) > 0$ implying that the maximum displacement occurs after the maximum force (see e.g., Sotin et al. 2009).

We note that when $a(\omega) > 0$ and $b(\omega) > 0$, condition $\delta(\omega) > 0$ is certainly met. Furthermore, in the high-frequency limit ($\omega \to \infty$), from (2.83) we have $a(\omega) \to x^e$ and $b(\omega) \to 0$, respectively, so that the tidal response of the Earth is essentially elastic, and there is no phase lag ($\delta(\omega) = 0$). In the low-frequency limit ($\omega \to 0$), we obtain $a(\omega) \to x^e + \sum_k \frac{x_k}{\gamma_k} = x^f$ where x^f is the fluid value of the TLN, and $b(\omega) \to 0$, so that the fluid response is in phase with the forcing. From the above, it turns out that function $b(\omega)$ attains a maximum, corresponding to frequency $\omega = \omega_M$ for which $\frac{d}{d\omega} \tan\delta = 0$ or equivalently $a \frac{db}{d\omega} = \frac{da}{d\omega} b$. At frequency $\omega = \omega_M$, which is only a function of the Earth's properties, a maximum phase lag $\delta = \delta_M$ is thus expected. The phase lag $\delta(\omega)$ is directly related to the dissipation rate of tidal energy in the viscoelastic mantle and, therefore, it has important implications for the study of the thermal evolution of planets (see, e.g., Efroimsky 2012, 2013).

2.4 Computing the Love Numbers

In the development of the GIA theory, a particular role has been played by spherically symmetric, incompressible Earth models, for which the density is not varying in consequence of deformation. In this case, analytical solutions exist for the LNs in the case of uniform or two-layer models (see Wu and Ni 1996). For more realistic

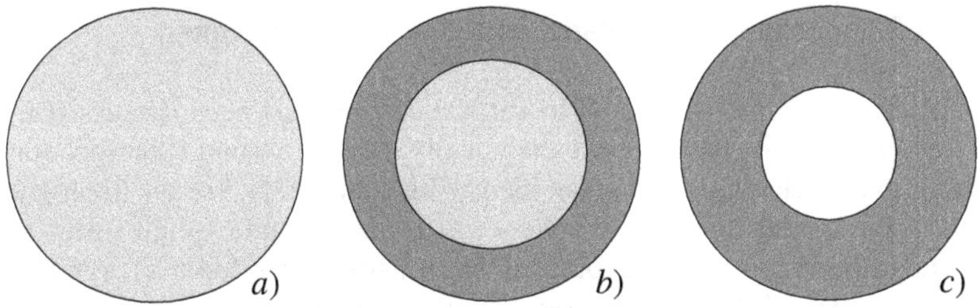

Fig. 2.5 **a** The homogeneous Earth model (the "Kelvin model"), for which simple expressions for the Love numbers exist, as shown in the text. **b** shows a two-layers planetary structure characterised by a mantle and a core while **c** represents an hollow sphere (a shell)

models, LNs can be computed only through a semi-analytical solution scheme (see Sect. C.7). In that case, at least for models characterised by a limited number of layers, the community has reached an agreed set of methods and numerical results[15] (Spada et al. 2011a).

2.4.1 *Kelvin Earth Model*

The so-called *Kelvin model* (Thomson 1863) had an important role in the development of modern global geodynamics.[16] It consists of a spherical, elastic, homogeneous, incompressible, non-rotating and self-gravitating sphere subject to tidal (or surface) forcing (see Fig. 2.5a). The Kelvin model is characterised by three independent parameters, namely radius a, density ρ_0 and shear modulus μ. We note that the mass of the model m^e and the gravity acceleration at its surface g_0 are not independent from each other; in fact, they are related with a and ρ_0, with $m^e = \frac{4\pi}{3}\rho_0 a^3$ and $g_0 = \frac{Gm^e}{a^2} = \frac{4\pi}{3}G\rho_0 a$, respectively. By the correspondence principle (see e.g., Fung 1965), the model can be extended to any linear viscoelasticity rheology (Spada 2013). In the particularly important case of the Maxwell rheology, there is one more free parameter, i.e., the dynamic viscosity η. The Kelvin model is somewhat unique, since tidal and loading Love numbers can be written in an appealingly simple analytical form, which makes this model a very useful tool in many applications.

---

[15] For compressible models such agreement has not been met yet.

[16] The model has been attributed to Lord Kelvin by Munk and MacDonald (1975), quoting Jeffreys (1952), pp. 211, 363–364. In his book on Earth rotation, Lambeck (1980) mentions the Kelvin Earth model at page 10ff but without quoting the original work of William Thomson (later knighted by Queen Victoria as Lord Kelvin).

2.4.1.1 LLNs for the Kelvin Model with Viscoelastic Rheology

The response of the Kelvin model to surface loading has been discussed by Wu and Peltier (1982), who have solved analytically the equilibrium equations for various Earth models, characterised by an increasing complexity. For the homogeneous incompressible model (the Kelvin sphere), they have found explicit forms for the loading Love numbers in the Laplace domain assuming a Maxwell rheology. In Sect. C.6, we give a full account about how the analytical expressions for the LNs of the Kelvin model can be obtained from the general solution of the equilibrium equations for a homogeneous shell.

After considerable algebra, the Laplace domain LLNs for a homogeneous sphere can be written in the following form:

$$\begin{Bmatrix} \tilde{h}_n^{\mathrm{L}} \\ \tilde{k}_n^{\mathrm{L}} \\ \tilde{\ell}_n^{\mathrm{L}} \end{Bmatrix}(s) = \frac{1}{1+\lambda_n \dfrac{\mu(s)}{\rho_0 g_0 a}} \begin{Bmatrix} h_n^{\mathrm{L,f}} \\ k_n^{\mathrm{L,f}} \\ \ell_n^{\mathrm{L,f}} \end{Bmatrix}, \tag{2.88}$$

where

$$\lambda_n = \frac{2n^2+4n+3}{n}, \tag{2.89}$$

and

$$\begin{aligned} h_n^{\mathrm{L,f}} &= -\frac{2n+1}{3} \\ k_n^{\mathrm{L,f}} &= -1 \\ \ell_n^{\mathrm{L,f}} &= -\frac{1}{n} \end{aligned} \tag{2.90}$$

represent the *fluid limits* of the LLNs, *i.e.*, the values assumed by the LLNs in the limit of vanishing frequency ($s \to 0$), while

$$\mu(s) = \frac{\mu s}{s+1/\tau_{\mathrm{M}}} \tag{2.91}$$

is the complex shear modulus appropriate for Maxwell rheology and

$$\tau_{\mathrm{M}} = \frac{\eta}{\mu} \tag{2.92}$$

is the Maxwell relaxation time, η being the dynamic viscosity and μ the elastic shear modulus of the sphere (see 1.76). Although generalised rheological laws (e.g., Burgers) involve more parameters (see e.g., Christensen 1982; Mainardi and Spada 2011), the analytical nature of the Kelvin model is preserved, provided that the appropriate complex shear modulus $\mu(s)$ is adopted. This is a direct consequence of the correspondence principle of linear viscoelasticity introduced in Sect. 1.3.9.

Of particular importance is the second of Eq. (2.90), since it implies that in the long-term limit the viscoelastic Kelvin model predicts a vanishing total variation of the gravitational potential. As we shall see in Sect. 3.4.2, this condition is not exactly met for models that include an elastic lithosphere, since the elastic strength of this layer can partially support the load even in the long-term (fluid) regime.

It is easy to recognise that the LLNs (2.88) are characterised, for a Maxwell rheology, by a single viscoelastic normal mode ($M = 1$) that for a 1-D GIA Earth model represents the simplest possible structure in the Laplace and in the time domains. Indeed, as we show in detail in Sect. C.6, by defining the non-dimensional number

$$\varepsilon_0 = \frac{\mu}{\rho_0 g_0 a}, \tag{2.93}$$

which represents a measure of the relative importance of elastic and gravitational forces (e.g., Francioni 2024), and the relaxation time scale

$$\tau_n = (1 + \lambda_n \varepsilon_0)\tau_{\mathrm{M}}, \tag{2.94}$$

any of the three LLNs (2.88) can be cast in the form

$$\tilde{x}_n^{\mathrm{L}}(s) = x_n^{\mathrm{L,e}} + \frac{x_n^{\mathrm{L,v}}}{s + 1/\tau_n}, \tag{2.95}$$

where the first and second term on *r.h.s.* correspond to the elastic and viscous responses, respectively,

$$x_n^{\mathrm{L,e}} = \frac{x_f}{1 + \lambda_n \varepsilon_0} x_n^{\mathrm{L,f}} \tag{2.96}$$

is the elastic limit of the LLN, and

$$x_n^{\mathrm{L,v}} = \left(\frac{1/\tau_{\mathrm{M}} - 1/\tau_n}{1 + \lambda_n \varepsilon_0}\right) x_n^{\mathrm{L,f}} \equiv \left(\frac{1}{\tau_{\mathrm{M}}} - \frac{1}{\tau_n}\right) x_n^{\mathrm{L,e}} \tag{2.97}$$

is the *viscoleastic residue*.[17] By comparing (2.95) with the general forms (2.68) and (2.77), it is easily seen that the spectrum of relaxation s_{ni} of the Kelvin model includes a single mode ($M = 1$) with $s_{n1} = -1/\tau_n$.

By a straightforward Laplace inversion of (2.95), the time-domain LLNs can be easily recovered. They can expressed by an elastic response and a single decaying exponential

$$x_n^{\mathrm{L}}(t) = x_n^{\mathrm{L,e}}\,\delta(t) + H(t)\, x_n^{\mathrm{L,v}} \mathrm{e}^{-t/\tau_n}, \tag{2.98}$$

[17] It is worth to note that the viscous part in (2.95) is also dependent upon the elastic rigidity through ε_0, hence it should be more appropriately referred to as *viscoelastic part*.

which following the terminology of Wu and Peltier (1982) is referred to as the M0 (or fundamental) viscoelastic mode of relaxation (see also Sabadini et al. 2016). The classical PGR theory for a flat viscous Earth similarly predicts a single exponentially decaying mode (see e.g., Ranalli 1995; Turcotte and Schubert 2014). Introducing other layers into the model like an elastic lithosphere and a fluid core would give rise to additional normal modes conventionally identified by L0, C0, *et cetera,* whose amplitude and time scale of relaxation are in general determined numerically (see e.g., Wu 1978; Wu and Peltier 1982; Spada et al. 1992b). A multi-mode structure would be also observed for the Kelvin model by introducing generalised Maxwell rheologies, however; for example, a two-mode solution would be found assuming a Burgers rheology (Spada 2013).

From (2.94) it is interesting to note that, for the Kelvin model with Maxwell rheology, the time scale of relaxation at harmonic degree n always exceeds the "natural" Maxwell relaxation time: $\tau_n > \tau_M$. Furthermore, for large values of n, τ_n increases with n, i.e., it *decreases* with *increasing* wavelength (see Jeans' rule, Eq. 2.39), in agreement with the classical PGR theory. The deviations from this law for small values of n can be seen as an effect of the sphericity of the Kelvin model and of its self-gravitation. Of course, since the homogeneous half-space model adopted in the classical PGR theory is purely viscous, it does not account for an elastic contribution to deformation, contrary to the Kelvin viscoelastic model (see 2.98).

As a final remark, it is worth to note that analytical LLNs for the Kelvin sphere are also available in the case self-gravitation is neglected. These are referred to as "non-self-gravitating solutions" by Wu and Ni (1996), and are obtained by assuming that *(i)* the gravitational field is due to the density structure of the unperturbed Earth only, with no contribution from the redistribution of mass within the Earth, and *(ii)* the upward attraction of the Earth due to the mass of the surface loads can be neglected. Since non-self-gravitation implies a constant gravitational field, $\tilde{k}_n(s) = 0$ and the solution is fully described by the displacement LLNs $\tilde{h}_n(s)$ and $\tilde{\ell}_n(s)$ (see their Eqs. 4). The analysis of Wu and Ni (1996) shows that self-gravitation has no effect on the number of eigenmodes (i.e., of viscoelastic normal modes). Interestingly, they also find that the solution for the non-self-gravitating case can be simply obtained by replacing all the g_0 terms with $g_0\,(2n+1)/(2n-2)$. As a consequence, non-self-gravitation results in a slightly larger downward gravitational pull, and the effect of self-gravitation decreases with increasing harmonic degree n. The readers are referred to Wu and Ni (1996) for a detailed inter-comparison between self-gravitating and non-self-gravitating results.

2.4.1.2 TLNs for the Kelvin Model with Viscoelastic Rheology

The TLNs for the Kelvin model have a structure that closely matches the one for the LLNs. As discussed in Sect. C.6, they are obtained by imposing tidal boundary conditions to the fundamental system of equations (see Sect. C.2.2 and Sect. C.5). Since tides mainly excite the $n = 2$ harmonic (Agnew 2007), a special focus shall be on this specific harmonic degree. Following Lambeck (1980) (see his p. 10), but using our notation, the TLNs for the Kelvin model are

$$\left\{\begin{array}{c}\tilde{h}_n^{\mathrm{T}}\\ \tilde{k}_n^{\mathrm{T}}\\ \tilde{\ell}_n^{\mathrm{T}}\end{array}\right\}(s)=\frac{1}{1+\lambda_n\dfrac{\mu(s)}{\rho_0 g_0 a}}\left\{\begin{array}{c}h_n^{\mathrm{T,f}}\\ k_n^{\mathrm{T,f}}\\ \ell_n^{\mathrm{T,f}}\end{array}\right\}, \tag{2.99}$$

where λ_n is given by (2.89) and

$$\begin{aligned} h_n^{\mathrm{T,f}} &= 1+\frac{3}{2(n-1)}\\ k_n^{\mathrm{T,f}} &= \frac{3}{2(n-1)}\\ \ell_n^{\mathrm{T,f}} &= \frac{3}{2n(n-1)} \end{aligned} \tag{2.100}$$

represent the *fluid limits* of the TLNs. We note that, for harmonic degree $n = 2$, we have:

$$\begin{aligned} h_2^{\mathrm{T,f}} &= 5/2\\ k_2^{\mathrm{T,f}} &= 3/2\\ \ell_2^{\mathrm{T,f}} &= 3/4, \end{aligned} \tag{2.101}$$

and

$$\lambda_2 = 19/2. \tag{2.102}$$

In the above equations, $\mu(s)$ represents the complex shear modulus corresponding to a general linear viscoelastic rheology. For a Maxwell rheology, manipulation of Eqs. (2.99) provides expressions like those we have obtained for the LLNs, but of course with different values for the elastic and viscous components in Eqs. (2.96) and (2.97). The relaxation time is however the same, with τ_2 given by (2.94).

2.4.2 *Other Analytical Models*

2.4.2.1 The Darwin Model

Contrary to the Kelvin model, the Darwin model is constituted by a homogeneous and incompressible *viscous newtonian* sphere (Darwin 1879). Peltier (1974) has reviewed Darwin's work in detail, exploiting a previous analysis reported in the PhD Thesis of Parsons (1972). The Darwin model has been often used to study the long-term obliquity variations forced by climate friction on Mars, since on the involved time scales elasticity plays no role (see Spada and Alfonsi 1998, and references therein).

In the limit of low frequencies ($s\tau_M \ll 1$, where τ_M is Maxwell time), the Maxwell complex modulus (2.91) approximates the newtonian limit

$$\mu_N(s) = \eta s\,, \tag{2.103}$$

where η is viscosity. By substituting $\mu_N(s)$ into the general expressions for the LLNs (2.88) or in those for the TLNs (2.99), after some straightforward algebra it is easily verified that the elastic components vanish, as expected due to the purely viscous nature of the Darwin model. In the time domain, any Love number for the Darwin model can be thus cast in the form

$$x_n(t) = \left(\frac{x_n^{\mathrm{f}}}{\tau_n^{\mathrm{D}}}\right) \mathrm{e}^{-t/\tau_n^{\mathrm{D}}}, \quad t \geq 0, \tag{2.104}$$

where x_n^{f} is the fluid Love number and, in agreement with Peltier (1974), the Darwin relaxation time is

$$\tau_n^{\mathrm{D}} = \lambda_n \frac{\eta}{\rho_0 g_0 a}\,, \tag{2.105}$$

where a is the radius of the sphere, ρ_0 is density, g_0 is gravity at the surface and λ_n is given by (2.89). When $x_n(t)$ denotes the vertical Love number, Eq. (2.104) describes the *free decay* of the initially perturbed surface of the sphere (Turcotte and Schubert 2014).

2.4.2.2 The Boussinesq's Flat Earth

Among simplified Earth's models for which an analytical expression of the LNs is available, it is of particular relevance the case of a non-gravitating, uniform elastic half-space subject to the pressure of a point load acting at the external surface. This model, first introduced by Boussinesq (1885), has been discussed in detail by Farrell (1972) and, more recently, revisited by Selvadurai (2001).

Since the wavelength of a Legendre polynomial approximately decreases with harmonic degree n as $\approx 2\pi a/n$ (Jean's rule, see Eq. 2.39), it is expected that at high harmonic degrees the response shall be sensitive only to the elastic parameters of the outer layer of the Earth model, and that effects due to sphericity will become progressively less significant. In this respect, the Boussinesq solutions may be regarded as an asymptotic limit for the response of an arbitrary spherical model to a surface load for large n, and they provide a reference solution against which the results of numerical computations may be benchmarked. By imposing equivalence between the asymptotic spherical solution and the exact half-space solution, Farrell (1972) has obtained the following expression for the asymptotic limit of LLNs:

$$\begin{Bmatrix} h_n^{\mathrm{L}} \\ n\ell_n^{\mathrm{L}} \\ nk_n^{\mathrm{L}} \end{Bmatrix} = \frac{m^{\mathrm{e}} g_0}{4\pi a^2(\lambda+\mu)} \begin{Bmatrix} -\dfrac{\lambda+2\mu}{\mu} \\ 1 \\ -\dfrac{3}{2}\dfrac{\rho}{\langle\rho\rangle}\dfrac{\lambda+\mu}{\mu} \end{Bmatrix}, \tag{2.106}$$

where ρ, λ and μ are the density and Lamé constants of the most external layer in the model depicted in Fig. 2.1, while $\langle\rho\rangle$ is the average model density, *i.e.* $\langle\rho\rangle = \frac{3m^{\mathrm{e}}}{4\pi a^3}$. Equations (2.106) provide a way of directly evaluating the h_∞, $(nk_n)_\infty$ and $(n\ell_n)_\infty$ terms appearing in (2.41), (2.43) and (2.46), respectively. For instance, assuming $\rho = 2600\,\mathrm{kg/m^3}$, $\lambda = 34.2\,\mathrm{GPa}$ and $\mu = 26.6\,\mathrm{GPa}$, corresponding to the outer crustal layer in model REF6371[18] of Kustowski et al. (2008), Eqs. (2.106) give $h_\infty \simeq -6.21$, $(n\ell_n)_\infty \simeq 1.89$ and $(nk_n)_\infty \simeq -3.05$, consistently with the LLNs shown in Fig. 2.3.

The Boussinesq's flat Earth model has an important role in the framework of geo-elasticity but it has seen a quite limited number of applications in geodynamics. However, it is noteworthy that a generation of analytical (or *quasi*-analytical) flat Earth models characterised by a viscoelastic rheology have been developed since the nineties, and had a role in the discipline of Post Glacial Rebound especially when dealing with surface loads of limited lateral extent. The readers are especially referred to the works of Detlef Wolf and collaborators, who have developed a number of flat models of increasing complexity, with some applications to PGR in northern European regions (Wolf 1985a, b, 1991a, b; Rümpker and Wolff 1996; Wolf and Kaufmann 2000; Klemann et al. 2003). Those readers interested in the topic can also benefit from the book of Cathles (1975) about the problem of the determination of mantle viscosity, who has employed a class of multi-layered PGR models in flat geometry, characterised by an elasto-newtonian rheology and having a quasi-analytical formulation (see also Cathles (2024)).

2.4.3 *Love Numbers for a 5-Layer Model*

Here we discuss some of the results of the benchmark study of Spada et al. (2011a), focusing on the relaxation spectrum and on the values of the elastic, fluid and viscous LLNs for a specific 1-D GIA model. These quantities have been introduced in Sect. 2.3.4 above and are further discussed in Appendix C, which deals with the response of a general, multi-layered spherically symmetric model to surface forcings. The results presented here, obtained by the Viscoelastic Normal Modes (VNM) method, shall be employed to accomplish a Post Glacial Rebound exercise further on in Sect. 3.4. The data reported in the figures that follow have been generated using the open source PGR solver `TABOO` and reproduce the benchmark results of Spada et al. (2011a) to a high degree of precision.

[18] See file `STW105.txt` from https://doi.org/10.17611/DP/9991858 [last accessed on February 5, 2025].

Table 2.1 Radial structure and parameters of the 1-D model B11, the reference model utilised in the benchmark study of Spada et al. (2011a). The model has five layers: the lithosphere (LT), the shallow upper mantle (SM), the mantle transition zone (TZ), the lower mantle (LM) and the core (C). At given radius r, the unperturbed gravity acceleration $g_0(r)$ is computed according to Eq. (2.6). The viscosity is expressed in Haskell units (Hu), with 1 Hu = 10^{21} Pa·s

Layer	r_{bot} km	r_{top} km	$r_{top} - r_{bot}$ km	ρ kg/m^3	$\mu \times 10^{11}$Pa	η Hu
LT	6301	6371	70	3037	0.50605	∞
SM	5951	6301	350	3438	0.70363	1
TZ	5701	5951	250	3871	1.05490	1
LM	3480	5701	2221	4978	2.28340	2
C	0	3480	3480	10750	0	0

The parameters of the 1-D GIA model adopted here—named B11—are shown in Table 2.1 (note that in Spada et al. (2011a) this same model was referred to as M3–L70–V01). Model B11 is incompressible and self–gravitating; the three mantle layers have a Maxwell rheology with a viscosity mildly increasing with depth, whereas the 70-km thick lithosphere is purely elastic. The core is homogeneous, fluid and inviscid. Due to the complexity of the model, no analytical solution exists for the Love numbers, so that here using program TABOO we have solved numerically the equilibrium equations adopting the VNM method, along the same lines outlined in Sect. C.7.

Figure 2.6 shows the *spectrum of relaxation* of model B11, obtained by solving numerically, for each harmonic degree n, the algebraic secular equation $N_n(s) = 0$ (see Eqs. 2.69–2.70 and Sect. C.7). The diagram shows the relaxation times $\tau_{nj} = -1/s_{nj}$ $(j = 1, \dots, M)$ as a function of the harmonic degree n, with s_{nj} being the M roots of the secular equations for a given harmonic degree. According to Eqs. (2.47), (2.52), and (2.56), the τ_{nj}'s are common to LLN and TLN since they are intrinsically dependent only on the model parameters and not on the boundary conditions imposed at the surface (see Sect. C.7.2). The spectrum shows nine branches, labelled according to the convention adopted within the VNM theory (Peltier 1985b; Spada et al. 1992b). These include four fast "transient modes" (modes T1–T4), a "core" mode (C0), a "lithospheric" mode (L0) and three "mantle" modes (M0, M1 and M2). Insight into the physical origin of these modes of relaxation, and on how their number depends upon the model complexity, is given by Wu (1978), Peltier (1985b), and Wu and Ni (1996). We note that, for a given harmonic degree n, the τ_{nj}'s span a very broad range of time scales, from the very fast "T" modes with time scales of a few centuries to values $> 10^7$ years for the mode M2 when $n = 2$. The M modes are associated with the very slow readjustment of internal interfaces that separate layers of different density (Peltier 1985b; Spada et al. 1992b).

The three frames of Fig. 2.7 show the values of the elastic ($h_n^{L,e}$, $\ell_n^{L,e}$, $k_n^{L,e}$) and the fluid ($h_n^{L,f}$, $\ell_n^{L,f}$, $k_n^{L,f}$) LLNs as a function of harmonic degree n. These represent the response immediately after loading ($t \to 0$) and for long times ($t \to \infty$), respectively (see Eq. 2.74). As such, they are completely unaffected by the viscosity profile and

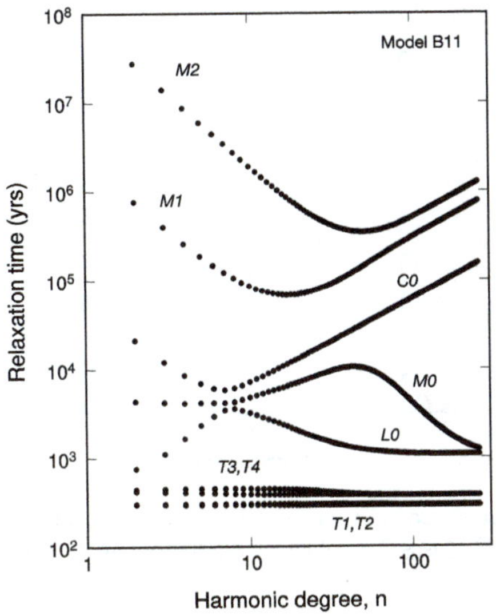

Fig. 2.6 Relaxation spectrum for model B11 in the range of harmonic degrees $2 \leq n \leq 256$. The results shown here, obtained by the TABOO PGR simulator, reproduce Fig. 2 of Spada et al. (2011a) very closely

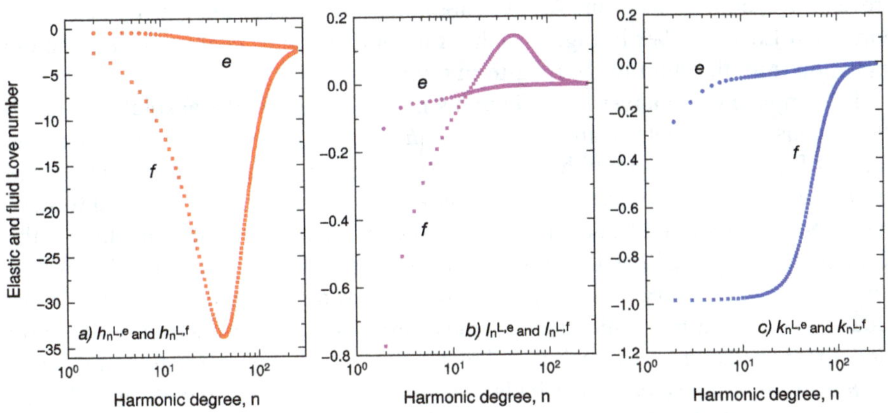

Fig. 2.7 Elastic (e) and fluid (f) LLNs h, ℓ and k for model B11, in the range of harmonic degrees $2 \leq n \leq 256$, obtained using program TABOO

are solely determined by the values of density and rigidity. The offset between the fluid and the elastic curves provide, at degree n, the amount of viscoelastic relaxation that each LLN undergoes. On one hand, for small n values (i.e., for loads of large lateral extent), the Earth is close to a state of perfect isostatic equilibrium in the long-term limit ($k_n^{f,L} \approx -1$ in Fig. 2.7c). In this case, the elastic lithosphere of model B11 is not sustaining efficiently the load, which is essentially floating on the underlying mantle, and the incremental gravitational potential almost exactly compensates the

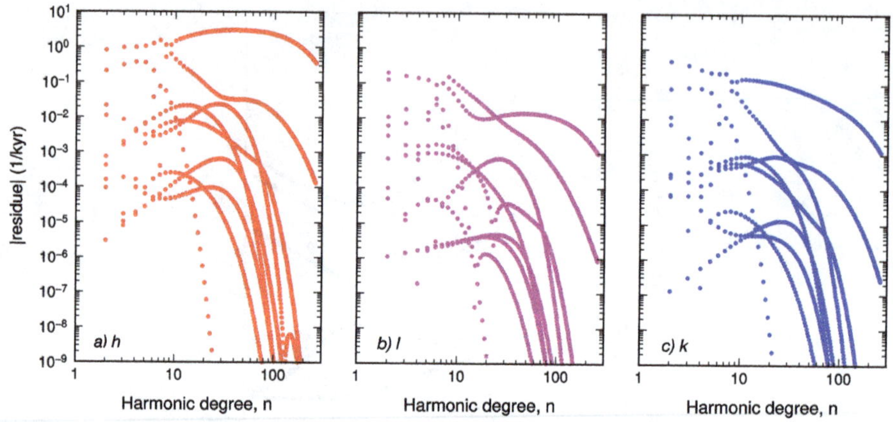

Fig. 2.8 Moduli of the viscous components of the h, ℓ and k LLN for the 1-D model B11, in the range of harmonic degrees $2 \leq n \leq 256$, obtained from program TABOO

direct contribution. On the other hand, for sufficiently large n values, elastic and fluid Love numbers reach a common asymptote. This is interpreted observing that loads of sufficiently short spatial scale are fully supported by the elastic lithosphere, thus preventing mantle relaxation. We also note that all LLNs are negative, except the horizontal Love number in Fig. 2.7b that for some n values changes its sign in the transition from the short to the long-term regime.

To complete our analysis, the diagrams in Fig. 2.8 show the absolute values of the viscous components of the LLNs, i.e., $|h_{nj}^{\mathrm{L}}|$, $|\ell_{nj}^{\mathrm{L}}|$ and $|k_{nj}^{\mathrm{L}}|$ that appear in the *r.h.s.* of Eq. (2.68). Along with relaxation spectrum in Fig. 2.6 and the elastic and fluid values in Fig. 2.7, these quantities characterise completely the spectral form of the LLNs for model B11 and can be used to construct, by a time convolution, the response to any type of load time history. From the figure, it is evident that the viscous components show a very complex pattern. As discussed by Brierley-Green and James (2024), for a given harmonic degree n these are related to the fraction of the Earth model response carried by each individual mode, which can be quantified through the *modal amplitude* (see, in particular, their Eqs. 8 and 9). In Spada et al. (2011a), the readers can find additional results, regarding the TLNs and the time-evolution of LLNs when a Heaviside load is imposed at the Earth's surface. The same work also reports a study of LNs of harmonic degree $n = 1$, tables containing numerical values of the LLNs and TLNs, and a detailed intercomparison among numerical results obtained by independent GIA research groups in various countries.

References

Agnew DC (2007) Earth tides. Treatise Geophys

Barletta VR, Ferrari C, Diolaiuti C, Carnielli T, Sabadini R, Smiraglia C (2006) Glacier shrinkage and modeled uplift of the Alps. Geophys Res Lett 33(14)

Bevis M, Melini D, Spada G (2016) On computing the geoelastic response to a disk load. Geophys J Int 205(1):1804–1812
Blewitt G (2003) Self-consistency in reference frames, geocenter definition, and surface loading of the solid Earth. J Geophys Res: Solid Earth 108(B2)
Boussinesq J (1885) Applications of potentials for the study of equilibrium and movement of elastic solids. Gautier-Villars, Paris
Brierley-Green C, James TS (2024) A hybrid hormal mode-collocation method for finding the response of laterally homogeneous compressible Maxwell viscoelastic Earth models. J Geophys Res: Solid Earth 129(7):e2023JB027238
Caron L, Métivier L, Greff-Lefftz M, Fleitout L, Rouby H (2017) Inverting glacial isostatic adjustment signal using Bayesian framework and two linearly relaxing rheologies. Geophys J Int 209(2):1126–1147
Caron L, Ivins E, Larour E, Adhikari S, Metivier L (2025) Love number computation within the Ice-Sheet and Sea-Level System Model (ISSM v4.24). EGUsphere 2025:1–39
Cathles L (1975) The viscosity of the Earth's Mantle. Princeton University Press
Cathles L (2024) On calculating glacial isostatic adjustment. Geod Geodyn 15(5):441–452
Chen W, Shen W, Dong X (2010) Atmospheric excitation of polar motion. Geo-Spat Inf Sci 13(2):130–136
Chen JY, Pan E, Bevis M (2017) Accurate computation of the elastic load Love numbers to high spectral degree for a finely layered, transversely isotropic and self-gravitating Earth. Geophys J Int 212(2):827–838
Christensen R (1982) Theory of viscoelasticity. Dover, Mineola, New York
Dahlen FA (1968) The normal modes of a rotating, elliptical Earth. Geophys J Int 16(4):329–367
Darwin GH (1879) I. On the bodily tides of viscous and semi-elastic spheroids, and on the ocean tides upon a yielding nucleus. Philos Trans R Soc Lond 170:1–35
Donnay JD (2011) Spherical trigonometry. Read Books Ltd
Dziewonski AM, Anderson DL (1981) Preliminary reference Earth model. Phys Earth Planet Inter 25(4):297–356
Efroimsky M (2012) Tidal dissipation compared to seismic dissipation: in small bodies, Earths, and super-Earths. Astrophys J 746(2):150
Efroimsky M (2013) Tidal dissipation compared to seismic dissipation: in small bodies, Earths, and super-Earths. Astrophys J 763(2):150
Farrell W (1972) Deformation of the Earth by surface loads. Rev Geophys 10(3):761–797
Fowler J, Ogle C, Bevis M (2019) An analytic method for computing the infinite sums occurring in the geoelastic disk load problem. Proc Tokyo Math-Phys Soc 124(2):2184–2201
Francioni L (2024) Tidal response and Love numbers of the moons of Uranus. Thesis, Department of Physics and Astronomy, University of Bologna
Fung Y (1965) Foundation of solid mechanics. Prentice-Hall, Englewood Cliffs, NJ
Giunchi C, Spada G, Sabadini R (1997) Lateral viscosity variations and post-glacial rebound: effects on present-day VLBI baseline deformations. Geophys Res Lett 24(1):13–16
Greff-Lefftz M (2000) Secular variation of the geocenter. J Geophys Res 25(B11):25,685–25,692
Greff-Lefftz M, Legros H (1997) Some remarks about the degree one deformations of the Earth. Geophys J Int 131:699–723
Hanyk L, Matyska C, Yuen D (1999) Secular gravitational instability of a compressible viscoelastic sphere. Geophys Res Lett 26(5):557–560
Hermans T, van der Wal W, Broerse T (2018) Reversal of the direction of horizontal velocities induced by GIA as a function of mantle viscosity. Geophys Res Lett 45(18):9597–9604
Hofmann-Wellenhof B, Moritz H (2006) Physical geodesy. Springer Science & Business Media
Huang P, Steffen R, Steffen H, Klemann V, Wu P, van der Wal W, Martinec Z, Tanaka Y (2023) A commercial finite element approach to modelling glacial isostatic adjustment on spherical self-gravitating compressible Earth models. Geophys J Int 235(3):2231–2256
James TS (1991) Postglacial deformation. PhD thesis, Princeton University

James TS, Lambert A (1993) A comparison of VLBI data with the ICE-3G glacial rebound model. Geophys Res Lett 20(9):871–874

James TS, Morgan WJ (1990) Horizontal motions due to post-glacial rebound. Geophys Res Lett 17:957–960

Jeffreys H (1952) The Earth. Cambridge University Press, Cambridge

Kachuck SB, Cathles LM (2018) GIAPY: glacial isostatic adjustment in PYthon: nondimensionalized relaxation method for computation of time-domain viscoelastic Love numbers. In: AGU fall meeting abstracts, vol 2018, pp DI31B–0008

Kachuck SB, Cathles LM (2019) Benchmarked computation of time-domain viscoelastic Love numbers for adiabatic mantles. Geophys J Int 218(3):2136–2149

King, M. A., Z. Altamimi, J. Boehm, M. Bos, R. Dach, P. Elosegui, F. Fund, M. Hernández-Pajares, D. Lavallee, P. J. M. Cerveira, N. Penna, R. E. Riva, P. Steigenberger, T. van Dam, L. Vittuari, S. Williams, and P. Willis (2010) Improved constraints on models of glacial isostatic adjustment: a review of the contribution of ground-based geodetic observations. Surv Geophys 31(5):465–507

Klemann V, Wu P, Wolf D (2003) Compressible viscoelasticity: stability of solutions for homogeneous plane-Earth models. Geophys J Int 153(3):569–585

Klemann V, Martinec Z (2011) Contribution of glacial-isostatic adjustment to the geocenter motion. Tectonophysics 511(3):99–108. Special Section on Observation and Modeling of Glacial Isostatic Adjustment

Kreemer C, Hammond WC, Blewitt G (2018) A robust estimation of the 3-D intraplate deformation of the North American plate from GPS. J Geophys Res: Solid Earth 123(5):4388–4412

Kustowski B, Ekström G, Dziewonski A (2008) Anisotropic shear-wave velocity structure of the Earth's mantle: a global model. J Geophys Res: Solid Earth 113(B6)

Lambeck K (1980) The Earth's variable rotation: geophysical causes and consequences. Cambridge University Press

Longman I (1962) A Green's function for determining the deformation of the Earth under surface mass loads: 1 Theory. J Geophys Res 67(2):845–850

Longman I (1963) A Green's function for determining the deformation of the Earth under surface mass loads. 2. Computations and numerical results. J Geophys Res 68:485–496

Love AEH (1909) The yielding of the Earth to disturbing forces. Proc R Soc London Ser A, Contain Pap Math Phys Character 82(551):73–88

Love AEH (1911) Some problems of geodynamics: being an essay to which the Adams prize in the University of Cambridge was Adjudged in 1911. CUP Archive

Love AEH (1944) A treatise on the mathematical theory of elasticity. Dover books on engineering series. Dover Publications

Mainardi F, Spada G (2011) Creep, relaxation and viscosity properties for basic fractional models in rheology. Eur Phys J Spec Top 193(1):133–160

Melchior P (1983) The tides of the planet Earth. Pergamon Press

Melini D, Gegout P, King M, Marzeion B, Spada G (2015) On the rebound: modeling Earth's ever-changing shape. Transactions American Geophysical Union, EOS, p 96

Melini D, Saliby C, Spada G (2022) On computing viscoelastic Love numbers for general planetary models: the ALMA^3 code. Geophys J Int 231(3):1502–1517

Michel A, Boy J-P (2021) Viscoelastic Love numbers and long period geophysical effects. Geophys J Int ggab369

Molodensky S (1977) Relation between Love numbers and load factors. Izv Phys Solid Earth 13:147–9

Munk WH, MacDonald GJ (1975) The rotation of the earth: a geophysical discussion. Cambridge University Press

Na S-H, Baek J (2011) Computation of the load Love number and the load Green's function for an elastic and spherically symmetric Earth. J Korean Phys Soc 58(5):1195–1205

Parsons B (1972) Changes in the Earth's shape. PhD thesis, Cambridge University, London

Peltier WR (1974) The impulse response of a Maxwell Earth. Rev Geophys Space Phys 12(4):649–669

Peltier W (1985) The LAGEOS constraint on deep mantle viscosity: results from a new normal mode method for the inversion of viscoelastic relaxation spectra. J Geophys Res: Solid Earth 90(B11):9411–9421
Peltier W (1995) VLBI baseline variations from the ICE-4G model of postglacial rebound. Geophys Res Lett 22(4):465–468
Peltier W (1998) A space geodetic target for mantle viscosity discrimination: horizontal motions induced by glacial isostatic adjustment. Geophys Res Lett 25(4):543–546
Ranalli G (1995) Rheology of the Earth. Chapman and Hall
Rümpker G, Wolff D (1996) Viscoelastic relaxation of a Burgers half-space: implications for the interpretation of the Fennoscandian uplift. Geophys J Int 124(2):541–555
Sabadini R, Yuen DA, Boschi E (1982) Polar wandering and the forced responses of a rotating, multilayered, viscoelastic planet. J Geophys Res: Solid Earth 87(B4):2885–2903
Sabadini R, Yuen D, Gasperini P (1985) The effects of transient rheology on the interpretation of lower mantle viscosity. Geophys Res Lett 12(6):361–364
Sabadini R, Vermeersen B, Cambiotti G (2016) Global dynamics of the Earth. Springer, Berlin
Saito M (1974) Some problems of static deformation of the Earth. J Phys Earth 22(1):123–140
Saito M (1978) Relationship between tidal and load Love numbers. J Phys Earth 26(1):13–16
Sella GF, Stein S, Dixon TH, Craymer M, James TS, Mazzotti S, Dokka RK (2007) Observation of glacial isostatic adjustment in 'stable' North America with GPS. Geophys Res Lett 34(2)
Selvadurai A (2001) On Boussinesq's problem. Int J Eng Sci 39(3):317–322
Shida T (1912a) On the body tides of the Earth, a proposal for the International Geodetic Association. Proc Tokyo Math-Phys Soc. 2nd Series 6(16):242–258
Shida T (1912b) On the elasticity of the Earth and the Earth's crust. Kyoto Imperial University
Sotin C, Tobie G, Wahr J, McKinnon W (2009) Tides and tidal heating on Europa. In: Pappalardo R, McKinnon W, Khurana K (eds), Europa. University of Arizona Press, pp 85–117
Spada G (2003) The theory behind TABOO. Samizdat Press. https://samizdat.mines.edu/home-old/
Spada G (2008) ALMA, a Fortran program for computing the viscoelastic Love numbers of a spherically symmetric planet. Comput Geosci 34(6):667–687
Spada G (2013) Love numbers of a generalized Maxwell sphere. Stud Geophys Geod 57(1):1–16
Spada G (2017) Glacial isostatic adjustment and contemporary sea level rise: an overview. Surv Geophys 38(1):1–33
Spada G, Boschi L (2006) Using the Post-Widder formula to compute the Earth's viscoelastic Love numbers. Geophys J Int 166(1):309–321
Spada G, Sabadini R, Yuen DA, Ricard Y (1992) Effects on post-glacial rebound from the hard rheology in the transition zone. Geophys J Int 109(3):683–700
Spada G, Barletta VR, Klemann V, Riva R, Martinec Z, Gasperini P, Lund B, Wolf D, Vermeersen L, King M (2011) A benchmark study for glacial isostatic adjustment codes. Geophys J Int 185(1):106–132
Spada G, Alfonsi L (1998) Obliquity variations and climate friction on Mars: Darwin versus layered models. J Geophys Res 103(E12):28,599–28,605
Spada G, Melini D (2019b) SELEN^4 (SELEN version 4.0): a Fortran program for solving the gravitationally and topographically self-consistent sea-level equation in glacial isostatic adjustment modeling. Geosci Model Dev 12(12):5055–5075
Spada G, Stocchi P (2006) The sea level equation, theory and numerical examples. Aracne Editrice
Sun W, Sjöberg L (1998) Gravitational potential changes in a spherically symmetric Earth model caused by a surface load. Phys Chem Earth 23(1):47–52
Thomson W (1863) XXVII. On the rigidity of the Earth. Philos Trans R Soc Lond 153:573–582
Tschoegl NW (2012) The phenomenological theory of linear viscoelastic behavior: an introduction. Springer Science & Business Media
Turcotte DL, Schubert G (2014) Geodynamics—applications of continuum physics to geological problems. Cambridge University Press

Van Hoolst T, Dehant V, Roosbeek F, Lognonné P (2003) Tidally induced surface displacements, external potential variations, and gravity variations on Mars. Icarus 161(2):281–296

Vermeersen L, Sabadini R (1997) A new class of stratified viscoelastic models by analytical techniques. Geophys J Int 129(3):531–570

Vermeersen LLA, Sabadini R, Spada G (1996) Compressible rotational deformation. Geophys J Int 126:735–761

Vermeersen LA, Sabadini R, Spada G (1996) Analytical visco-elastic relaxation models. Geophys Res Lett 23(7):697–700

Wahr J, Zuber M, Smith D, Lunine J (2006) Tides on Europa, and the thickness of Europa's icy shell. J Geophys Res: Planets 111(E12)

Wang H, Xiang L, Jia L, Jiang L, Wang Z, Hu B, Gao P (2012) Load Love numbers and Green's functions for elastic Earth models PREM, iasp91, ak135, and modified models with refined crustal structure from Crust 2.0. Comput Geosci 49:190–199

Whitehouse PL (2009) Glacial isostatic adjustment and sea-level change. Svensk Kärnbränslehantering AB, Swedish Nuclear Fuel and Waste Management Co., Stockholm, State of the art report, p 105

Whitehouse PL (2018) Glacial isostatic adjustment modelling: historical perspectives, recent advances, and future directions. Earth Surf Dyn 6(2):401–429

Widder DV (2015) Laplace transform (PMS-6). Princeton University Press, Princeton

Wolf D (1985) On Boussinesq's problem for Maxwell continua subject to an external gravity field. Geophys J Int 80(1):275–279

Wolf D (1985) The normal modes of a uniform, compressible Maxwell half-space. J Geophys 56(1):100–105

Wolf D (1991) Boussinesq's problem of viscoelasticity. Terra Nova 3(4):401–407

Wolf D (1991) Viscoelastodynamics of a stratified, compressible planet: incremental field equations and short-and long-time asymptotes. Geophys J Int 104(2):401–417

Wolf D, Kaufmann G (2000) Effects due to compressional and compositional density stratification on load-induced Maxwell viscoelastic perturbations. Geophys J Int 140(1):51–62

Wu P (1978) The response of a Maxwell Earth to applied surface mass loads: glacial isostatic adjustment. Master's thesis, University of Toronto, Toronto, Ontario

Wu P, Ni Z (1996) Some analytical solutions for the viscoelastic gravitational relaxation of a two-layer non-self-gravitating incompressible spherical Earth. Geophys J Int 126(2):413–436

Wu P, Peltier W (1982) Viscous gravitational relaxation. Geophys J Int 70(2):435–485

Wu P, Peltier W (1984) Pleistocene deglaciation and the Earth's rotation: a new analysis. Geophys J Int 76(3):753–791

Yuen DA, Sabadini RCA, Gasperini P, Boschi E (1986) On transient rheology and glacial isostasy. J Geophys Res 91:11,420–11,438

Yuen DA, Peltier WR (1982) Normal modes of the viscoelastic Earth. Geophys J Int 69(2):495–526

Zhang C (1992) Love numbers of the Moon and of the terrestrial planets. Earth Moon Planet 56(3):193–207

Chapter 3
The Earth's Response to Surface Loads

All art is at once surface and symbol.
Those who go beneath the surface do so at their peril.

(Oscar Wilde)

3.1 Introduction

This chapter deals with the Earth's response to finite-sized, time-varying surface loads, introducing various *surface response functions* (SRFs). Mathematically, the SRFs are defined by 3-D spatio-temporal convolutions involving a Green's function (GF, constructed in Chap. 2) and a surface load function, introduced in Chap. 1. Note that here and in the following we use the superscript *sur* to remark that we are dealing with *surface* response functions. *Rotational* response functions, associated with variations of the Earth's centrifugal potential, shall be introduced in the next chapter.

The *geoid SRF* $\mathcal{G}^{\rm sur}(\gamma, t)$ accounts for variations of the geoid height, caused by both the direct and the indirect gravitational effects of the surface load. The coefficients of the expansion of $\mathcal{G}^{\rm sur}(\gamma, t)$ in series of CSH, denoted by $\mathcal{G}_{nm}(t)$, shall be constructed using the $k_n^{\rm L}(t)$ LLN. More SRFs shall be introduced, associated with the three components of the displacement vector in spherical coordinates, which shall be referred to as *displacement SRFs*. The SRF for the vertical component is denoted by $\mathcal{U}^{\rm sur}$, whereas those for the horizontal components are $\mathcal{V}_\theta^{\rm sur}$ and $\mathcal{V}_\lambda^{\rm sur}$. These shall be defined by means of the LLNs $h_n^{\rm L}(t)$ and $\ell_n^{\rm L}(t)$, respectively. Invoking the properties of solenoidal fields, the displacement SRFs shall be expanded in terms of the CSHs and their surface gradients, with coefficients $\mathcal{U}_{nm}(t)$ (vertical) and $\mathcal{V}_{nm}(t)$ (horizontal). The *sea-level SRF* $\mathcal{R}^{\rm sur}$ is constructed from the geoid and the vertical

G. Spada and D. Melini, *Glacial Isostatic Adjustment*, Springer Geophysics,
https://doi.org/10.1007/978-3-031-90538-4_3

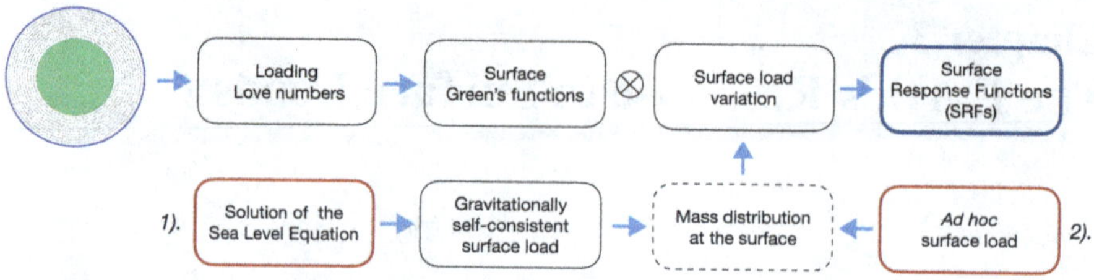

Fig. 3.1 Logical steps followed to obtain the SRFs starting from the 1-D Earth model on the left. Two alternative approaches are outlined. In approach *(1)*, a fully gravitationally self-consistent surface load is available from the SLE whereas in *(2)*, a simplified method is adopted, based upon an *ad hoc* eustatic approximation of the ocean load. Symbol ⊗ denotes the 3-D spatio-temporal convolution defined by (3.2)

displacement SRFs, being simply defined as the difference $\mathcal{R}^{\mathrm{sur}}(\gamma, t) = \mathcal{G}^{\mathrm{sur}} - \mathcal{U}^{\mathrm{sur}}$ and having coefficients $\mathcal{R}_{nm}(t)$. This SRF is of particular importance since it constitutes an essential ingredient of the Sea Level Equation that we shall discuss later in Chap. 5.

The approach that we shall follow in this chapter is sketched in Fig. 3.1. In Sect. 3.2.1 we define the SRFs in terms of spatio-temporal convolution between a GF and a general surface load. In Sect. 3.2.2 explicit forms are obtained for the harmonic coefficients $(\mathcal{R}_{nm}, \mathcal{G}_{nm}, \mathcal{U}_{nm}, \mathcal{V}_{nm})^{\mathrm{sur}}$, useful for further numerical applications in the framework of the SLE in Chap. 6. It shall be assumed that the loads are arbitrarily distributed over the Earth's surface, and evolving in a *piecewise constant* manner. A simplified surface loading problem shall be solved in Sect. 3.3, were we assume that the ice load distributed over the continents is compensated by a uniform (*eustatic*) ocean load. Finally, adopting the open source program TABOO, in Sect. 3.4 we will study quantitatively the SRFs obtained in a Post Glacial Rebound (PGR) simulation, employing a realistic axisymmetric ice cap model with a simplified time history.

3.2 Convolutions, Response Functions, *and All that*

3.2.1 *Three-Dimensional Convolution*

All the SRFs have the following form:

$$\mathrm{SRF}(\gamma, t) = \Gamma \otimes \mathcal{L}, \tag{3.1}$$

where $\Gamma = \Gamma(\gamma, t)$ is the viscoelastic Green's function (GF) associated to the SRF, $\mathcal{L}(\gamma, t)$ is surface load variation (1.119), and symbol ⊗ denotes the 3-D convolution

$$(\Gamma \otimes \mathcal{L})(\gamma, t) \equiv \int_{-\infty}^{+\infty} \mathrm{d}t' \int_{\mathrm{e}} \Gamma(\alpha, t - t')\, \mathcal{L}(\gamma', t')\, \mathrm{d}A', \tag{3.2}$$

which implies a 1-D time convolution and a 2-D spatial convolution over the Earth's surface, where we recall that the area element is $dA = a^2 \sin\theta d\theta d\lambda$, a being Earth's radius. In Eq. (3.2), α denotes the angular distance separating the *observer*, located at point $\gamma = (\theta, \lambda)$, from an arbitrary point (the *source*) that belongs to the surface mass distribution, at coordinates $\gamma' = (\theta', \lambda')$ (see Fig. 2.2). By the cosines rule for sides in spherical trigonometry (see e.g., Donnay 2011), the angular separation between the source and the observer is

$$\alpha = \alpha(\gamma, \gamma'), \tag{3.3}$$

with

$$\cos\alpha = \cos\theta\cos\theta' + \sin\theta\sin\theta'\cos(\lambda - \lambda'). \tag{3.4}$$

It is important to remark that in Eq. (3.1) the GF $\Gamma(\gamma, t)$ depends upon the properties of the Earth model adopted, whereas $\mathcal{L}(\gamma, t)$ only depends upon the geometry and the spatio-temporal variability of the surface load. Since the 3-D convolution is performed over the whole surface of the Earth e, it accounts for contributions from both ice and melt-water sources and, through the GF, it incorporates both direct and indirect loading effects. In particular, since the convolution is extended over the oceans, these are referred to as *gravitating* oceans; otherwise, the oceans are *non-gravitating* (see pages 649ff of Farrell and Clark 1976).

The physical meaning of the GFs has been already described in Sect. 2.2.1. A further insight is gained by considering a source localised in space and time, expressed by a load function of the type:

$$\mathcal{L}^{\Delta}(\gamma, t) = \frac{m_L \delta(t)}{a^2} \frac{\delta(\theta - \theta_0)\delta(\lambda - \lambda_0)}{\sin\theta}, \tag{3.5}$$

where δ is Dirac's delta, (θ_0, λ_0) are the source coordinates and without loss of generality the source mass is $m_L = 1$ kg. By substituting (3.5) into (3.1) using (3.2), one obtains the corresponding SRF as:

$$\text{SRF}^{\Delta}(\gamma, t) = \Gamma(\gamma, \gamma_0, t), \tag{3.6}$$

showing that for a localised, impulsive, unit load, the SRF coincides with the GF.

3.2.2 *Surface Response Functions*

In Sect. 2.3, we have progressively constructed the GFs pertaining to a minimal set of geophysical variables. In particular, we have found that once expanded in series of CSHs, the GFs share a common structure, which here is summarised for the readers convenience.

The geoid, vertical displacement and sea level GFs take the form:

$$\begin{Bmatrix} \Gamma^g \\ \Gamma^u \\ \Gamma^s \end{Bmatrix} (\gamma, \gamma', t) \equiv \frac{a}{m^e} \sum_{nm} \begin{Bmatrix} \Gamma^g_{nm} \\ \Gamma^u_{nm} \\ \Gamma^s_{nm} \end{Bmatrix} (\gamma', t)\, Y_{nm}(\gamma), \tag{3.7}$$

with harmonic coefficients

$$\begin{Bmatrix} \Gamma^g_{nm} \\ \Gamma^u_{nm} \\ \Gamma^s_{nm} \end{Bmatrix} (\gamma', t) \equiv \frac{1}{2n+1} \begin{Bmatrix} \delta(t) + k^L_n(t) \\ h^L_n(t) \\ \eta^L_n(t) \end{Bmatrix} Y^*_{nm}(\gamma'), \tag{3.8}$$

whereas the GFs for the two components of horizontal displacement have a slightly different form, since these involve the horizontal gradient of the CSHs:

$$\begin{Bmatrix} \Gamma^{v_\theta} \\ \Gamma^{v_\lambda} \end{Bmatrix} (\gamma, \gamma', t) \equiv \frac{a}{m^e} \sum_{nm} \Gamma^v_{nm}(\gamma', t) \begin{Bmatrix} \nabla_\theta \\ \nabla_\lambda \end{Bmatrix} Y_{nm}(\gamma), \tag{3.9}$$

with coefficients:

$$\Gamma^v_{nm}(\gamma', t) = \frac{\ell^L_n(t)}{2n+1} Y^*_{nm}(\gamma'). \tag{3.10}$$

Following the general definition (3.1), the corresponding SRFs are:

$$\begin{Bmatrix} \mathcal{G} \\ \mathcal{U} \\ \mathcal{R} \end{Bmatrix}^{\text{sur}} (\gamma, t) \equiv \begin{Bmatrix} \Gamma^g \\ \Gamma^u \\ \Gamma^s \end{Bmatrix} \otimes \mathcal{L}, \tag{3.11}$$

and

$$\begin{Bmatrix} \mathcal{V}_\theta \\ \mathcal{V}_\lambda \end{Bmatrix}^{\text{sur}} (\gamma, t) \equiv \begin{Bmatrix} \Gamma^{v_\theta} \\ \Gamma^{v_\lambda} \end{Bmatrix} \otimes \mathcal{L}, \tag{3.12}$$

respectively.

Additional SRFs could be considered as well, like those associated to gravity anomalies and tilts of the Earth's surface induced by the load variations. These can be constructed using GFs which entail different combinations of the three LLNs (h^L_n, ℓ^L_n, k^L_n); for these, the readers are referred to Longman (1962, 1963); Farrell (1972); Peltier (1974) and Sun and Sjöberg (1998).

3.2.3 Sea Level SRF

Here we aim at obtaining an explicit expression for the harmonic components $\mathcal{R}^{\text{sur}}_{nm}(t)$ of the sea-level SRF $\mathcal{R}^{\text{sur}}$, defined as the difference between the geoid and the vertical displacement SRFs:

$$\mathcal{R}^{\text{sur}}(\gamma, t) = \mathcal{G}^{\text{sur}} - \mathcal{U}^{\text{sur}}. \tag{3.13}$$

The computations are fully detailed, since they shall serve as a basis for the evaluation of the other SRFs in the sections that follow.

According to the last of Eq. (3.11) and to definition of 3-D convolution (3.2), we have:

$$\mathcal{R}^{\text{sur}}(\gamma, t) = \Gamma^s \otimes \mathcal{L} = \int_{-\infty}^{+\infty} \mathrm{d}t' \int_{\text{e}} \Gamma^s(\gamma, \gamma', t - t')\, \mathcal{L}(\gamma', t')\, \mathrm{d}A', \tag{3.14}$$

where the sea-level GF has the spectral form

$$\Gamma^s(\gamma, \gamma', t) = \frac{a}{m^{\text{e}}} \sum_{nm} \Gamma^s_{nm}(\gamma', t)\, Y_{nm}(\gamma), \tag{3.15}$$

with harmonic coefficients:

$$\Gamma^s_{nm}(\gamma', t) = \frac{\eta^{\text{L}}_n(t)}{2n+1} Y^*_{nm}(\gamma'), \tag{3.16}$$

and $\eta^{\text{L}}_n(t) = \delta(t) + k^{\text{L}}_n(t) - h^{\text{L}}_n(t)$.

In Eq. (3.14), there are no restrictions upon the spatial dependence of $\mathcal{L}(\gamma, t)$, which in general may represent a spatially variable ice or water load. However, henceforth it is assumed that $\mathcal{L}(\gamma, t)$ is a piecewise constant (PC) function of time (see Sect. E.1). The outcomes of the ensuing computations shall be useful in view of the numerical solution of the SLE in Chap. 6 and Appendix E, which is based on the same time-discretisation scheme. According to Eq. (E.11), the CSH coefficients of the load function expansion

$$\mathcal{L}(\gamma, t) = \sum_{nm} \mathcal{L}_{nm}(t)\, Y_{nm}(\gamma), \tag{3.17}$$

are step-wise as well, with:

$$\mathcal{L}_{nm}(t) = \sum_{k=0}^{N} \Delta\mathcal{L}_{nm,k}\, H(t - t_k), \tag{3.18}$$

where $H(t)$ is the unit step function, $\Delta\mathcal{L}_{nm,k} \equiv \mathcal{L}_{nm,k+1} - \mathcal{L}_{nm,k}$ for $k = 0, \ldots N$, where N is the number of time steps, and $\mathcal{L}_{nm,k} \equiv \mathcal{L}_{nm}(t_k)$.

The computations are straightforward, although a bit cumbersome. For ease of presentation, the whole material has been organised into four steps; the final result can be found at Sect. 3.2.3.4.

3.2.3.1 Step 1

Direct substitution of (3.15) into (3.14), using (3.16) and (3.17), recalling that $dA = a^2 \, d\gamma$ and exploiting of the orthogonality of the spherical harmonics, gives

$$
\begin{aligned}
\mathcal{R}^{\mathrm{sur}}(\gamma, t) &= \frac{a^3}{m^{\mathrm{e}}} \int_{-\infty}^{+\infty} \mathrm{d}t' \int_{\gamma} \left(\sum_{nm} \frac{\eta_n^{\mathrm{L}}(t-t')}{2n+1} Y_{nm}^*(\gamma') Y_{nm}(\gamma) \right) \cdot \\
&\qquad \cdot \left(\sum_{n'm'} \mathcal{L}_{n'm'}(t') \, Y_{n'm'}(\gamma') \right) \mathrm{d}\gamma' \\
&= \frac{a^3}{m^{\mathrm{e}}} \sum_{nm} \frac{Y_{nm}(\gamma)}{2n+1} \cdot \\
&\qquad \cdot \int_{-\infty}^{+\infty} \mathrm{d}t' \eta_n^{\mathrm{L}}(t-t') \sum_{n'm'} \mathcal{L}_{n'm'}(t') \underbrace{\int_{\gamma} Y_{nm}^*(\gamma') Y_{n'm'}(\gamma') \mathrm{d}\gamma'}_{4\pi \, \delta_{nn'} \, \delta_{mm'}} \\
&= \frac{4\pi a^3}{m^{\mathrm{e}}} \sum_{nm} \frac{Y_{nm}(\gamma)}{2n+1} \int_{-\infty}^{+\infty} \eta_n^{\mathrm{L}}(t-t') \, \mathcal{L}_{nm}(t') \, \mathrm{d}t',
\end{aligned}
\tag{3.19}
$$

which can be recognised, for the *sea-level* SRF, as a CSH expansion

$$
\mathcal{R}^{\mathrm{sur}}(\gamma, t) = \sum_{nm} \mathcal{R}_{nm}^{\mathrm{sur}}(t) Y_{nm}(\gamma),
\tag{3.20}
$$

with harmonic coefficients

$$
\mathcal{R}_{nm}^{\mathrm{sur}}(t) = \frac{3}{\rho^{\mathrm{e}}} \frac{\eta_n^{\mathrm{L}}(t) * \mathcal{L}_{nm}(t)}{2n+1},
\tag{3.21}
$$

where $*$ is time convolution and

$$
\rho^{\mathrm{e}} = \frac{3 \, m^{\mathrm{e}}}{4\pi a^3}
\tag{3.22}
$$

is the Earth's average density.

3.2.3.2 Step 2

We now transform $\mathcal{R}^{\mathrm{sur}}_{nm}(t)$ using the definition (2.65) of $\eta^{\mathrm{L}}_n(t)$, also recalling the multi-exponential form of the LLNs in the time domain, given by (2.47) and (2.52), respectively. Thus from (3.21) we obtain:

$$\begin{aligned}\mathcal{R}^{\mathrm{sur}}_{nm}(t) &= \frac{3}{\rho^{\mathrm{e}}}\left[\left(\frac{1+k^{\mathrm{L,e}}_n - h^{\mathrm{L,e}}_n}{2n+1}\right)\delta(t) + H(t)\sum_{i=1}^{M}\frac{k^{\mathrm{L}}_{ni} - h^{\mathrm{L}}_{ni}}{2n+1}\,\mathrm{e}^{s_{ni}t}\right] * \mathcal{L}_{nm}(t)\\ &= \frac{3}{\rho^{\mathrm{e}}}\left[\left(\frac{1+k^{\mathrm{L,e}}_n - h^{\mathrm{L,e}}_n}{2n+1}\right)\mathcal{L}_{nm}(t) + \sum_{i=1}^{M}\frac{k^{\mathrm{L}}_{ni} - h^{\mathrm{L}}_{ni}}{2n+1}\left(H(t)\,\mathrm{e}^{s_{ni}t}\right) * \mathcal{L}_{nm}(t)\right]\\ &= \mathcal{R}^{\mathrm{sur,e}}_{nm} + \mathcal{R}^{\mathrm{sur,v}}_{nm},\end{aligned} \tag{3.23}$$

where we have used the sifting property of Dirac's delta, and we have decomposed $\mathcal{R}^{\mathrm{sur}}_{nm}$ into an elastic and a viscous part, with

$$\mathcal{R}^{\mathrm{sur,e}}_{nm}(t) = \frac{3}{\rho^{\mathrm{e}}}\left(\frac{1+k^{\mathrm{L,e}}_n - h^{\mathrm{L,e}}_n}{2n+1}\right)\mathcal{L}_{nm}(t), \tag{3.24}$$

and

$$\mathcal{R}^{\mathrm{sur,v}}_{nm}(t) = \frac{3}{\rho^{\mathrm{e}}}\sum_{i=1}^{M}\frac{k^{\mathrm{L}}_{ni} - h^{\mathrm{L}}_{ni}}{2n+1}\left(H(t)\,\mathrm{e}^{s_{ni}t}\right) * \mathcal{L}_{nm}(t), \tag{3.25}$$

respectively. We note that these two last expressions are completely general, being valid for surface loads characterised by an arbitrary (non necessarily PC) time history.

3.2.3.3 Step 3

Using the PC expression of the load function CSH coefficients (3.18), the elastic part of $\mathcal{R}^{\mathrm{sur}}_{nm}$ reads

$$\mathcal{R}^{\mathrm{sur,e}}_{nm}(t) = \frac{3}{\rho^{\mathrm{e}}}\,e^{s}_{n}\sum_{k=0}^{N}\Delta\mathcal{L}_{nm,k}\,H(t-t_k), \tag{3.26}$$

where we have defined the constants

$$e^{s}_{n} = \frac{1+k^{\mathrm{L,e}}_n - h^{\mathrm{L,e}}_n}{2n+1}. \tag{3.27}$$

Using the definition of time convolution, the viscous part is:

$$
\begin{aligned}
\mathcal{R}_{nm}^{\mathrm{sur,v}}(t) &= \frac{3}{\rho^{\mathrm{e}}} \sum_{i=1}^{M} \frac{k_{ni}^{\mathrm{L}} - h_{ni}^{\mathrm{L}}}{2n+1} \int_{-\infty}^{+\infty} H(t-t')\, \mathrm{e}^{s_{ni}(t-t')}\, \mathcal{L}_{nm}(t')\, \mathrm{d}t' \\
&= \frac{3}{\rho^{\mathrm{e}}} \sum_{i=1}^{M} \frac{k_{ni}^{\mathrm{L}} - h_{ni}^{\mathrm{L}}}{2n+1} \int_{-\infty}^{t} \mathrm{e}^{s_{ni}(t-t')} \left(\sum_{k=0}^{N} \Delta\mathcal{L}_{nm,k}\, H(t'-t_k) \right) \mathrm{d}t' \\
&= \frac{3}{\rho^{\mathrm{e}}} \sum_{k=0}^{N} \Delta\mathcal{L}_{nm,k} \sum_{i=1}^{M} \frac{k_{ni}^{\mathrm{L}} - h_{ni}^{\mathrm{L}}}{2n+1} \int_{-\infty}^{t} \mathrm{e}^{s_{ni}(t-t')} H(t'-t_k)\, \mathrm{d}t' \\
&= \frac{3}{\rho^{\mathrm{e}}} \sum_{k=0}^{N} \Delta\mathcal{L}_{nm,k} \sum_{i=1}^{M} \frac{k_{ni}^{\mathrm{L}} - h_{ni}^{\mathrm{L}}}{2n+1} H(t-t_k)\, \mathrm{e}^{s_{ni}t} \int_{t_k}^{t} \mathrm{e}^{-s_{ni}t'} \mathrm{d}t' \\
&= \frac{3}{\rho^{\mathrm{e}}} \sum_{k=0}^{N} \Delta\mathcal{L}_{nm,k} \sum_{i=1}^{M} \frac{k_{ni}^{\mathrm{L}} - h_{ni}^{\mathrm{L}}}{2n+1} \frac{\mathrm{e}^{s_{ni}(t-t_k)} - 1}{s_{ni}} H(t-t_k),
\end{aligned}
\tag{3.28}
$$

which can be expressed as follows

$$
\mathcal{R}_{nm}^{\mathrm{sur},v}(t) = \frac{3}{\rho^{\mathrm{e}}} \sum_{k=0}^{N} \Delta\mathcal{L}_{nm,k}\, \beta_n^s(t-t_k)\, H(t-t_k), \tag{3.29}
$$

where we have defined the functions

$$
\beta_n^s(t) = \frac{1}{2n+1} \sum_{i=1}^{M} \frac{k_{ni}^{\mathrm{L}} - h_{ni}^{\mathrm{L}}}{s_{ni}} \left(\mathrm{e}^{s_{ni}t} - 1 \right). \tag{3.30}
$$

Combining the elastic (3.26) and the viscous parts (3.29), we arrive at the following expression for the CSH coefficients of the sea-level SRF:

$$
\mathcal{R}_{nm}^{\mathrm{sur}}(t) = \frac{3}{\rho^{\mathrm{e}}} \sum_{k=0}^{N} \Delta\mathcal{L}_{nm,k} \left(e_n^s + \beta_n^s(t-t_k) \right) H(t-t_k). \tag{3.31}
$$

3.2.3.4 Step 4

We recall, from Sect. 1.4.3.3, that the general expression for the surface load variation $\mathcal{L}(\gamma, t)$ stems from three components, associated with variations of the ice thickness (superscript i), of sea-level (s) and of the ocean function (o); namely:

$$
\mathcal{L}(\gamma, t) = \mathcal{L}^{\mathrm{i}} + \mathcal{L}^{\mathrm{s}} + \mathcal{L}^{\mathrm{o}}, \tag{3.32}
$$

where $\mathcal{L}^{\rm i} = \rho^{\rm i}\mathcal{W}$, $\mathcal{L}^{\rm s} = \rho^{\rm w}\mathcal{Z}$ and $\mathcal{L}^{\rm o} = \rho^{\rm r}\mathcal{X}$, hence

$$\mathcal{L}(\gamma, t) = \rho^{\rm i}\mathcal{W} + \rho^{\rm w}\mathcal{Z} + \rho^{\rm r}\mathcal{X}. \tag{3.33}$$

Thus, substituting

$$\Delta\mathcal{L}_{nm,k} = \rho^{\rm i}\,\Delta\mathcal{W}_{nm,k} + \rho^{\rm w}\,\Delta\mathcal{Z}_{nm,k} + \rho^{\rm r}\,\Delta\mathcal{X}_{nm,k} \tag{3.34}$$

into Eq. (3.31) we find:

$$\mathcal{R}^{\rm sur}_{nm}(t) = \mathcal{R}^{\rm i}_{nm} + \mathcal{R}^{\rm s}_{nm} + \mathcal{R}^{\rm o}_{nm}, \tag{3.35}$$

where superscripts (i,s,o) have the same meaning as in (3.32), and

$$\left\{\begin{array}{c} \mathcal{R}^{\rm i}_{nm} \\ \mathcal{R}^{\rm s}_{nm} \\ \mathcal{R}^{\rm o}_{nm} \end{array}\right\}(t) = \frac{3}{\rho^{\rm e}}\sum_{k=0}^{N}\left\{\begin{array}{c} \rho^{\rm i}\Delta\mathcal{W}_{nm,k} \\ \rho^{\rm w}\Delta\mathcal{Z}_{nm,k} \\ \rho^{\rm r}\Delta\mathcal{X}_{nm,k} \end{array}\right\}\left(e^{s}_{n} + \beta^{s}_{n}(t - t_k)\right) H(t - t_k), \tag{3.36}$$

which completely determines the sea-level SRF for a surface load characterised by a PC time evolution.

3.2.4 *Geoid and Vertical Displacement SRFs*

By their definition, the GFs for the geoid and for vertical displacement have the same structure of the sea-level GF, as it is apparent from Eq. (3.7). Indeed, these three GFs only differ for the particular combinations of LLNs involved.

Following the scheme outlined for the sea-level SRF, for the geoid and vertical displacement SRFs we have the spectral forms

$$\mathcal{G}^{\rm sur}(\gamma, t) = \sum_{nm}\mathcal{G}^{\rm sur}_{nm}(t)\,Y_{nm}(\gamma), \tag{3.37}$$

and

$$\mathcal{U}^{\rm sur}(\gamma, t) = \sum_{nm}\mathcal{U}^{\rm sur}_{nm}(t)\,Y_{nm}(\gamma), \tag{3.38}$$

where the coefficients are

$$\mathcal{G}^{\rm sur}_{nm}(t) = \frac{3}{\rho^{\rm e}}\frac{\left(\delta(t) + k^{\rm L}_{n}(t)\right) * \mathcal{L}_{nm}(t)}{2n+1} \tag{3.39}$$

and

$$\mathcal{U}_{nm}^{\text{sur}}(t) = \frac{3}{\rho^{\text{e}}} \frac{h_n^{\text{L}}(t) * \mathcal{L}_{nm}(t)}{2n+1}, \tag{3.40}$$

respectively.

For a PC load, the coefficients can be written as

$$\mathcal{G}_{nm}^{\text{sur}}(t) = \mathcal{G}_{nm}^{\text{i}} + \mathcal{G}_{nm}^{\text{s}} + \mathcal{G}_{nm}^{\text{o}} \tag{3.41}$$

$$\mathcal{U}_{nm}^{\text{sur}}(t) = \mathcal{U}_{nm}^{\text{i}} + \mathcal{U}_{nm}^{\text{s}} + \mathcal{U}_{nm}^{\text{o}}, \tag{3.42}$$

with

$$\begin{Bmatrix} \mathcal{G}_{nm}^{\text{i}} \\ \mathcal{G}_{nm}^{\text{s}} \\ \mathcal{G}_{nm}^{\text{o}} \end{Bmatrix}(t) = \frac{3}{\rho^{\text{e}}} \sum_{k=0}^{N} \begin{Bmatrix} \rho^{\text{i}} \Delta \mathcal{W}_{nm,k} \\ \rho^{\text{w}} \Delta \mathcal{Z}_{nm,k} \\ \rho^{\text{r}} \Delta \mathcal{X}_{nm,k} \end{Bmatrix} \left(e_n^g + \beta_n^g(t - t_k)\right) H(t - t_k), \tag{3.43}$$

and, similarly

$$\begin{Bmatrix} \mathcal{U}_{nm}^{\text{i}} \\ \mathcal{U}_{nm}^{\text{s}} \\ \mathcal{U}_{nm}^{\text{o}} \end{Bmatrix}(t) = \frac{3}{\rho^{\text{e}}} \sum_{k=0}^{N} \begin{Bmatrix} \rho^{\text{i}} \Delta \mathcal{W}_{nm,k} \\ \rho^{\text{w}} \Delta \mathcal{Z}_{nm,k} \\ \rho^{\text{r}} \Delta \mathcal{X}_{nm,k} \end{Bmatrix} \left(e_n^u + \beta_n^u(t - t_k)\right) H(t - t_k), \tag{3.44}$$

where as usual, superscripts (i,s,o) stand for *ice*, *sea level* and *ocean*, respectively, and where we have defined the constants:

$$e_n^g \equiv \frac{1 + k_n^{\text{L,e}}}{2n+1} \tag{3.45}$$

$$e_n^u \equiv \frac{h_n^{\text{L,e}}}{2n+1}, \tag{3.46}$$

and the auxiliary functions

$$\beta_n^g(t) \equiv \frac{1}{2n+1} \sum_{i=1}^{M} \frac{k_{ni}^{\text{L}}}{s_{ni}} \left(\text{e}^{s_{ni} t} - 1\right) \tag{3.47}$$

$$\beta_n^u(t) \equiv \frac{1}{2n+1} \sum_{i=1}^{M} \frac{h_{ni}^{\text{L}}}{s_{ni}} \left(\text{e}^{s_{ni} t} - 1\right). \tag{3.48}$$

3.2.5 Horizontal Displacement SRFs

The horizontal displacement SRFs can be analysed following the same scheme adopted for $\mathcal{R}^{\rm sur}$, $\mathcal{G}^{\rm sur}$ and $\mathcal{U}^{\rm sur}$, starting from the convolutions (3.12) and using the GFs given by Eq. (3.9).

After some simple algebra the two horizontal displacement SRFs can be cast in the form

$$\mathcal{V}_{\theta}^{\rm sur}(\gamma, t) = \sum_{nm} \mathcal{V}_{nm}^{\rm sur}(t)\, \nabla_{\theta}\, Y_{nm}(\gamma) \tag{3.49}$$

$$\mathcal{V}_{\lambda}^{\rm sur}(\gamma, t) = \sum_{nm} \mathcal{V}_{nm}^{\rm sur}(t)\, \nabla_{\lambda}\, Y_{nm}(\gamma), \tag{3.50}$$

which share the same CSH coefficients

$$\mathcal{V}_{nm}^{\rm sur}(t) = \mathcal{V}_{nm}^{\rm i} + \mathcal{V}_{nm}^{\rm s} + \mathcal{V}_{nm}^{\rm o}, \tag{3.51}$$

where superscripts (i,s,o) stand for *ice*, *sea level* and *ocean*, respectively, with

$$\begin{Bmatrix} \mathcal{V}_{nm}^{\rm i} \\ \mathcal{V}_{nm}^{\rm s} \\ \mathcal{V}_{nm}^{\rm o} \end{Bmatrix}(t) = \frac{3}{\rho^{\rm e}} \sum_{k=0}^{N} \begin{Bmatrix} \rho^{\rm i} \Delta \mathcal{W}_{nm,k} \\ \rho^{\rm w} \Delta \mathcal{Z}_{nm,k} \\ \rho^{\rm r} \Delta \mathcal{X}_{nm,k} \end{Bmatrix} \left(e_n^{v} + \beta_n^{v}(t - t_k)\right) H(t - t_k), \tag{3.52}$$

and where we have defined the constants

$$e_n^{v} = \frac{\ell_n^{\rm L,e}}{2n+1}, \tag{3.53}$$

and the auxiliary function:

$$\beta_n^{v}(t) \equiv \frac{1}{2n+1} \sum_{i=1}^{M} \frac{\ell_{ni}^{\rm L}}{s_{ni}} \left(e^{s_{ni} t} - 1\right). \tag{3.54}$$

3.3 Simplified Surface Loading Problem

Although the SLE represents the fundamental equation of GIA, it is sometimes useful to consider a *simplified loading problem* that only involves the ice load and a spatially uniform meltwater load, thus neglecting other complexities implied in the SLE. Such approach has been adopted in the benchmark paper of Spada et al. (2011a) and it has been numerically implemented in the open source Post Glacial Rebound (PGR) solver TABOO.[1] Due to the assumptions made, the simplified loading problem can be solved in a closed form, in terms of vertical displacements and other fundamental quantities. No iterative numerical scheme is required to obtain the solution, as it is indeed necessary in the case of the SLE. The solution of the simplified loading problem is useful to obtain a first estimate of the geophysical variables involved in GIA and to model PGR in a very realistic way. Furthermore, the solutions obtained are useful to calibrate fully numerical methods sometimes employed in GIA modelling (see Spada et al. 2011a, for a few examples).

In previous section, we have expressed the surface load variation $\mathcal{L}$ as a function of variables $\mathcal{W}$, $\mathcal{Z}$, and $\mathcal{X}$, which entail time variations of the ice thickness, of sea level and of the ocean function, respectively. However, even assuming that the history of the ice thickness is known a priori, it is apparent that $\mathcal{L}$ is not fully determined until the SLE is numerically solved, which shall be done in Chap. 6. Thus, at the present stage of our discussion, the response of the Earth's surface can only be obtained by making *ad hoc* approximations that are common in PGR modelling (see e.g., Spada et al. 2011a). Indeed, here we will assume that the surface load is given over a subset i of the whole Earth's surface where its thickness varies in space and time in a known fashion, and over a disjoint subset w, where it takes—at a given time—a spatially uniform value. The former contribution, which shall be referred to as the *ice load*, mimics the waning and waxing of a continental ice sheet while the latter—the *meltwater load*—represents a uniform compensating mass distribution across a hypothetical ocean with fixed shorelines. For *compensated surface loads*, both contributions are present, while if only the first contribution is taken into account, the surface load is *uncompensated.*

As we will see below in detail, by solving the simplified loading problem it is possible to obtain expressions for vertical displacements and geoid height changes, without invoking the *gravitationally self-consistent* approach that is implicit in the SLE.

3.3.1 *Ice and Meltwater Loads*

The surface mass distribution is modelled as the linear combination of two functions, $\sigma^{\mathrm{i}}(\gamma, t)$ and $\sigma^{\mathrm{w}}(\gamma, t)$, describing the ice and the meltwater components of the total load, respectively.

[1] TABOO (Spada 2003) is available from https://github.com/danielemelini/TABOO. It was originally distributed by the Samizdat Press and now it can also be found at the page https://samizdat.mines.edu/.

3.3.1.1 Ice Load

In continental areas, we define the *ice load* by

$$\sigma^{\mathrm{i}}(\gamma, t) = \rho^{\mathrm{i}} d^{\mathrm{i}}(\gamma)\, \chi^{\mathrm{i}}(\gamma)\, f^{\mathrm{i}}(t), \tag{3.55}$$

where ρ^{i} is the ice density, d^{i} ($d^{\mathrm{i}} \geq 0$) is the ice thickness at a given location of spherical coordinates $\gamma = (\theta, \lambda)$, $\chi^{\mathrm{i}}(\gamma)$ is the *ice characteristic function*

$$\chi^{\mathrm{i}}(\gamma) \equiv \begin{cases} 1, \text{ if } \gamma \in i \\ 0, \text{ otherwise,} \end{cases} \tag{3.56}$$

and $f^{\mathrm{i}}(t) \geq 0$ is non-dimensional modulating function describing the ice thickness time history. No marine based ice is considered.

3.3.1.2 Meltwater Load

The *meltwater load* is

$$\sigma^{\mathrm{w}}(\gamma, t) = \rho^{\mathrm{w}} d^{\mathrm{w}}(t)\, \chi^{\mathrm{w}}(\gamma), \tag{3.57}$$

where ρ^{w} is the density of water, we have assumed that the water layer is spatially uniform (i.e., its thickness d^{w} only depends on time), and $\chi^{\mathrm{w}}(\gamma)$ is the *ocean characteristic function*

$$\chi^{\mathrm{w}}(\gamma) \equiv \begin{cases} 1, \text{ if } \gamma \in w \\ 0, \text{ otherwise.} \end{cases} \tag{3.58}$$

Since the contour of domain w is assumed to be not time dependent, in this approach the fixed shorelines approximation is implicit.

3.3.2 *Simplified Loading Problem*

The problem consists in evaluating the Earth's response to the total surface load

$$\sigma^{\mathrm{iw}}(\gamma, t) = \sigma^{\mathrm{i}} + \sigma^{\mathrm{w}}, \tag{3.59}$$

where according to (3.55) and (3.57), we have

$$\sigma^{\mathrm{iw}}(\gamma, t) = \rho^{\mathrm{i}} d^{\mathrm{i}}(\gamma)\, \chi^{\mathrm{i}}(\gamma)\, f^{\mathrm{i}}(t) + \rho^{\mathrm{w}} d^{\mathrm{w}}(t)\, \chi^{\mathrm{w}}(\gamma). \tag{3.60}$$

3.3.3 Complex Loads

For a single ice load, the geometry of the simplified loading problem is sketched in Fig. 3.2a. Expression (3.59) can be generalised to the case of $N_{\rm i}$ individual ice elements:

$$\sigma^{\rm iw}(\gamma, t) = \rho^{\rm i} \sum_{k=1}^{N_{\rm i}} \left(d^{{\rm i},k}(\gamma)\, \chi^{{\rm i},k}(\gamma)\, f^{{\rm i},k}(t) \right) + \rho^{\rm w} d^{\rm w}(t)\, \chi^{\rm w}(\gamma), \tag{3.61}$$

where it can be assumed (although this is not mandatory) that all the ice elements are disjoint and that the ocean and the ice elements are not overlapping, so that there is no marine ice and all ice is grounded above sea level (see Fig. 3.2b). Since in Eq. (3.55) we have separated the space and time dependence of the ice loads into functions $d^{\rm i}(\gamma)$ and $f^{\rm i}(t)$, respectively, Eq. (3.61) is useful to describe complex ice loads by combining the effects of geometrically simple (i.e., disc–shaped) elements that evolve in time according to different histories. This approach, which relies upon the superposition principle, has been also adopted in the work of Bevis et al. (2016) and exploited in the Post Glacial Rebound calculator TABOO (Spada et al. 2011a).

We note that in Eqs. (3.59) and (3.61) the water term is sometimes neglected ($d^{\rm w}(t) \approx 0$), since the ice loads are generally expected to have a dominating role, especially across the glaciated areas and in the surroundings. This approximation has been adopted, for example, in the benchmark tests published in Spada et al. (2011a). However, neglecting the meltwater contribution would violate the mass conservation principle (the surface load would be, in this case, *physically implausible*). Of course, assuming a uniform meltwater load as done in Eq. (3.57) constitutes sometimes a crude approximation, since it is known that due to gravitational, deformational and rotational effects the meltwater load is effectively not uniform, as will be discussed

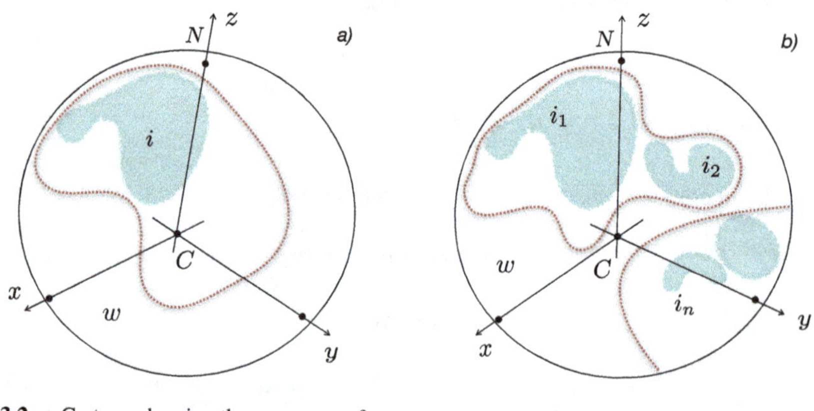

Fig. 3.2 **a** Cartoon showing the geometry of a system composed by an ice load i and by an ocean w of arbitrary geometry. In **b** the case of multiple ice loads is considered. The (fixed) shorelines are shown by dashed contours

in Chap. 5. We finally note that in Eqs. (3.59) and (3.61) we are neglecting rotational contributions, which in the *rotationally self-consistent* approach discussed in the next chapter are implicitly taken into account.

3.3.4 Mass Conservation

Since mass is conserved, according to Eq. (1.135) the following condition is met[2]

$$\langle \sigma^{\mathrm{iw}}(\gamma, t) \rangle^{\mathrm{e}} = 0, \tag{3.62}$$

which allows to determine, from (3.59), the thickness variation of the meltwater layer

$$d^{\mathrm{w}}(t) = -\frac{\rho^{\mathrm{i}}}{\rho^{\mathrm{w}}} \frac{\langle d^{\mathrm{i}}(\gamma)\chi^{\mathrm{i}}(\gamma)\rangle^{\mathrm{e}}}{\langle \chi^{\mathrm{w}}(\gamma)\rangle^{\mathrm{e}}} f^{\mathrm{i}}(t). \tag{3.63}$$

We note that, using the definition (1.136) of whole-Earth average $\langle \cdot \rangle^{\mathrm{e}}$, it is easy to obtain a more physically meaningful expression for $d^{\mathrm{w}}(t)$. Indeed, $\langle d^{\mathrm{i}}(\gamma)\chi^{\mathrm{i}}(\gamma)\rangle^{\mathrm{e}} = (A^{\mathrm{i}}/A^{\mathrm{e}})\ \langle d^{\mathrm{i}}(\gamma)\rangle^{\mathrm{i}}$, where A^{i} is the area of the surface of the ice element and $\langle \cdot \rangle^{\mathrm{i}}$ denotes the average over the ice domain. Since the mass of the ice is $m^{\mathrm{i}}(t) = \rho^{\mathrm{i}} \int_i d^{\mathrm{i}}(\gamma) f^{\mathrm{i}}(t) \mathrm{d}A = A^{\mathrm{i}} \rho^{\mathrm{i}} f^{\mathrm{i}}(t) \langle d^{\mathrm{i}}(\gamma)\rangle^{\mathrm{i}}$, where $\int_i$ is the integral over the ice distribution, we have $\langle d^{\mathrm{i}}(\gamma)\chi^{\mathrm{i}}(\gamma)\rangle^{\mathrm{e}} = m^{\mathrm{i}}(t)/(\rho^{\mathrm{i}} A^{\mathrm{e}} f^{\mathrm{i}}(t))$. Furthermore, because $\langle \chi^{\mathrm{w}}(\gamma)\rangle^{\mathrm{e}} = A^{\mathrm{w}}/A^{\mathrm{e}}$, where A^{w} is the area of the surface of the oceans, (3.63) gives

$$d^{\mathrm{w}}(t) = -\frac{m^{\mathrm{i}}(t)}{\rho^{\mathrm{w}} A^{\mathrm{w}}}, \tag{3.64}$$

which matches the expression of the *eustatic* (or *barystatic*, according to Gregory et al. 2019) sea-level change defined in (5.49) in the framework of the classical theory of the SLE by Farrell and Clark (1976). Of course, this is a direct consequence of our assumption of a uniform sea-level rise in the simplified loading problem. The negative sign indicates that, as expected, an increase in the ice thickness ($m^{\mathrm{i}}(t) > 0$) is compensated by a uniform sea-level fall and *vice versa*. Finally, by substitution of (3.63) into (3.60), we obtain the final form of the mass-conserving surface load:

$$\sigma^{\mathrm{iw}}(\gamma, t) = \rho^{\mathrm{i}} \left(d^{\mathrm{i}}(\gamma)\chi^{\mathrm{i}}(\gamma) - \frac{\langle d^{\mathrm{i}}(\gamma)\chi^{\mathrm{i}}(\gamma)\rangle^{\mathrm{e}}}{\langle \chi^{\mathrm{w}}(\gamma)\rangle^{\mathrm{e}}} \chi^{\mathrm{w}}(\gamma) \right) f^{\mathrm{i}}(t). \tag{3.65}$$

---

[2] We note that the mass conservation condition (1.135) actually prescribes a vanishing average over the Earth's surface for the *variation* of the load with respect to a reference state, and not for the load itself. Hence in what follows we are implicitly assuming, without loss of generality, that the surface load in the reference state is zero.

3.3.5 Expansion in Spherical Harmonics

The mass-conserving surface load (3.65) can be easily expanded in series of spherical harmonics, which facilitates the computation of the surface response functions further on. First, we expand the ice component, with

$$d^{\mathrm{i}}(\gamma)\chi^{\mathrm{i}}(\gamma) = \sum_{nm} d^{\mathrm{i}}_{nm}\, Y_{nm}(\gamma), \tag{3.66}$$

where, using Eq. (A.36) and the definition of $\chi^{\mathrm{i}}(\gamma)$, the coefficients are

$$d^{\mathrm{i}}_{nm} \equiv \frac{1}{4\pi}\int_{\gamma} \chi^{\mathrm{i}}(\gamma)\, d^{\mathrm{i}}(\gamma)\, Y^{*}_{nm}(\gamma)\mathrm{d}\gamma = \frac{1}{4\pi}\int_{\mathrm{i}} d^{\mathrm{i}}(\gamma)\, Y^{*}_{nm}(\gamma)\mathrm{d}\gamma, \tag{3.67}$$

and where the integral $\int_{\mathrm{i}}$ is over the ice load. In a similar way, we expand the ocean characteristic function

$$\chi^{\mathrm{w}}(\gamma) = \sum_{nm} \chi^{\mathrm{w}}_{nm}\, Y_{nm}(\gamma), \tag{3.68}$$

with coefficients

$$\chi^{\mathrm{w}}_{nm} \equiv \frac{1}{4\pi}\int_{\gamma} \chi^{\mathrm{w}}(\gamma)\, Y^{*}_{nm}(\gamma)\mathrm{d}\gamma = \frac{1}{4\pi}\int_{\mathrm{o}} Y^{*}_{nm}(\gamma)\mathrm{d}\gamma, \tag{3.69}$$

where the second integral is over the oceans. Last, since according to (A.38) the two spatial averages in Eq. (3.65) can be expressed in terms of degree $n = 0$ coefficients, we obtain the expansion

$$\sigma^{\mathrm{iw}}(\gamma, t) = \rho^{\mathrm{i}}\, f^{\mathrm{i}}(t) \sum_{nm} d^{\mathrm{iw}}_{nm}\, Y_{nm}(\gamma), \tag{3.70}$$

with new coefficients

$$d^{\mathrm{iw}}_{nm} = d^{\mathrm{i}}_{nm} - \left(\frac{d^{\mathrm{i}}_{00}}{\chi^{\mathrm{w}}_{00}}\right) \chi^{\mathrm{w}}_{nm}, \tag{3.71}$$

where we note that, as a consequence of mass conservation, $d^{\mathrm{iw}}_{00} = 0$. Neglecting the second term on the *r.h.s.* of Eq. (3.71) would imply a violation of the mass conservation principle.

3.3.6 General Solution

In Sect. 3.2.2, we have obtained the Earth's response to a spatially variable loading function, characterised by a piecewise constant time history. In order to solve the simplified loading problem, here we adopt the same principles, by performing a 3-D convolution involving appropriate GFs and the *ad hoc* load function σ^{iw} introduced in previous section.

Specifically, the SRFs for the simplified loading problem are

$$\begin{Bmatrix} \mathcal{U} \\ \mathcal{V}_\theta \\ \mathcal{V}_\lambda \\ \mathcal{G} \end{Bmatrix} (\gamma, t) \equiv \begin{Bmatrix} \Gamma^u \\ \Gamma^{v_\theta} \\ \Gamma^{v_\lambda} \\ \Gamma^g \end{Bmatrix} \otimes \sigma^{\mathrm{iw}}, \tag{3.72}$$

where $\mathcal{U}$, $\mathcal{V}_\theta$, and $\mathcal{V}_\lambda$ are the components, in a spherical coordinates system, of the displacement vector at the Earth's surface, and $\mathcal{G}$ is the change in the geoid height.[3] The corresponding GFs Γ^u, Γ^{v_θ}, Γ^{v_λ} and Γ^g are given by Eqs. (3.7) and (3.9), whereas σ^{iw} is expressed by (3.65).

To evaluate the spatiotemporal 3-D convolutions in Eq. (3.72), we follow the same approach outlined in Sect. 3.2.3 to determine the solution of the general loading problem. For the sake of the readers convenience, we solve this exercise in detail for the vertical displacement SRF, obtaining:

$$\begin{aligned}
\mathcal{U}(\gamma, t) \equiv\ & \Gamma^u \otimes \sigma^{\mathrm{iw}} = \\
=\ & \int_{-\infty}^{+\infty} \mathrm{d}t' \int_e \Gamma^u(\gamma, \gamma', t - t') \sigma^{\mathrm{iw}}(\gamma', t')\, \mathrm{d}A' \\
=\ & \frac{a^3 \rho^{\mathrm{i}}}{m^{\mathrm{e}}} \int_{-\infty}^{+\infty} \mathrm{d}t' f^{\mathrm{i}}(t') \int_\gamma \left(\sum_{nm} \frac{h_n^{\mathrm{L}}(t - t')}{2n + 1} Y_{nm}^*(\gamma') Y_{nm}(\gamma) \right) \cdot \\
& \cdot \left(\sum_{n'm'} d_{n'm'}^{\mathrm{iw}} Y_{n'm'}(\gamma') \right) \mathrm{d}\gamma' \\
=\ & \frac{a^3 \rho^{\mathrm{i}}}{m^{\mathrm{e}}} \sum_{nm} \frac{Y_{nm}(\gamma)}{2n + 1} \cdot \\
& \cdot \int_{-\infty}^{+\infty} \mathrm{d}t' h_n^{\mathrm{L}}(t - t') f^{\mathrm{i}}(t') \sum_{n'm'} d_{n'm'}^{\mathrm{iw}} \underbrace{\int_\gamma Y_{nm}^*(\gamma') Y_{n'm'}(\gamma') \mathrm{d}\gamma'}_{4\pi\, \delta_{nn'}\, \delta_{mm'}} \\
=\ & \frac{3 \rho^{\mathrm{i}}}{\rho^{\mathrm{e}}} \sum_{nm} d_{nm}^{\mathrm{iw}}\, c_n^h(t)\, Y_{nm}(\gamma),
\end{aligned} \tag{3.73}$$

where we have used the spectral form of the GF Γ^u (2.54), coefficients d_{nm}^{iw} are given by Eq. (3.71), we have exploited the orthogonality property of the CSHs, ρ^{e} is the

---

[3] To simplify the notation, here we do not use superscript *sur* to mark these SRFs, as we have done also in Sect. 3.2.2. This is not necessary, since in the simplified loading problem we *only* deal with surface loading, while in the general GIA problem *rotational response functions* (RRFs) shall be needed and labelled by *rot* in later sections.

average density of the Earth, $f^{\mathrm{i}}(t)$ is time history of the ice load, and function $c_n^h(t)$ is defined in Eq. (3.75) below.

Along these same lines, after some algebra, the full set of the four fundamental solutions of the simplified loading problem is easily established, with

$$\begin{Bmatrix} \mathcal{U} \\ \mathcal{V}_\theta \\ \mathcal{V}_\lambda \\ \mathcal{G} \end{Bmatrix} (\gamma, t) \equiv \frac{3\rho^{\mathrm{i}}}{\rho^{\mathrm{e}}} \sum_{nm} d_{nm}^{\mathrm{iw}} \begin{Bmatrix} c_n^h \\ c_n^\ell \\ c_n^\ell \\ c_n^k \end{Bmatrix} (t) \cdot \begin{Bmatrix} 1 \\ \nabla_\theta \\ \nabla_\lambda \\ 1 \end{Bmatrix} Y_{nm}(\gamma), \tag{3.74}$$

where $\nabla_\theta \equiv \frac{\partial}{\partial\theta}$ and $\nabla_\lambda \equiv \frac{1}{\sin\theta}\frac{\partial}{\partial\lambda}$ are the $\hat{\theta}$ and $\hat{\lambda}$ components of the horizontal gradient operator, respectively, and where we have defined the convolved LLNs as

$$\begin{Bmatrix} c_n^h \\ c_n^\ell \\ c_n^k \end{Bmatrix} (t) \equiv \frac{1}{2n+1} \int_{-\infty}^{+\infty} \mathrm{d}t' \begin{Bmatrix} h_n^{\mathrm{L}} \\ \ell_n^{\mathrm{L}} \\ \delta + k_n^{\mathrm{L}} \end{Bmatrix} (t - t') f^{\mathrm{i}}(t'), \tag{3.75}$$

where we see that the δ term in the third of Eq. (3.75) accounts for the direct effect of the load on gravitational potential, which is also present for a rigid Earth (i.e., for $h_n^{\mathrm{L}} = \ell_n^{\mathrm{L}} = k_n^{\mathrm{L}} = 0$).

We remark that solutions (3.74) hold for a general surface load, without any particular symmetry. However, in the following Sect. 3.3.7 and Sect. 3.3.8, we shall consider simple surface loads characterized by a strong symmetry, for which the response can be expressed by expansions in series of Legendre polynomials. A particularly important asymmetric load, the *rectangular load*, shall be considered in Sect. 3.3.9.

3.3.7 Axis-Symmetric Loads

Here we consider the particularly important case of axis-symmetric surface loads, i.e., surface loads for which regions i and w are symmetric for rotations around the polar z-axis. A cartoon showing examples of axis-symmetric loads is shown in Fig. 3.3. More specifically, we find it convenient to consider the particular axis-symmetric loads for which, regardless the value of longitude in the range $(0 \le \lambda \le 2\pi)$, we have

$$\chi^{\mathrm{i}}(\theta) \equiv \begin{cases} 1, & \text{if } 0 \le \theta \le \alpha \\ 0, & \text{if } \alpha < \theta \le \pi, \end{cases} \tag{3.76}$$

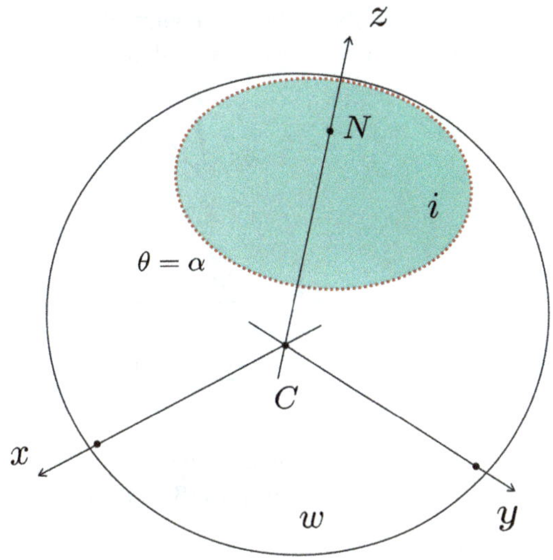

Fig. 3.3 A load characterised by an axial symmetry with respect to the z axis of the cartesian reference frame. Labels i and w denote the ice and the meltwater components of the load, respectively. The ice margin, which follows the dashed shoreline, is located at co-latitude $\theta = \alpha$

where α is a fixed angle (the load *half-amplitude*). The characteristic ocean function $\chi^{\mathrm{w}}(\theta)$ is chosen to be complementary to $\chi^{\mathrm{i}}(\theta)$, with

$$\chi^{\mathrm{w}}(\theta) \equiv \begin{cases} 0, \text{ if } 0 \le \theta \le \alpha \\ 1, \text{ if } \alpha < \theta \le \pi. \end{cases} \tag{3.77}$$

According to these definitions, the ice margin is located at co-latitude $\theta = \alpha$, which also represents the location of the hypothetical shorelines.[4] The geometry of the ice load is still specified by the ice thickness d^{i} which, by virtue of axial symmetry, is only a function of colatitude θ.

For axis-symmetric loads, it can be easily verified that coefficients d^{i}_{nm} and χ^{w}_{nm} in Eq. (3.71) turn out to be purely zonal (i.e., they vanish except for the rotationally-symmetric modes with harmonic order $m = 0$). This can be seen using (3.67) and the definition of CSHs (A.22), which provide

$$d^{\mathrm{i}}_{nm} = \frac{1}{4\pi} \int_{\gamma} d^{\mathrm{i}}(\theta) \chi^{\mathrm{i}}(\theta) Y^{*}_{nm}(\gamma) \mathrm{d}\gamma = \frac{\delta_{m0}\, \mu_{nm}}{2n+1}\, d^{\mathrm{i}}_{n}, \tag{3.78}$$

where μ_{nm} is the normalization constant defined by Eq. (A.23), and

$$\begin{aligned} d^{\mathrm{i}}_{n} &= \frac{2n+1}{2} \int_{0}^{\pi} d^{\mathrm{i}}(\theta) \chi^{\mathrm{i}}(\theta) P_n(\cos\theta) \sin\theta \, \mathrm{d}\theta \\ &= \frac{2n+1}{2} \int_{0}^{\alpha} d^{\mathrm{i}}(\theta) P_n(\cos\theta) \sin\theta \, \mathrm{d}\theta \end{aligned} \tag{3.79}$$

[4] Of course, more general axis-symmetric definitions are possible, in which a zonal ocean only partially fills the region $\theta \ge \alpha$.

are the coefficients of the expansion of $d^{\mathrm{i}}(\theta)\chi^{\mathrm{i}}(\theta)$ in series of LPs (see Eq. A.14). Recalling (3.69), we further note that

$$\chi^{\mathrm{w}}_{nm} = \frac{1}{4\pi}\int_{\gamma} \chi^{\mathrm{w}}(\theta) Y^{*}_{nm}(\gamma)\mathrm{d}\gamma = \frac{\delta_{m0}\,\mu_{nm}}{2n+1}\,\chi^{\mathrm{w}}_{n}, \tag{3.80}$$

where

$$\begin{aligned} \chi^{\mathrm{w}}_{n} &= \frac{2n+1}{2}\int_{0}^{\pi} \chi^{\mathrm{w}}(\theta) P_n(\cos\theta)\sin\theta\,\mathrm{d}\theta \\ &= \frac{2n+1}{2}\int_{\alpha}^{\pi} P_n(\cos\theta)\sin\theta\,\mathrm{d}\theta \end{aligned} \tag{3.81}$$

are the coefficients of the expansion of $\chi^{\mathrm{w}}(\theta)$ in series of LPs. Hence, upon substitution of (3.81) and (3.79) into (3.71), for axis-symmetric loads we have

$$d^{\mathrm{iw}}_{nm} = \frac{\delta_{m0}\,\mu_{nm}}{2n+1}\,d^{\mathrm{iw}}_{n}, \tag{3.82}$$

with

$$d^{\mathrm{iw}}_{n} = d^{\mathrm{i}}_{n} - \left(\frac{d^{\mathrm{i}}_{0}}{\chi^{\mathrm{w}}_{0}}\right)\chi^{\mathrm{w}}_{n}, \tag{3.83}$$

where constants d^{i}_{n} and χ^{w}_{n} shall depend on the geometry of the specific axis-symmetric surface load adopted. In Sect. 3.3.8 below these will be explicitly evaluated for two very important load types, namely the *disc load* and the *cap load.*

By substituting (3.83) into the general solutions of the simplified loading problem (3.74), we obtain particular solutions for axis-symmetric loads, which are now simply expressed in series of Legendre polynomials. For the components of surface displacement and the geoid height we have

$$\begin{Bmatrix} \mathcal{U} \\ \mathcal{V}_{\theta} \\ \mathcal{V}_{\lambda} \\ \mathcal{G} \end{Bmatrix}(\theta,t) \equiv \frac{3\rho^{\mathrm{i}}}{\rho^{\mathrm{e}}}\sum_{n=n_{\min}}^{n_{\max}} d^{\mathrm{iw}}_{n} \begin{Bmatrix} c^{h}_{n} \\ c^{\ell}_{n} \\ c^{\ell}_{n} \\ c^{k}_{n} \end{Bmatrix}(t) \cdot \begin{Bmatrix} 1 \\ \partial_{\theta} \\ 0 \\ 1 \end{Bmatrix} P_n(\cos\theta), \tag{3.84}$$

where we note that the time convolutions $c^{h}_{n}(t)$, $c^{\ell}_{n}(t)$, and $c^{k}_{n}(t)$ are still given by Eq. (3.75) and that condition $\mathcal{V}_{\lambda} = 0$ is a consequence of the axial symmetry of the load. In (3.84), we have made explicit the minimum and maximum degrees of the harmonic synthesis (i.e., $n_{\min}$ and $n_{\max}$, respectively), which are important parameters in all numerical applications.

We note that by a rotation of the reference frame, the surface response to axis-symmetric loads placed at an arbitrary position of the Earth's surface can be easily obtained from the solutions above. This is sometimes exploited in numerical tools that take advantage of the simple solution available for axis-symmetric loads, like programs REAR (Melini et al. 2015) and TABOO (Spada et al. 2011a).

3.3.8 *Disc and Cap Loads*

We now turn our attention to particular axis-symmetric surface loads that are commonly employed in PGR modelling: the *disc load* and the *cap load*, respectively.

3.3.8.1 Disc Load

For this particular axis-symmetric surface load, depicted in Fig. 3.4b, the ice thickness is

$$d^{\mathrm{i}}(\theta) = d^{\mathrm{i}} = \text{constant} \tag{3.85}$$

across region i. The assumption may appear grossly unrealistic, since it is known that under perfect plastic conditions, the equilibrium profile of an ice sheet is parabolic (see, e.g., Oerlemans 2008). However, based on the superposition principle expressed by Eq. (3.61), in PGR modelling the disc load is useful to build more complex loads composed by several elements, whose margins can be made time-dependent by choosing appropriately the individual ice height time histories (see e.g., Tushingham and Peltier 1991).

An illustration of this concept is shown in Fig. 3.5. Loads of non-glacial origin like those associated to sedimentary basins can be modeled as combinations of disc loads as well (see e.g., Ivins et al. 2007; Karpytchev et al. 2018). Furthermore, due to its simple geometry and mathematical expression, the disc load is routinely employed to model the Earth's response to the melting of the present ice sheets and glaciers due to global warming (see e.g., Spada et al. 2012b; Memin et al. 2014; Melini et al. 2015; Bevis et al. 2016). Disc loads are also employed to study the flexure of the planetary lithospheres under the load exerted by mascons (see e.g., Melosh 1978; Pullan and Lambeck 1982), based on the classical result of Brotchie (1971) in planar geometry. In the example of Fig. 3.6, discs of variable size are used as *finite elements* to model

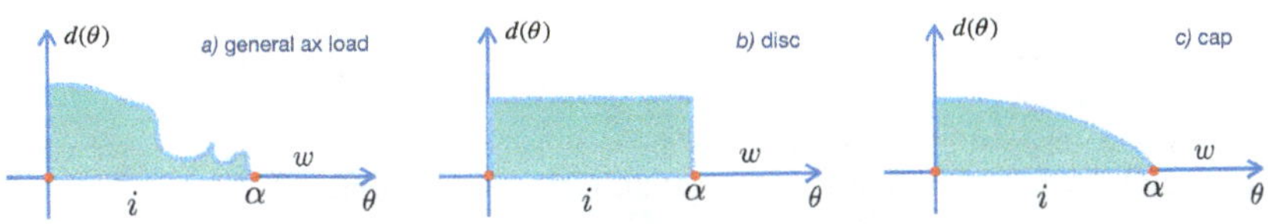

Fig. 3.4 Sketch showing the thickness $d = d(\theta)$ for **a** a general axis-symmetric ice load, **b** a disc load and **c** a cap load. The load margin is located at $\theta = \alpha$ and it is coincident with the shoreline. The ice and the ocean domains are labelled by i and w, respectively. The axis of symmetry of the loads is the z axis of the cartesian reference frame

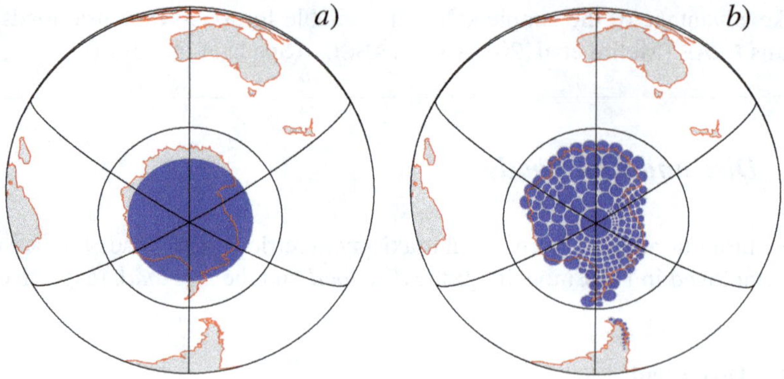

Fig. 3.5 A bed of discs like those depicted here, each characterised by a different thickness, resolution and time-history, can be employed to describe a complex ice load with evolving boundaries. Interstices or overlaps between the discs, however, cannot be avoided. In **a** a single disc is shown over Antarctica, with a poor coverage. The discs in **b** represent the ice sheet in Antarctica, according to the ICE-3G model of Tushingham and Peltier (1991). The *fixed shorelines* are marked in red

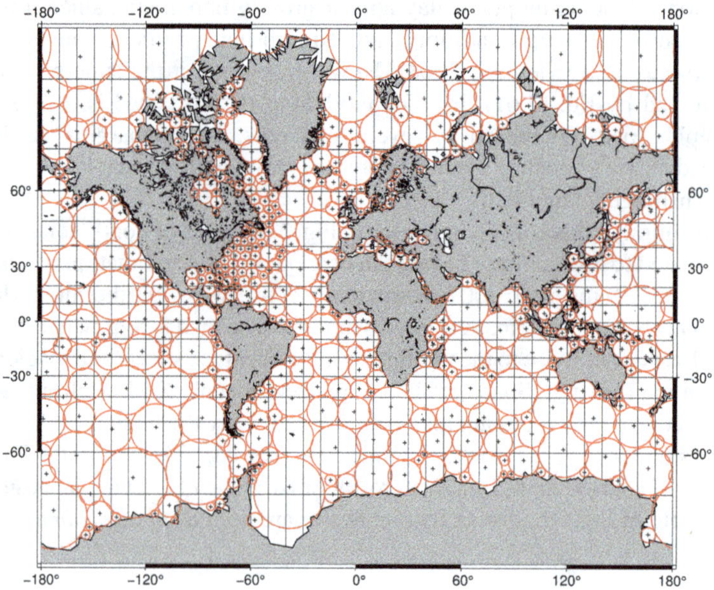

Fig. 3.6 Axis-symmetric disc loads distributed over the world's oceans, in order to describe the meltwater load. The height of each disk is determined by solving the gravitationally self-consistent SLE. The image, inspired by the work of Tushingham and Peltier (1991), has been reconstructed and drawn by Veronica Guerra during a stage at the University of Urbino *Carlo Bo* in 2016

the non-uniform meltwater load distributed over the oceans in the context of the solution of the SLE (Tushingham and Peltier 1991).

In the case of the disc load, the evaluation of the various terms in Eq. (3.83) is quite straightforward, since one just needs to take into account the properties of the LPs. In particular, the "parity" (A.4) and the "shifted derivative" properties (A.5) are particularly useful. After some algebra one obtains

$$d_n^{\mathrm{iw}} \text{ for the disc load} = d^{\mathrm{i}} \cdot \begin{cases} 0 & \text{if } n = 0 \\ -\dfrac{P_{n+1}(\cos\alpha) - P_{n-1}(\cos\alpha)}{1+\cos\alpha} & \text{if } n \geq 1, \end{cases} \tag{3.86}$$

in accord with Eq. (8) of Bevis et al. (2016). By substitution of (3.86) into (3.84) the response of the Earth to an axis-symmetric disc load is thus fully determined.

3.3.8.2 Cap Load

The cap load captures the geometry of actual ice sheets better than an individual disc load. For this particular axis-symmetric surface load, the ice thickness is *not* constant across region i, but it varies according to

$$d^{\mathrm{i}}(\theta) = d^{\mathrm{i,max}} \sqrt{\frac{\cos\theta - \cos\alpha}{1-\cos\alpha}}, \quad 0 \leq \theta \leq \alpha, \tag{3.87}$$

where the parallel defined by $\theta = \alpha$ still represents the margin of the ice load *and* the shoreline at the same time, and parameter $d^{\mathrm{i,max}}$ denotes the maximum thickness of the cap load, attained along the axis of symmetry ($\theta = 0$, see Fig. 3.4c).

As far as we know, expression (3.87) was first introduced, in the context of GIA, by Yuen et al. (1986, see their Appendix A). The rationale for introducing this particular surface load is that it qualitatively matches the profile of an ice sheet at equilibrium, assuming a perfect plastic behavior (see e.g., Oerlemans 2008). We note that a more general expression for the profile of an ice sheet is presented by Nye (1959), and more recently by Faraoni and Vokey (2015). Indeed, the nature of the profile described by Eq. (3.87) can be better appreciated considering the limit for small α, and consequently small θ, which is appropriate in the case we neglect the Earth's sphericity. With $\cos\theta \approx 1 - \theta^2/2$ and $\cos\alpha \approx 1 - \alpha^2/2$, we find $d^{\mathrm{i}}(\theta) \approx d^{\mathrm{i,max}}\sqrt{1 - (\theta/\alpha)^2}$, which effectively describes an elliptic profile, and not a parabolic one as proposed by Yuen et al. (1986). In any instance, Eq. (3.87) constitutes an excellent alternative to the disc load since, as discussed by Yuen et al. (1986), it minimises the distortions at the border of the ice sheet due to edge effects, which instead are more pronounced for a disc of constant thickness.

Due to the θ-dependence of d^{i} in Eq. (3.87), the evaluation of the various terms in (3.83) is somewhat cumbersome, contrary to the case of the disc load considered in previous section. Nevertheless, due to the particular form of $d^{\mathrm{i}}(\theta)$, it is possible to obtain the harmonic coefficients of the surface load in an analytical form. We only give the final result, leaving the readers to Spada (2003) for a detailed derivation:

$$d_n^{\mathrm{iw}} \text{ for the cap load} = d^{\mathrm{i,max}} \cdot \begin{cases} 0 & \text{if } n = 0 \\ c_n^{(1)}(\alpha) + c_n^{(2)}(\alpha) & \text{if } n \geq 1, \end{cases} \tag{3.88}$$

where the two constants $c_n^{(1,2)}$ depend upon the load half-amplitude, with

$$c_n^{(1)}(\alpha) = -\frac{1/4}{1-\cos\alpha}\left(\frac{T_{n+1}(\alpha) - T_{n+2}(\alpha)}{n + 3/2} - \frac{T_{n-1}(\alpha) - T_n(\alpha)}{n - 1/2}\right) \tag{3.89}$$

and

$$c_n^{(2)}(\alpha) = -1/3\left(\frac{1-\cos\alpha}{1+\cos\alpha}\right)\left(P_{n+1}(\cos\alpha) - P_{n-1}(\cos\alpha)\right), \tag{3.90}$$

respectively, where $T_n(\alpha) = \cos(n\alpha)$ are the Čebyšëv polynomials of 1st kind and we have used property (A.6) of the Legendre polynomials.

3.3.9 Non Axis-Symmetric Loads

As an example of non-axis-symmetric surface load, we consider the "rectangular" load, which is depicted in Fig. 3.7. As it is clear from the figure, the load is bounded by two parallels and two meridians. Of course, since it is defined on the spherical surface of the Earth, it would be more appropriate to refer to this load to as a *spherical rectangle*.

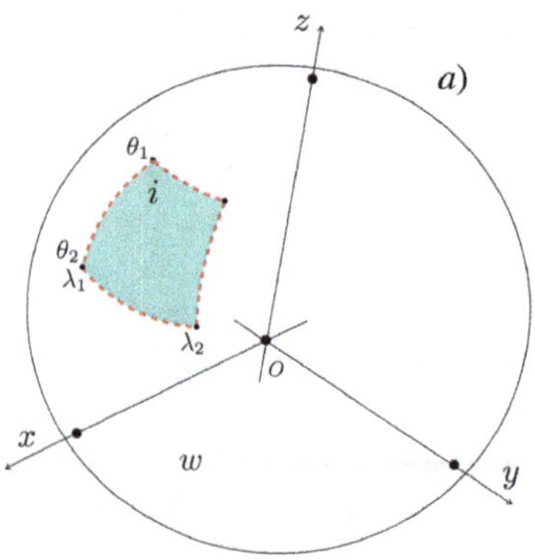

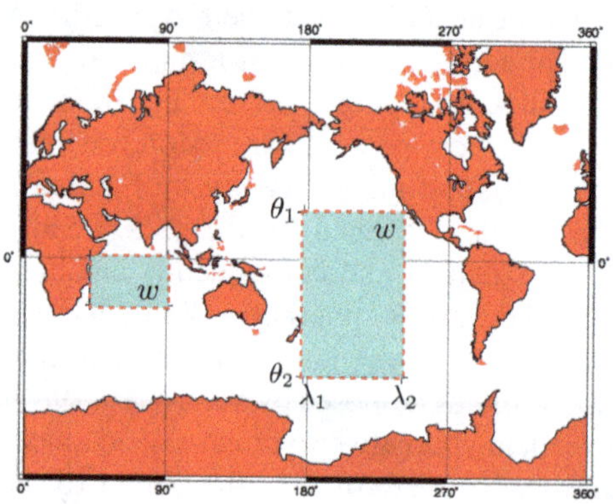

Fig. 3.7 Frame **a** shows a rectangular ice load (i) bounded by lines of latitude and longitude. The margins follow the (dashed) shorelines and an ocean of uniform thickness (w) is complementary to the ice load. Frame **b** shows, in a different geographical projection, two rectangles that mimic water loads over the oceans

The rectangular load is a valid alternative to the axis-symmetric disc load, especially when we consider complex aggregates combining individual loads. Indeed, the advantage with respect to the disc loads is that using the rectangles the interstices and the overlaps can be avoided. However, contrary to disc loads, the spherical harmonic coefficients of the rectangular loads are not fully analytical and their numerical computation poses some difficulties.

A single rectangular element is defined by the ice characteristic function

$$\chi^{\mathrm{i}}(\gamma) \equiv \begin{cases} 1, \text{ if } (\theta_1 \leq \theta \leq \theta_2) \text{ and } (\lambda_1 \leq \lambda \leq \lambda_2) \\ 0, \text{ elsewhere}, \end{cases} \tag{3.91}$$

where $0 < \theta_i < \pi$ and $0 \leq \lambda_i \leq 2\pi$ $(i = 1, 2)$ are the bounding co-latitudes and longitudes shown in Fig. 3.7, respectively, and we assume that the load thickness is

$$d^{\mathrm{i}}(\gamma) = d^{\mathrm{i}} = \text{constant}, \tag{3.92}$$

as in the case of the disc load. Of course, the choice of the geometry of the "ocean" that compensates the excess mass of the rectangular load is somewhat arbitrary. For the sake of simplicity, in the following we choose the ocean geometry by defining a characteristic function

$$\chi^{\mathrm{w}}(\gamma) \equiv \begin{cases} 0, \text{ if } (\theta_1 \leq \theta \leq \theta_2) \text{ and } (\lambda_1 \leq \lambda \leq \lambda_2) \\ 1, \text{ elsewhere}, \end{cases} \tag{3.93}$$

which is complementary to the load characteristic function (3.91), i.e., $\chi^{\mathrm{w}}(\gamma) + \chi^{\mathrm{i}}(\gamma) = 1$. Other more realistic choices are of course possible, such as the one of using an *ocean function* that closely follows the actual contours of the coastlines.

We now turn our attention to the evaluation of the harmonic coefficients d^{i}_{nm} and of χ^{w}_{nm}, which are both required in the general expression for the total surface load d^{iw}_{nm} (Eq. 3.71). From (3.67), also using the definition of CSH (A.22), the harmonic coefficients of the rectangular load are

$$\begin{aligned} d^{\mathrm{i}}_{nm} &= \frac{1}{4\pi} \int_{\gamma} \chi^{\mathrm{i}}(\gamma)\, d^{\mathrm{i}}(\gamma)\, Y^*_{nm}(\gamma) \mathrm{d}\gamma = \frac{d^{\mathrm{i}}}{4\pi} \int_{i} Y^*_{nm}(\gamma)\, \mathrm{d}\gamma \\ &= \frac{d^{\mathrm{i}}}{4\pi} \mu_{nm} \int_{\lambda_1}^{\lambda_2} \mathrm{e}^{-im\lambda} \mathrm{d}\lambda \int_{\theta_1}^{\theta_2} P_{nm}(\cos\theta) \sin\theta \mathrm{d}\theta \\ &= \frac{d^{\mathrm{i}}}{4\pi} \mu_{nm}\, \sigma_{nm}\, a_m, \end{aligned} \tag{3.94}$$

where the normalization factor μ_{lm} is given by (A.23) and we have defined the constants

$$\sigma_{nm} \equiv - \int_{x_1}^{x_2} P_{nm}(x)\, \mathrm{d}x, \tag{3.95}$$

with $x_1 = \cos\theta_1$ and $x_2 = \cos\theta_2$, and

$$a_m \equiv \int_{\lambda_1}^{\lambda_2} \mathrm{e}^{-im\lambda} \mathrm{d}\lambda. \tag{3.96}$$

Due to the arbitrary interval of integration, σ_{nm} cannot be expressed in a closed form, although useful recursive relations exist for integrals of associated Legendre polynomials (Paul 1978). According to our experience, standard integration routines in double precision (for example those proposed by Press 2007) provide satisfactory results in spite of the oscillatory nature of the integrand. As expression (3.96) suggests, the evaluation of a_m is however fully analytical but the two cases $m = 0$ and $m \neq 0$ must be dealt with separately. By omitting the straightforward details, one obtains:

$$a_m \equiv \begin{cases} \lambda_2 - \lambda_1, & m = 0 \\ \dfrac{1}{m}\Big[(\sin m\lambda_2 - \sin m\lambda_1) + i\,(\cos m\lambda_2 - \cos m\lambda_1)\Big], & m \neq 0. \end{cases} \tag{3.97}$$

For the evaluation of χ^{w}_{nm}, we start from the general definition (3.69) and we use the property of additivity of the integrals to obtain

$$\begin{aligned} \chi^{\mathrm{w}}_{nm} &= \frac{1}{4\pi}\int_\gamma \chi^{\mathrm{w}}(\gamma)\, Y^*_{nm}(\gamma)\mathrm{d}\gamma = \frac{1}{4\pi}\int_w Y^*_{nm}(\gamma)\mathrm{d}\gamma, \\ &= \frac{1}{4\pi}\left(\int_\gamma Y^*_{nm}(\gamma)\mathrm{d}\gamma - \int_i Y^*_{nm}(\gamma)\mathrm{d}\gamma\right) \\ &= \delta_{n0}\delta_{m0} - \frac{d^{\mathrm{i}}_{nm}}{d^{\mathrm{i}}}, \end{aligned} \tag{3.98}$$

where we have also taken advantage of property (A.32) and of the first of Eq. (3.94).

The evaluation of the harmonic coefficients of the total surface load is now straightforward. Starting from the definition (3.71) and using the results above for d^{i}_{nm} and χ^{w}_{nm}, for degrees and orders $(n, m) \neq (0, 0)$ after some algebra we obtain

$$d^{\mathrm{iw}}_{nm} = d^{\mathrm{i}}_{nm}\left(\frac{1}{1 - \dfrac{d^{\mathrm{i}}_{00}}{d^{\mathrm{i}}}}\right), \tag{3.99}$$

which, using $\frac{d^{\mathrm{i}}_{00}}{d^{\mathrm{i}}} = \frac{1}{4\pi}\mu_{00}\,\sigma_{00}\,a_0 = \frac{1}{4\pi}(\lambda_2 - \lambda_1)(\cos\theta_1 - \cos\theta_2)$ can be cast in the equivalent form

$$d^{\mathrm{iw}}_{nm} = \frac{d^{\mathrm{i}}}{4\pi}\,\mu_{nm}\,\sigma_{nm}\,a_m\left(\frac{1}{1 - A^{\mathrm{r}}/A^{\mathrm{e}}}\right), \tag{3.100}$$

Fig. 3.8 **a** A bed of rectangular ice loads that describes the ice elements according to model Ice-1 of Peltier and Andrews (1976). Frame **b** shows a uniform coverage of the world's oceans by means of rectangular loads of size 5° × 5° and uniform thickness that *compensate* the ice load in (**a**). Note the poor approximation of the contour of the shorelines attained with such a coarse grid size

where

$$A^{\mathrm{r}} = a^2 \int_{\lambda_1}^{\lambda_2} \mathrm{d}\lambda \int_{\theta_1}^{\theta_2} \sin\theta \mathrm{d}\theta = a^2(\lambda_2 - \lambda_1)(\cos\theta_1 - \cos\theta_2) \tag{3.101}$$

is the area of the rectangle and $A^{\mathrm{e}} = 4\pi a^2$ is the area of the Earth's surface. From result (3.100), it is clear that unless the area of the rectangle is a significant fraction of A^{e}, we have $d_{nm}^{\mathrm{iw}} \approx d_{nm}^{\mathrm{i}}$, meaning that the harmonic coefficients of the rectangle can be assimilated, in practice, to those obtained by violating the mass conservation principle.

The rectangular loads have been often employed in the GIA literature. An example is shown in Fig. 3.8a, which shows the structure of the global ice model called Ice-1, characterised by a juxtaposition of large rectangular loads, with a size of 5° × 5°, all distributed over the northern hemisphere (Peltier and Andrews 1976). A possible *compensating* water load is considered in Fig. 3.8b.

3.4 A Post Glacial Rebound Simulation

Post Glacial Rebound (or PGR) represents the regional response of the solid Earth to the disintegration of the continental ice sheets at the end of last glaciation, which culminated about 20, 000 years ago. During the last great ice age, large portions of the Earth's surface (e.g., Scandinavia and North America) were covered by extensive ice sheets. Their weight was so large to cause a significant central subsidence (sinking of the ground) and a lateral uplift (peripheral bulge). These topographic features *rebounded* to the equilibrium configuration when the ice sheets started melting.

The process is illustrated in the cartoons of Fig. 3.9. The rebound has not been instantaneous, and it still continues in the formerly glaciated regions, which is a clear manifestation of the anelasticity of the Earth. Since the *post-glacial uplift* and its rate

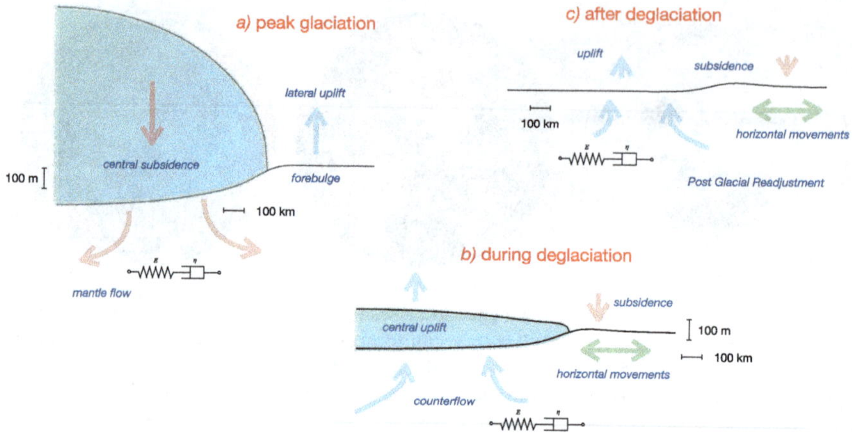

Fig. 3.9 Cartoon illustrating the physics of PGR, progressing from **a** peak glaciation to **b** deglaciation and **c** after deglaciation. Note the central subsidence and the slight lateral uplift (*forebulge* or *lateral bulge*) in response to the ice load. The flow and the counter-flow induced in the mantle are sensitive to rheology, sketched by the Maxwellian spring-dashpot model. The uplift and subsidence caused by the surface load are accompanied by horizontal displacements and geoid height variations, discussed in the text. The figure is obtained from an actual PGR simulation using idealised ice and Earth models, by means of program `TABOO`. The horizontal and vertical scales are also shown. The figure is inspired to an artwork of Thomas Sinclair James, available from the page: https://www.antarcticglaciers.org/glaciers-and-climate/sea-level-rise-2/recovering-from-an-ice-age/ [last accessed on August 11, 2024]

of variation are sensitive to the mantle viscosity, past and present uplift data can be employed to constrain the rheological properties of the Earth's interior. As pointed out by Spada (2017) and Gregory et al. (2019), PGR should not be confused with GIA (Glacial Isostatic Adjustment). Indeed, the latter is a global phenomenon that is not restricted to deglaciated areas and that is principally tied to sea-level variations (Whitehouse 2018). While for studying GIA it is necessary to solve the Sea Level Equation, for simulating regional PGR it is possible to adopt the simplified approach to the surface loading problem outlined in Sect. 3.3, neglecting the self-gravitation of the oceans, variations in the rotational state of the Earth, and the horizontal migration of the shorelines. As a matter of fact, however, in some studies PGR and GIA have been regarded as synonyms (see e.g., Sabadini et al. 2016).

Due to the broad subject, it is very difficult to provide a comprehensive review of PGR. We leave the readers to the bibliography of the work of Whitehouse (2018) for a list of important contributions to the subject (see also Spada 2017). Here we just mention a few fundamental works and reviews on this topic, essential for the comprehension of the material presented in this section. The physics of PGR has been first disclosed by Haskell (1935, 1936) who studied the *motion of a viscous fluid under a surface load*, providing a first estimate of the Earth's Newtonian viscosity[5] ($\eta =$

[5] Haskell's result have been subsequently revisited by Mitrovica (1996).

10^{21} Pa·s). A fundamental overview of PGR (or *glacio-isostatic rebound*) has been given by Walcott (1973), where the geological evidence is presented and some basic PGR modelling is proposed, with a discussion of the implications on the knowledge of the deep Earth properties. The book of Cathles (1975) deals with all aspects of PGR, ranging from the observations to the theories and provides a very broad presentation of the subject. However, the Sea Level Equation is not introduced and the attention is mainly devoted to near-field data and regional modelling efforts, aimed at estimating the viscosity of the Earth's mantle. A nice and succinct introduction to PGR is given by Turcotte and Schubert (2014), who has adopted Earth models in flat geometry in order to illustrate the basic physics of this process. The PhD dissertation of James (1991) constitutes a valuable introduction to PGR, since in addition to a qualitative description, it presents an overview of the development of the physical models of PGR, both in flat and in spherical geometry. An important section devoted to PGR can also found in the book of Ranalli (1995), who has emphasised the rheological aspects of the problem. Finally, it is worth to mention the work of Steffen and Wu (2011) for the in-depth review of the data and modelling for the region of Fennoscandia, the region where the PGR phenomenon has been first discovered and documented (see Ekman 2009, for an historical perspective), also providing a comprehensive review of the existing literature.

3.4.1 Setting up the PGR Model

Here we illustrate the basic settings of the model employed in this guided PGR exercise. We first describe the Earth model, the ice load geometry and its time-history. Then, we simulate PGR and we discuss the patterns of surface displacements obtained. Finally, we briefly focus on the present-day rates of displacement at the Earth's surface, in view of their importance from a geodetic standpoint. This exercise has mainly pedagogical purposes, aimed at providing an approximate estimate of the signals associated with PGR. The numerical computations have been carried out by means of the computer program `TABOO` (Spada 2003), which can be employed to design more realistic case studies.[6]

3.4.1.1 Earth Model

In the following PGR exercise, we shall employ the rheological five layer model B11,[7] the same that has been used as a case study for the evaluation of the viscoelastic normal modes in Sect. 2.4.3. The readers are therefore referred to Table 2.1 for the

---

[6] Program `TABOO` is now available from https://github.com/danielemelini/TABOO. It was originally distributed by the Samizdat Press, founded in the 1990s by John Scales and Martin Smith, and it can be also found at https://samizdat.mines.edu/ [last accessed on August 18, 2024].

[7] This same model was referred to as "M3-L70-V01" in the benchmark paper of Spada et al. (2011a).

values of the elastic moduli and for the viscosity profile of the model, and to Fig. 1 of Spada et al. (2011a) for a sketch of its radial structure. Loading Love numbers have been presented in Figs. 2.7 and 2.8, respectively. Here it is only useful to recall that model B11 is characterised by a inviscid homogeneous core, a three-layer mantle with Maxwell incompressible rheology, and a 70 km thick uniform elastic lithosphere. In the following PGR simulation, we are employing model B11 because it represents a good compromise between simplicity and realism. Furthermore, in the community there is a consensus about its spectral properties and these have been successfully reproduced using different numerical tools (see Spada et al. 2011a), something that has not been yet achieved neither for more complex models nor for others characterised by a compressible rheology.

3.4.1.2 Ice Load Geometry

We adopt an axis-symmetric ice load characterised by a quasi-elliptical cross-section (a *cap load*), whose general geometrical features have been illustrated in the second part of Sect. 3.3.8. The profile of the cap load is shown in Fig. 3.10. Mass conservation is ensured introducing a "complementary ocean" of uniform thickness that compensates the ice load, according to Eqs. (3.88)–(3.90). In all simulations, the half-amplitude of the cap load is fixed to $\alpha = 10°$ and its maximum thickness is $d^{i,\max} = 2,000$ m. The ice load is centred at the North Pole ($\theta = 0$) and its size is very roughly comparable to that of the major northern hemisphere ice sheets at the Last Glacial Maximum. We note that the assumption of a fixed margin is rather unrealistic, and possibly prevents a correct modelisation of the lateral bulge (see Fig. 3.9) and of its horizontal migration (Cathles 1975).

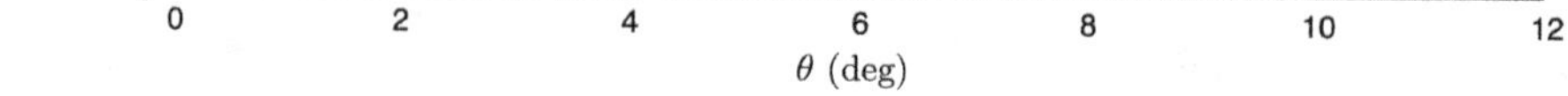

Fig. 3.10 Profile of the *quasi elliptical cap ice model* adopted in the PGR simulation performed in this section. The edge of the cap is fixed at $\theta = \alpha$, and the maximum thickness is attained along the axis of symmetry. At a given co-latitude θ, the thickness is given by Eq. (3.87). Note that at the edge the inclination of the cap is vertical, whereas it is horizontal at the summit ($\theta = 0$)

3.4.1.3 Ice Load Time History

The time variation of the cap load thickness mimics a simple deglaciation, schematically drawn in Fig. 3.11. It includes three phases:

1. an infinitely long-lasting *isostatic equilibrium phase*, during which the load keeps a constant thickness,
2. a *deglaciation phase*, during which the ice load thickness decreases at a constant rate while keeping the margins fixed,
3. a *post-glacial rebound phase*, during which the Earth's surface is free from loads, and subject to readjustment.

This sequence grossly mimics the time-history of the large continental ice sheets since the Last Glacial Maximum (see e.g., Walcott 1973; Lambeck et al. 2014). Of course, more detailed time histories can be introduced, possibly including previous loading and unloading phases or complex deglaciation histories. An account is given in the TABOO theory booklet (Spada 2003). However, the one adopted here captures the main events during the evolution of the ice sheets and it shows a sufficient realism, introducing a minimum number of parameters.

Mathematically, the time history considered here and shown in Fig. 3.11 can be expressed by

$$f^{\mathrm{i}}(t) = \begin{cases} 1, & \text{if } -\infty \leq t \leq 0 \\ 1 - \dfrac{t}{\tau}, & \text{if } 0 < t \leq \tau \\ 0, & \text{if } t > \tau, \end{cases} \tag{3.102}$$

where τ is the length of the deglaciation phase, and we have conventionally set the origin of times ($t = 0$) at the onset of deglaciation. In the following, we shall refer to Eq. (3.102) as to *simple deglaciation* time history. We note that it can be written in a more compact and convenient form as

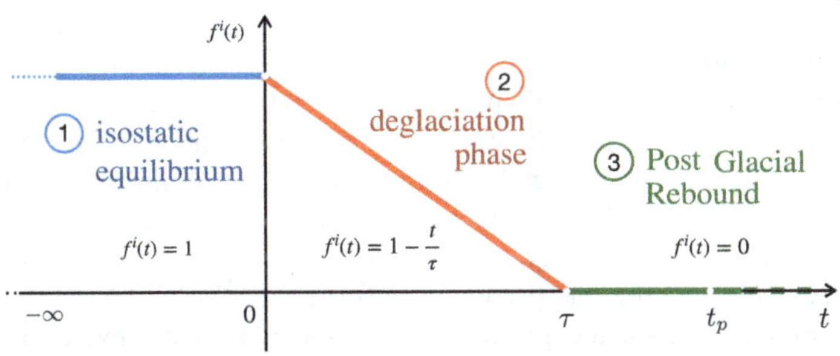

Fig. 3.11 Schematic time-history $f^{\mathrm{i}}(t)$ of the ice source considered in the PGR simulation (*simple deglaciation*), showing three distinct phases. The simple deglaciation history is characterised by a single parameter, i.e., the duration of the deglaciation phase τ. With t_p we denote present time

$$f^{\rm i}(t) = [1 - H(t)] + \left(1 - \frac{t}{\tau}\right)[H(t) - H(t-\tau)], \tag{3.103}$$

where we have used the properties of the Heaviside step function (see Eq. E.1). Henceforth, the duration of the deglaciation phase is assumed to be $\tau = 10$ kyr; present time corresponds to $t_p = 15$ kyrs, i.e., the deglaciation phase is assumed to have ended 5 kyr before present.

3.4.1.4 Time Convolutions

According to Eq. (3.84), which gives the response to a general axis-symmetric surface load in the framework of the *simplified surface loading problem*, in order to solve the PGR problem we need to evaluate the time convolutions $c_n^h(t)$, $c_n^k(t)$, and $c_n^\ell(t)$ for the specific time history chosen. The computation is not difficult, but it requires some attention. Using the expression (2.52) for the Love number $h_n^{\rm L}(t)$ in the first of Eq. (3.75), and performing the required integrations term-by-term taking into account the properties of the step function $H(t)$, after some algebra we obtain:

$$c_n^h(t) = \frac{1}{2n+1}\left[h_n^{\rm L,f}\, f^{\rm i}(t) + \left(H(t)\, G_n^h(t) - H(t-\tau)\, G_n^h(t-\tau)\right)\right], \tag{3.104}$$

where the ice time history $f_i(t)$ is defined by (3.103), $h_n^{\rm L,f}$ is the fluid LLN, we have defined the auxiliary multi-exponential function

$$G_n^h(t) = \sum_{i=1}^{M} \frac{h_{ni}^{\rm L}}{s_{ni}}\, \frac{1 - {\rm e}^{s_{ni} t}}{s_{ni}\tau}, \tag{3.105}$$

where $h_{ni}^{\rm L}$ are the viscous components of the LLNs, s_{ni} are the inverse relaxation times, and $M \equiv 4N_{\rm V} = 12$ is the total number of viscoelastic normal modes for the present case study (Spada et al. 2011a). It is useful to remark that $c_n^h(t)$ is only a function of the rheological profile adopted, and does not depend upon the geometrical features of the ice model.

For the convolution associated with the geoid height $c_n^k(t)$, an expression similar to (3.104) holds, with $h_n^{\rm L,f}$ substituted by $1 + k_n^{\rm L,f}$ (the term "1" stems from the δ-term in Eq. 3.84) and with residues $h_{ni}^{\rm L}$ substituted by $k_{ni}^{\rm L}$. Similarly, the convolution associated with horizontal displacements $c_n^\ell(t)$ is obtained by substituting $h_n^{\rm L,f}$ with $\ell_n^{\rm L,f}$ and $h_{ni}^{\rm L}$ with $\ell_{ni}^{\rm L}$, respectively. Although stated in a different form, expression (3.104) is equivalent to the one presented in Spada (2003). In that same work the readers can find the analytical expressions for a number of time convolutions relative to other schematic time histories, useful for pedagogical purposes, implemented in the Post Glacial Rebound simulator TABOO (see Spada et al. 2011a). Equation (3.104) can be easily interpreted from a physical standpoint. The first term in parentheses, proportional to $h_n^{\rm L,f}$, describes the Earth's response *as if* the mantle would behave as a perfectly inviscid (i.e., zero-viscosity) fluid. The response of the Earth is, in this case, exactly in phase with the load time history $f^{\mathrm{i}}(t)$. The second term manifests corrections that account for the finite viscosity of the Earth. The first of these acts immediately after the inception of the deglaciation phase ($t \geq 0$) and accounts for the gradual readjustment of the Earth's surface to the progressive unloading. The second only acts after the end of deglaciation ($t \geq \tau$), when according to (3.103) the external surface of the Earth is free. For $t \geq \tau$, the combination of these two (multi)-exponentially decaying terms is responsible for PGR. For a perfectly elastic Earth, the viscous part of the LLN is $h^{\mathrm{L}}_{ni} \equiv 0$, so that according to (3.105) we would have $G^h_n(t) \equiv 0$. Thus, the PGR that occurs for $t \geq \tau$ is a direct manifestation of the viscoelastic properties of the Earth.

The multi-exponential *relaxation* of the free surface of the Earth occurring for $t \geq \tau$ has a simpler, single-exponential counterpart in the basic PGR models illustrated by Ranalli (1995) and Turcotte and Schubert (2014), in which the Earth is uniform and viscous, and its sphericity and self-gravitation are not taken into consideration. These simplified treatments of the PGR problem are ultimately based on the seminal works of Haskell (1935, 1936) about the *motion of a viscous fluid under a surface load*.

3.4.2 Surface Displacements and Geoid

3.4.2.1 Vertical Component

Figure 3.12 shows the surface vertical displacement $\mathcal{U}(\theta, t)$, evaluated by means of Eq. (3.84), using the model parameters listed in Table 3.1 and the load described in section above. The vertical displacement is evaluated in the geocentric reference frame, i.e., in the frame with origin in the Earth's centre of mass, and this also applies to all quantities considered in the following.

The displacement is shown along a meridian as a function of θ, at various times. In the isostatic equilibrium configuration ($t \leq 0$, blue curve), the displacement along the axis of symmetry of the cap load is ~ -540 m, i.e. $\sim 1/4$ of the ice load thickness. This value approximately matches the ratio between the ice density and the upper mantle density (see Table 2.1), as we would expect according to the Airy isostatic compensation model (see e.g., Turcotte and Schubert 2014). However, as we shall discuss in the following, a perfect isostatic state is not reached and the analogy with Airy compensation mechanism is not fully appropriate, since in our modelling the lithosphere has a finite strength. At equilibrium, an uplifted region outside the ice margins (*peripheral bulge*) is clearly visible for $\theta \approx 13°$, with an amplitude of ~ 10 m. Since our ice model has fixed margins, however, it is possible that the shape of the peripheral bulge and its lateral movements are not reproduced realistically.

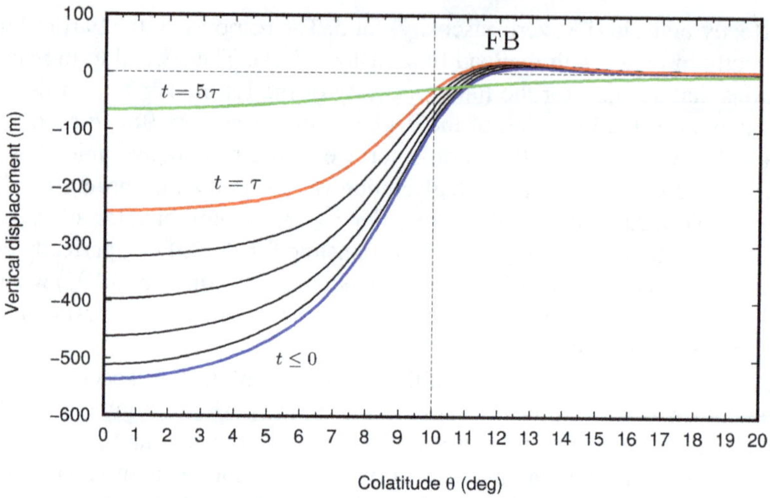

Fig. 3.12 Surface vertical displacement $\mathcal{U}(\theta, t)$ obtained in the PGR simulation, as a function of co-latitude θ, for various times t. The horizontal dashed line shows the unperturbed Earth's surface and the vertical dashed segment marks the (fixed) load margin. During the deglaciation phase ($0 \leq t \leq \tau$), curves are plotted with a time increment of 2 kyr. FB marks the *forebulge* (or *lateral bulge*)

Table 3.1 Numerical values of the essential parameters employed in the PGR simulation illustrated in the text. For Earth model B11, the rheological parameters are given in Table 2.1

PGR model parameter	Symbol	Value
Length of deglaciation	τ	10 kyr
Present time	t_p	15 kyr
Load half-amplitude	α	10°
Maximum load thickness	$d^{\mathrm{i,max}}$	2000 m
Minimum harmonic degree	n_{min}	1
Maximum harmonic degree	n_{max}	512
Ice density	ρ^{i}	931 kg m^{-3}
Earth's average density	ρ^{e}	5512 kg m^{-3}

The phenomenon of *post glacial rebound* is clearly illustrated by the sequence of curves in the time range $0 \leq t \leq \tau$; at the end of deglaciation ($t = \tau$, red curve), the upward movement of the Earth's surface has recovered $\approx 50\%$ of the initial topography. During deglaciation, the peripheral bulge further develops and it is slightly displaced inward. Remarkably, the state of isostatic disequilibrium is still significant at time $t = 5\tau$, i.e., long after the end of deglaciation, with a residual topography of ≈ -70 m in the central region.

3.4.2.2 Horizontal Component

The spatio-temporal pattern of horizontal displacements $\mathcal{V}(\theta, t)$ is shown in Fig. 3.13, where positive and negative values denote movements away from the ice load and towards the ice load, respectively. Contrary to vertical displacement $\mathcal{U}(\theta, t)$, considered in Fig. 3.12, $\mathcal{V}(\theta, t)$ vanishes along the polar axis $\theta = 0°$, as a consequence of the axial symmetry of the surface load and of the spherical symmetry of the model adopted here. It is also apparent that horizontal displacements are relatively small compared to vertical displacement, with an amplitude reduced by approximately one order of magnitude. At isostatic equilibrium ($t \leq 0$, blue curve), the maximum horizontal displacement (≈ 40 m) is found just outside the margin of the load, but the response curve is not monotonous and another maximum is reached within the margins of the glaciated region. During deglaciation, when PGR is manifest, the amplitude of $\mathcal{V}(\theta, t)$ progressively decays until it reaches negative values for time $t = \tau$ (red curve), denoting that the Earth surface has moved to the left with respect to the unperturbed state (dashed horizontal line). Similar to vertical displacement, we note that a significant amount of isostatic disequilibrium remains for $t = 5\tau$, well after the disappearance of the surface load.

It is well known that the pattern of $\mathcal{V}(\theta, t)$ is more difficult to interpret compared with that of vertical displacement $\mathcal{U}(\theta, t)$ (see e.g., James and Morgan 1990; Spada et al. 1992b; James and Lambert 1993; Giunchi et al. 1997; Sella et al. 2007; Lidberg et al. 2010). More recent work (see Hermans et al. 2018) has shown an enhanced sensitivity of horizontal movements to the rheological layering in the lithosphere and in the shallow upper mantle. It is possible that, adopting viscosity profiles that

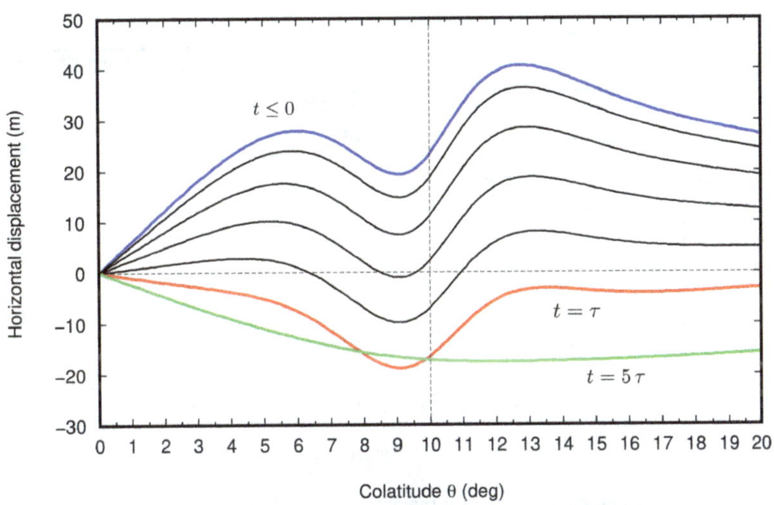

Fig. 3.13 Surface horizontal displacement $\mathcal{V}(\theta, t)$ obtained in the PGR simulation, as a function of the co-latitude of the observer θ, for various times t. For other details, see caption of Fig. 3.12

differ from that of model B11 (see Table 3.1), the patterns of horizontal displacement shown in Fig. 3.13 could well be modified or even reversed. The readers can make their own numerical experiments using program TABOO.

3.4.2.3 Geoid Displacement

In Fig. 3.14 we consider the displacement of the geoid $\mathcal{G}(\theta, t)$ as a function of co-latitude θ, for various times t. At equilibrium ($t \leq 0$), we note that the geoid is only displaced by $\sim$1 m, negligible compared with the subsidence caused by the load (see the blue curve in Fig. 3.12). This indicates that in isostatic conditions the gravitational effects due to the excess mass of the ice sheet are *almost* exactly cancelled by those induced by surface displacement, shown by the blue curve in Fig. 3.12. However, the cancellation is not complete because we have assumed finite elastic strength for the lithosphere, which partly supports the surface load even on very long time scales; in the case of a viscous lithosphere, the cancellation would be effectively complete (see e.g., Ricard et al. 1992). During deglaciation (black curves), $\mathcal{G}(\theta, t)$ becomes more and more negative, due to the negative contribution of the net mass anomaly associated with surface displacement, which progressively dominates the vanishing positive effect of the surface load. In analogy with the case of vertical deformation, isostatic disequilibrium is still manifest at $t = 5\tau$ as a negative displacement of the geoid with a maximum amplitude of $\approx$ 3 m under the centre of the former load.

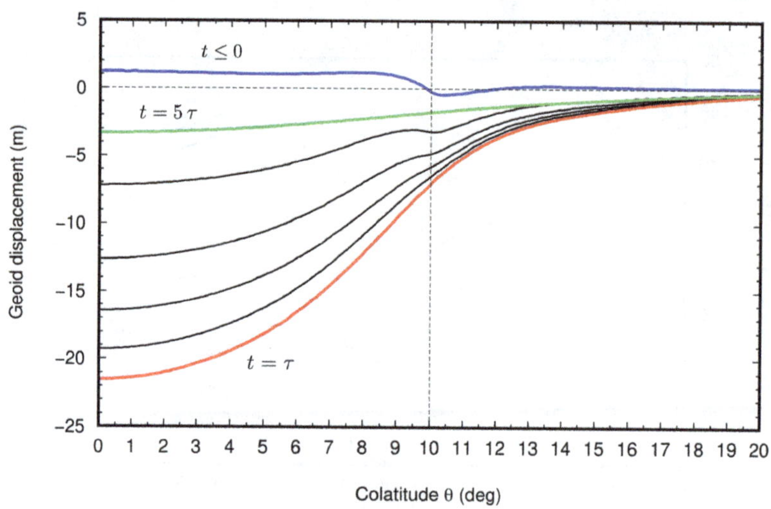

Fig. 3.14 Geoid displacement $\mathcal{G}(\theta, t)$, as a function of co-latitude θ, for various times t. See also caption of Fig. 3.12

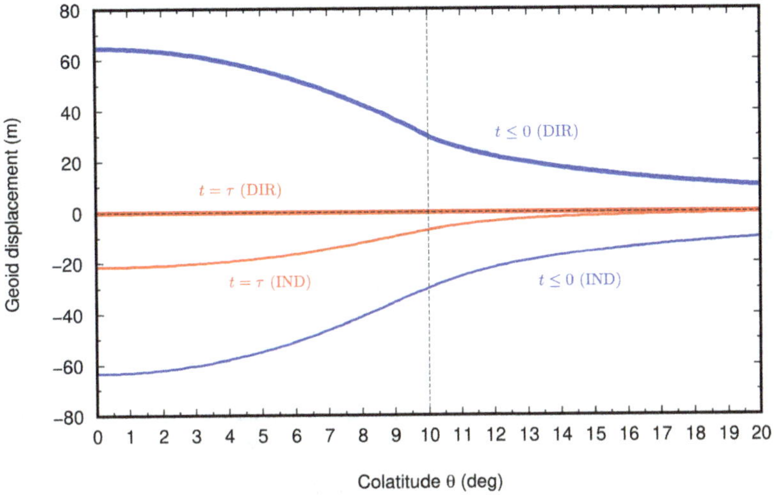

Fig. 3.15 Direct (thick curves) and indirect (thin) contributions to the total geoid variation $\mathcal{G}(\theta, t)$ at equilibrium (blue) and at the end of deglaciation (red). Note that here the vertical scale is enlarged with respect to Fig. 3.14. The vertical dashed segment marks the load margin

More insight into the results obtained for $\mathcal{G}(\theta, t)$ is gained from Fig. 3.15, where thick and thin curves show the direct (DIR) and the indirect (IND) contributions to $\mathcal{G}(\theta, t)$. From Chap. 2, we recall that the former is evaluated in the assumption of a rigid Earth and that it only accounts for the gravitational attraction of the surface load, while the latter is associated with the deformation-induced gravitational effects. Practically, $\mathcal{G}^{\mathrm{DIR}}(\theta, t)$ is evaluated only keeping the δ term in the expression of $c_n^k(t)$ (see Eq. 3.75), while $\mathcal{G}^{\mathrm{IND}}(\theta, t)$ only contains contributions from the elastic and viscous components of the Love number $k_n^{\mathrm{L}}(t)$. Of course, $\mathcal{G}^{\mathrm{DIR}} + \mathcal{G}^{\mathrm{IND}} \equiv \mathcal{G}$. On one hand, the results of Fig. 3.15 justify the small values found in Fig. 3.14 for $\mathcal{G}(\theta, t)$ in the equilibrium state $t \leq \tau$, because of condition $\mathcal{G}^{\mathrm{IND}} \approx -\mathcal{G}^{\mathrm{DIR}}$ (compare the thick with the thin blue curves). On the other hand, at the end of deglaciation ($t = \tau$), because of the disappearance of the ice load ($\mathcal{G}^{\mathrm{DIR}} = 0$, see red thick curve), the negative contribution due to the still depressed central topography dominates, giving $\mathcal{G} = \mathcal{G}^{\mathrm{IND}} < 0$ (red thin curve).

3.4.3 *Present–Day Isostatic Disequilibrium*

The state of present-day disequilibrium in the formerly glaciated region is studied in Fig. 3.16, where the rates of change $(\dot{\mathcal{U}}, \dot{\mathcal{V}}, \dot{\mathcal{G}})$ are evaluated at time $t = t_p = 15$ kyr after the inception of deglaciation, i.e., 5 kyr after the end of melting. The rates

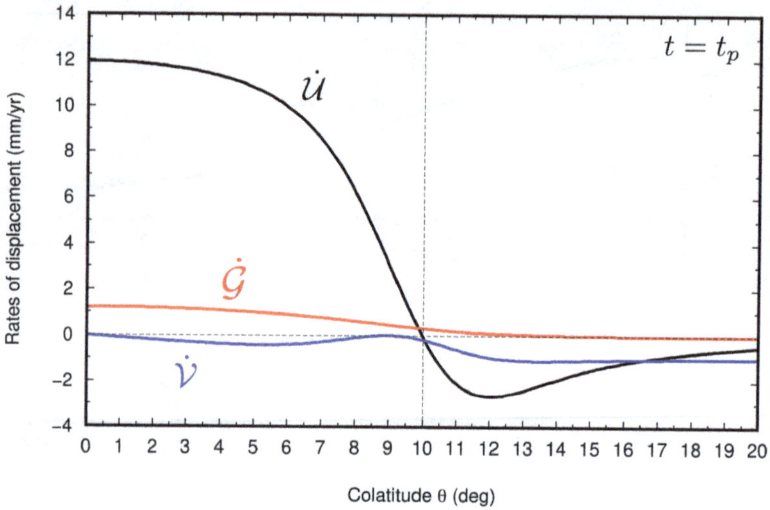

Fig. 3.16 Present-day ($t = t_p$) vertical velocity ($\dot{\mathcal{U}}$), horizontal velocity ($\dot{\mathcal{V}}$) and rate of change of the geoid height ($\dot{\mathcal{G}}$) as a function of colatitude θ

have been obtained taking the time derivative of Eq. (3.84), and noting that for time $t = t_p > \tau$ we have $\dot{f}^{\mathrm{i}}(t) = 0$, so that Eqs. (3.104) and (3.105) give

$$\dot{c}_n^h(t) = -\frac{1}{2n+1} \sum_{i=1}^{M} h_{ni}^{\mathrm{L}} \left(\frac{1 - \mathrm{e}^{-s_{ni}\tau}}{s_{ni}\tau} \right) \mathrm{e}^{s_{ni}t}, \tag{3.106}$$

with similar results for the convolutions $\dot{c}_n^\ell$ and $\dot{c}_n^k$, provided that h_{ni}^{L} is substituted by ℓ_{ni}^{L} and k_{ni}^{L}, respectively. Equation (3.106) shows immediately that, for an rigid Earth ($h_{ni}^{\mathrm{L}} = 0$) $\dot{c}_n^h$ vanishes. This simple consideration shows the importance of geodetic observations of PGR signals in deglaciated areas, which have the potential to provide constraints on the viscosity profile of the mantle and on the evolution of the past ice sheets (see e.g., King et al. 2010).

Figure 3.16 confirms that the dominating signal across de-glaciated areas is vertical uplift ($\dot{\mathcal{U}} > 0$), with values as large as $\sim$ 12 mm yr^{-1} close to the centre of the former ice load. At the margins, $\dot{\mathcal{U}}$ is close to zero. We note that PGR is causing a subsidence ($\dot{\mathcal{U}} < 0$) in the region of the peripheral bulges, which are collapsing at the significant rate of $\sim$ 2 mm yr^{-1}. The pattern of the rate of change of the geoid height ($\dot{\mathcal{G}}$, red curve) is broadly reproducing that of vertical velocity, but its amplitude is $\approx$ 10 times smaller. We observe that the collapse of the peripheral bulge is not affecting, for this PGR model, the values of $\dot{\mathcal{G}}$. The horizontal velocity field $\dot{\mathcal{V}}$ amounts to fractions of mm yr^{-1} and tend to be slightly negative across the formerly glaciated area, thus indicating a moderate inward motion (Hermans et al. 2018). Outside the glaciated area, $\dot{\mathcal{V}}$ is close to -1 mm yr^{-1} across a broad range of co-latitudes.

It useful to remark that the numerical values above and the details of the spatial patterns can vary significantly according to the assumptions made about the viscosity of the mantle and to the characteristics of the ice load. Readers wishing to pursue this matter further can study these sensitivities using the publicly available TABOO PGR simulator (Spada 2003) or similar tools developed within the GIA community (see Table 1 in Spada et al. 2011a).

References

Bevis M, Melini D, Spada G (2016) On computing the geoelastic response to a disk load. Geophys J Int 205(1):1804–1812
Brotchie J (1971) Flexure of a liquid-filled spherical shell in a radial gravity field. Mod Geol 3:15–23
Cathles L (1975) The viscosity of the Earth's Mantle. Princeton University Press
Donnay JD (2011) Spherical trigonometry. Read Books Ltd
Ekman M (2009) The changing level of the Baltic Sea during 300 years: a clue to understanding the Earth. Summer Institute for Historical Geophysics Åland Islands
Faraoni V, Vokey MW (2015) The thickness of glaciers. Eur J Phys 36(5):055031
Farrell W (1972) Deformation of the Earth by surface loads. Rev Geophys 10(3):761–797
Farrell W, Clark J (1976) On postglacial sea level. Geophys J Roy Astron Soc 46(3):647–667
Giunchi C, Spada G, Sabadini R (1997) Lateral viscosity variations and post-glacial rebound: effects on present-day VLBI baseline deformations. Geophys Res Lett 24(1):13–16
Gregory JM, Griffies SM, Hughes CW, Lowe JA, Church JA, Fukimori I, Gomez N, Kopp RE, Landerer F, Le Cozannet G et al (2019) Concepts and terminology for sea level: mean, variability and change, both local and global. Surv Geophys 1–39
Haskell N (1935) The motion of a viscous fluid under a surface load. Physics 6(8):265–269
Haskell N (1936) The motion of a viscous fluid under a surface load. Part II. J Appl Phys 7(2):56–61
Hermans T, van der Wal W, Broerse T (2018) Reversal of the direction of horizontal velocities induced by GIA as a function of mantle viscosity. Geophys Res Lett 45(18):9597–9604
Ivins ER, Dokka RK, Blom RG (2007) Post-glacial sediment load and subsidence in coastal Louisiana. Geophys Res Lett 34(16)
James TS (1991) Postglacial deformation. PhD thesis, Princeton University
James TS, Lambert A (1993) A comparison of VLBI data with the ICE-3G glacial rebound model. Geophys Res Lett 20(9):871–874
James TS, Morgan WJ (1990) Horizontal motions due to post-glacial rebound. Geophys Res Lett 17:957–960
Karpytchev M, Ballu V, Krien Y, Becker M, Goodbred S, Spada G, Calmant S, Shum C, Khan Z (2018) Contributions of a strengthened early Holocene monsoon and sediment loading to present-day subsidence of the Ganges-Brahmaputra delta. Geophys Res Lett 45(3):1433–1442
King MA, Altamimi Z, Boehm J, Bos M, Dach R, Elosegui P, Fund F, Hernández-Pajares M, Lavallee D, Cerveira PJM, Penna N, Riva RE, Steigenberger P, van Dam T, Vittuari L, Williams S, Willis P (2010) Improved constraints on models of glacial isostatic adjustment: a review of the contribution of ground-based geodetic observations. Surv Geophys 31(5):465–507
Lambeck K, Rouby H, Purcell A, Sun Y, Sambridge M (2014) Sea level and global ice volumes from the last glacial maximum to the Holocene. Proc Natl Acad Sci 111(43):15,296–15,303
Lidberg M, Johansson JM, Scherneck H-G, Milne GA (2010) Recent results based on continuous GPS observations of the GIA process in Fennoscandia from BIFROST. J Geodyn 50(1):8–18
Longman I (1962) A Green's function for determining the deformation of the Earth under surface mass loads: 1. Theory. J Geophys Res 67(2):845–850

Longman I (1963) A Green's function for determining the deformation of the Earth under surface mass loads. 2. Computations and numerical results. J Geophys Res 68:485–496
Melini D, Gegout P, Spada G, King M (2015) REAR—a regional ElAstic rebound calculator—user manual for version 1.0. http://danielemelini.github.io/rear/
Melosh H (1978) The tectonics of mascon loading. In: Lunar and planetary science conference proceedings, vol 9, pp 3513–3525
Memin A, Spada G, Boy J-P, Rogister Y, Hinderer J (2014) Decadal geodetic variations in Ny-Ålesund (Svalbard): role of past and present ice-mass changes. Geophys J Int 198(1):285–297
Mitrovica JX (1996) Haskell [1935] revisited. J Geophys Res: Solid Earth 101(B1):555–569
Nye J (1959) The motion of ice sheets and glaciers. J Glaciol 3(26):493–507
Oerlemans J (2008) Minimal glacier models. Utrecht Publishing & Archiving Service
Paul M (1978) Recurrence relations for integrals of associated Legendre functions. Bulletin Geodesique 52(3):177–190
Peltier WR (1974) The impulse response of a Maxwell Earth. Review of Geophysics and Space Physics 12(4):649–669
Peltier W, Andrews J (1976) Glacial-isostatic adjustment I. The forward problem. Geophys J Int 46(3):605–646
Press WH (2007) Numerical recipes, 3rd edn. The art of scientific computing. Cambridge University Press
Pullan S, Lambeck K (1982) Mascons and loading of the lunar lithosphere. In: Lunar and planetary science conference proceedings, vol 12, pp 853–865
Ranalli G (1995) Rheology of the Earth. Chapman and Hall
Ricard Y, Sabadini R, Spada G (1992) Isostatic deformations and polar wander induced by redistribution of mass within the Earth. J Geophys Res: Solid Earth 97(B10):14223–14236
Sabadini R, Vermeersen B, Cambiotti G (2016) Global dynamics of the Earth. Springer, Berlin
Sella GF, Stein S, Dixon TH, Craymer M, James TS, Mazzotti S, Dokka RK (2007) Observation of glacial isostatic adjustment in "stable" North America with GPS. Geophys Res Lett 34(2)
Spada G (2003) The theory behind TABOO. Samizdat Press. http://samizdat.mines.edu/taboo/teoria.pdf
Spada G (2017) Glacial isostatic adjustment and contemporary sea level rise: an overview. Surv Geophys 38(1):1–33
Spada G, Sabadini R, Yuen DA, Ricard Y (1992) Effects on post-glacial rebound from the hard rheology in the transition zone. Geophys J Int 109(3):683–700
Spada G, Barletta VR, Klemann V, Riva R, Martinec Z, Gasperini P, Lund B, Wolf D, Vermeersen L, King M (2011) A benchmark study for glacial isostatic adjustment codes. Geophys J Int 185(1):106–132
Spada G, Ruggieri G, Sørensen LS, Nielsen K, Melini D, Colleoni F (2012) Greenland uplift and regional sea level changes from ICESat observations and GIA modelling. Geophys J Int 189(3):1457–1474
Steffen H, Wu P (2011) Glacial isostatic adjustment in Fennoscandia—a review of data and modeling. J Geodyn 52(3):169–204
Sun W, Sjöberg L (1998) Gravitational potential changes in a spherically symmetric Earth model caused by a surface load. Phys Chem Earth 23(1):47–52
Turcotte DL, Schubert G (2014) Geodynamics—applications of continuum physics to geological problems. Cambridge University Press
Tushingham A, Peltier W (1991) ICE-3G—a new global model of late Pleistocene deglaciation based upon geophysical predictions of post-glacial relative sea level change. J Geophys Res 96(B3):4497–4523
Walcott RI (1973) Structure of the Earth from glacio-isostatic rebound. Annu Rev Earth Planet Sci 1:15

Whitehouse PL (2018) Glacial isostatic adjustment modelling: historical perspectives, recent advances, and future directions. Earth Surf Dyn 6(2):401–429
Yuen DA, Sabadini RCA, Gasperini P, Boschi E (1986) On transient rheology and glacial isostasy. J Geophys Res 91:11,420–11,438

Chapter 4
Rotational Response of the Earth

Round about the cauldron go;
In the poisoned entrails throw.

(Macbeth: Act IV)

4.1 Introduction

In GIA modelling, rotational effects constitute a significant contribution to long-wavelength and regional patterns of sea-level change, together with gravitational and deformational effects. In its most advanced version, the Sea Level Equation accounts for surface loading due to the waning and waxing continental ice sheets but also for fluctuations of Earth's rotation axis. The former have been presented in previous chapter, the latter are illustrated and discussed herein. In its pristine form (see Farrell and Clark 1976), the Sea level Equation did not account for rotational effects at all, being limited to the isostatic response arising from surface loading. Earth rotation has been introduced in Post Glacial Rebound studies by Han and Wahr (1989) and shortly after by Bills and James (1996), who adopted simplified models in which only the major Laurentide ice sheet was considered, and eustatic (i.e., spatially uniform) sea-level variations were assumed. Later on, with the contributions of Milne and Mitrovica (1996, 1998) and Milne (1998), rotational effects have been included into the framework of the SLE, opening the route to a number of subsequent developments, which have been reviewed by Spada (2017) and Whitehouse (2018).

In the following, due to the very broad subject, we shall not describe the Earth rotation theory in detail. Indeed, we shall limit our attention to basic concepts and topics that are strictly necessary in the context of GIA and in the development of a general, *rotationally self-consistent* Sea Level Equation. The concept of *polar*

G. Spada and D. Melini, *Glacial Isostatic Adjustment*, Springer Geophysics,
https://doi.org/10.1007/978-3-031-90538-4_4

motion (or wobble), which represents the motion of the rotation axis with respect to the Earth's crust (Lambeck 1980), is central. The readers are referred to fundamental contributions (see e.g., Routh 1905; Munk and MacDonald 1975; Lambeck 1980) for a full account of the problem of Earth rotation, including observational aspects and modelling. The readers can also take advantage from the comprehensive works of Terpos (1973), Rochester (1973), Barnes et al. (1983) and Zharkov et al. (1996) for insightful discussions. The treatise of Sabadini et al. (2016) deals with GIA and Earth rotation in detail, also providing a thorough historical perspective and discussing previous debates about the interpretation of some terms of the Liouville equations for GIA.

The chapter is organised into three sections. In Sect. 4.2 we present some basic concepts, and starting from rigid rotation we arrive to a suitable form of the linearised Liouville equations describing polar motion and variations in the length of day (LOD) for a deformable planet. Then, we deal with the polar motion and LOD excitation functions (EFs), which represent the forcing terms of the linearised Liouville equations. The EFs for rotational deformation and those for surface loading are derived in detail separately. A form of the Liouville equations suitable for GIA modelling is obtained in Sect. 4.3, where two distinct rotational theories that have been proposed in the literature are also discussed, in view of their implementation in the SLE. The concept of Polar Motion Transfer Function (PMTF) is also introduced, and a general solution of Liouville equations is presented. Finally, in Sect. 4.4, an account is given of the rotational response functions (RRFs), which represent an essential ingredient of the Sea Level Equation and constitute the counterpart of the surface response functions (SRFs) introduced in previous chapter. The workflow of the rotational theory discussed in this chapter is summarised in Fig. 4.1.

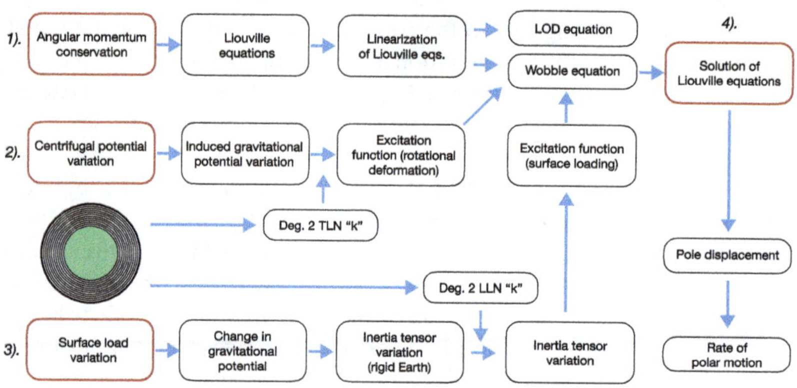

Fig. 4.1 Scheme showing the workflow of this chapter, with four major steps enlightened

4.2 Earth Rotation and Excitation Functions

The two following subsections summarise basic concepts regarding rigid and non-rigid body rotation. In our presentation, we have essentially followed the treaties of Munk and MacDonald (1975) and Lambeck (1980).

4.2.1 *Rigid Rotation*

Before introducing the equations governing the rotation of a deformable body, it is necessary to recall the basic facts regarding the rotation of a perfectly rigid body. Here we largely follow Lambeck (1980), but we only report the essential concepts useful for the application to the GIA problem. For a more general formulation of the theory of rigid bodies rotation, the readers are referred to the classical essay by Goldstein (1950).

4.2.1.1 Euler's Equations

According to the second cardinal law of mechanics, the rate of change of the angular momentum $\vec{L}$ of a body equals the externally applied torque $\vec{N}$. Using the index notation, it reads:

$$\frac{\mathrm{d}}{\mathrm{d}t} L_i = N_i \,, \tag{4.1}$$

where both vectors are expressed in an inertial reference frame. However, in view of the applications in geodynamics, it is more convenient to write these equations in a rotating (non inertial) reference frame, in which the axes (x_1, x_2, x_3) are rigidly connected to the rotating body in some prescribed way. Written in this frame, Eq. (4.1) are known as *Euler's equations*, and take the form:

$$\frac{\mathrm{d}}{\mathrm{d}t} L_i + \varepsilon_{ijk}\, \omega_j\, L_k = N_i \,, \tag{4.2}$$

where

$$\omega_i = (\omega_1, \omega_2, \omega_3) \tag{4.3}$$

is the angular velocity vector and ε_{ijk} is the Ricci Levi-Civita alternating symbol.[1]

[1] By definition, $\varepsilon_{ijk} = 0$ when at least two indices coincide, $\varepsilon_{ijk} = +1$ when *ijk* is an even permutation of 123 (e.g., 231), and $\varepsilon_{ijk} = -1$ when the permutation is odd (e.g., 132). In this way, the vector product between vectors ω_i and L_i is $\varepsilon_{ijk}\omega_j L_k$.

Following e.g., Goldstein (1950), the angular momentum components are linear combinations of the angular velocity vector components as

$$L_i = I_{ij}\,\omega_j\,, \tag{4.4}$$

where I_{ij} is the rank two, symmetric inertia tensor

$$I_{ij} = \int_V \left(x_l' x_l'\,\delta_{ij} - x_i' x_j'\right) \mathrm{d}m\,, \tag{4.5}$$

and where the integration is over the volume V of the body, $\mathrm{d}m = \rho\, dV$ is the mass element, $\rho = \rho(\vec{x})$ is density, and due to the assumption of rigidity the elements I_{ij} are constants that only depend upon the internal mass distribution (see Sect. 1.2.1.3). By substituting (4.4) into (4.2), Euler's equations are transformed into

$$\frac{\mathrm{d}}{\mathrm{d}t} I_{ij}\,\omega_j + \varepsilon_{ijk}\,\omega_j\, I_{kl}\,\omega_l = N_i\,, \tag{4.6}$$

which constitute a set of three coupled, non-linear equations with unknowns ω_i.

4.2.1.2 Solution of Euler's Equations

Due to the symmetry of the inertia tensor, it is always possible to choose the axes x_i in such a way that the products of inertia I_{ij} with $i \neq j$ vanish. In such a frame, the tensor is diagonal (see Sect. 1.2.1.3). A solution of Euler's equations can be found assuming that the body is axially symmetric about the x_3 axis, with principal moments of inertia $I_{11} = I_{22} = A$ and $I_{33} = C > A$, so that:

$$I_{ij} = \begin{pmatrix} A & 0 & 0 \\ 0 & A & 0 \\ 0 & 0 & C \end{pmatrix}\,. \tag{4.7}$$

Also assuming that the body is free from external torques ($N_i = 0$), Euler's equations (4.6) reduce to

$$\dot{\omega}_1 + \left(\frac{C-A}{A}\right) \omega_2\,\omega_3 = 0 \tag{4.8}$$

$$\dot{\omega}_2 - \left(\frac{C-A}{A}\right) \omega_1\,\omega_3 = 0 \tag{4.9}$$

$$\dot{\omega}_3 = 0\,, \tag{4.10}$$

where the dot indicates the time derivative. These equations can be integrated assuming initial conditions $\omega_i(0) = \omega_{i0}$, obtaining:

$$\omega_1(t) = \omega_{10} \cos \sigma_r t - \omega_{20} \sin \sigma_r t \tag{4.11}$$

$$\omega_2(t) = \omega_{10} \sin \sigma_r t + \omega_{20} \cos \sigma_r t \tag{4.12}$$

$$\omega_3(t) = \omega_{30} = \Omega\,, \tag{4.13}$$

where Ω is a constant, and the Euler frequency is:

$$\sigma_r = \left(\frac{C - A}{A}\right) \Omega\,, \tag{4.14}$$

where r stands for *rigid*.

4.2.1.3 Free Nutation

According to (4.11–4.13) and (4.14), the angular velocity vector ω_i maintains a constant length and rotates stably around the principal axis of inertia (the *figure axis*) x_3, describing a circle with a radius determined by the initial conditions. The angular frequency of motion σ_r depends upon the principal moments of inertia of the body C and A, hence it is determined by the internal mass distribution. Such motion is known as *free Eulerian precession* or *free nutation* (Lambeck 1980).

Since for the Earth[2]

$$\frac{C - A}{A} \approx \frac{1}{300}\,, \tag{4.15}$$

and the mean angular velocity is

$$\Omega = 7.29 \times 10^{-5}\,\mathrm{s}^{-1}\,, \tag{4.16}$$

from (4.14) we have

$$\sigma_r \approx \frac{1}{306}\,\mathrm{d}^{-1}\,, \tag{4.17}$$

which corresponds to the eulerian period

$$T_r = \frac{2\pi}{\sigma_r} \simeq 10 \text{ months.} \tag{4.18}$$

It is important to note that the eulerian period T_r is significantly shorter than the observed *Chandlerian wobble* of $\approx$ 14 months, named after the astronomer Seth Carlo Chandler who discovered it in 1891 (see e.g., Lambeck 1980; Okubo 1982; Barnes et al. 1983; Carter and Carter 2000; Gross 2000). The discrepancy was attributed to possible influences of air mass movements on the axis of rotation, or to

---

[2] Observed values of constants defining the form and the gravity of the Earth are given in Table 2.5 of Lambeck (1980).

the elastic yielding of the Earth to the varying centrifugal forces. For a geophysical discussion, the readers are referred to Chap. 8 of Lambeck (1980).

4.2.2 *Non-rigid Rotation*

Euler's equations (4.6) and their solutions for free nutation (4.11–4.13) constitute a valid zeroth-order approximation of the rotational behaviour of the Earth. However, these equations are neglecting two very important aspects: *i)* the time-variations of the inertia tensor due to the Earth's deformation in response to e.g., surface loads, mantle convection or large earthquakes and *(ii)* the motion relative to the axes of the rotating reference frame, associated to ocean currents and atmospheric circulation (Lambeck 1980; Barnes et al. 1983).

4.2.2.1 Liouville Equations

To study how relevant geophysical processes can affect Earth rotation on various time-scales, it is necessary to generalise Euler's equations appropriately. Following Barnes et al. (1983), we extend Eq. (4.4) by writing the angular momentum as

$$L_i(t) = I_{ij}(t)\,\omega_j(t) + h_i(t)\,, \tag{4.19}$$

where the time-dependence of the inertia tensor I_{ij} is shown explicitly and

$$h_i(t) = \int_V \varepsilon_{ijk}\, x_j\, v_k\, \mathrm{d}m \tag{4.20}$$

where v_i is the velocity field of the geophysical fluids (Munk and MacDonald 1975; Lambeck 1980). Hence, from (4.2), the momentum equations take the final form:

$$\frac{\mathrm{d}}{\mathrm{d}t}\big(I_{ij}(t)\,\omega_j + h_i(t)\big) + \varepsilon_{ijk}\,\omega_j\,\left(I_{kl}(t)\,\omega_l + h_k(t)\right) = N_i\,, \tag{4.21}$$

known as *Liouville equations* for polar motion. As mentioned by Munk and MacDonald (1975) quoting the work of Routh (1905), they were first obtained by Joseph Liouville in 1858 (see Fang et al. 2022, for a discussion).

4.2.2.2 Linearised Liouville Equations

Liouville equations constitute a set of three coupled, non-linear ODEs in the unknowns ω_i, for which a closed form solution is generally not available (see e.g., Spada et al. 1993). For many problems of non-rigid rotation, including GIA, it is possible to consider small displacements of the rotation axis relative to an initial state of uniform rotation, obtaining a linearised set of equations (Sabadini et al. 1982; Wu and Peltier 1984). For a more general non-linear approach, valid for large displacements of the axis of rotation and suitable to describe the long-term rotational dynamics of the Earth, the readers are referred to Lefftz (1991), Spada (1992), Spada et al. (1992a, 1993), Ricard et al. (1993), Steinberger and O'Connell (1997), and and to citations therein. As we shall see in the following, for small excursions of the rotation axis, the three Liouville equations reduce to a set of two linear equations for polar motion, and one decoupled equation describing the variations of the LOD, respectively (Lambeck 1980).

The instantaneous pole of rotation is marked by P in Fig. 4.2, whereas the cartesian components of the angular velocity vector are

$$\begin{aligned} \omega_1(t) &\equiv \omega \sin\theta_p \cos\lambda_p \\ \omega_2(t) &= \omega \sin\theta_p \sin\lambda_p \\ \omega_3(t) &= \omega \cos\theta_p \,, \end{aligned} \tag{4.22}$$

where θ_p and λ_p are the co-latitude and longitude of P, respectively, and where $\omega \equiv \sqrt{\omega_k \omega_k}$ is the modulus of ω_i. These equations hold in general, for any location of the pole with respect to axis x_3. However, in our ensuing discussion, it is more convenient to consider small excursions of the axis of rotation and to write the angular velocity components as follows:

$$\begin{aligned} \omega_1(t) &= \Omega\, m_1 \\ \omega_2(t) &= \Omega\, m_2 \\ \omega_3(t) &= \Omega\,(1 + m_3)\,, \end{aligned} \tag{4.23}$$

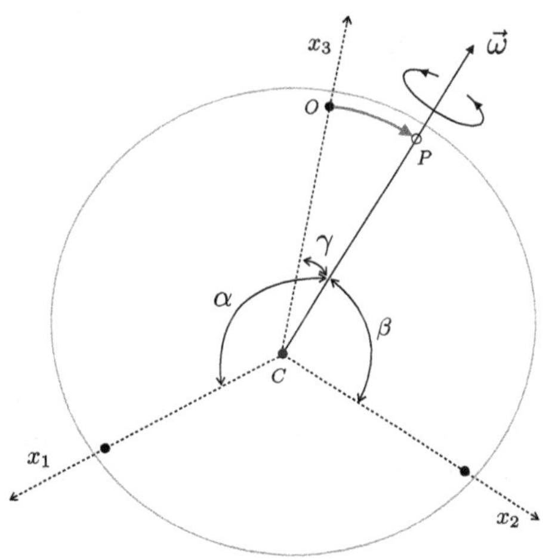

Fig. 4.2 In response to a surficial or endogenous geophysical process, the instantaneous pole of rotation slightly moves from O to P. At the same time the LOD changes, due to variations in the spin rate

or

$$\omega_i(t) = \Omega(m_i + \delta_{i3})\,, \qquad (i = 1, 2, 3)\,, \tag{4.24}$$

where Ω is a reference angular velocity that can be identified with the mean angular velocity of the Earth, and where $m_i = m_i(t)$ denote small, non-dimensional quantities ($|m_i| \ll 1$). They have a direct physical meaning, since $m_1(t)$ and $m_2(t)$ describe the *polar motion* (or *wobble*) of the pole of rotation and its erratic motion on the Earth's surface, while $m_3(t)$ is associated with variations of the axial rotation with

$$m_3(t) = -\frac{\Delta\text{LOD}}{\text{LOD}}\,, \tag{4.25}$$

where LOD stands for *length of day* (Munk and MacDonald 1975; Lambeck 1980).

Astronomical observation of Earth rotation during last century confirm that quantities m_i are effectively small, which fully justifies approximations (4.23). Between 1900 and 1972, the observations have revealed that the quasi-periodic Chandler wobble had an average amplitude of $\sim 0.2''$ (arc seconds,[3] see e.g., Guinot 1972), corresponding to values

$$(|m_1|, |m_2|)^{\text{obs}} \approx 10^{-6}\,. \tag{4.26}$$

Furthermore, according to Morrison (1973), values of m_3 from 1820 to 1975 have been subject to significant temporal variations, but their amplitude has never exceeded a few units in 10^{-9}, so that we can safely assume that

$$|m_3|^{\text{obs}} \lesssim 10^{-8}\,. \tag{4.27}$$

From a geometrical standpoint (see Fig. 4.2), m_1, m_2 and $1 + m_3$ approximately represent the direction cosines of ω_i relative to axes x_1, x_2, and x_3, respectively, i.e., they are cosines of the angles (α, β, γ) that the angular velocity vector forms *w.r.t.* the directions of axes (x_1, x_2, x_3) (Gross 1992). This can be appreciated first observing that, to first order, the modulus of ω_i is $\omega \equiv \sqrt{\omega_k \omega_k} \sim \Omega(1 + m_3) = \omega_3$. We note also that since $\omega_1 = \omega \cos\alpha$, where α is the angle between axis x_1 and the angular velocity vector, we have $\cos\alpha = \omega_1/\omega = (\omega_1/\Omega)(\Omega/\omega) \sim m_1(\Omega/\omega_3) = m_1/(1 + m_3) \sim m_1$. In a similar way, we see that $\cos\beta \sim m_2$ and $\cos\gamma \sim 1 + m_3$. We further note that, in general, $\omega \neq \Omega$; indeed, from Eqs. (4.23–4.24), we have

$$\omega^2 = \Omega^2\left(1 + 2m_3 + m^2\right)\,, \tag{4.28}$$

with $m^2 = m_1^2 + m_2^2 + m_3^2$.

Within the linearization scheme of the Liouville equations, a further assumption is that the variations of the inertia tensor can be regarded as small quantities, if compared to the diagonal terms. This is justified for a broad range of

---

[3] 1 arc second ($1''$) corresponds to 1/3600 of a degree.

geophysical phenomena (Munk and MacDonald 1975; Lambeck 1980). Hence, following a perturbative approach, we write

$$I_{ij}(t) = \begin{pmatrix} A & 0 & 0 \\ 0 & A & 0 \\ 0 & 0 & C \end{pmatrix} + \Delta I_{ij}(t), \tag{4.29}$$

where $\Delta I_{ij}(t)$ is a small time-dependent perturbation, such that $|\Delta I_{ij}|/C \ll 1$; similarly, it is assumed that the relative momentum and the torque are small quantities as well, with $|h_i| \ll \Omega C$ and $|N_i| \ll \Omega^2 C$ (Barnes et al. 1983).

A direct consequence of Eq. (4.29) is that x_3 is not the axis of maximum inertia of the rotating body, which shall be slightly displaced in consequence of the inertia perturbations ΔI_{ij}. In particular, it is not difficult to show that the direction cosines of the figure axis (i.e., the axis of maximum inertia) with respect to axes x_1 and x_2 are approximately $\Delta I_{13}/(C - A)$ and $\Delta I_{23}/(C - A)$, respectively and that the corresponding moment of inertia is $C + \Delta I_{33}$ (see e.g., Terpos 1973).

By substituting Eq. (4.24) and (4.29) into (4.21), and ignoring the products of small quantities, the Liouville equations reduce to a coupled set of two linear non-homogeneous ODEs for the *wobble* (or *polar motion*):

$$\frac{\dot{m}_1}{\sigma_\mathrm{r}} + m_2 = +\psi_2 \tag{4.30}$$

$$\frac{\dot{m}_2}{\sigma_\mathrm{r}} - m_1 = -\psi_1, \tag{4.31}$$

and to a decoupled LOD equation:

$$\dot{m}_3 = \dot{\psi}_3, \tag{4.32}$$

where the two forcing terms

$$\psi_1(t) = \frac{\Omega^2 \Delta I_{13} + \Omega^2 \Delta \dot{I}_{23} + \Omega h_1 + \dot{h}_2 - N_2}{\Omega^2 (C - A)} \tag{4.33}$$

$$\psi_2(t) = \frac{\Omega^2 \Delta I_{23} - \Omega^2 \Delta \dot{I}_{13} + \Omega h_2 - \dot{h}_1 + N_1}{\Omega^2 (C - A)}, \tag{4.34}$$

are the *polar motion excitation functions*, whereas

$$\psi_3(t) = \frac{-\Omega^2 \Delta I_{33} - \Omega h_3 + \Omega \int_0^t N_3(t')\mathrm{d}t'}{\Omega^2 C} \tag{4.35}$$

is the *LOD excitation function* (Terpos 1973; Rochester 1973; Munk and MacDonald 1975; Lambeck 1980; Barnes et al. 1983; Gross 1992; Zharkov et al. 1996).

As it is clear from their definition, the three excitation functions (EFs) ψ_i describe all factors perturbing the rotational state of the Earth, such as for example changes in the inertia tensor and in relative momentum associated with the atmospheric and oceanic circulation, global mass redistributions due to convection currents in the mantle, variations in the centrifugal forces and tides, inertia variations due to large earthquakes and global surface loading processes such as GIA or the melting of contemporary ice sheets and glaciers, core-mantle coupling, and possibly other so far unknown processes (see e.g., Munk and MacDonald 1975; Lambeck 1980; Barnes et al. 1983). As we shall see more in detail in Sect. 4.2.3, the EFs can be conveniently classified according to their physical meaning.

4.2.2.3 General Solution of the Linearised Liouville Equations

Using complex variables, the two linearised Liouville equations for polar motion (4.30–4.31) can be written in the compact form:

$$\mathrm{i}\,\frac{\dot{\mathbf{m}}}{\sigma_{\mathrm{r}}} + \mathbf{m} = \boldsymbol{\Psi}, \tag{4.36}$$

where the *polar motion vector* is:

$$\mathbf{m}(t) = m_1 + \mathrm{i}\, m_2\,, \tag{4.37}$$

and the *complex EF for polar motion* is:

$$\boldsymbol{\Psi}(t) = \frac{\Omega^2 \Delta\mathbf{I} - \mathrm{i}\,\Omega\Delta\dot{\mathbf{I}} + \Omega\mathbf{h} - \mathrm{i}\,\dot{\mathbf{h}} + \mathrm{i}\,\mathbf{N}}{\Omega^2\,(C-A)}\,, \tag{4.38}$$

with:

$$\boldsymbol{\Psi}(t) = \psi_1 + \mathrm{i}\,\psi_2 \tag{4.39}$$

$$\Delta\mathbf{I}(t) = \Delta I_{13} + \mathrm{i}\,\Delta I_{23} \tag{4.40}$$

$$\mathbf{h}(t) = h_1 + \mathrm{i}\,h_2 \tag{4.41}$$

$$\mathbf{N}(t) = N_1 + \mathrm{i}\,N_2\,. \tag{4.42}$$

Assuming that all the perturbing geophysical phenomena can be neglected so that $\boldsymbol{\Psi}(t) = 0$, the system (4.36) is homogeneous, and the solution is a free oscillation

$$\mathbf{m} = \mathbf{m}_0\, \mathrm{e}^{\mathrm{i}\sigma_{\mathrm{r}} t}\,, \tag{4.43}$$

where $\mathbf{m}_0 = m_{10} + \mathrm{i}\, m_{20}$ is the initial value of $\mathbf{m}$. Furthermore, with $\psi_3 = 0$ the LOD equation (4.32) gives

$$m_3 = m_{30} = \text{constant}\,, \tag{4.44}$$

so that we easily recognise that (4.43) and (4.44) describe, as expected, the free nutation of a rigid Earth (4.11–4.13).

In the more geophysically interesting case of a excitation $\boldsymbol{\Psi}(t) \neq 0$, the general solution of the non-homogeneous wobble equation (4.36) can be found, for instance, using the method of Laplace transforms. Assuming that $\boldsymbol{\Psi}(t)$ is a causal function (*i.e.*, that it vanishes for $t < 0$), the solution of (4.36) can be cast in the form:

$$\mathbf{m}(t) = \mathbf{m}_0\, \mathrm{e}^{\mathrm{i}\sigma_\mathrm{r} t} - i\sigma_\mathrm{r} \int_0^t \boldsymbol{\Psi}(t')\mathrm{e}^{-\mathrm{i}\sigma_\mathrm{r}(t-t')}\mathrm{d}t'\,, \tag{4.45}$$

where we recognise that the integral represents the time-convolution between functions $\mathrm{e}^{-\mathrm{i}\sigma_\mathrm{r} t}$ and $\boldsymbol{\Psi}(t)$ (see B.68). The problem of obtaining information about the geophysical processes perturbing the Earth rotation using astronomical observations of m_i is thus one of deconvolution (e.g., Lambeck 1980).

4.2.3 Excitation Functions

It is customary to classify the EFs into three families, according to the physical origin of the excitation, namely:

$$\psi_i(t) = \psi_i^{\text{matter}} + \psi_i^{\text{motion}} + \psi_i^{\text{torque}} \quad (i = 1, 2, 3)\,, \tag{4.46}$$

where the *matter EF* depend upon variations of the inertia tensor $\Delta\mathbf{I}$ and ΔI_{33}, the *motion EF* is a linear combination of $\Delta\dot{\mathbf{I}}$, $\mathbf{h}$, $\dot{\mathbf{h}}$ and h_3, while the *torque EF* is only dependent upon $\mathbf{N}$ and N_3 (Lambeck 1980). We note that ψ_i^{motion} could be further decomposed into velocity and acceleration contributions (see Lambeck 1980). For explicit expressions of the EFs, both in cartesian and in spherical coordinates, the readers are referred to Eqs. (4.1.1) and (4.1.2) of Lambeck (1980) and to his Table 4.1.

Since GIA mainly involves the solid Earth, the relative angular momentum and the external torques can be disregarded (Peltier 1974; Wu and Peltier 1982; Sabadini et al. 1982; Wu and Peltier 1984); so that ψ_i^{motion} and ψ_i^{torque} shall not be considered further. According to Eqs. (4.33–4.34) and (4.35) above, the matter EFs are:

$$\boldsymbol{\Psi}^{\text{matter}}(t) = \frac{\Delta\mathbf{I}}{C - A} \tag{4.47}$$

$$\psi_3^{\text{matter}}(t) = -\frac{\Delta I_{33}}{C}\,, \tag{4.48}$$

where, recalling that $\Delta\mathbf{I} = \Delta I_{13} + i\,\Delta I_{23}$, we note that these EFs only involve three out of the six independent components of the inertia tensor variation.

To describe the effects of GIA on Earth rotation, two types of matter EFs are required. The first (the *rotational deformation EF*) is associated with the deformations induced by the centrifugal force; the second one (*surface loading EF*) is associated with loads acting at the Earth's surface. Hence, the total matter EFs are

$$\mathbf{\Psi}^{\text{matter}}(t) = \mathbf{\Psi}^{\text{rot}} + \mathbf{\Psi}^{\text{sur}} \tag{4.49}$$

$$\psi_3^{\text{matter}}(t) = \psi_3^{\text{rot}} + \psi_3^{\text{sur}}, \tag{4.50}$$

with

$$\mathbf{\Psi}^{\text{rot}}(t) = \frac{\Delta \mathbf{I}^{\text{rot}}}{C - A} \tag{4.51}$$

$$\mathbf{\Psi}^{\text{sur}}(t) = \frac{\Delta \mathbf{I}^{\text{sur}}}{C - A}, \tag{4.52}$$

and

$$\psi_3^{\text{rot}}(t) = -\frac{\Delta I_{33}^{\text{rot}}}{C} \tag{4.53}$$

$$\psi_3^{\text{sur}}(t) = -\frac{\Delta I_{33}^{\text{sur}}}{C}, \tag{4.54}$$

where it is useful to remark that $\mathbf{\Psi}^{\text{rot}}$ and ψ_3^{rot} depend upon the polar motion vector $\mathbf{m}$, while $\mathbf{\Psi}^{\text{sur}}$ and ψ_3^{sur} do not, as we will see further on.

We note that for the purpose of modelling rotational effects of GIA on sea-level change, it is not necessary to consider the problem of the length-of-day variations (see e.g., Mitrovica and Wahr 2011). The physical reason is that, to first order, the centrifugal potential variations only depend upon the wobble components m_1 and m_2, and not on the LOD variable m_3, as we will see in next section.

4.2.4 Excitation Function for Rotational Deformation

We first find a convenient expression for the centrifugal potential variation ΔW^c induced by polar motion, which is a prerequisite for the evaluation of the excitation functions $\mathbf{\Psi}^{\text{rot}}$ and ψ_3^{rot}.

4.2.4.1 Centrifugal Potential Variations

The centrifugal potential variation ΔW^c in response to variations of the angular velocity can be evaluated in various ways. Here we shall follow an elementary approach; a more formal derivation is reported in Sect. D.1, leading to the same result, and fully consistent with relevant works in the literature (see e.g., Milne and Mitrovica 1998).

A further method exists, followed by e.g., Lambeck (1980), Sabadini and Peltier (1981), Peltier (2015) and Sabadini et al. (2016) which exploits the MacCullagh's formula (Munk and MacDonald 1975). Since the centrifugal potential variations are directly caused by Earth rotation fluctuations, we expect that ΔW^c is a function of m_i. Indeed we shall see that, to first order in small quantities, the relation is one of linear proportionality and only involves m_1 and m_2, i.e., the two polar motion components.

We assume a reference state in which the Earth has been rotating in equilibrium until time $t_0 = 0$, with its angular velocity vector constantly aligned with the z axis, i.e., $\omega_{i0} = \Omega\, \delta_{i3}$, where Ω is a constant. Subsequently, for $t > 0$, a new dynamical state is attained in response to some geophysical forcing like e.g., GIA. In this state, the rotation axis is slightly displaced with respect to the equilibrium state, so that $\omega_i = \Omega(m_i + \delta_{i3})$.

At an arbitrary point $A = (a, \theta, \lambda)$ of the Earth's surface, in the general state the centrifugal potential is $W^c = \frac{1}{2}\,\omega^2 d^2$, where d is the distance that separates A from the rotation axis (see Fig. 4.3). But since $d = a \sin\Theta$, we have

$$W^c(\Theta, t) = \frac{1}{2}\,\omega^2\, a^2 \left(1 - \cos^2\Theta\right), \tag{4.55}$$

and similarly, in the reference state

$$W^{c0}(\theta) = \frac{1}{2}\,\Omega^2\, a^2 \left(1 - \cos^2\theta\right), \tag{4.56}$$

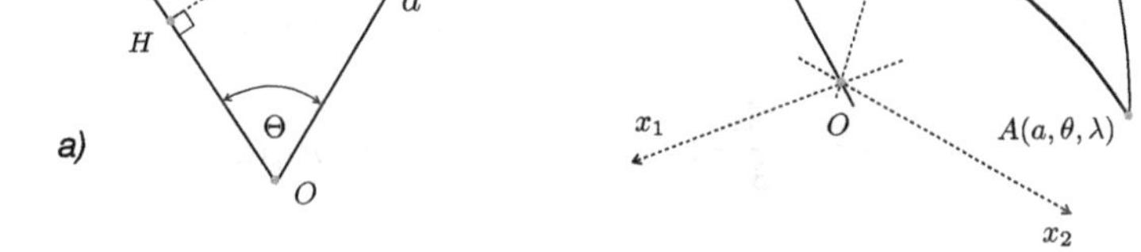

Fig. 4.3 Geometry for the study of variations of the centrifugal potential induced by movements of the pole of rotation. Point A denotes an Earth-fixed observer at (a, θ, λ), $P = (a, \theta_p, \lambda_p)$ is the instantaneous position of the pole of rotation (IPR), and $\vec{\omega}$ is the angular velocity vector. Θ is the angular separation between A and P, with $\cos\Theta = \cos\theta \cos\theta_p + \sin\theta \sin\theta_p \cos(\lambda - \lambda_p)$

so that the potential variation in A is

$$\Delta W^{c}(\gamma, t) = W^{c} - W^{c0} = \frac{a^2}{2}\left(\omega^2 - \Omega^2\right) - \frac{a^2}{2}\left(\omega^2 \cos^2\Theta - \Omega^2\cos^2\theta\right) . \tag{4.57}$$

From Eq. (4.28), we note that the condition $\omega^2 \approx \Omega^2$ is attained with a precision of a few parts in 10^6, due to the small observed values of the m_i's (see 4.26 and 4.27). Hence, the first term on the *r.h.s.* of (4.57), of harmonic degree $n = 0$, can be neglected (Lambeck and Burša 1973; Lambeck 1980). Thus, the centrifugal potential variation can be approximated by

$$\Delta W^{c}(\gamma, t) = -\frac{a^2\Omega^2}{2}\left(\cos^2\Theta - \cos^2\theta\right) . \tag{4.58}$$

The idea is now to transform this expression assuming a small value of θ_p, the co-latitude of the pole of rotation. This is consistent with the assumption that the axis of rotation is displaced only slightly in consequence of some geophysical excitation, coherent with the linearization scheme for the Liouville equations adopted. Applying the cosine rule for sides (A.29) to spherical triangle $\widehat{APQ}$ in Fig. 4.3, we have

$$\begin{aligned}
\Delta W^{c}(\gamma, t) &= -\frac{a^2\Omega^2}{2}\left(\left(\cos\theta_p\cos\theta + \sin\theta_p\sin\theta\cos(\lambda-\lambda_p)\right)^2 - \cos^2\theta\right)\\
&= -\frac{a^2\Omega^2}{2}\Big(-\sin^2\theta_p\cos^2\theta + \sin^2\theta_p\sin^2\theta\cos^2(\lambda-\lambda_p) +\\
&\qquad +2\sin\theta_p\cos\theta_p\sin\theta\cos\theta\cos(\lambda-\lambda_p)\Big)\\
&\simeq -a^2\Omega^2\sin\theta\cos\theta\left(\sin\theta_p\cos\lambda_p\cos\lambda + \sin\theta_p\sin\lambda_p\sin\lambda\right)\\
&= -\frac{a^2\Omega^2}{2}\left(m_1\cos\lambda + m_2\sin\lambda\right)\sin 2\theta ,
\end{aligned} \tag{4.59}$$

where we have approximated $\sin^2\theta_p \approx 0$ and $\cos\theta_p \approx 1$, and we have used (4.22). Result (4.59) agrees with the formula (D.41), established by means of rigorous methods.[4]

By recalling the definition of $P_{21}(\cos\theta)$ in Table A.1, we have

$$\Delta W^{c}(\gamma, t) = \frac{a^2\Omega^2}{3}\left(m_1\cos\lambda + m_2\sin\lambda\right) P_{21}(\cos\theta) , \tag{4.60}$$

showing that, to a high degree of precision, the variation of centrifugal potential due to a small displacement of the Earth's rotation axis is a pure harmonic function of degree $n = 2$ and order $m = 1$. This potential does not load the Earth but it has an important indirect effect, since the induced deformations will cause a variation of

[4] Note that in Appendix D.1 we use symbol $\Lambda(\gamma, t)$ to indicate the centrifugal potential variation.

the gravitational potential and of the inertia tensor of the rotating body. We note that result (4.60), obtained for $r = a$, would be also valid for other values of radius r.

4.2.4.2 Rotational Deformation EF for Polar Motion

By application of the concept of Love number,[5] the induced variation of gravitational potential $\Delta W^{\rm ind}$ (*the effect*) is linearly proportional to $\Delta W^{\rm c}(t)$ (*the cause*):

$$\Delta W^{\rm ind}(\gamma, t) = k_2^{\rm T}(t) * \Delta W^{\rm c}(\gamma, t)\,, \tag{4.61}$$

where $k_2^{\rm T}(t)$ is the tidal Love number of degree $n = 2$, and the time convolution $(*)$ is an expression of the principle of additivity in linear systems[6] (Tschoegl 2012). The use of the tidal Love number here – and not of an hypothetical *centrifugal Love number* $k_2^{\rm C}(t)$ – is motivated by the fact that the tidal potential and the centrifugal potential have the same radial dependence $\sim r^2$ and are both harmonic functions (see page 9 of Lambeck 1980), although the latter has a spectrum limited to harmonic degree $n = 2$ whereas the former contains, in principle, all contributions of degrees $n = 2, 3, \ldots$ (see page 23 of Sabadini et al. 2016).

On one hand, combining (4.61) with (4.60) gives:

$$\Delta W^{\rm ind}(\gamma, t) = \frac{a^2\Omega^2}{3}\left[k_2^{\rm T} * (m_1 \cos\lambda + m_2 \sin\lambda)\right] P_{21}(\cos\theta)\,, \tag{4.62}$$

but on the other hand, according to (1.27), we also have

$$\Delta W^{\rm ind}(\gamma, t) = \frac{Gm^{\rm e}}{a}\left(\Delta c_{21}^{\rm rot} \cos\lambda + \Delta s_{21}^{\rm rot} \sin\lambda\right) P_{21}(\cos\theta)\,, \tag{4.63}$$

where a is Earth's radius, $m^{\rm e}$ is its mass and $(\Delta c_{21}^{\rm rot}, \Delta s_{21}^{\rm rot})$ are appropriate Stokes' coefficients of harmonic degree 2 and order 1, so that by comparing the two last equations we get:

$$\Delta c_{21}^{\rm rot}(t) + {\rm i}\,\Delta s_{21}^{\rm rot}(t) = \frac{a^3\Omega^2}{3Gm^{\rm e}}\left(k_2^{\rm T} * \mathbf{m}\right)\,. \tag{4.64}$$

But according to Eqs. (1.42) and (1.43)

$$\Delta c_{21}^{\rm rot}(t) + {\rm i}\,\Delta s_{21}^{\rm rot}(t) = \frac{\Delta I_{13}^{\rm rot} + {\rm i}\,\Delta I_{23}^{\rm rot}}{Ma^2}\,, \tag{4.65}$$

[5] In Chen et al. (2010) this concept is referred to as *Love's principle*.

[6] We note that the Laplace transform of (4.61) is $\Delta\tilde{W}^{\rm ind}(\gamma, s) = \tilde{k}_2^{\rm T}(s)\Delta\tilde{W}^{\rm c}(\gamma, s)$, where $\tilde{k}_2^{\rm T}(s)$ and $\Delta\tilde{W}^{\rm c}(\gamma, s)$ are the Laplace transforms of $k_2^{\rm T}(t)$ and $\Delta W^{\rm c}(\gamma, t)$, respectively. In the frequency domain, this is effectively a relation of simple proportionality.

so that

$$\Delta \mathbf{I}^{\mathrm{rot}}(t) = \frac{a^5\Omega^2}{3G}\left(k_2^{\mathrm{T}} * \mathbf{m}\right) \tag{4.66}$$

represents the inertia variation in response to the change of centrifugal potential $\Delta W^{\mathrm{c}}(t)$. By the definition (4.51), we therefore conclude that the rotational deformation EF is

$$\boldsymbol{\psi}^{\mathrm{rot}}(t) = \frac{1}{k^{\mathrm{s}}}\left(k_2^{\mathrm{T}} * \mathbf{m}\right), \tag{4.67}$$

where we have defined the non-dimensional constant:

$$k^{\mathrm{s}} = \frac{3\,G}{a^5\,\Omega^2}(C - A)\,. \tag{4.68}$$

We note that k^{s} was introduced by Munk and MacDonald (1975) to characterise the long-term yielding of the Earth in response to centrifugal force. We shall refer to k^{s} to as *secular Love number*[7] (we observe that in Lambeck (1980), it is considered *merely a convenient constant* and it is denoted by k_{o}). Others, like e.g., Han and Wahr (1989), Mitrovica and Wahr (2011) and Peltier (2015) call it *fluid Love number*, and denote it by k_{f}. Using observed values of the constants involved, Lambeck (1980) finds that k^{s} is close to unity, with:

$$k^{\mathrm{s}} = 0.942\,. \tag{4.69}$$

It is worth noting that for an elastic Earth, the TLN is impulsive, with: $k_2^{\mathrm{T}}(t) = k_2^{\mathrm{T,e}}\,\delta(t)$. In this case, the convolutions are transformed into a simple linear proportionality, and the rotational EF turns out to be

$$\boldsymbol{\psi}^{\mathrm{rot}}(t) = \frac{1}{k^{\mathrm{s}}}\left(k_2^{\mathrm{T,e}}\,\mathbf{m}\right). \tag{4.70}$$

4.2.4.3 Rotational Deformation EF for the Length of Day

Since the potential $\Delta W^{\mathrm{ind}}(\gamma, t)$ is a pure harmonic of degree $n = 2$ and order $m = 1$, it is uniquely responsible for time variations of the products of inertia $\Delta I_{13}^{\mathrm{rot}}$ and $\Delta I_{23}^{\mathrm{rot}}$, and not for variations of the axial component $\Delta I_{33}^{\mathrm{rot}}$ (see e.g., Lambeck 1980;

[7] Following Burša (1992), it may be interesting to observe that the secular Love number can be equivalently written as $k^{\mathrm{s}} = -3\,J_2/m^{\mathrm{c}}$ where $J_2 = -(C - A)/Ma^2$ is the dynamical form factor of the Earth and m^{c}, which also appears in Clairaut's equation (see e.g., Hofmann-Wellenhof and Moritz 2006) is $m^{\mathrm{c}} = a\omega^2/g_0(a)$, the ratio between the centrifugal force and the gravitational acceleration at the equator.

Barnes et al. 1983; Han and Wahr 1989; Sidorenkov 2009; Peltier 2015). Hence from Eq. (4.53), the LOD EF associated with centrifugal deformation is simply:

$$\psi_3^{\mathrm{rot}}(t) = 0\,. \tag{4.71}$$

4.2.5 *Excitation Function for Surface Loading*

Here we study the surface loading EFs $\boldsymbol{\Psi}^{\mathrm{sur}}$ and ψ_3^{sur}, which along with the rotational deformation EF obtained in previous section, constitute an essential part of the Liouville equations. We find it convenient to start from the general expression of the surface response function $\mathcal{G}^{\mathrm{sur}}(\gamma, t)$ obtained in Sect. 3.2.4, which represents the variation of the geoid height in response to surface loading. Following, e.g., Seitz and Schuh (2010), the idea is to take advantage of the relation between the degree $n = 2$ Stokes coefficients of the gravitational potential and the elements of the inertia tensor.

By Bruns formula (see e.g., Hofmann-Wellenhof and Moritz 2006), the variation of gravitational potential due to surface loading is $\Delta W^{\mathrm{sur}} = g_0\, \mathcal{G}^{\mathrm{sur}}$, where $g_0 = Gm^{\mathrm{e}}/a^2$ is the reference surface gravity. On one hand, according to (3.39) the coefficients of the CSH expansion of the potential are

$$\Delta W_{nm}^{\mathrm{sur}}(t) = g_0 \frac{3}{\rho^{\mathrm{e}}} \frac{\left[\delta(t) + k_n^{\mathrm{L}}(t)\right] * \mathcal{L}_{nm}(t)}{2n+1}\,, \tag{4.72}$$

where $\mathcal{L}_{nm}(t)$ are the coefficients of the surface load variation and $k_n^{\mathrm{L}}(t)$ is the degree n LLN for the potential. On the other hand, following (1.27), the gravitational potential can be expanded in series of ALPs according to:

$$\Delta W^{\mathrm{sur}}(\gamma, t) = \frac{Gm^{\mathrm{e}}}{a} \sum_{nm} \left(\Delta c_{nm}^{\mathrm{sur}} \cos m\lambda + \Delta s_{nm}^{\mathrm{sur}} \sin m\lambda\right) P_{nm}(\cos\theta)\,, \tag{4.73}$$

where the sum is limited to positive orders and $(\Delta c_{nm}^{\mathrm{sur}}, \Delta s_{nm}^{\mathrm{sur}})$ represent the corresponding variation of the cosine and sine Stokes coefficients, respectively. Taking advantage of the transformation formula (A.47), we find

$$\Delta c_{nm}^{\mathrm{sur}}(t) + \mathrm{i}\,\Delta s_{nm}^{\mathrm{sur}}(t) = \frac{3}{\rho^{\mathrm{e}} a^2} \frac{2 - \delta_{0m}}{\sqrt{2n+1}} \sqrt{\frac{(n-m)!}{(n+m)!}} \left[\delta(t) + k_n^{\mathrm{L}}(t)\right] * \mathcal{L}_{nm}^{*}(t)\,. \tag{4.74}$$

4.2.5.1 Surface Loading EF for Polar Motion

Using Eqs. (1.42) and (1.43), the inertia variation due to surface loading is

$$\Delta \mathbf{I}^{\text{sur}}(t) = Ma^2 \left(\Delta c_{21}^{\text{sur}} + \mathrm{i}\, \Delta s_{21}^{\text{sur}}\right), \tag{4.75}$$

where $\Delta c_{21}^{\text{sur}}$ and $\Delta s_{21}^{\text{sur}}$ can be obtained from (4.74), giving:

$$\Delta \mathbf{I}^{\text{sur}}(t) = \left[\delta(t) + k_2^{\mathrm{L}}(t)\right] * \Delta \mathbf{I}^{\text{rig}}, \tag{4.76}$$

where

$$\Delta \mathbf{I}^{\text{rig}}(t) = \frac{4\pi}{3}\sqrt{\frac{6}{5}}\, a^4\, \mathcal{L}_{21}^*(t) \tag{4.77}$$

is the inertia variation that would be observed if the Earth was rigid.[8] The EF for surface loading follows immediately from (4.76) and (4.52), with

$$\mathbf{\Psi}^{\text{sur}}(t) = \left[\delta(t) + k_2^{\mathrm{L}}(t)\right] * \mathbf{\Psi}^{\text{rig}}(t), \tag{4.81}$$

---

[8] That this interpretation is physically correct, it can be seen by the following argument, along the lines of Wu and Peltier (1984). For a rigid Earth, $\Delta \mathbf{I}^{\text{rig}}$ can only depend upon the surface mass redistribution and therefore it should be possible to write it as a surface integral. By definition of inertia tensor (1.8), we have

$$\Delta \mathbf{I}^{\text{rig}}(t) = -\int_e (x_1 x_3 + i\, x_2 x_3)\, dM, \tag{4.78}$$

where dM is the elementary change of surface mass. But recalling from Eq. (1.119) that $\mathcal{L}(\gamma, t) = dM/dA$, we obtain:

$$\Delta \mathbf{I}^{\text{rig}}(t) = -a^2 \int_\gamma (x_1 x_3 + i x_2 x_3) \mathcal{L}(\gamma, t)\, \mathrm{d}\gamma = -a^4 \int_\gamma \sin\theta \cos\theta\, \mathrm{e}^{i\lambda} \mathcal{L}(\gamma, t)\, \mathrm{d}\gamma =$$

$$= \frac{1}{3}\sqrt{\frac{6}{5}}\, a^4 \int_\gamma \mathcal{L}(\gamma, t) Y_{21}(\gamma)\, \mathrm{d}\gamma, \tag{4.79}$$

where we have written the x_i's in terms of spherical coordinates and we have used the expression of $Y_{21}(\gamma)$ in Table A.1. We now expand the scalar field $\mathcal{L}(\gamma, t)$ in series of CSHs and use the orthogonality condition, obtaining:

$$\begin{aligned}
\Delta \mathbf{I}^{\text{rig}}(t) &= \frac{a^4}{3}\sqrt{\frac{6}{5}} \int_\gamma \Big(\sum_{nm} \mathcal{L}_{nm}^*(t) Y_{nm}^*(\gamma)\Big) Y_{21}(\gamma)\, \mathrm{d}\gamma \\
&= \frac{1}{3}\sqrt{\frac{6}{5}}\, a^4 \sum_{nm} \mathcal{L}_{nm}^*(t) \int_\gamma Y_{nm}^*(\gamma) Y_{21}(\gamma)\, \mathrm{d}\gamma =
\end{aligned}$$

$$= \frac{4\pi}{3}\sqrt{\frac{6}{5}}\, a^4 \sum_{nm} \mathcal{L}_{nm}^*(t)\, \delta_{l2}\delta_{m1} = \frac{4\pi}{3}\sqrt{\frac{6}{5}}\, a^4\, \mathcal{L}_{21}^*(t), \tag{4.80}$$

which agrees with (4.77) in the text.

where the polar motion excitation function for a rigid Earth is

$$\boldsymbol{\Psi}^{\mathrm{rig}}(t) = c_{21}^{\psi} \, \mathcal{L}_{21}^{*}(t) \,, \tag{4.82}$$

and we have defined the constant

$$c_{21}^{\psi} = \frac{4\pi}{3} \sqrt{\frac{6}{5}} \frac{a^4}{C - A} \,. \tag{4.83}$$

4.2.5.2 Surface Loading EF for the Length of Day

According to (1.47), the axial inertia variation is

$$\Delta I_{33}^{\mathrm{sur}}(t) = -\frac{2}{3} M a^2 \Delta c_{20}^{\mathrm{sur}} \,, \tag{4.84}$$

where, following Rochester and Smylie (1974), we have taken advantage of the invariance of the trace of inertia tensor, since we are considering solenoidal displacement fields (see Eq. 1.47). The zonal coefficient $\Delta c_{20}^{\mathrm{sur}}$ is obtained setting $n = 2$, and $m = 0$ in (4.74), giving

$$\Delta I_{33}^{\mathrm{sur}}(t) = \left[\delta(t) + k_2^{\mathrm{L}}(t) \right] * \Delta I_{33}^{\mathrm{rig}} \,, \tag{4.85}$$

where

$$\Delta I_{33}^{\mathrm{rig}}(t) = -\frac{4\pi}{3} \sqrt{\frac{2}{5}} \, a^4 \, \mathcal{L}_{20}(t) \tag{4.86}$$

is the axial inertia variation for a rigid Earth. The surface loading EF for the LOD follows immediately from (4.85) and (4.54), with

$$\psi_3^{\mathrm{sur}}(t) = \left[\delta(t) + k_2^{\mathrm{L}}(t) \right] * \psi_3^{\mathrm{rig}}(t) \,, \tag{4.87}$$

where

$$\psi_3^{\mathrm{rig}}(t) = c_{20}^{\psi} \, \mathcal{L}_{20}(t) \,, \tag{4.88}$$

and we have defined the constant:

$$c_{20}^{\psi} = -\frac{4\pi}{3} \sqrt{\frac{2}{5}} \frac{a^4}{C} \,. \tag{4.89}$$

4.2.5.3 A Note on the Inertia Variations Due to Surface Loading

We have seen that the total variation of the inertia tensor due to surface loading is

$$\Delta I_{ij}^{\mathrm{sur}}(t) = \left[\delta(t) + k_2^{\mathrm{L}}(t)\right] * \Delta I_{ij}^{\mathrm{rig}}(t)\,, \tag{4.90}$$

where $\Delta I_{ij}^{\mathrm{rig}}$ ($ij = 12, 13, 33$) is the rigid Earth contribution and the degree 2 Love number $k_2^{\mathrm{L}}(t)$ accounts for the effects of the loading-induced deformations.

We remark that $k_2^{\mathrm{L}}(t)$ has been introduced when dealing with the *indirect effect* of surface loading on the *gravitational potential* in Sect. 2.2.3. Indeed, the relationship between inertia and the gravitational potential is the premise to result (4.90), which has been obtained starting from the expression of the change in the geoid height due to surface loading ($\mathcal{G}^{\mathrm{sur}}(\gamma, t)$), and exploiting the linear relationship between inertia and the degree $n = 2$ harmonics of the gravitational potential. Since the expression of $\mathcal{G}^{\mathrm{sur}}(\gamma, t)$ is ultimately based on "Love's principle" described by Eq. (2.25), Eq. (4.90) is itself a direct consequence of that same principle.

In GIA studies that appeared in the literature since the 1980s, Eq. (4.90) has been adopted extensively (see e.g., Sabadini and Peltier 1981; Peltier 1982; Wu and Peltier 1984; Ricard et al. 1992, 1993; Spada et al. 1992a, 2011a; Vermeersen et al. 1996b), although its relationship with "Love's principle" has not always been acknowledged. For example, in the treatise of Sabadini et al. (2016), the Laplace transform of Eq. (4.90) is given without any further comment (see their Eq. 3.66).

By analytical methods, Sabadini and Peltier (1981) and later Peltier (1982) have shown that (4.90) effectively holds for the viscoelastic Kelvin sphere considered in Sect. C.6. Assuming that the surface load induces a displacement field of spherical components (u_r, u_θ, u_λ), they have used directly the definition of inertia tensor and have integrated radially the required functionals, assuming incompressibility. We note that a similar approach was also followed, for an elastic Earth with a fluid core, by Smylie and Mansinha (1971a), in order to determine the inertia variations due to internal dislocations in the context of the problem of the seismic excitation of the Chandler wobble. In the same framework, Spada (1995) has pointed that a correction to Eq. (4.90) is needed to account for the special conditions at the solid-fluid core-mantle boundary (CMB). However, such correction, anticipated by Smylie and Mansinha (1971a), has never been estimated numerically and it is possibly unimportant.

Last, we note that Wu and Peltier (1984) have assumed that Eq. (4.90) holds for arbitrary linearly viscoelastic and radially stratified Earth (1-D) models, leaving a simple proof of its validity to an application of Green's identity but without giving any account of it.

4.3 Liouville Equations for GIA

For the purpose of modelling rotational effects of GIA on sea-level change, it is not necessary to consider the problem of the length-of-day variations (see e.g., Milne and Mitrovica 1998). Hence, in the following, we only deal with the Liouville equations for polar motion, which constitute a set of two coupled linear ODEs.

In previous sections, we have written the linearised Liouville equations for polar motion in the complex form:

$$\mathrm{i}\,\frac{\dot{\mathbf{m}}}{\sigma_{\mathrm{r}}} + \mathbf{m} = \boldsymbol{\Psi}\,, \tag{4.91}$$

and we have split the total matter excitation function into two components:

$$\boldsymbol{\Psi}(t) = \boldsymbol{\Psi}^{\mathrm{rot}} + \boldsymbol{\Psi}^{\mathrm{sur}}\,, \tag{4.92}$$

where $\boldsymbol{\Psi}^{\mathrm{rot}}$ and $\boldsymbol{\Psi}^{\mathrm{sur}}$ represent the contributions due to the surface loading and to the rotationally-induced deformations, respectively. We note that the linearised Liouville equation can be equivalently written as

$$\dot{\mathbf{m}} = \mathrm{i}\,\sigma_{\mathrm{r}}(\mathbf{m} - \boldsymbol{\Psi})\,, \tag{4.93}$$

showing that the first term—the velocity of wandering of the pole of rotation (see Terpos 1973)—prevents the condition $\mathbf{m} \equiv \boldsymbol{\Psi}$ to be met. This can be also stated saying that in the equilibrium position $\dot{\mathbf{m}} = 0$, the rotation axis is aligned with the total excitation $\boldsymbol{\Psi}$. Hence, the location of the instantaneous pole of rotation does not coincide, in general, with the location of the *excitation pole* (the point where the direction of the total excitation function pierces the surface of the Earth). In physical terms, this means that during polar motion in response to some geophysical excitation—either due to rotational deformation or to surface loads, or both—the Earth's rotation axis does not necessarily maintain itself perpendicular to the plane containing the equatorial bulge, i.e., the axis of rotation is not coincident with the figure axis (see Ricard et al. 1993).

For both EFs, an explicit form has been found in Sect. 4.2.3. However, it is convenient to remind that the EF for rotational deformation is

$$\boldsymbol{\Psi}^{\mathrm{rot}}(t) = \frac{k_2^{\mathrm{T}}(t)}{k^{\mathrm{s}}} * \mathbf{m}(t), \tag{4.94}$$

where $k_2^{\mathrm{T}}(t)$ is the harmonic degree 2 'k' TLN (see Sect. 2.3.4.2) and k^{s} is the secular Love number. Using Eq. (4.94) into (4.91) we obtain the Liouville equations for polar motion in the new form

$$\mathrm{i}\frac{\dot{\mathbf{m}}}{\sigma_{\mathrm{r}}} + \left(\delta(t) - \frac{k_2^{\mathrm{T}}(t)}{k^{\mathrm{s}}}\right) * \mathbf{m} = \boldsymbol{\Psi}^{\mathrm{sur}}\,, \tag{4.95}$$

where we note that, for the sifting property of Dirac's delta, we have $\mathbf{m}(t) = \delta(t) * \mathbf{m}(t)$.

We observe that the first term in Eq. (4.95) physically represents the rate of change of polar motion normalised by the Eulerian frequency σ_r. In GIA studies, this term can be neglected since the late-Pleistocene ice sheets are evolving on time scales of centuries or millennia, or even longer, largely exceeding the Eulerian period $2\pi/\sigma_r \approx 10$ months. This does not imply that the pole is not moving, but only that in GIA studies we can neglect the high-frequency component of the polar wandering velocity, just keeping the component responsible for the slow drift of the axis of rotation (the *secular polar motion* of the pole, see Dickman 1977). Hence, by making this approximation, from (4.95) we obtain a form of the Liouville equations which is suitable in the framework of GIA:

$$\left(\delta(t) - \frac{k_2^{\mathrm{T}}(t)}{k^{\mathrm{s}}}\right) * \mathbf{m} = \mathbf{\Psi}^{\mathrm{sur}}, \tag{4.96}$$

and that shall be adopted henceforth.

Dealing with the Liouville equations, it is often convenient to work in the Laplace domain, since using the normal modes method the Love numbers are directly computed as a function of complex frequency s (see, in particular, Sect. 2.3.4). In this context, this is also convenient since in this way the polar motion equation is transformed into an algebraic equation (e.g., Tschoegl 2012). Thus, taking the fundamental property of the time convolution into account (see Table B.1), the LT of the Liouville equation (4.96) is

$$\left(1 - \frac{\tilde{k}_2^{\mathrm{T}}(s)}{k^{\mathrm{s}}}\right) \tilde{\mathbf{m}}(s) = \tilde{\mathbf{\Psi}}^{\mathrm{sur}}(s), \tag{4.97}$$

where $\tilde{k}_2^{\mathrm{T}}(s)$, $\tilde{\mathbf{m}}(s)$ and $\tilde{\mathbf{\Psi}}^{\mathrm{sur}}(s)$ are the LTs of $k_2^{\mathrm{T}}(t)$, $\mathbf{m}(t)$, and $\mathbf{\Psi}^{\mathrm{sur}}(t)$, respectively.

4.3.1 Earth Rotation Theories

As first pointed out by Mitrovica et al. (2005) and later re-discussed in depth by Peltier (2015) and Mitrovica and Wahr (2011), in GIA modelling it is possible to approach the solution of Liouville equations (4.97) into two distinct ways. This is related with the interpretation of the secular Love number k^{s}, which we have estimated using observed values of the relevant geodetic quantities (see Eqs. 4.68 and 4.69).

A problem arises with the use of k^{s} in GIA applications. Indeed, in GIA, it is customary to adopt Earth models characterised by an elastic, unbroken lithosphere of finite strength, something that is generally considered appropriate in view of the relatively short time scale of GIA (thousands of years), compared to the time scale of mantle convection (million of years). However, for Earth models that include such a feature, the long-term equilibrium rotational shape may well differ from the actual observed shape described by k^s, due to the elastic stress "frozen" in the elastic lithosphere, even assuming that the Earth has been spinning for a very long time.

According to the possible interpretations of k^s, two distinct theories have been proposed, which are briefly summarised below. As discussed in detail in Mitrovica and Wahr (2011), the two theories predict responses that differ qualitatively and quantitatively, with the revised one implying a less significant impact of GIA long-wavelength (harmonic degree $n = 2$) geodetic signatures compared to the traditional theory, and a reduced rate of polar drift. It is convenient to remark that in both theories the "fast" Chandler wobble is filtered out since the onset, in agreement with Eq. (4.96).

4.3.1.1 Traditional Rotation Theory

Many previous studies have been based on the so-called *traditional rotation theory*, illustrated by Peltier (1982), Wu and Peltier (1984), Peltier (2015) and also adopted in the benchmark study of Spada et al. (2011a). In this framework, ultimately based upon the concept of "equivalent Earth model" of Munk and MacDonald (1975), it is assumed that

$$k^s \equiv k_2^{T,f} = \lim_{s \to 0} \tilde{k}_2^T(s)\,, \tag{4.98}$$

i.e., that the secular Love number coincides with the fluid 'k' TLN $k_2^{T,f}$ characterizing the specific Earth model employed (see Eq. 2.78), possibly including an elastic lithosphere having a long-term finite strength. According to Mitrovica et al. (2005) and Mitrovica and Wahr (2011), in conjunction with a coarse description of the Earth's density profile that is sometimes adopted in GIA modelling (e.g., Spada et al. 1992b, 2011a) using Eq. (4.98) could lead to significant inaccuracies in the prediction of the present-day rate of polar motion.

4.3.1.2 New (or Revised) Rotation Theory

The *new (or revised) rotation theory* proposed by Mitrovica et al. (2005) and Mitrovica and Wahr (2011), suggests that *(i)* the k^s values employed in GIA modelling should not depend upon the Earth's elasticity, and that *(ii)* the density profile of the Earth model should closely match the seismically determined profile (e.g., the PREM profile, see Dziewonski and Anderson 1981) in order to accurately describe the Earth's rotational response. Furthermore, *(iii)* the k^s value should also account for the Earth's excess ellipticity. Accordingly, Mitrovica et al. (2005) and Mitrovica and Wahr (2011) have proposed to model the secular Love number as

$$k^s \equiv k_2^{T,f}(LT = 0) + k^{ee}\,, \tag{4.99}$$

where[9] $k_2^{\mathrm{T,f}}(LT = 0)$ is the 'k' TLN of an effectively fluid Earth, characterised by a viscous lithosphere and a PREM density profile, and k^{ee} is a small correction associated to the excess ellipticity of the Earth. Following Nakiboglu (1982) and Mitrovica and Wahr (2011), we adopt the numerical value $k^{\mathrm{ee}} = 0.0082$.

4.3.2 Polar Motion Transfer Function

Regardless the rotational theory adopted – either traditional or revised—the Liouville equations in the Laplace domain (4.97) can be formally solved (or *inverted*) to obtain

$$\tilde{\mathbf{m}}(s) = \tilde{\mathcal{A}}(s)\,\tilde{\mathbf{\Psi}}^{\mathrm{sur}}(s)\,, \tag{4.100}$$

where using a terminology akin to that of Munk and MacDonald (1975), the function

$$\tilde{\mathcal{A}}(s) = \left(1 - \frac{\tilde{k}_2^{\mathrm{T}}(s)}{k^{\mathrm{s}}}\right)^{-1} \tag{4.101}$$

shall be referred to as *polar motion transfer function* (PMTF). We note that, in the time domain, Eq. (4.100) would simply correspond to the time convolution

$$\mathbf{m}(t) = \mathcal{A}(t) * \mathbf{\Psi}^{\mathrm{sur}}(t)\,. \tag{4.102}$$

Equation (4.100) can be used to provide a physical interpretation for the PMTF. Indeed, assuming an impulsive surface loading EF $\mathbf{\Psi}^{\mathrm{sur}}(t) \sim \delta(t)$ and recalling that the Laplace transform of a Dirac delta is 1 (see Table B.1), from (4.102) we would obtain $\mathbf{m}(t) \sim \mathcal{A}(t)$, showing that the PMTF physically represents the polar motion driven by an impulsive (δ-like) surface loading EF. Due to the nature of (4.102), we also observe that obtaining $\mathbf{\Psi}^{\mathrm{sur}}(t)$ from the knowledge of the PMTF $\mathcal{A}(t)$ and some astrogeodetic observation of $\mathbf{m}(t)$ constitutes a classical deconvolution problem (see e.g., Lambeck 1980).

The structure of the PMTF is studied in detail in Sect. D.2, where we show that it can be reduced to the general form

$$\tilde{\mathcal{A}}(s) = A^{\mathrm{e}} + \frac{A^{\mathrm{s}}}{s} + \sum_{i=1}^{M'} \frac{A_i}{s - a_i}\,, \tag{4.103}$$

where the value of M' depends upon the Earth's rotation theory adopted (either traditional or revised, as discussed below), and constants A^{e}, A^{s}, A_i and a_i, which

[9] Note that here LT stands for "lithospheric thickness" and should not be confused with "Laplace transform".

depend upon the rheological layering of the Earth, are known within the GIA community as *elastic rotational residue*, *secular rotational residue*, *rotational residues*, and *rotational roots*, respectively (e.g., Spada et al. 2011a). As detailed in Appendix D.2, constants A^{e}, A^{s}, and A_i are numerically found to be real-valued (i.e., their imaginary parts are negligible in front of their real parts).

According to Eq. (4.103), using Table B.1, in the time domain the PMTF takes the following form

$$\mathcal{A}(t) = A^{\mathrm{e}}\,\delta(t) + A^{\mathrm{s}}\,H(t) + H(t)\sum_{i=1}^{M'} A_i\,\mathrm{e}^{a_i t}\,, \tag{4.104}$$

and according to all the numerical experiments performed so far in the framework of incompressible models (see e.g., Spada et al. 2011a) all the a_i's are real-valued and negative, which ensures the asymptotic stability of the PMTF regardless the rotation theory adopted (Spada and Melini 2019).

Furthermore, in Appendix D.2 it is also shown that when the *traditional rotation theory* is employed, one has

$$M' = M - 1 \quad \text{and} \quad A^{\mathrm{s}} \neq 0\,, \tag{4.105}$$

where M is the number of VNMs (Spada et al. 2011a), while adopting the *new (or revised) rotation theory* one obtains

$$M' = M \quad \text{and} \quad A^{\mathrm{s}} = 0\,, \tag{4.106}$$

in agreement with the analysis of Mitrovica and Wahr (2011) and Peltier (2015).

It is worth to remark that the different value of the secular rotational residue A^{s} in the two theories has implications in the long-term behavior of the PMTF, and consequently, of polar motion. In fact, from Eq. (4.100), it is clear that for a stepwise surface excitation (i.e., $\tilde{\boldsymbol{\Psi}}^{\mathrm{sur}}(s) \sim {}^{1}\!/\!_{s}$) that mimics an abrupt deglaciation episode, the long-term polar motion shall increase linearly with time for $A^{\mathrm{s}} \neq 0$ (traditional theory), and it shall decay exponentially for $A^{\mathrm{s}} = 0$ (revised theory). A physical interpretation of this distinct behaviour has been detailed by Mitrovica and Wahr (2011) and discussed by Peltier (2015). We should be aware, however, that we have assumed infinitesimal excursions of the pole of rotation, hence both theories are doomed to become inaccurate when employed to describe polar motion on the time-scales of plate tectonics, since large excursions of the rotation axis are expected (Ricard et al. 1992, 1993; Spada et al. 1992a, 1993; Steinberger and O'Connell 1997; Sabadini et al. 2016).

4.3.3 Solution of Liouville Equations

In previous sections, we have gathered all the elements needed to obtain a computationally convenient form of the Liouville equations for polar motion, suitable for applications in the framework of GIA. Here they are used to describe the response of the rotation axis to a arbitrary surface mass distribution that evolves in time in a stepwise manner.

Using Eq. (4.81) into (4.102), we obtain the polar motion displacement in the form of a double convolution product

$$\mathbf{m}(t) = \mathcal{A}(t) * \left[\delta(t) + k_2^{\mathrm{L}}(t)\right] * \mathbf{\Psi}^{\mathrm{rig}}(t)\,, \tag{4.107}$$

where $\mathcal{A}(t)$ is the PMTF and $\mathbf{\Psi}^{\mathrm{rig}}(t)$ is the surface loading EF. To avoid the cumbersome computations implied by the convolutions, we take the LT of (4.107) exploiting property *8)* of Table B.1, so that we obtain

$$\tilde{\mathbf{m}}(s) = \tilde{\mathcal{A}}(s)\left[1 + \tilde{k}_2^{\mathrm{L}}(s)\right]\tilde{\mathbf{\Psi}}^{\mathrm{rig}}(s)\,, \tag{4.108}$$

where $\tilde{\mathcal{A}}(s)$ and $\tilde{\mathbf{\Psi}}^{\mathrm{rig}}(s)$ are the LTs of $\mathcal{A}(t)$ and $\mathbf{\Psi}^{\mathrm{rig}}(t)$, respectively. In an equivalent way, we can write

$$\tilde{\mathbf{m}}(s) = \tilde{\mathcal{A}}'(s)\,\tilde{\mathbf{\Psi}}^{\mathrm{rig}}(s)\,, \tag{4.109}$$

where following Spada et al. (2011a) we have defined the *modified form of the PMTF* as

$$\tilde{\mathcal{A}}'(s) = \left[1 + \tilde{k}_2^{\mathrm{L}}(s)\right]\tilde{\mathbf{\Psi}}^{\mathrm{rig}}(s)\,. \tag{4.110}$$

The lengthy analytical computations presented in Sect. D.2.3 indicate that, in close analogy with the PMTF $\tilde{\mathcal{A}}(s)$, for both rotation theories also $\tilde{\mathcal{A}}'(s)$ can be cast in the spectral form

$$\tilde{\mathcal{A}}'(s) = A'^{\mathrm{e}} + \frac{A'^{\mathrm{s}}}{s} + \sum_{i=1}^{M'} \frac{A_i'}{s - a_i}\,, \tag{4.111}$$

where the primed residues A'^{e}, A'^{s}, and A_i' are appropriate combinations of the constants that appear in the PMTF and in elastic and viscous parts of $\tilde{k}_2^{\mathrm{L}}(s)$, and the value of M' depends upon the rotation theory adopted.

The LT of the surface loading EF, required in (4.108), can be obtained recalling from Sect. 4.2.5 that, in the time domain, we have

$$\mathbf{\Psi}^{\mathrm{rig}}(t) = c_{21}^{\psi}\,\mathcal{L}_{21}^{*}(t)\,, \tag{4.112}$$

with constant c_{21}^{ψ} given by (4.83). We note that Eq. (4.112) holds for a general surface load EF. However, in view of the numerical solution of the SLE, it is now convenient to assume a PC-evolving surface load. This is not implying a loss of generality, since any geophysically plausible time history can be approximated by a stepwise function, provided that a suitable number of steps N is adopted (of course, this may come at a cost, in numerical applications). According to Eq. (E.8) and to those that follow, the harmonic coefficients of a PC load function are PC as well, so that the harmonic coefficient of degree $n = 2$ and order $m = 1$ is

$$\mathcal{L}_{21}(t) = \sum_{k=0}^{N} \Delta\mathcal{L}_{21,k}\, H(t - t_k)\,, \tag{4.113}$$

where the increments are defined as $\Delta\mathcal{L}_{21,k} = \mathcal{L}_{21,k+1} - \mathcal{L}_{21,k}$ ($k = 0, 1, \ldots, N$). Thus, with

$$\boldsymbol{\Psi}_k^{\mathrm{rig}} \equiv c_{21}^{\psi} \Delta\mathcal{L}_{21,k}^{*}\,, \tag{4.114}$$

using (4.113) into (4.112) gives

$$\boldsymbol{\Psi}^{\mathrm{rig}}(t) = \sum_{k=0}^{N} \boldsymbol{\Psi}_k^{\mathrm{rig}}\, H(t - t_k)\,, \tag{4.115}$$

which by property *5)* of Table B.1, is Laplace-transformed into

$$\tilde{\boldsymbol{\Psi}}^{\mathrm{rig}}(s) = \sum_{k=0}^{N} \boldsymbol{\Psi}_k^{\mathrm{rig}}\, \frac{\mathrm{e}^{-st_k}}{s}\,. \tag{4.116}$$

Hence, substituting Eqs. (4.116) and (4.111) into (4.109) yields the LT of the polar motion displacement in the form

$$\tilde{\mathbf{m}}(s) = \sum_{k=0}^{N} \boldsymbol{\Psi}_k^{\mathrm{rig}} \left(A'^{\mathrm{e}} + \frac{A'^{\mathrm{s}}}{s} + \sum_{i=1}^{M'} \frac{A'_i}{s - a_i} \right) \frac{\mathrm{e}^{-st_k}}{s}\,, \tag{4.117}$$

which can be easily inverted back into the time domain using again property *(5)* of Table B.1 and obtaining

$$\mathbf{m}(t) = \sum_{k=0}^{N} \boldsymbol{\Psi}_k^{\mathrm{rig}} \left(A'^{\mathrm{e}} + A'^{\mathrm{s}}(t - t_k) + \sum_{i=1}^{M'} \frac{A'_i}{a_i} \left(\mathrm{e}^{a_i(t - t_k)} - 1 \right) \right) H(t - t_k)\,, \tag{4.118}$$

which represents the sought explicit solution of the Liouville equations for polar motion, in the special case of a surface load excitation that evolves in a PC manner. We note that, taking the time derivative of Eq. (4.118) and ignoring δ-like terms arising from the time-derivative of the unit step function $H(t)$, the *rate of polar motion* is:

$$\dot{\mathbf{m}}(t) = \sum_{k=0}^{N} \boldsymbol{\Psi}_k^{\text{rig}} \left(A'^s + \sum_{i=1}^{M'} A'_i e^{a_i(t-t_k)} \right) H(t-t_k) \,. \qquad (4.119)$$

4.4 Rotation Response Functions

Solving the Liouville equations, in previous section we have evaluated the displacement of the Earth's pole of rotation $\mathbf{m}(t)$ in response to geophysical excitations. Polar wandering and its rate of variation, described by Eqs. (4.118) and (4.119) respectively, induce a change in the centrifugal potential that, in turn, deforms the Earth vertically and horizontally and is responsible for changes in the gravitational potential and sea level. In this section, these effects are expressed in terms of rotation response functions (RRFs), which shall be computed with the aid of appropriate rotational GFs. There are analogies with the theory framework adopted for the SRFs in Sect. 3.2.2, but also some important differences.

The first RRF accounts for variations of the geoid height, caused by the direct and the indirect effects of the change of centrifugal potential. This SRF shall be referred to as the *geoid* RRF, and denoted by $\mathcal{G}^{\text{rot}}(\gamma, t)$. Its CSH coefficients $\mathcal{G}_{nm}^{\text{rot}}$ shall be constructed using an appropriate GF, by means of the $k_n^{\text{T}}(t)$ TLN.

Three other RRFs are associated with the three components of the displacement field in spherical coordinates and shall be referred to as *displacement RRFs* ($\mathcal{U}^{\text{rot}}$, $\mathcal{V}_\theta^{\text{rot}}$ and $\mathcal{V}_\lambda^{\text{rot}}$). Invoking the Mie representation for solenoidal displacement fields (Sect. B.2), these RRFs shall be expanded in series of CSH ($\mathcal{U}^{\text{rot}}$) and of their surface gradients ($\mathcal{V}_\theta^{\text{rot}}$ and $\mathcal{V}_\lambda^{\text{rot}}$) utilising the scalar coefficient $\mathcal{U}_{nm}^{\text{rot}}$ for the vertical component and one single coefficient $\mathcal{V}_{nm}^{\text{rot}}$ for the two horizontals. The RRFs shall be expressed in terms of TLNs $h_n^{\text{T}}(t)$ and $\ell_n^{\text{T}}(t)$ for vertical and horizontal components, respectively. A fourth RRF shall also be considered, the *sea-level* RRF, defined as the difference between the RRFs for vertical displacement and that for the geoid: $\mathcal{R}^{\text{rot}}(\gamma, t) = \mathcal{G}^{\text{rot}} - \mathcal{U}^{\text{rot}}$, with harmonic coefficients $\mathcal{R}_{nm}^{\text{rot}} = \mathcal{G}_{nm}^{\text{rot}} - \mathcal{U}_{nm}^{\text{rot}}$. The sea-level RRF constitutes an essential ingredient of the SLE, along with the sea-level SRF considered in previous chapters.

As we did previously for the SRFs, our purpose is to determine explicit expressions for the three RRFs, providing in particular the CSH coefficients $\mathcal{G}_{nm}^{\text{rot}}$, $\mathcal{U}_{nm}^{\text{rot}}$, $\mathcal{V}_{nm}^{\text{rot}}$ and $\mathcal{R}_{nm}^{\text{rot}}$, necessary for the numerical implementation of the SLE. Our approach is sketched in Fig. 4.4.

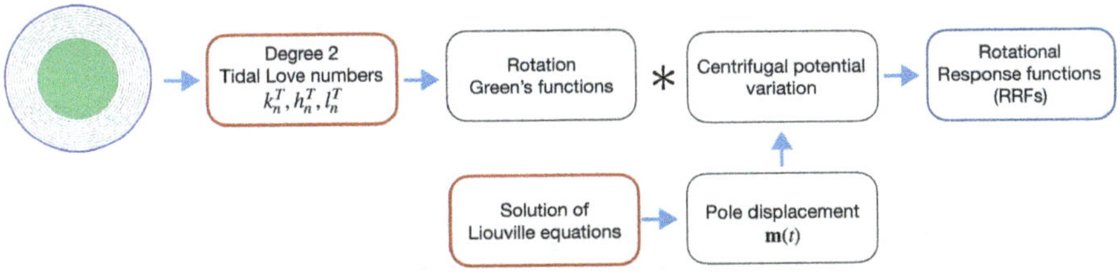

Fig. 4.4 Steps necessary for the computation of the spectral coefficients of the RRFs, starting from the 1-D Earth model sketched on the left. The asterisk denotes time convolution

4.4.1 *Rotational Green's Functions*

The harmonic components of the four RRFs considered here (i.e., $\mathcal{G}^{\rm rot}$, $\mathcal{U}^{\rm rot}$, $\mathcal{V}^{\rm rot}$ and $\mathcal{R}^{\rm rot}$) have all the general form

$$\mathcal{RRF}_{nm}(t) = \mathrm{GF}_n * \Lambda_{nm} \,, \tag{4.120}$$

where here $\mathrm{GF}_n(t)$ denotes the corresponding *rotation Green's function*, $\Lambda_{nm}(t)$ is degree n and order m harmonic component of the change of centrifugal potential associated with variations in the Earth's angular velocity (see Sect. D.1), and $*$ denotes the 1-D temporal convolution, i.e.,

$$(\mathrm{GF}_n * \Lambda_{nm})\,(t) \equiv \int_{-\infty}^{+\infty} \mathrm{GF}_n(t-t')\,\Lambda_{nm}(t')\,\mathrm{d}t', \tag{4.121}$$

as opposed to the 3-D spatio-temporal convolution employed in the construction of the SRFs (see Eq. 3.2). Indeed, since we are dealing with a perturbing potential that does not load the Earth's surface, a spatial integration will not be necessary. For a further clarification about this important difference, the readers are referred to Milne and Mitrovica (1998). It is also important to remark that, according to Eq. (4.120), the RRFs spectrum could possess, in principle, all the harmonic degrees and orders. However, recalling the main result of Sect. D.1, we know that $\Lambda_{nm}(t)$ is essentially a harmonic degree $n = 2$ and order $m = 1$ function, so that all the RRFs discussed in this section shall have this same characteristic.

Based upon Eq. (4.120), the harmonic coefficients of the three fundamental RRFs can be expressed by

$$\left\{\begin{matrix} \mathcal{G}_{nm} \\ \mathcal{U}_{nm} \\ \mathcal{V}_{nm} \end{matrix}\right\}^{\rm rot}(t) \equiv \left\{\begin{matrix} \Upsilon_n^g \\ \Upsilon_n^u \\ \Upsilon_n^v \end{matrix}\right\}(t) * \Lambda_{nm}(t)\,, \tag{4.122}$$

whereas the sealevel RRF is a combination of the geoid and vertical displacement RRFs, with

$$\mathcal{R}_{nm}^{\text{rot}}(t) = \mathcal{G}_{nm}^{\text{rot}} - \mathcal{U}_{nm}^{\text{rot}} = \Upsilon_n^s(t) * \Lambda_{nm}(t)\,, \tag{4.123}$$

where we have used symbols Υ_n to denote the corresponding GFs, which following Milne and Mitrovica (1998) are defined by:

$$\begin{Bmatrix} \Upsilon_n^g \\ \Upsilon_n^u \\ \Upsilon_n^v \end{Bmatrix}(t) \equiv \frac{1}{g_0}\begin{Bmatrix} \delta + k_n^{\text{T}} \\ h_n^{\text{T}} \\ \ell_n^{\text{T}} \end{Bmatrix}(t)\,, \tag{4.124}$$

and

$$\Upsilon_n^s(t) \equiv \Upsilon_n^g - \Upsilon_n^u\,, \tag{4.125}$$

where g_0 is the reference surface gravity acceleration, and $(k_n^{\text{T}}, h_n^{\text{T}}, \ell_n^{\text{T}})$ are the TLNs of harmonic degree n.

We note that the impulsive term $\delta(t)$ in Eq. (4.124) represents the direct effect of the change in centrifugal potential upon the gravitational potential $\mathcal{G}^{\text{rot}}$, while $k_n^{\text{T}}(t)$ describes the indirect effect induced by rotational deformation. Thus, $\mathcal{G}^{\text{rot}}$ would not vanish in the case of a rigid Earth (i.e., for $k_n^{\text{T}} = 0$). We also note that, contrary to the SRFs (see Eq. 3.7), the rotation GFs do not imply a sum over the harmonic degrees; for this reason, their n-dependence remains explicit (Milne and Mitrovica 1998). Furthermore, the GFs associated to the rotational response do not depend upon the coordinates of the "observer" $\gamma = (\theta, \lambda)$, as it occurs for the SRFs. We also mention that the GFs Υ have dimensions of $[\text{TL}^{-1}]$; hence, all the RRFs have dimensions of [L], consistently with the SRFs.

As a final remark, it is worth to mention that using property *(8)* of Table B.1 and taking (4.124) into account, the LT of Eq. (4.122) is

$$\begin{Bmatrix} \tilde{\mathcal{G}}_{nm} \\ \tilde{\mathcal{U}}_{nm} \\ \tilde{\mathcal{V}}_{nm} \end{Bmatrix}^{\text{rot}}(s) \equiv \frac{1}{g_0}\begin{Bmatrix} 1 + \tilde{k}_n^{\text{T}} \\ \tilde{h}_n^{\text{T}} \\ \tilde{\ell}_n^{\text{T}} \end{Bmatrix}(s)\,\tilde{\Lambda}_{nm}(s)\,, \tag{4.126}$$

where $\tilde{\Lambda}_{nm}(s)$ is the LT of $\Lambda_{nm}(t)$. Furthermore, from (4.123), for the sea-level RRF we have

$$\tilde{\mathcal{R}}_{nm}^{\text{rot}}(s) = \frac{1}{g_0}\left(1 + \tilde{k}_n^{\text{T}}(s) - \tilde{h}_n^{\text{T}}(s)\right)\tilde{\Lambda}_{nm}(s)\,. \tag{4.127}$$

4.4.2 Geoid, Displacement and Sealevel RRFs

Obtaining an explicit form for the RRFs would demand the evaluation of the convolution integral in Eq. (4.121). However, we shall find it more convenient to work with the Laplace-transformed equations (4.126). This can be done in a few steps, taking advantage of some previous results. Here we consider in detail the RRF for the geoid and give summary formulae for the other RRFs.

4.4.2.1 Geoid RRF

According to the first of Eqs. (4.126), the geoid RRF in the Laplace domain reads

$$\tilde{\mathcal{G}}^{\mathrm{rot}}_{nm}(s) = \frac{1}{g_0}\left(1 + \tilde{k}^{\mathrm{T}}_n(s)\right)\tilde{\Lambda}_{nm}(s), \tag{4.128}$$

but from Eq. (D.39) we note that $\tilde{\Lambda}_{nm}(s)$ is a degree $n = 2$ and order $m = 1$ harmonic function with

$$\tilde{\Lambda}_{nm}(s) = \delta_{n2}\,\delta_{m1}\,c^{\lambda}_{nm}\,\tilde{\mathbf{m}}^*(s), \tag{4.129}$$

so that

$$\tilde{\mathcal{G}}^{\mathrm{rot}}_{nm}(s) = \frac{c^{\lambda}_{nm}\,\delta_{n2}\,\delta_{m1}}{g_0}\left(1 + \tilde{k}^{\mathrm{T}}_n(s)\right)\tilde{\mathbf{m}}^*(s), \tag{4.130}$$

which, using the complex conjugate of the solution of the Liouville equations for polar motion (4.117), taking (4.114) into account, and taking advantage of the fact that A'^{e}, A'^{s}, A'_i and a_i are real-valued constants, becomes

$$\tilde{\mathcal{G}}^{\mathrm{rot}}_{21}(s) = \frac{c^{\lambda}_{21}c^{\psi}_{21}}{g_0}\sum_{k=0}^{N}\Delta\mathcal{L}_{21,k}\left(1 + \tilde{k}^{\mathrm{T}}_2(s)\right)\left(A'^{\mathrm{e}} + \frac{A'^{\mathrm{s}}}{s} + \sum_{i=1}^{M'}\frac{A'_i}{s - a_i}\right)\frac{\mathrm{e}^{-st_k}}{s}, \tag{4.131}$$

We now note that the product of the two factors in parentheses above is analogous to the product $(1 + \tilde{k}^{\mathrm{L}}_2(s))\tilde{\mathcal{A}}(s)$ that defines the modified form of the PMTF in Sect. D.2.3. Thus, without the need of further computations, we arrive at

$$\tilde{\mathcal{G}}^{\mathrm{rot}}_{21}(s) = \frac{c^{\psi}_{21}c^{\lambda}_{21}}{g_0}\sum_{k=0}^{N}\Delta\mathcal{L}_{21,k}\left(g'^{\mathrm{e}} + \frac{g'^{\mathrm{s}}}{s} + \sum_{i=1}^{M'}\frac{g'_i}{s - a_i} + \sum_{i=1}^{M}\frac{g''_i}{s - s_i}\right)\frac{\mathrm{e}^{-st_k}}{s}, \tag{4.132}$$

where, according to Eqs. (D.87–D.90) and (D.84–D.85), we have introduced the new constants

$$\begin{aligned}
g'^{\mathrm{e}} &= \left(1 + k^{\mathrm{T,e}}_2\right)A'^{\mathrm{e}},\\
g'^{\mathrm{s}} &= \left(1 + k^{\mathrm{T,f}}_2\right)A'^{\mathrm{s}},\\
g'_i &= \left(1 + k^{\mathrm{T,e}}_2 + \sum_{i'=1}^{M}\frac{k^{\mathrm{T}}_{2i'}}{a_i - s_{i'}}\right)A'_i \equiv \left(1 + k^{\mathrm{T}}_2(a_i)\right)A'_i, \quad (i = 1, \ldots, M')\\
g''_i &= \left(A'^{\mathrm{e}} + \frac{A'^{\mathrm{s}}}{s_i} + \sum_{i'=1}^{M'}\frac{A'_{i'}}{s_i - a_{i'}}\right)k^{\mathrm{T}}_{2i} \equiv \mathcal{A}'(s_i)\,k^{\mathrm{T}}_{2i}, \quad (i = 1, \ldots, M).
\end{aligned}$$

Using Table B.1, the inverse LT of Eq. (4.132) is now readily found, giving the time-domain CSH coefficient of the geoid RRF:

$$\mathcal{G}_{21}^{\rm rot}(t) = \frac{c_{21}^{\psi} c_{21}^{\lambda}}{g_0} \sum_{k=0}^{N} \Delta\mathcal{L}_{21,k}\, \gamma^{g}(t - t_k)\, H(t - t_k), \tag{4.133}$$

where all the other coefficients are identically zero, except $\mathcal{G}_{2-1}^{\rm rot} = -(\mathcal{G}_{21}^{\rm rot})^*$, and where we have defined the function

$$\gamma^{g}(t) \equiv g'^{\rm e} + g'^{\rm s} t + \sum_{i=1}^{M'} \frac{g_i'}{a_i} \left(e^{a_i t} - 1\right) + \sum_{i=1}^{M} \frac{g_i''}{s_i} \left(e^{s_i t} - 1\right). \tag{4.134}$$

In view of their own definition, the explicit expressions for the other RRFs follow from the one found for the geoid RRF in a straightforward way, employing the appropriate combinations of Love numbers. The relevant formulae are listed below.

4.4.2.2 Vertical Displacement RRF

For the vertical displacement RRF we have

$$\mathcal{U}_{21}^{\rm rot}(t) = \frac{c_{21}^{\psi} c_{21}^{\lambda}}{g_0} \sum_{k=0}^{N} \Delta\mathcal{L}_{21,k}\, \gamma^{u}(t - t_k)\, H(t - t_k), \tag{4.135}$$

where

$$\gamma^{u}(t) \equiv u'^{\rm e} + u'^{\rm s} t + \sum_{i=1}^{M'} \frac{u_i'}{a_i} \left(e^{a_i t} - 1\right) + \sum_{i=1}^{M} \frac{u_i''}{s_i} \left(e^{s_i t} - 1\right), \tag{4.136}$$

with

$$\begin{aligned}
u'^{\rm e} &= h_2^{\rm T,e} A'^{e}, \\
u'^{\rm s} &= h_2^{\rm T,f} A'^{s}, \\
u_i' &= \left(h_2^{\rm T,e} + \sum_{i'=1}^{M} \frac{h_{2i'}^{\rm T}}{a_i - s_{i'}} \right) A_i' \equiv h_2^{\rm T}(a_i)\, A_i', \quad (i = 1, \ldots, M') \\
u_i'' &= \left(A'^{e} + \frac{A'^{s}}{s_i} + \sum_{i'=1}^{M'} \frac{A_{i'}'}{s_i - a_{i'}} \right) h_{2i}^{\rm T} \equiv \mathcal{A}'(s_i)\, h_{2i}^{\rm T}, \quad (i = 1, \ldots, M).
\end{aligned}$$

4.4.2.3 Horizontal Displacement RRF

For the horizontal displacement RRF we have

$$\mathcal{V}_{21}^{\text{rot}}(t) = \frac{c_{21}^{\psi} c_{21}^{\lambda}}{g_0} \sum_{k=0}^{N} \Delta \mathcal{L}_{21,k}\, \gamma^{v}(t - t_k)\, H(t - t_k)\,, \tag{4.137}$$

where

$$\gamma^{v}(t) \equiv v'^{\text{e}} + v'^{\text{s}} t + \sum_{i=1}^{M'} \frac{v_i'}{a_i} \left(\text{e}^{a_i t} - 1\right) + \sum_{i=1}^{M} \frac{v_i''}{s_i} \left(\text{e}^{s_i t} - 1\right)\,, \tag{4.138}$$

with

$$\begin{aligned}
v'^{\text{e}} &= \ell_2^{Te}\, A'^{e}, \\
v'^{\text{s}} &= \ell_2^{Tf}\, A'^{s}, \\
v_i' &= \left(\ell_2^{Te} + \sum_{i'=1}^{M} \frac{\ell_{2i'}^{\text{T}}}{a_i - s_{i'}}\right) A_i' \equiv \ell_2^{\text{T}}(a_i)\, A_i'\,, \quad (i = 1, \ldots, M') \\
v_i'' &= \left(A'^{e} + \frac{A'^{s}}{s_i} + \sum_{i'=1}^{M'} \frac{A_{i'}'}{s_i - a_{i'}}\right) \ell_{2i}^{\text{T}} \equiv \mathcal{A}'(s_i)\, \ell_{2i}^{\text{T}}, \quad (i = 1, \ldots, M)\,.
\end{aligned}$$

4.4.2.4 Sea Level RRF

Recalling that, by definition,

$$\mathcal{R}^{\text{rot}}(\gamma, t) = \mathcal{G}^{\text{rot}} - \mathcal{U}^{\text{rot}}\,, \tag{4.139}$$

from Eqs. (4.133) and (4.135) we conclude that

$$\mathcal{R}_{21}^{\text{rot}}(t) = \frac{c_{21}^{\psi} c_{21}^{\lambda}}{g_0} \sum_{k=0}^{N} \Delta \mathcal{L}_{21,k}\, \gamma^{s}(t - t_k) H(t - t_k), \tag{4.140}$$

where we have set

$$\gamma^{s}(t) \equiv \gamma^{g} - \gamma^{u}\,, \tag{4.141}$$

with functions γ^{g} and γ^{u} given by Eqs. (4.134) and (4.136), respectively.

References

Barnes RTH, Hide R, White AA, Wilson CA (1983) Atmospheric angular momentum fluctuations, length-of-day changes and polar motion. Proc R Soc Lond. Ser A. Math Phys Sci 387:31–73
Bills BG, James TS (1996) Late quaternary variations in relative sea level due to glacial cycle polar wander. Geophys Res Lett 23(21):3023–3026
Burša M (1992) Secular Loave numbers of the major planets. Earth Moon Planets 59(3):239–244
Carter M, Carter W (2000) Seth Carlo Chandler Jr.: the discovery of variation of latitude. Int Astron Union Colloq 178:109–119
Chen W, Shen W, Dong X (2010) Atmospheric excitation of polar motion. Geo-spatial Inf Sci 13(2):130–136
Dickman S (1977) Secular trend of the Earth's rotation pole: consideration of motion of the latitude observatories. Geophys J R Astron Soc 51(1):229–244
Dziewonski AM, Anderson DL (1981) Preliminary reference Earth model. Phys Earth Planet Inter 25(4):297–356
Fang M, Hager BH, Liao X, Zhou Y, Xu X (2022) Revisit the theory of Earth rotation-anatomy of the Liouville equation. Geophys J Int 229(3):2175–2191
Farrell W, Clark J (1976) On postglacial sea level. Geophys J R Astron Soc 46(3):647–667
Goldstein H (1950) Classical mechanics. Addison-Wesley
Gross RS (1992) Correspondence between theory and observations of polar motion. Geophys J Int 109(1):162–170
Gross RS (2000) The excitation of the Chandler wobble. Geophys Res Lett 27(15):2329–2332
Guinot B (1972) The Chandlerian Wobble from 1900 and 1970. Astron Astrophys 19:207–214
Han D, Wahr J (1989) Post-glacial rebound analysis for a rotating Earth. In: Cohen S, Vanicek P (eds) Slow deformations and transmission of stress in the Earth, Geophysical monograph series
Hofmann-Wellenhof B, Moritz H (2006) Physical geodesy. Springer
Lambeck K (1980) The Earth's variable rotation: geophysical causes and consequences. Cambridge University Press
Lambeck K, Burša M (1973) Temporal variations of rotational origin in the absolute value of gravity. Studia Geophysica et Geodaetica 17(4):269–271
Lefftz M (1991) Aspects théoriques de la rotation de la Terre et de son noyau: influence de la viscosité. Univesite de Strasbourg, Strasbourg, France, PhD Thése
Milne G (1998) Refining models of the glacial isostatic adjustment process, Canadian theses. Thesis (Ph.D.)–University of Toronto
Milne G, Mitrovica J (1996) Postglacial sea-level change on a rotating Earth: first results from a gravitationally self-consistent sea-level equation. Geophys J Int 126(3):F13–F20
Milne GA, Mitrovica JX (1998) Postglacial sea-level change on a rotating Earth. Geophys J Int 133(1):1–19
Mitrovica JX, Wahr J (2011) Ice age Earth rotation. Annu Rev Earth Planet Sci 39:577–616
Mitrovica JX, Wahr J, Matsuyama I, Paulson A (2005) The rotational stability of an ice-age earth. Geophys J Int 161(2):491–506
Morrison L (1973) Rotation of the Earth from AD 1663–1972 and the Constancy of G. Nature 241:519–520
Munk WH, MacDonald GJ (1975) The rotation of the Earth: a geophysical discussion. Cambridge University Press
Nakiboglu S (1982) Hydrostatic theory of the Earth and its mechanical implications. Phys Earth Planet Inter 28(4):302–311
Okubo S (1982) Is the Chandler period variable? Geophys J Int 71(3):629–646
Peltier W (2015) The history of the Earth's rotation: impacts of deep Earth physics and surface climate variability. In: Treatise on Geophysics: Second Edition, vol 9, pp 221–279
Peltier WR (1982) Dynamics of the ice age Earth. In: Advances in geophysics, vol 24, pp. 1–146. Elsevier

Peltier WR (1974) The impulse response of a Maxwell Earth. Rev Geophys Space Phys 12(4):649–669

Ricard Y, Sabadini R, Spada G (1992) Isostatic deformations and polar wander induced by redistribution of mass within the Earth. J Geophys Res Solid Earth 97(B10):14223–14236

Ricard Y, Spada G, Sabadini R (1993) Polar wandering of a dynamic Earth. Geophys J Int 113(2):284–298

Rochester MG (1973) The Earth's rotation. EOS, Trans Am Geophys Union 54(8):769–780

Rochester M, Smylie D (1974) On changes in the trace of the Earth's inertia tensor. J Geophys Res 79(32):4948–4951

Routh EJ (1905) The advanced Part of A treatise on the dynamics of a system of rigid bodies: being Part II of a treatise on the whole subject. With numerous examples. MacMillan and Company

Sabadini R, Peltier W (1981) Pleistocene deglaciation and the Earth's rotation: implications for mantle viscosity. Geophys J Int 66(3):553–578

Sabadini R, Yuen DA, Boschi E (1982) Polar wandering and the forced responses of a rotating, multilayered, viscoelastic planet. J Geophys Res Solid Earth 87(B4):2885–2903

Sabadini R, Vermeersen B, Cambiotti G (2016) Global dynamics of the Earth. Springer

Seitz F, Schuh H (2010) Earth rotation. In: Sciences of geodesy-I, pp 185–227. Springer

Sidorenkov NS (2009) The interaction between Earth's rotation and geophysical processes. Wiley

Smylie D, Mansinha L (1971) The elasticity theory of dislocations in real Earth models and changes in the rotation of the Earth. Geophys J Int 23(3):329–354

Spada G (1992) Rebound Post-glaciale e Dinamica Rotazionale di un Pianeta Viscoelastico Stratificato. PhD thesis, University of Bologna, Bologna, Italy

Spada G (1995) Changes in the Earth inertia tensor: the role of boundary conditions at the core-mantle interface. Geophys Res Lett 22(24):3557–3560

Spada G (2017) Glacial isostatic adjustment and contemporary sea level rise: an overview. Surv Geophys 38(1):1–33

Spada G, Ricard Y, Sabadini R (1992a) Excitation of true polar wander by subduction. Nature 360(6403):452–454

Spada G, Sabadini R, Yuen DA, Ricard Y (1992b) Effects on post-glacial rebound from the hard rheology in the transition zone. Geophys J Int 109(3):683–700

Spada G, Sabadini R, Ricard Y (1993) On a particular solution of the non-linear Liouville equations. Geophys J Int 114:399–404

Spada G, Barletta VR, Klemann V, Riva R, Martinec Z, Gasperini P, Lund B, Wolf D, Vermeersen L, King M (2011) A benchmark study for glacial isostatic adjustment codes. Geophys J Int 185(1):106–132

Spada G, Melini D (2019) SELEN^4 (SELEN version 4.0): a Fortran program for solving the gravitationally and topographically self-consistent sea-level equation in glacial isostatic adjustment modeling. Geosci Model Dev 12(12):5055–5075

Steinberger B, O'Connell RJ (1997) Changes of the Earth's rotation axis owing to advection of mantle density heterogeneities. Nature 387(6629):169–173

Terpos A (1973) A study of polar motion. Master's thesis, University of Thessaloniki (NASA 15066)

Tschoegl NW (2012) The phenomenological theory of linear viscoelastic behavior: an introduction. Springer

Vermeersen LLA, Sabadini R, Spada G (1996) Compressible rotational deformation. Geophys J Int 126:735–761

Whitehouse PL (2018) Glacial isostatic adjustment modelling: historical perspectives, recent advances, and future directions. Earth Surface Dyn 6(2):401–429

Wu P, Peltier W (1982) Viscous gravitational relaxation. Geophys J Int 70(2):435–485

Wu P, Peltier W (1984) Pleistocene deglaciation and the Earth's rotation: a new analysis. Geophys J Int 76(3):753–791

Zharkov VN, Molodensky SM, Brzeziński A, Groten E, Varga P (1996) The Earth and its rotation. Low frequency geodynamics. Wichmann, Heidelberg

Chapter 5
The Sea Level Equation

Circumferentia aquae et circumferentia terrae inequaliter distanti.

(Dante Alighieri, Quaestio de aqua et terra)

5.1 Introduction

The Sea Level Equation (SLE) is the equation that governs the physics of GIA. The SLE is consistent with the laws of continuum mechanics, with the mass conservation principle and with Newton's law of universal gravitation. For this reason, the SLE is referred to as a *gravitationally self-consistent* equation. Contrary to many other important equations in physics, the SLE is not a partial differential equation, rather it is a *three-dimensional non-linear integral equation* that demands a special approach and particular methods of solution.

In this chapter, we first describe the essentials of the theory behind the SLE from a qualitative standpoint (Sect. 5.2). Then, in Sect. 5.3, we consider the various forms that the SLE takes, from the most basic to the final general *pseudo-spectral* form. Due to its important role in the literature, the Farrell and Clark (1976) original version of the SLE shall be discussed in detail. Attention shall be also paid to the implicit nature of the SLE, which suggests an iterative approach to its solution, and its non-linearity shall be discussed. In Sect. 5.4, special solutions of the SLE that can be studied by means of analytical methods shall also be discussed, including the *eustatic* solution due to Suess (1906), which has an important role in understanding global sea-level variations, and the Woodward's (1888) solution, which neatly illustrates the importance of gravitational attraction in the SLE. The approach to the numerical solution of the SLE is not of concern here, and will be presented in Appendix E.

G. Spada and D. Melini, *Glacial Isostatic Adjustment*, Springer Geophysics,
https://doi.org/10.1007/978-3-031-90538-4_5

5.2 What is the Sea Level Equation?

To understand the physical nature of the SLE, it is convenient to define a *reference state* in which the solid Earth, the surface loads (i.e., the ice sheets) and the oceans are in equilibrium until a certain time t_0 (Fig. 5.1a), and to compare it to a *perturbed state* at time $t > t_0$ (Fig. 5.1b). Essentially, this was the approach originally proposed by Farrell and Clark (1976, FC76 hereinafter) in their seminal work where the SLE was introduced first. Although the reference configuration can be chosen arbitrarily, it may be convenient to identify it with the Last Glacial Maximum (LGM, $\sim 21,000$ years ago). In the reference state, at point $\gamma = (\theta, \lambda)$ we denote by $L_0(\gamma)$ the load acting on the Earth's surface and by $I_0(\gamma)$ the ice thickness. By definition, in this state, the ocean surface is a level surface of the total gravity potential of the Earth.

The purpose of the SLE is to describe how sea level and other variables shall change when the configuration of the system evolves in the *perturbed state*, where the surface load and the ice thickness are $L(\gamma, t)$ and $I(\gamma, t)$, respectively. Despite the global variations observed in the new state, there are two fundamental constraints: *i)* the mass of the (ice+oceans) system must be conserved, and *ii)* the new sea surface must remain an equipotential for the total gravity field. In the perturbed state, the

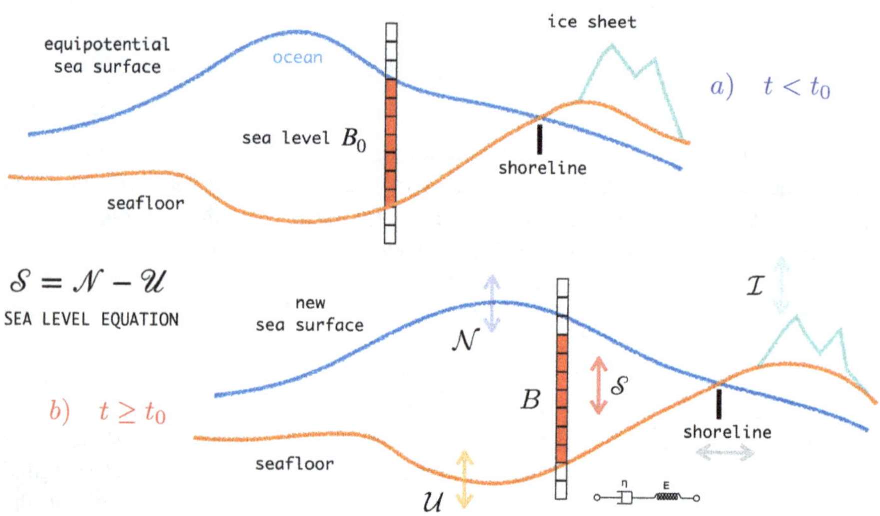

Fig. 5.1 Sketches of **a** the reference equilibrium state at time $t < t_0$ and **b** of the general configuration for $t \geq t_0$, also showing the three components of the Earth's system that are interacting in the SLE: the solid Earth, the oceans and the ice sheets. Changes in sea level $\mathcal{S}$ relative to the solid Earth are observed by the stick meter (or *tide gauge*). With $\mathcal{N}$ and $\mathcal{U}$ we denote the sea surface variation and the vertical displacement of the solid Earth, respectively. The sea surface is equipotential in (**a**) but also in (**b**), after the ice sheets have lost (or gained) mass and the oceans mass is varied to compensate exactly that loss (gain). In (**b**), the arrows indicate that the sea surface and the solid Earth have moved relative to the geocentre, and that the shorelines have migrated laterally relative to (**a**)

Earth shall be deformed and its gravitational field shall be modified, too, along with its rotational state. These changes depend on the surface mass variation but also on the rheology of the Earth's mantle, illustrated by a Maxwell spring-dashpot system in Fig. 5.1b. Absolute sea level and the pattern of the world's shorelines shall be altered, too. Ultimately, the SLE aims at describing these complex changes occurring in the *Earth's system* in consequence to the surface mass redistribution.

The physical interactions responsible for the changes observed in the perturbed state are qualitatively sketched in the diagram of Fig. 5.2, inspired to the work of Clark et al. (1978). Since the interactions shown here are operating simultaneously and at all spatial wavelengths, their individual contributions cannot be disentangled, which makes the interpretation of the GIA effects particularly challenging. The top triangle shows the three basic elements of the SLE, i.e., the ice sheets, the solid Earth, and the oceans (Conrad 2013). As indicated by the arrows, these elements are interacting by two mechanisms: *i)* surface loading and *ii)* mutual gravitational attraction. The waxing and waning ice sheets exert a load at the surface of the solid Earth (*glacial or ice loading*, related to *glacio-isostasy*), but the mass variation of the oceans is also loading the Earth, modifying the equilibrium on the seafloor (*melt-water loading*, associated to *hydro-isostasy*). These two spatially non-uniform loads are tightly interconnected, since the mass conservation of the (ice+oceans) system imposes that the

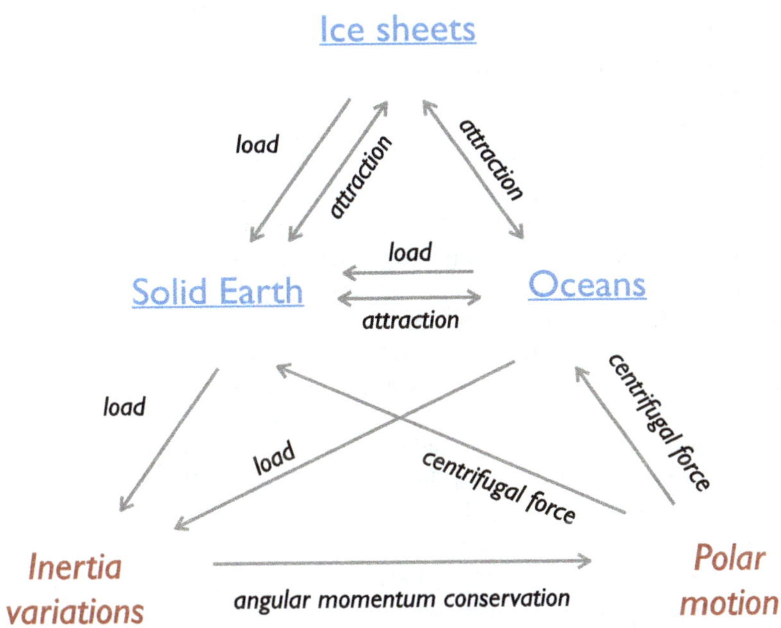

Fig. 5.2 In the top part, the triangle shows the elements that are perpetually interacting in the SLE (the solid Earth, the ice sheets and the oceans) through surface loading and mutual gravitational attraction. The bottom part qualitatively shows how Earth rotational effects are coming into play. Inspired by diagram "G" that originally was published in Fig. 1 of Clark et al. (1978)

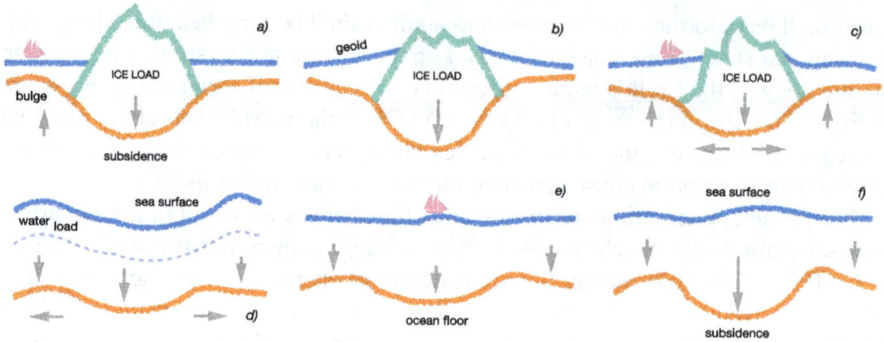

Fig. 5.3 Physical interactions among ice loads, the melt-water load, glacial history, and non-rigid solid Earth: **a** The weight of the ice deforms the Earth and **b** the ice mass attracts the water. **c** The transfer of mass distorts the geoid. **d** Similarly, the weight of melt-water deforms the seabed and **e** more water flows into this depression, increasing the melt-water load and **f** causing further subsidence of the ocean floor. These processes are interrelated, as the diagram of Fig. 5.2 suggests. The figure is inspired by the work of Clark et al. (1978)

average load variation vanishes across the whole Earth's surface. Due to imperfections in the mantle elasticity, the past loads induce delayed effects that are manifest as a global state of isostatic disequilibrium until present. Furthermore, the equi-potential surfaces of the Earth's gravity field are twisted by the mass redistributed over the Earth surface and in the oceans, changing the geoid shape.

The three elements that enter into the SLE are all affected by gravitational attraction. In particular, the sea surface is warped by the attraction of the continental ice sheets, but at the same time the geoid variations caused by the solid Earth deformation concur to modify the shape of the oceans surface. The bottom part of Fig. 5.3 considers further physical interactions, driven by the Earth's irregular rotation. Inertia perturbations, associated to long wavelength deformations and sea-level variations can drive excursions of the rotation axis and variations of the rate of rotation in order to conserve the Earth's angular momentum (see e.g., Lambeck 1980). The consequent variation of the centrifugal potential alters, in turn, both the solid Earth and the sea surface (*rotational feedback* on sea level, see Peltier 1999).

The *inextricably related interactions* first acknowledged by Clark et al. (1978) and illustrated in Fig. 5.3 are responsible for the regional imprints of GIA. As first noted by Woodward (1888) and later discussed by Daly (1925), Walcott (1972) and Farrell and Clark (1976), the sea-level variations associated with glacial isostasy depart significantly from the spatially uniform pattern that we would observe for a rigid, non-gravitating and non-rotating Earth (i.e., ignoring the interactions). Often, in the geological literature the spatially uniform sea-level change is referred to as

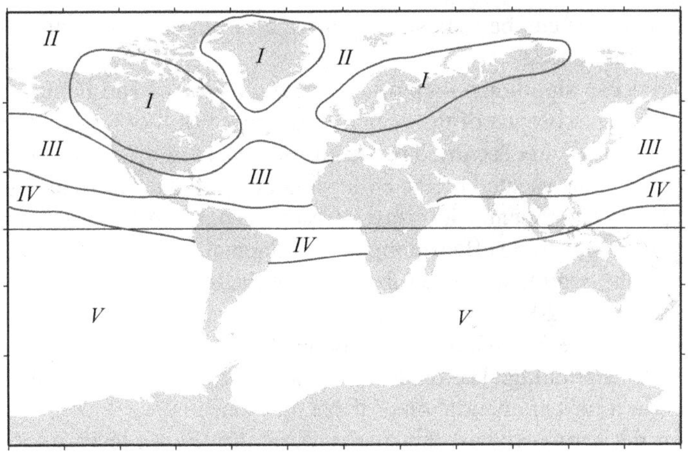

Fig. 5.4 Sketch of the extents of RSL zones of post-glacial sea-level change from *I* to *V*, redrawn and adapted from Clark et al. (1978). *"Within each zone, the sea-level signatures at all locations are similar to one another"*. This figure shows the layout of zones for a realistic deglaciation history in the northern hemisphere, assuming no contributions from Antarctica. Zone *VI*, associated with continental shorelines, is not explicitly marked on the map

eustatic, a word attributed to Suess (1906); eustatic[1] variations only depend on the history of the grounded ice volume (Milne and Mitrovica 2008; Borreggine et al. 2022).

The interactions are responsible for a global pattern of relative sea level (RSL) variations during the melting of the late-Pleistocene ice sheets, which Clark et al. (1978) have characterised by defining six *RSL zones*, labelled from *I* to *VI* and displayed in Fig. 5.4. The RSL zones encompass the glaciated areas (zone *I*), the region of the collapsing fore-bulge (*II*), the time-dependent emergence (*III*) and the oceanic submergence region (*IV*), the oceanic emergence region (*V*), and the continental shorelines (*VI*). The extent and time evolution of the RSL zones depend, in general, by the chronology of the ice sheets, by the rheological layering of the Earth, and by the presence or not of the rotational feedback.

The various RSL zones reflect the physical interactions embedded into the SLE. A major effort to establish the origin of the RSL patterns has been made by Mitrovica and Milne (2002), who have studied the nature of the RSL zones in connection with the various terms of the SLE, describing the physical mechanisms responsible for their establishment and unveiling, in particular, the enigmatic far-field processes of *equatorial ocean siphoning* and *continental levering*. Note that Mitrovica and Milne (2002) have used a variant of model ICE-3G of Tushingham and Peltier (1991), without including rotational effects in their simulations and assuming, for simplicity, fixed shorelines. It is reasonable that the identified RSL patterns and the associated

[1] The term *eustatic* has received many qualifications, so far. Today, the term *barystatic* seems to be preferred (Gregory et al. 2019).

physical mechanisms can be only slightly affected from these assumptions, but the details may be influenced.

Equatorial ocean siphoning—a term coined by Mitrovica and Peltier (1991)—is responsible for the existence of the oceanic emergence (fall of RSL) region in zone *V* of Fig. 5.4. The process is caused by the displacement of meltwater from far-field into near-field regions vacated by the collapsing lateral bulges in zone *II* (Mitrovica and Milne 2002). Continental levering, already recognised by Walcott (1972), is caused by the ocean loading effect along the continental margins (zone *VI*), inducing a "levering" of the continents and a subsidence of the offshore regions. As shown by Mitrovica and Milne (2002) by analysing the SLE, this subsidence acts to reinforce the sea-level fall in the far-field regions, producing effects comparable to those of the equatorial ocean siphoning. The readers are also referred to the work of Borreggine et al. (2022) for a further elucidation of these mechanisms.

Following the above studies, the spatial variability in sea level associated with GIA has been widely investigated with the aid of the SLE, aiming at reconstructing the history of deglaciation since the LGM (see e.g., Clark et al. 2002; Kendall et al. 2008; Mitrovica et al. 2011). On a more limited spatial scale, the concept of RSL zone has also been useful to interpret the Holocene sea-level variations across the Mediterranean Sea (Stocchi and Spada 2007; Mauz et al. 2015). During last decades, the SLE has been the subject of a number of further investigations, aimed *i)* at disclosing its physical meaning and complexity, *ii)* at extending its validity for a 3-D Earth and *iii)* at discussing the relation between rotational variations and sea-level change. Interested readers are referred to the works of Johnston (1993); Milne and Mitrovica (1996, 1998); Milne et al. (1999); Wolf et al. (2002); Mitrovica and Milne (2002); Lambeck et al. (2003); Mitrovica and Milne (2003); Wu (2004); Kendall et al. (2005); Spada and Stocchi (2006, 2007); Melini et al. (2010); Dal Forno et al. (2012); Broerse et al. (2014); Spada (2017); Martinec et al. (2018); Hartmann et al. (2020); Coulson et al. (2021) but this list is certainly not exhaustive. These studies have also emphasised the physical reasons of the spatial variability in sea level predicted by the SLE. Furthermore, they have given insight into the method of solution, on which we shall focus in detail in Appendix E.

5.3 Theory of the Sea Level Equation

The Sea Level Equation (SLE) manifests the relation between sea-level change, the sea surface variation, and the vertical displacement of the solid surface of the Earth. In the following, also for pedagogical purposes, various but equivalent forms of the SLE are introduced and discussed, characterised by an increasing level of complexity. They range from the most basic form to a few intermediate forms, and lastly to the most complete pseudo-spectral form that is currently employed in numerical applications. The FC76 form of the SLE is of particular historical importance and it is considered as a special case of the complete pseudo-spectral form.

5.3.1 Basic Form of the SLE

Sea level can be defined as the offset between the sea surface and the surface of the solid Earth. Based upon this concept, in order to establish a basic form of the SLE, we compare the two states of the system (ice+oceans+solid Earth) depicted in Fig. 5.1. In the reference state, sea level is

$$B_0(\gamma) = r_0^{ss} - r_0^{se}, \tag{5.1}$$

where $r_0^{ss}(\gamma)$ and $r_0^{se}(\gamma)$ are the initial values of the radii of the equipotential sea surface (ss) and of the solid Earth (se) in a geocentric reference frame, respectively. As shown in Fig. 5.1a, B_0 would be directly measured by a stick-meter (or a *tide gauge*) placed at a point of coordinates $\gamma = (\theta, \lambda)$. In response to the melting (or accretion) of the continental ice sheets, to consequent variations of the volume of the oceans, and to rotational variations, at time $t \geq t_0$ the system evolves into the perturbed state of Fig. 5.1b. The rheology of the solid Earth, illustrated by the Maxwell spring-dashpot mechanical analogue, has a fundamental role in establishing the new state. The new sea level is

$$B(\gamma, t) = r^{ss} - r^{se}, \tag{5.2}$$

where $r^{ss}(\gamma, t)$ and $r^{se}(\gamma, t)$ are the perturbed values of the radii of the sea surface and of the solid Earth, respectively. At γ, we define *sea-level change* as the difference

$$\mathcal{S}(\gamma, t) = B(\gamma, t) - B_0(\gamma), \tag{5.3}$$

which is equivalent, by the same definition of B and B_0 given above, to the double difference

$$\mathcal{S}(\gamma, t) = \left(r^{ss} - r_0^{ss}\right) - \left(r^{se} - r_0^{se}\right). \tag{5.4}$$

Introducing the *sea surface variation* (or *absolute sea-level change*) by

$$\mathcal{N}(\gamma, t) = r^{ss} - r_0^{ss}, \tag{5.5}$$

and the *vertical displacement* of the solid surface of the Earth

$$\mathcal{U}(\gamma, t) = r^{se} - r_0^{se}, \tag{5.6}$$

from Eq. (5.4), we obtain the SLE in its most basic form

$$\mathcal{S}(\gamma, t) = \mathcal{N} - \mathcal{U}, \tag{5.7}$$

expressing the height variation of the water column bounded by the sea surface and the ocean floor. It is apparent that $\mathcal{S}$ could only be determined once $\mathcal{N}$ and $\mathcal{U}$ are simultaneously known.

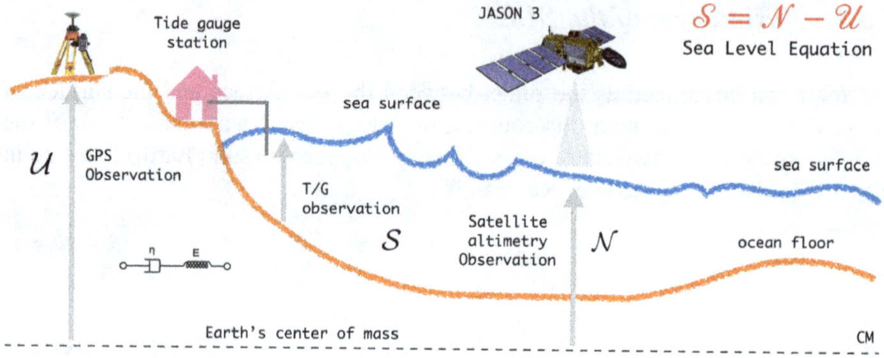

Fig. 5.5 Elements that define the SLE, i.e., equation $\mathcal{S} = \mathcal{N} - \mathcal{U}$, and the geodetic observations that make variables $\mathcal{S}$, $\mathcal{N}$, and $\mathcal{U}$ accessible. This figure, drawn by GS, is inspired by the works of Wöppelmann and Marcos (2016) and Pfeffer and Allemand (2016). See also Pfeffer et al. (2017)

In the way it has been constructed, illustrated in Fig. 5.1, the SLE (5.7) could be also referred to as the *tide gauge equation*. Indeed, the SLE indicates that an observation of sea-level change performed at a tide gauge ($\mathcal{S}$) is equivalent to the difference between two observations made at the same location, performed by satellite altimetry (absolute sea-level change $\mathcal{N}$) and by geodetic techniques such as GNSS (vertical displacement of the Earth's solid surface $\mathcal{U}$), respectively. This basic principle is discussed in Pfeffer and Allemand (2016) and Wöppelmann and Marcos (2016), and illustrated in Fig. 5.5.

5.3.2 *A Suite of Other Forms of the SLE*

5.3.2.1 First Form of the SLE

As shown by FC76, the sea surface variation $\mathcal{N}(\gamma, t)$ is tightly associated to the variation of the geoid height, but it is not *the* variation of the geoid height. Indeed, while for a rigid Earth model there is no distinction between changes in geoid radius and changes in sea level, these two quantities become different if the Earth is deformable. A further problem arises from the fact that, in the new state illustrated in Fig. 5.1b, the volume of the oceans is varied relative to the reference state in Fig. 5.1a, due to changes in the ice thickness.

Taking the factors above into account, FC76 have shown[2] that the sea surface variation can be expressed as

$$\mathcal{N}(\gamma, t) = \mathcal{G} + c, \tag{5.8}$$

where in agreement with Bruns formula (see e.g., Hofmann-Wellenhof and Moritz 2006) the variation of the geoid radius relative to the reference state is

$$\mathcal{G}(\gamma, t) = \frac{\Delta W}{g_0}, \tag{5.9}$$

where $\Delta W(\gamma, t)$ the variation of the total gravity potential of the Earth system either caused by surface loading or rotational effects or both, g_0 is the reference surface gravity acceleration and c is well known within the GIA community as the "Farrell and Clark's c constant".

As pointed out by Tamisiea (2011), some confusion arose recently about Eqs. (5.8) and (5.9), since sometimes the sea surface variation has been used as a synonymous of geoid height variation (i.e., it has been erroneously assumed that $\mathcal{N} = \mathcal{G}$ thus ignoring the c constant). From our discussion above, neglecting c would certainly imply a conceptual error. Beside this, the numerical value of the present-day rate of variation of the c-constant, $\dot{c}$, which can be estimated in the range between 0.2 and 0.3 mm yr^{-1} according to the GIA model adopted and to the approximations made, is indeed not negligible in modern geodetic applications (see Tamisiea 2011; Spada 2017).

Using Eqs. (5.8) and (5.9), the SLE (5.7) becomes

$$\mathcal{S}(\gamma, t) = \mathcal{G} - \mathcal{U} + c, \tag{5.10}$$

---

[2] The FC76 argument about the geoid and the sea surface, and the way they have introduced the c constant is somewhat burdensome but since it is important in the development of the SLE, it is worth to revisit it here. In the reference state, we define the *geoid* as the surface $r = r_0^{ss}(\gamma)$, upon which the gravity potential $W_0(\gamma)$ assumes a constant value. However, in the new state, the gravity potential $W(\gamma, t)$ shall not be, in general, constant over $r = r_0^{ss}(\gamma)$. Nevertheless, it is possible to shift this surface by a small amount $\varepsilon(\gamma, t)$ such that over the surface $r = r_0^{ss} + \varepsilon$ the gravity potential shall assume that same value that it had in the reference state, i.e., $W(r_0^{ss} + \varepsilon) = W_0(r_0^{ss})$. Hence $\varepsilon(\gamma, t)$ exactly represents the variation in the geoid radius. In fact, expanding $W(r_0^{ss} + \varepsilon)$ in Taylor series about $r = r_0^{ss}$ and approximating, to first order, the surface gravity as $g \approx -\frac{\partial W}{\partial r}|_{r=r_0^{ss}} = \text{constant}$, one easily obtains $\varepsilon(\gamma, t) = \mathcal{G}$ with $\mathcal{G}$ given by Eq. (5.9), in which $\Delta W(\gamma, t) = W(r_0^{ss}) - W_0(r_0^{ss})$. Although $r = r_0^{ss} + \varepsilon$ is equipotential, it does not necessarily represent the new sea surface, since changes in the oceans volume caused by ice melt or accretion have not been taken into account. At this point, FC76 argue that if $r = r_0^{ss} + \varepsilon$ is an equipotential, another one can be found by shifting this surface by a small amount $c \ll r_0^{ss}$ (indeed, $W(r_0^{ss} + \varepsilon + c) \approx W(r_0^{ss} + \epsilon) - cg = \text{constant}$). Thus, $r_0^{ss} + \varepsilon + c$ is the radius of another equipotential, where c is to be ultimately determined by the constraint of mass conservation. When this is done, the new sea surface is found at radius $r^{ss} = r_0^{ss} + \varepsilon + c$. Recalling Eq. (5.5), this is equivalent to Eq. (5.8) since above we have seen that $\varepsilon(\gamma, t) = \mathcal{G}(\gamma, t)$.

which represents the *first form* of the SLE, written in terms of the yet unknown constant c.

5.3.2.2 Second Form of the SLE

Starting from the first form, the SLE can be easily cast in a new symmetric form, which we shall refer to as *second form*. To see this, it is now convenient to compute the average of the SLE over the oceans, where we recall that the ocean-average of a scalar function $F(\gamma, t)$ is defined as

$$\langle F \rangle^{\mathrm{o}}(t) \equiv \frac{1}{A^{\mathrm{o}}} \int_{\mathrm{o}} F(\gamma, t)\, \mathrm{d}A, \tag{5.11}$$

where $A^{\mathrm{o}}(t) = \int_{\mathrm{o}} \mathrm{d}A$ is the area of the oceans, $\mathrm{d}A = a^2 \sin\theta \mathrm{d}\theta \mathrm{d}\lambda$ is the element of area, and a the average Earth's radius. Of course, the shape of the shorelines (hence the ocean function) is yet to be determined at this stage, since the SLE has not been actually solved.

By ocean-averaging both sides of the first form of the SLE (5.10), we obtain

$$\langle \mathcal{S} \rangle^{\mathrm{o}}(t) = \langle \mathcal{G} - \mathcal{U} \rangle^{\mathrm{o}} + \langle c \rangle^{\mathrm{o}}, \tag{5.12}$$

hence

$$c(t) = \langle \mathcal{S} \rangle^{\mathrm{o}} - \langle \mathcal{G} - \mathcal{U} \rangle^{\mathrm{o}}, \tag{5.13}$$

where have set $\langle c \rangle^{\mathrm{o}} = c(t)$ since c is a spatially invariant constant. By substituting $c(t)$ into the first form (5.10), we obtain the symmetric *second form* of the SLE:

$$\mathcal{S}(\gamma, t) = (\mathcal{G} - \mathcal{U}) + \mathcal{S}^{\mathrm{ave}} - \langle \mathcal{G} - \mathcal{U} \rangle^{\mathrm{o}}. \tag{5.14}$$

We note that the yet undetermined ocean average

$$\mathcal{S}^{\mathrm{ave}} \equiv \langle \mathcal{S} \rangle^{\mathrm{o}} \tag{5.15}$$

can be obtained by the mass conservation constraint given by (1.147), providing

$$\mathcal{S}^{\mathrm{ave}}(t) = \mathcal{S}^{\mathrm{equ}} + \mathcal{S}^{\mathrm{ofu}}, \tag{5.16}$$

where $\mathcal{S}^{\mathrm{equ}}$ and $\mathcal{S}^{\mathrm{ofu}}$ are two spatially invariant terms. The first, henceforth referred to as *equivalent sea-level change*, is

$$\mathcal{S}^{\mathrm{equ}}(t) \equiv -\frac{\mu}{\rho^{\mathrm{w}} A^{\mathrm{o}}}, \tag{5.17}$$

where μ is the mass variation of the grounded ice and A^{o} is the time evolving area of the oceans, whereas the second

$$\mathcal{S}^{\mathrm{ofu}}(t) \equiv \frac{A^{\mathrm{e}}}{A^{\mathrm{o}}}\langle T_0 O\rangle^{\mathrm{e}} \tag{5.18}$$

depends explicitly upon variations of the OF, either due to the horizontal migration of the shorelines or to transitions from grounded to floating ice (or *vice versa*). It is important to note that $\mathcal{S}^{\mathrm{equ}}(t)$ and $\mathcal{S}^{\mathrm{ofu}}(t)$ are dynamic variables, since $A^{\mathrm{o}}(t)$ and $O(t)$ shall depend upon the spatio-temporal variations of sea level (i.e., on $\mathcal{S}$ itself), and therefore also from the rheological properties of the Earth.

5.3.2.3 Third Form of the SLE

In Eq. (5.14), the change of geoid height $\mathcal{G}$ and the vertical displacement $\mathcal{U}$ contain contributions from surface loading and rotational deformation, which have been considered separately in Chaps. 3 and 4, respectively. By combining the two contributions linearly, we have

$$\mathcal{G}(\gamma, t) = \mathcal{G}^{\mathrm{sur}} + \mathcal{G}^{\mathrm{rot}}, \tag{5.19}$$

and

$$\mathcal{U}(\gamma, t) = \mathcal{U}^{\mathrm{sur}} + \mathcal{U}^{\mathrm{rot}}, \tag{5.20}$$

where $\mathcal{G}^{\mathrm{sur}}$ and $\mathcal{U}^{\mathrm{sur}}$ are given explicitly by Eqs. (3.11) and the harmonics of $\mathcal{G}^{\mathrm{rot}}$ and $\mathcal{U}^{\mathrm{rot}}$ are expressed by (4.122), respectively. Hence, introducing a *total sea-level response function*

$$\mathcal{R}(\gamma, t) = \mathcal{G} - \mathcal{U}, \tag{5.21}$$

we obtain

$$\mathcal{R}(\gamma, t) = \mathcal{R}^{\mathrm{sur}} + \mathcal{R}^{\mathrm{rot}}, \tag{5.22}$$

where the contribution of surface loading is

$$\mathcal{R}^{\mathrm{sur}}(\gamma, t) = \mathcal{G}^{\mathrm{sur}} - \mathcal{U}^{\mathrm{sur}}, \tag{5.23}$$

while the one arising from rotational effects is

$$\mathcal{R}^{\mathrm{rot}}(\gamma, t) = \mathcal{G}^{\mathrm{rot}} - \mathcal{U}^{\mathrm{rot}}. \tag{5.24}$$

The previous equations can be easily rearranged in terms of the $\mathcal{R}$ variables. In particular, the first form of the SLE (5.10) becomes

$$\mathcal{S}(\gamma, t) = \mathcal{R} + c, \tag{5.25}$$

with the c constant (see 5.13) given by

$$c(t) = \mathcal{S}^{\mathrm{ave}} - \langle \mathcal{R}^{\mathrm{sur}} \rangle^{\mathrm{o}} - \langle \mathcal{R}^{\mathrm{rot}} \rangle^{\mathrm{o}}\,; \tag{5.26}$$

while the symmetric second form (5.14) becomes:

$$\mathcal{S}(\gamma, t) = \mathcal{R} + \mathcal{S}^{\mathrm{ave}} - \langle \mathcal{R} \rangle^{\mathrm{o}}, \tag{5.27}$$

or, equivalently:

$$\mathcal{S}(\gamma, t) = (\mathcal{R}^{\mathrm{sur}} - \langle \mathcal{R}^{\mathrm{sur}} \rangle^{\mathrm{o}}) + \mathcal{S}^{\mathrm{ave}} + \left(\mathcal{R}^{\mathrm{rot}} - \langle \mathcal{R}^{\mathrm{rot}} \rangle^{\mathrm{o}}\right), \tag{5.28}$$

which represents the *third form* of the SLE.

5.3.3 Pseudospectral Form of the SLE

The third form of the SLE (5.28) contains the surface response function $\mathcal{R}^{\mathrm{sur}}$. From Sect. 3.2.3, we recall that $\mathcal{R}^{\mathrm{sur}}$ is obtained by a 3-D *spatio-temporal* convolution

$$\mathcal{R}^{\mathrm{sur}}(\gamma, t) = \Gamma^{s} \otimes \mathcal{L}, \tag{5.29}$$

where $\Gamma^{s}(\gamma, t)$ is the surface *sea-level Green's function* given by (3.15). Using the decomposition (3.32) of the surface load, we can similarly split the sea-level surface response function according to:

$$\mathcal{R}^{\mathrm{sur}}(\gamma, t) = \mathcal{R}^{\mathrm{i}} + \mathcal{R}^{\mathrm{s}} + \mathcal{R}^{\mathrm{o}}, \tag{5.30}$$

where superscripts *i*, *s* and *o* refer to *ice*, *sea level* and *ocean*, respectively, and the three terms are obtained by convolving Γ^{s} with the three components of the load function:

$$\mathcal{L}^{\mathrm{i}}(\gamma, t) = \rho^{\mathrm{i}}\ \mathcal{C}\mathcal{I} = \rho^{\mathrm{i}}\,\mathcal{W} \tag{5.31}$$

$$\mathcal{L}^{\mathrm{s}}(\gamma, t) = \rho^{\mathrm{w}}\ \mathcal{O}\mathcal{S} = \rho^{\mathrm{w}}\mathcal{Z} \tag{5.32}$$

$$\mathcal{L}^{\mathrm{o}}(\gamma, t) = \rho^{\mathrm{r}}\ \mathcal{Q}\mathcal{O} = \rho^{\mathrm{r}}\mathcal{X}, \tag{5.33}$$

respectively. For a piecewise constant surface load, the harmonic coefficients of the three components $\mathcal{R}^{\mathrm{i,s,o}}$ are given by Eqs. (3.36).

The SLE in the third form (5.28) also contains $\mathcal{R}^{\mathrm{rot}}$, the rotational response function for sea level; its harmonic coefficients are directly obtained by a 1-D *time* convolution

$$\mathcal{R}^{\mathrm{rot}}_{nm}(t) = \Upsilon^{s}_{n} * \Lambda_{nm}, \tag{5.34}$$

where $\Upsilon_n^s(t)$ is the *rotation sea-level Green's function* and $\Lambda_{nm}(t)$ are the coefficients of degree n and order m of the expansion of the *variation of centrifugal potential* $\Lambda(\gamma, t)$. Following Milne and Mitrovica (1998), in Sect. 4.4.2 we have shown that $\Lambda(\gamma, t)$ is essentially a spherical harmonic function of degree $n = 2$ and order $m = \pm 1$ (see also Martinec and Hagedoorn 2014).

Thus, using (5.30), the SLE (5.28) is rearranged as follows

$$\mathcal{S}(\gamma, t) = \left(\widehat{\mathcal{R}}^{\mathrm{i}} + \widehat{\mathcal{R}}^{\mathrm{s}} + \widehat{\mathcal{R}}^{\mathrm{o}}\right) + \mathcal{S}^{\mathrm{ave}} + \widehat{\mathcal{R}}^{\mathrm{rot}}, \tag{5.35}$$

where the three terms in parentheses originate from surface loading and the *modified response functions* $\widehat{\mathcal{R}}$ are defined as

$$\widehat{\mathcal{R}}^{x}(\gamma, t) = \mathcal{R}^{x} - \langle \mathcal{R}^{x} \rangle^{\mathrm{o}}, \quad \text{for} : x = \mathrm{i, s, o}, \tag{5.36}$$

and

$$\widehat{\mathcal{R}}^{\mathrm{rot}}(\gamma, t) = \mathcal{R}^{\mathrm{rot}} - \langle \mathcal{R}^{\mathrm{rot}} \rangle^{\mathrm{o}}, \tag{5.37}$$

respectively.

Now it is worth to note that $\widehat{\mathcal{R}}^{\mathrm{s}}$ depends upon variable $\mathcal{O}\mathcal{S}$ through the component $\mathcal{L}^{\mathrm{s}}$ of the surface load variation (see, in particular, Eq. 5.32). According to Eq. (4.140), this holds for $\widehat{\mathcal{R}}^{\mathrm{rot}}$ as well. Following Mitrovica and Peltier (1991), in view of the numerical solution of the SLE it is convenient to transform Eq. (5.35) in such a way that the new variable $\mathcal{O}\mathcal{S}$ becomes the unknown *in lieu* of $\mathcal{S}$. This is accomplished by "projecting" Eq. (5.35) on the OF (i.e., multiplying both sides by the ocean function $\mathcal{O}$), which provides the final pseudospectral form of the SLE:

$$\mathcal{Z}(\gamma, t) = \mathcal{K}^{\mathrm{i}} + \mathcal{K}^{\mathrm{s}}(\mathcal{Z}) + \mathcal{K}^{\mathrm{o}} + \mathcal{Z}^{\mathrm{ave}} + \mathcal{K}^{\mathrm{rot}}(\mathcal{Z}), \tag{5.38}$$

with the new unknown function:

$$\mathcal{Z}(\gamma, t) \equiv \mathcal{O}\,\mathcal{S}, \tag{5.39}$$

and

$$\mathcal{Z}^{\mathrm{ave}}(\gamma, t) \equiv \mathcal{O}\,\mathcal{S}^{\mathrm{ave}} = \mathcal{O}\left(\mathcal{S}^{\mathrm{equ}} + \mathcal{S}^{\mathrm{ofu}}\right), \tag{5.40}$$

$$\mathcal{K}^{\mathrm{i,s,o}}(\gamma, t) \equiv \mathcal{O}\,\widehat{\mathcal{R}}^{\mathrm{i,s,o}}, \tag{5.41}$$

$$\mathcal{K}^{\mathrm{rot}}(\gamma, t) \equiv \mathcal{O}\,\widehat{\mathcal{R}}^{\mathrm{rot}}. \tag{5.42}$$

The ocean-projected SLE (5.38) can be approached successfully by means of spectral methods, i.e., it can be expanded in series of spherical harmonics following a procedure that is detailed in Appendix E. However, since the solution of the SLE is performed with the aid of a suitable numerical integration on a spherical grid, the term *pseudo*-spectral approach is certainly more appropriate than *spectral* (Mitrovica and Peltier 1991; Mitrovica et al. 1994). We note that the adoption of a fully spectral approach would impose the explicit computation of Wigner 3-j symbols known from the theory of angular momentum in quantum mechanics, as done by Plag and Jüettner (2001).

5.3.4 The Farrell and Clark's SLE

The SLE was first introduced in the seminal paper of FC76, although a similar equation was previously discussed by Farrell (1973) in the context of tidal changes in sea level (see also Woodward 1888). The work of FC76 reports the SLE in three different forms, characterised by increasing complexity, in which the Earth is assumed to be rigid, elastic and viscoelastic, respectively. Since the latter ensures realistic results and meaningful comparisons with field data, the classical theory of FC76 has strongly influenced all the literature that followed. In fact, a number of numerical implementations of the SLE published in the last two decades have adopted the FC76 theory (see e.g., Spada and Stocchi 2007; Adhikari et al. 2016; Kachuck and Cathles 2018).

In their work, FC76 have made three fundamental assumptions that we have relaxed in the general SLE theory illustrated in previous sections. First, they have assumed that the effects of Earth rotation on sealevel can be neglected. Second, their theory assumes that a steep, cliff-like shoreline is running along the present-day boundary between the oceans and the continental masses, thus prohibiting the horizontal migration of the shorelines. Third, FC76 neglected the grounded marine-based ice and floating ice, thus assuming that all the ice masses are grounded above sea level. For an in-depth discussion of these approximations and of the errors implied, the readers are referred to the works of Mitrovica and Milne (2003) and Kendall et al. (2005), where details and inter-comparisons between the classic theory and its subsequent extensions can be also found.

Our purpose here is simply to recover the FC76 formulation as a particular case of the general equations introduced in previous sections. It is convenient to start from the symmetric second form of the SLE, which reads

$$\mathcal{S}(\gamma, t) = (\mathcal{G} - \mathcal{U}) + \mathcal{S}^{\text{ave}} - \langle \mathcal{G} - \mathcal{U} \rangle^{\circ}, \tag{5.43}$$

where according to (3.13) and (3.14), we have

$$(\mathcal{G} - \mathcal{U})\,(\gamma, t) \equiv \mathcal{R} = \Gamma^{s} \otimes \mathcal{L}, \tag{5.44}$$

with the sea-level GF given by

$$\Gamma^{s}(\gamma, t) = \Gamma^{g} - \Gamma^{u}, \tag{5.45}$$

where Γ^g and Γ^u are the GFs for the geoid height and for vertical displacement, respectively, and we have neglected the rotational contributions, in accord with the first of FC76 hypotheses. Since in FC76 it is also assumed that the ocean function is constant, with $O(\gamma, t) = O^{\mathrm{p}}(\gamma)$ where O^{p} is the present-day OF, according to Eq. (1.124) and to those that immediately follow, the surface load variation can be approximated by

$$\mathcal{L}(\gamma, t) = \rho^{\mathrm{i}}\, C^{\mathrm{p}}\, \mathcal{I} + \rho^{\mathrm{w}}\, O^{\mathrm{p}}\, \mathcal{S}, \tag{5.46}$$

where the continent function $C^{\mathrm{p}}(\gamma) = 1 - O^{\mathrm{p}}(\gamma)$ is a constant as well. By substituting (5.46) into (5.44) we obtain

$$(\mathcal{G} - \mathcal{U})\,(\gamma, t) = \rho^{\mathrm{i}}\, \Gamma^s \otimes_{\mathrm{i}} \mathcal{I} + \rho^{\mathrm{w}}\, \Gamma^s \otimes_{\mathrm{o}} \mathcal{S}, \tag{5.47}$$

where symbols $\otimes_{\mathrm{i}}$ and $\otimes_{\mathrm{o}}$ indicate a 3-D spatio-temporal convolution like the one defined by Eq. (3.2) but limited to the region covered by ice and by the oceans, respectively. Thus, by inserting (5.47) into (5.43), we obtain the SLE in the FC76 form[3]

$$\begin{aligned} \mathcal{S}(\gamma, t) = &+\rho^{\mathrm{i}}\, \Gamma^s \otimes_{\mathrm{i}} \mathcal{I} + \rho^{\mathrm{w}}\, \Gamma^s \otimes_{\mathrm{o}} \mathcal{S} + \\ &+\mathcal{S}^{\mathrm{eus}} + \\ &-\rho^{\mathrm{i}} \langle \Gamma^s \otimes_{\mathrm{i}} \mathcal{I} \rangle^{\mathrm{o}} - \rho^{\mathrm{w}} \langle \Gamma^s \otimes_{\mathrm{o}} \mathcal{S} \rangle^{\mathrm{o}}, \end{aligned} \tag{5.48}$$

where

$$\mathcal{S}^{\mathrm{eus}}(t) = -\frac{\mu}{\rho^{\mathrm{w}}\, A^{\mathrm{op}}}, \tag{5.49}$$

and where the constant

$$A^{\mathrm{op}} = \int_{\mathrm{e}} O^{\mathrm{p}}\, \mathrm{d}A \tag{5.50}$$

represents the present-day area of the surface of the oceans. In obtaining (5.48), we have taken advantage of the fact that $\mathcal{S}^{\mathrm{ofu}} = 0$ since according to FC76 the ocean function is not allowed to vary (see Eq. 5.18). Within the FC76 theory, the ice mass variation that appears in Eq. (5.49) can be written as

$$\mu(t) = \rho^{\mathrm{i}} \int_{\mathrm{i}} \mathcal{I}\, \mathrm{d}A, \tag{5.51}$$

where in the general expression (1.140) we have set $C{=}C_0{=}C^{\mathrm{p}}$ in agreement with the fundamental FC76 assumption of a constant ocean function. We remark that, in the FC76 formulation, the bathymetry of the oceans has no influence on the change in sea level, hence no information about topography is needed in order to solve the SLE.

---

[3] Note that with respect to FC76, in Eq. (5.48) a factor $^1/_g$ is missing, where g is surface gravity, because we are using a different normalization for the sea-level GF.

The spatially invariant term $\mathcal{S}^{\text{eus}}(t)$, which is the value taken by $\mathcal{S}^{\text{equ}}(t)$ for constant OF (see Eq. 5.17), has an important role in GIA studies. It is sometimes referred to as *eustatic sea-level change*. Following FC76 (see their p. 648), it is worth to observe that the term *eustatic* has received many qualifications (see Dott 1992) since the word has been introduced by Eduard Suess in 1906. For this reason, some authors prefer to avoid using this term. We note that in the current standard terminology for sea level the term *barystatic* should be preferred to *eustatic* (Gregory et al. 2019). Besides the issue of terminology, the $\mathcal{S}^{\text{eus}}(t)$ term constitutes one very special solution of SLE, on which we shall return in Sect. 5.4.1 below. Here we only note that, in the general formulation of the SLE (Eq. 5.38), $\mathcal{S}^{\text{eus}}(t)$ does not appear, hence this quantity has a physical sense only within the framework of the classical FC76 theory, in which the *oceans are bounded by vertical edges*.

5.3.5 Implicit Nature of the SLE

In the framework of the FC76 classical theory, the SLE (5.48) has the form of an implicit (or integral) equation, since the unknown function $\mathcal{S}$ appears on the *l.h.s.* but it is also embedded in the spatio-temporal convolution integral $\otimes_o$ on the *r.h.s.* The problem is three-dimensional, since it involves the two spatial variables (θ, λ) and time t. The implicit nature remains, of course, when one considers the more general version of the SLE (5.38); in this case, however, the solution is considerably more challenging due to rotational effects, transition between marine and ground-based ice and the horizontal migration of the shorelines (Mitrovica and Milne 2003; Lambeck et al. 2003; Kendall et al. 2005).

WR Peltier has first noted that the SLE is similar, in some respects, to an inhomogeneous Fredholm equation of the second kind (see e.g., Jerri 1999). For this reason, the SLE cannot be solved explicitly unless drastic simplifying assumptions are made, which shall be discussed below. Because of the integral nature of the SLE, a successive approximations approach to its solution is generally viable. Such method, already suggested by FC76, is borrowed from the theory of linear integral equations in one variable and it is usually referred to as Neumann series method (Jerri 1999).

To qualitatively illustrate the approach to the solution of the SLE, we consider an integral, non-homogeneous Fredholm equation of the second kind, whose general form is

$$u(x) = f(x) + \int_a^b K(x, x')\, u(x')\, \mathrm{d}x', \tag{5.52}$$

where a and b are constants, function $u(x)$ plays the role of the unknown $\mathcal{S}$ in the SLE (5.48), $f(x)$ represents all the terms on the right hand side of the SLE not depending upon $\mathcal{S}$, and the kernel $K(x, x')$ mimics the sea-level GF, Γ^s. Although the SLE is actually three-dimensional, while (5.52) is one-dimensional, the approach to the solution is similar in the two cases.

As a first approximation, a reasonable guess $u = u^{(0)}(x)$ is assumed and substituted on the right hand side, thus giving

$$u^{(1)}(x) = f(x) + \int_a^b K(x, x')\, u^{(0)}(x')\, \mathrm{d}x', \tag{5.53}$$

where the integral is evaluated analytically or numerically. The process is iterated obtaining, in this way, various approximations of the actual solution $u(x)$, based upon the recursion

$$u^{(n)}(x) = f(x) + \int_a^b K(x, x')\, u^{(n-1)}(x')\, \mathrm{d}x', \tag{5.54}$$

where $u^{(n-1)}(x)$ is the estimate of the solution obtained in previous step. The iteration can be stopped when current solution (say, $u^{(N)}(x)$) does not differ significantly from the previous one, e.g., when

$$\frac{\displaystyle\int_a^b \left|u^{(N)}(x) - u^{(N-1)}(x)\right| \mathrm{d}x}{\displaystyle\int_a^b \left|u^{(N)}(x)\right| \mathrm{d}x} \leq \varepsilon, \tag{5.55}$$

where $\varepsilon > 0$ is a pre-established threshold (e.g., $\varepsilon = 0.01$ or $\varepsilon = 0.05$).

When the FC76 theory is adopted, the most natural choice is to use $\mathcal{S}^{\mathrm{eus}}$ as an initial guess. Indeed, $\mathcal{S}^{\mathrm{eus}}$ is not *the* solution of the SLE, but it is certainly a valid solution *on average*, as we shall discuss in Sect. 5.4.1 below. For the FC76 version of the SLE, the convergence of the iterative approach to the true solution is rather fast, since normally three to five iterations are enough to ensure sufficiently small values of ε (e.g., 5%; see Spada and Stocchi 2007; Spada et al. 2012). However, when the general form of the SLE is considered, the natural first guess would be $\mathcal{S}^{\mathrm{ave}}$, since in this case $\mathcal{S}^{\mathrm{eus}}$ is a meaningless quantity. A further complication arises, however, since due to the migration of the shorelines the Earth's topography itself is evolving, so that an external iteration scheme is necessary, possibly characterised by a different convergence rate. These issues shall be addressed in Sect. 6.3, where we will discuss in detail the numerical convergence of the iterative SLE solution scheme for an example case.

5.3.6 *Non-linearity of the SLE*

As far as we know, the "degree of non-linearity" of the SLE has not been the subject of investigation. Here we sketch some ideas about this problem.

Within the framework of the FC76 theory, for which Eq. (5.48) holds, the linearity of the SLE can be easily demonstrated observing that the conditions of *homogeneity* and *additivity* are both met, as it is required for linear systems (*superposition principle*). Regarding homogeneity, let us suppose that $\mathcal{S}$ is a solution of the SLE corresponding to a certain ice source $\mathcal{I}$. Function $k\mathcal{S}$, where $k \neq 0$, is a solution as well, corresponding to source $k\mathcal{I}$. The statement is immediately demonstrated using Eq. (5.48), due to the linearity of the operations of convolution $\otimes_{\mathrm{i}}$ and $\otimes_{\mathrm{o}}$, and to the fact that $\mathcal{S}^{\mathrm{eus}}$ defined by (5.49) is a linear function of $\mathcal{I}$ in view of (5.51). Now, suppose that $\mathcal{S}^a$ and $\mathcal{S}^b$ are solutions corresponding to sources $\mathcal{I}^a$ and $\mathcal{I}^b$, respectively. By the same arguments adopted to test the condition of homogeneity, it is easily demonstrated that $\left(\mathcal{S}^a + \mathcal{S}^b\right)$ is also a solution of the SLE, corresponding to source $\left(\mathcal{I}^a + \mathcal{I}^b\right)$; thus, the condition of additivity is also met.

However, when the FC76 assumptions are relaxed, homogeneity and additivity are not ensured, and the SLE cannot be classified as a *linear* equation. The reason can be understood by considering the term $\mathcal{S}^{\mathrm{ave}}$ that appears in the third form of the SLE (see 5.28). According to its own definition (see Eq. 5.16), $\mathcal{S}^{\mathrm{ave}}$ is not merely a geometrical quantity like $\mathcal{S}^{\mathrm{eus}}$, since it depends implicitly (and dynamically) upon $\mathcal{S}$ through terms A^{o} at the denominator of Eqs. (5.17) and (5.18) and through the ocean function variation $\mathcal{O}$, thus violating the conditions of homogeneity and additivity that would ensure linearity.

The strength of the non-linearity of the SLE should be ultimately tested numerically running a solver like SELEN[4] (Spada and Melini 2019b) using different configurations of the ice sources. Due to non-additivity, the solution of the SLE assuming that deglaciation occurs simultaneously in e.g., Antarctica and Greenland should differ somewhat from the solution obtained considering individual deglaciations over the two regions separately. Hopefully, we shall return on this issue in future work.

5.4 Some Special Versions of the SLE

In some conditions, the SLE takes forms having a particular significance, also from an historical viewpoint. In the following, some of these special versions of the SLE are briefly discussed, and the solutions are presented, when available.

5.4.1 *Suess's Solution*

Within the FC76 theory, a closed-form solution of the SLE is possible in a very special but very important case. Assuming for one moment that all the four terms containing the sea-level Green's function Γ^s are omitted in Eq. (5.48), the SLE is immediately solved, with:

$$\mathcal{S}(\gamma, t) = \mathcal{S}^{\mathrm{eus}}(t) \equiv -\frac{\mu}{\rho^{\mathrm{w}} A^{\mathrm{op}}}, \tag{5.56}$$

showing that, in this case, sea-level change would match the spatially uniform but time-dependent *eustatic* value (5.49) everywhere. In fact, recalling from Eq. (2.63) the definition of sea-level GF, a sufficient condition for $\Gamma^s = 0$ is $\Gamma^u = \Gamma^g = 0$. From a physical stand point, this is equivalent to assume *i)* a perfectly rigid Earth ($\mathcal{U} = 0$), and *ii)* that the mass redistribution at the surface is not perturbing the gravitational potential ($\mathcal{G} = 0$). The first condition would be met by neglecting the Earth's rheological properties, while the second would be met assuming $G = 0$, where G is Newton's constant (*non-gravitating* Earth). From the basic form of the SLE (5.7), we note that condition $\mathcal{U} = 0$ implies

$$\mathcal{S}(\gamma, t) = \mathcal{N} = \mathcal{S}^{\text{eus}}(t), \tag{5.57}$$

meaning that *in eustatic conditions relative sea-level variations are not distinguished from absolute sea-level variations*, which are spatially uniform as well. Furthermore, setting $\mathcal{G} = 0$ in (5.8) and using (5.57) we obtain

$$c(t) = \mathcal{S}^{\text{eus}}(t). \tag{5.58}$$

Indeed, the *eustatic solution* (5.56) generally constitutes a very poor approximation to the actual solution of the SLE. However, it is worth noting that a general and elegant relation exists between $\mathcal{S}^{\text{eus}}$ and the effective solution of the SLE in non-eustatic conditions, which has been explicitly illustrated in FC76. This allows to better appreciate the actual meaning of *eustasy*. In fact, taking the ocean-average of both sides of Eq. (5.48) gives

$$\begin{aligned} \langle \mathcal{S}(\gamma, t)\rangle^{\text{o}} &= +\rho_{\text{i}}\langle \Gamma^s \otimes_{\text{i}} \mathcal{I}\rangle^{\text{o}} + \rho_{\text{w}}\langle \Gamma^s \otimes_{\text{o}} \mathcal{S}\rangle^{\text{o}} + \\ &\quad +\langle \mathcal{S}^{\text{eus}}\rangle^{\text{o}} + \\ &\quad -\rho_{\text{i}}\langle\langle \Gamma^s \otimes_{\text{i}} \mathcal{I}\rangle^{\text{o}}\rangle^{\text{o}} - \rho_{\text{w}}\langle\langle \Gamma^s \otimes_{\text{o}} \mathcal{S}\rangle^{\text{o}}\rangle^{\text{o}}, \end{aligned} \tag{5.59}$$

which, since $\langle\langle f\rangle^{\text{o}}\rangle^{\text{o}} = \langle f\rangle^{\text{o}}$ for any function $f = f(\gamma, t)$ and $\langle g(t)\rangle^{\text{o}} = g(t)$ for any spatially invariant function $g(t)$, provides

$$\langle \mathcal{S}(\gamma, t)\rangle^{\text{o}} = \mathcal{S}^{\text{eus}}. \tag{5.60}$$

This last important result tells us that the *ocean-averaged* sea-level variation in general dynamic (*i.e.*, non-eustatic) conditions (*i.e.*, non-rigid and self-gravitating Earth), matches the eustatic value that one would observe in eustatic conditions (*i.e.*, rigid and non-gravitating Earth). Equation (5.60) is a simple consequence of mass conservation. Remarkably, it is independent from the ice sheets chronology and from their spatial distribution, from the rheology of the Earth's mantle, from the shape of the coastlines, and would still hold true in presence of a rotational feedback on sea level. Furthermore, it holds regardless the time-scale of the GIA process. However, it is important to remark that (5.60) is only valid within the framework of the FC76 classical theory. Adopting the general formulation of the SLE (5.28), by ocean-averaging we would obtain

$$\langle \mathcal{S}(\gamma, t) \rangle^{\mathrm{o}} = \mathcal{S}^{\mathrm{ave}} \neq \mathcal{S}^{\mathrm{eus}}, \tag{5.61}$$

where the dynamic term $\mathcal{S}^{\mathrm{ave}}$ is given by Eq. (5.16).

Recalling the definition of eustatic sea-level change (5.49), result (5.60) implies that from the knowledge of the time-dependent spatial average $\langle \mathcal{S} \rangle^{\mathrm{o}}$, one would immediately obtain the time-history of the mass balance of the ice sheets (*i.e.,* function $\mu(t)$), without invoking any further geophysical observation. This argument assumes that the area of the oceans did not change significantly since the LGM, in accord with the FC76 assumptions. Unfortunately, the geographically sparse sea-level observations from the Holocene are far from allowing for a reliable estimate of the spatial average $\langle \mathcal{S} \rangle^{\mathrm{o}}$, which prevents a straightforward assessment of the time evolution of the major continental ice sheets during the last millennia (see e.g., Tushingham and Peltier 1991). This is inferred, instead, using evidence from locations in the far-field of the regions that were formerly deeply covered by ice at the Last Glacial Maximum (LGM, $\sim$ 21 kyrs BP), where it can be assumed that the effects of isostatic disequilibrium, governed by the SLE terms that contain the sea-level Green's function Γ^s, are small compared to the eustatic component of sea-level rise (Peltier 2004; Milne and Mitrovica 2008).

5.4.2 *Woodward's Solution*

In his pioneering work, Woodward (1888) considered the problem of the determination of the shape of the oceans surface due to gravitational attraction between a localised ice cap and the surrounding water. Here we consider a very particular case, in which the lateral extension of the ice cap is neglected. As we shall see, in this case the solution of the SLE can be found in a closed form, which is referred to as the *Woodward's solution.* Such solution clearly illustrates the importance of gravitational effects upon the pattern of sea-level change. We assume a point-like island surrounded by a global ocean, and we consider the instantaneous freezing of a unit ice mass on the island, neglecting the deformations of the solid Earth in response to variations of the surface load. The geometry of the problem is sketched in Fig. 5.6.

The Woodward SLE is immediately obtained from the FC76 form (Eq. 5.48), by dropping the terms describing the oceans self-attraction, expressed by the spatio-temporal convolutions involving $\mathcal{S}(\gamma, t)$. The hypothesis of non-gravitating oceans implies that a local change in the water column does not alter the global mass distribution of the whole ocean. This gives

$$\mathcal{S}^{\mathrm{w}}(\gamma, t) = \rho^{\mathrm{i}}\, \Gamma^s \otimes_{\mathrm{i}} \mathcal{I} + \mathcal{S}^{\mathrm{eus}} - \rho^{\mathrm{i}} \langle \Gamma^s \otimes_{\mathrm{i}} \mathcal{I} \rangle^{\mathrm{o}}, \tag{5.62}$$

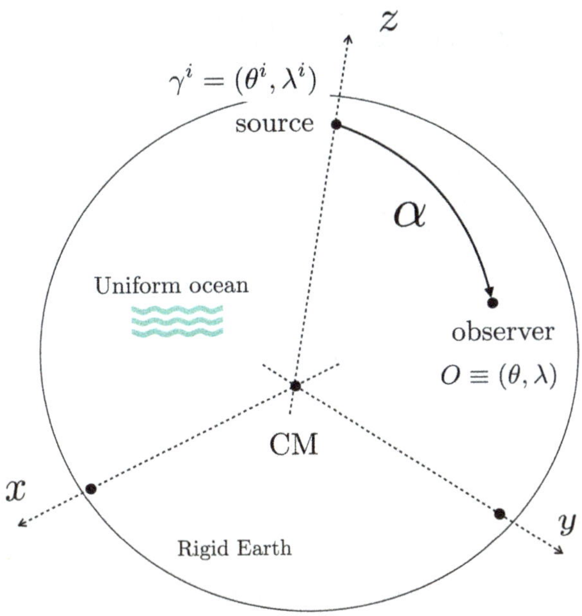

Fig. 5.6 Source-observer geometry for the Woodward problem. A point mass freezes (melts) instantaneously at $\gamma^{\mathrm{i}} = (\theta^{\mathrm{i}}, \lambda^{\mathrm{i}})$ and sea-level falls (rises) by $\mathcal{S}^{\mathrm{w}}$ at the observer location $O = (\theta, \lambda)$

where w stands for Woodward, and the terms containing ice-convolutions describe the sea-level change associated to the gravitational attraction between the ice and the water masses. Along with the constraint of mass conservation, this constitutes the only physical ingredient that enters into the SLE in Woodward's approach. We note that, in such a form, the SLE has lost its implicit character and the solution can be found without invoking the recursive methods that are required when the oceans self-gravitation is taken into account.

In order to solve the SLE in the Woodward form (5.62), we need to make the sea-level GF $\Gamma^s(\gamma, t)$ and the source term $\mathcal{I}(\gamma, t)$ explicit. Concerning $\Gamma^s(\gamma, t)$, in the general expression (2.64) we impose $k^{\mathrm{L}}(t) = h^{\mathrm{L}}(t) = 0$, thus neglecting the indirect gravitational potential variation and vertical displacement, only keeping the $\delta(t)$ term that describes the direct gravitational effect of the source. Also taking advantage of the Legendre sum (2.19), we find

$$\Gamma^s(\alpha, t) \equiv \frac{a}{m^{\mathrm{e}}} \frac{1}{2\sin(\alpha/2)} \delta(t), \tag{5.63}$$

where α is the angular distance along the great circle connecting the source with the observer, given by (2.13). For a spatially localised (delta-like) source, switched on at time $t = 0$ at point $\gamma^{\mathrm{i}} = (\theta^{\mathrm{i}}, \lambda^{\mathrm{i}})$, we have

$$\mathcal{I}(\gamma, t) = \Delta\mathcal{I}\,\frac{\delta(\theta - \theta^{\mathrm{i}})\delta(\lambda - \lambda^{\mathrm{i}})}{\sin\theta} H(t), \tag{5.64}$$

where $H(t)$ is the Heaviside step function and with the instantaneous freezing and melting corresponding to $\Delta\mathcal{I} > 0$ and $\Delta\mathcal{I} < 0$, respectively. However, given the axial symmetry of the problem, it is convenient to consider the particular case of a source at the north pole ($\theta^{\mathrm{i}} = \lambda^{\mathrm{i}} = 0$), with

$$\mathcal{I}(\gamma, t) = \Delta\mathcal{I}\,\frac{\delta(\theta)\delta(\lambda)}{\sin\theta} H(t). \tag{5.65}$$

Then, the first term on the *r.h.s.* of (5.62) is

$$\begin{aligned}\rho^{\mathrm{i}}\,\Gamma^{s} \otimes_{\mathrm{i}} \mathcal{I} &= \rho^{\mathrm{i}}\Delta\mathcal{I}\frac{a^3}{m^{\mathrm{e}}}\int_{-\infty}^{\infty}\mathrm{d}t'\int_{\mathrm{e}}\frac{1}{2\sin\alpha/2}\,\delta(t-t')\,\delta(\theta')\,\delta(\lambda')\,H(t')\,\mathrm{d}\theta'\,\mathrm{d}\lambda' \\ &= \rho^{\mathrm{i}}\Delta\mathcal{I}\frac{a^3}{m^{\mathrm{e}}}\,H(t)\frac{1}{2\sin\theta/2},\end{aligned} \tag{5.66}$$

where we have used (5.63) and (5.65), the sifting property of the Dirac's delta and we have recalled that $\mathrm{d}A = a^2\sin\theta\mathrm{d}\theta\mathrm{d}\lambda$.

The second term of (5.62), i.e., the eustatic term, is

$$\begin{aligned}\mathcal{S}^{\mathrm{eus}}(t) &= -\frac{\mu(t)}{\rho^{\mathrm{w}} A^{\mathrm{o}}} \\ &= -\frac{\rho^{\mathrm{i}}}{\rho^{\mathrm{w}} A^{\mathrm{o}}}\int_{\mathrm{e}}\mathcal{I}(\gamma, t)\mathrm{d}A \\ &= -\frac{\rho^{\mathrm{i}}}{\rho^{\mathrm{w}}}\frac{a^2}{A_{\mathrm{o}}}\Delta\mathcal{I}\int_{\mathrm{e}}\delta(\theta)\delta(\lambda)H(t)\mathrm{d}\theta\mathrm{d}\lambda \\ &= -\frac{\rho^{\mathrm{i}}}{\rho^{\mathrm{w}}}\frac{\Delta\mathcal{I}}{4\pi}H(t),\end{aligned} \tag{5.67}$$

where we have used the definition of eustatic sea-level change (5.49), the expression giving the mass variation of the ice load (5.51), and we have set $A^{\mathrm{o}} = 4\pi a^2$ since in Woodward model the oceans cover the whole Earth's surface, with the exception of the micro-continent of vanishing area where the ice source is located.

The third and last term of (5.62) is the ocean-average of the first; hence:

$$\begin{aligned}\rho^{\mathrm{i}}\,\langle\Gamma^{s} \otimes_{\mathrm{i}} \mathcal{I}\rangle^{\mathrm{o}} &\equiv \rho^{\mathrm{i}}\frac{1}{A^{\mathrm{o}}}\int_{\mathrm{o}}\left(\Gamma^{s} \otimes_{\mathrm{i}} \mathcal{I}\right)\mathrm{d}A \\ &= \rho^{\mathrm{i}}\,\Delta\mathcal{I}\,\frac{a^3}{m^{\mathrm{e}}}\,H(t)\,\frac{a^2}{A^{\mathrm{o}}}\int_0^{2\pi}\mathrm{d}\lambda\int_0^{\pi}\frac{\sin\theta\,\mathrm{d}\theta}{2\sin\theta/2} \\ &= \rho^{\mathrm{i}}\,\Delta\mathcal{I}\,\frac{a^3}{m^{\mathrm{e}}}\,H(t)\,\frac{a^2}{4\pi a^2}2\pi\int_0^{\pi}\frac{2\sin\theta/2\,\cos\theta/2\,\mathrm{d}\theta}{2\sin\theta/2}\end{aligned}$$

$$
\begin{aligned}
&= \rho^{\mathrm{i}}\,\Delta\mathcal{I}\,\frac{a^3}{m^{\mathrm{e}}}\,H(t)\,\frac{1}{2}\int_0^{\pi}\cos\theta/2\,\mathrm{d}\theta \\
&= \rho^{\mathrm{i}}\,\Delta\mathcal{I}\,\frac{a^3}{m^{\mathrm{e}}}\,H(t)\int_0^{\pi/2}\cos x\,\mathrm{d}x \\
&= \rho^{\mathrm{i}}\,\Delta\mathcal{I}\,\frac{a^3}{m^{\mathrm{e}}}\,H(t),
\end{aligned}
\tag{5.68}
$$

where we have integrated over the whole water-covered Earth surface.

Hence, substituting (5.66), (5.67) and (5.68) into the SLE (5.62), the solution of the Woodward problem reads

$$
\mathcal{S}^{\mathrm{w}}(\theta, t) = \frac{\rho^{\mathrm{i}} a^3\,\Delta\mathcal{I}}{m^{\mathrm{e}}}\,w(\theta)\,H(t), \tag{5.69}
$$

with the θ-dependence described by function

$$
w(\theta) = \left(\frac{1}{2\sin(\theta/2)} - 1\right) - \frac{\rho^{\mathrm{e}}}{3\,\rho^{\mathrm{w}}}, \tag{5.70}
$$

where the term in parentheses accounts for the contribution of gravitational attraction, and the other represents the eustatic term. Note that, since in constructing the Woodward solution we have dealt with point sources, $w(\theta)$ has a singularity for $\theta \to 0$, which however disappears as soon as finite-sized sources are considered.

To illustrate the meaning of the result obtained, we consider the plot of $w(\theta)$ displayed in Fig. 5.7. In this case, we have considered the instantaneous freezing of a point mass ($\Delta\mathcal{I} > 0$) adopting the numerical values $\rho^{\mathrm{e}} = 5513$ kg m^{-3} and $\rho^{\mathrm{w}} = 1000\,\mathrm{kg\,m^{-3}}$. Equations (5.69) and (5.70) imply a sea-level rise ($\mathcal{S}^{\mathrm{w}} > 0$) in the neighbourhood of the source, for $\sin\theta/2 \le (2(1 + \rho^{\mathrm{e}}/3\rho^{\mathrm{w}}))^{-1}$, i.e., for $\theta \lesssim 20.2°$ (*vice versa,* a sea-level fall is expected for $\theta \gtrsim 20.2°$). This result, which may appear as counter-intuitive, is simply a consequence of Newton's law of gravitation. This implies that in the neighbours of an ice mass subject to *deglaciation* ($\Delta\mathcal{I} < 0$), the Woodward theory predicts a sea-level fall, due to the loss of the gravitational attraction between the ice and the ocean mass. The effect would be enhanced due to elastic (or viscoelastic) rebound of the solid Earth, which however is not considered here since Woodward model is rigid (the case of an elastic Earth was later considered by Daly 1925). We further note that according to (5.69) and (5.70) sea level attains the eustatic value for $1/(2\sin\theta/2) = 1$, hence only along the parallel of co-latitude $\theta = 60°$; as a consequence, super- and sub-eustatic values are obtained for $\theta \le 60°$ and for $\theta > 60°$, respectively. However, according to the general result (5.60), *on average*, sea-level change will be equal to the eustatic value $\mathcal{S}^{\mathrm{eus}} = \rho^{\mathrm{i}} a^3 \Delta\mathcal{I}\langle w(\theta)\rangle^{\mathrm{o}}/m^{\mathrm{e}}$, with $\langle w(\theta)\rangle^{\mathrm{o}} = -\rho^{\mathrm{e}}/(3\,\rho^{\mathrm{w}}) \approx -1.84$.

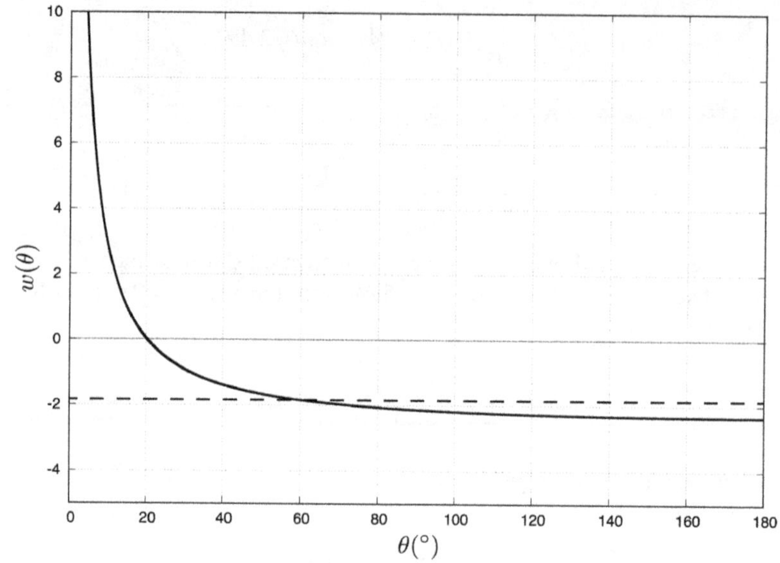

Fig. 5.7 Solution of the Woodward problem $w = w(\theta)$ for an instantaneously freezing ice mass (solid curve), according to Eq. (5.70). The dashed line represents the uniform sea-level drop that we would observe in eustatic conditions, neglecting the gravitational attraction between the ice and the water mass

5.4.3 Farrell and Clark's Solution Neglecting Ice Loads

At the end of this chapter we mention another special variant of the SLE, which is of some interest. In their work, FC76 considered the special case in which the ice sources are *added from outside the Earth, rather than coming from a melting ice sheet.* They used this case to illustrate the convergence of the iterative solution for sea level change and to discuss the errors introduced by the traditional assumption that when a mass M is added to the ocean, it causes a uniform rise in sea level $M/A^{o}\rho^{w}$. In fact, even neglecting the load exerted by the ice sources, the SLE predicts a geographically variable differential sea-level change $\mathcal{S}$ that stems from an eustatic contribution plus a contribution due to the self-attraction of the ocean masses and to the yielding of the ocean floor.

With the source term $\mathcal{I} = 0$, the FC76 form of the SLE (5.48) becomes

$$\mathcal{S}^{\mathrm{nic}}(\gamma, t) = \rho^{\mathrm{w}}\,\Gamma^{s} \otimes_{\mathrm{o}} \mathcal{S} + \mathcal{S}^{\mathrm{eus}} - \rho^{\mathrm{w}}\langle \Gamma^{s} \otimes_{\mathrm{o}} \mathcal{S}\rangle^{\mathrm{o}}, \tag{5.71}$$

where *nic* stands for *no ice* and we note that the SLE retains, in this case, its implicit nature, so that no analytical solution can be determined and an iterative approach needs to be employed. Using elastic Love numbers, FC76 have shown that the added mass tends to collect into the largest basin (the Pacific basin) in the expense of the North Atlantic. There, FC76 report a $\sim$5% *difference between the true sea level change and the uniform approximation*, solely based on the eustatic solution given by $\mathcal{S}^{eus}$ in Eq. (5.71).

References

Adhikari S, Ivins ER, Larour E (2016) ISSM-SESAW v1. 0: mesh-based computation of gravitationally consistent sea-level and geodetic signatures caused by cryosphere and climate driven mass change. Geosci Model Dev 9(3)

Borreggine M, Powell E, Pico T, Mitrovica JX, Meadow R, Tryon C (2022) Not a bathtub: a consideration of sea-level physics for archaeological models of human migration. J Archaeol Sci 137:105507

Broerse T, Riva R, Vermeersen B (2014) Ocean contribution to seismic gravity changes: the sea level equation for seismic perturbations revisited. Geophys J Int 199(2):1094–1109

Clark JA, Farrell WE, Peltier WR (1978) Global changes in postglacial sea level: a numerical calculation. Quatern Res 9(3):265–287

Clark PU, Mitrovica J, Milne G, Tamisiea M (2002) Sea-level fingerprinting as a direct test for the source of global meltwater pulse IA. Science 295(5564):2438–2441

Conrad C (2013) The solid Earth's influence on sea level. Bull Geol Soc Am 125(7–8):1027–1052

Coulson S, Al-Attar D, Mitrovica J (2021) An extended ice-age sea-level equation: incorporating water flux across sills. Geophys J Int 225(1):236–252

Dal Forno G, Gasperini P, Spada G (2012) Implementation of the complete Sea Level Equation in a 3D finite elements scheme: a validation study. In: VII Hotine-Marussi symposium on mathematical geodesy: proceedings of the symposium in Rome, 6–10 June, 2009. Springer, Berlin, pp 393–397

Daly RA (1925) Pleistocene changes of level. Am J Sci 10(58):281–313

Dott RH (1992) An introduction to the ups and downs of eustasy. Geol Soc Am Mem 180:1–16

Farrell W (1973) A discussion on the measurement and interpretation of changes of strain in the Earth—Earth tides, ocean tides and tidal loading. Philos Trans R Soc London Ser A, Math Phys Sci 274(1239):253–259

Farrell W, Clark J (1976) On postglacial sea level. Geophys J Roy Astron Soc 46(3):647–667

Gregory JM, Griffies SM, Hughes CW, Lowe JA, Church JA, Fukimori I, Gomez N, Kopp RE, Landerer F, Cozannet GL et al (2019) Concepts and terminology for sea level: mean, variability and change, both local and global. Surv Geophys 1–39

Hartmann R, Ebbing J, Conrad CP (2020) A Multiple 1D Earth Approach (M1DEA) to account for lateral viscosity variations in solutions of the sea level equation: an application for glacial isostatic adjustment by Antarctic deglaciation. J Geodyn 135:101695

Hofmann-Wellenhof B, Moritz H (2006) Physical geodesy. Springer Science & Business Media

Jerri A (1999) Introduction to integral equations with applications. Wiley

Johnston P (1993) The effect of spatially non-uniform water loads on prediction of sea-level change. Geophys J Int 114(3):615–634

Kachuck SB, Cathles LM (2018) giapy: glacial isostatic adjustment in PYthon: nondimensionalized relaxation method for computation of time-domain viscoelastic Love numbers. In: AGU fall meeting abstracts, vol 2018, pp DI31B–0008

Kendall RA, Mitrovica JX, Milne GA (2005) On post-glacial sea level–II. Numerical formulation and comparative results on spherically symmetric models. Geophys J Int 161(3):679–706

Kendall RA, Mitrovica JX, Milne GA, Törnqvist TE, Li Y (2008) The sea-level fingerprint of the 8.2 ka climate event. Geology 36(5):423–426

Lambeck K (1980) The Earth's variable rotation: geophysical causes and consequences. Cambridge University Press

Lambeck K, Purcell A, Johnston P, Nakada M, Yokoyama Y (2003) Water-load definition in the glacio-hydro-isostatic sea-level equation. Quatern Sci Rev 22(2):309–318

Martinec Z, Hagedoorn J (2014) The rotational feedback on linear-momentum balance in glacial isostatic adjustment. Geophys J Int 199(3):1823–1846

Martinec Z, Klemann V, van der Wal W, Riva R, Spada G, Sun Y, Melini D, Kachuck S, Barletta V, Simon K (2018) A benchmark study of numerical implementations of the sea level equation in GIA modelling. Geophys J Int 215(1):389–414

Mauz B, Ruggieri G, Spada G (2015) Terminal Antarctic melting inferred from a far-field coastal site. Quatern Sci Rev 116:122–132

Melini D, Spada G, Piersanti A (2010) A sea level equation for seismic perturbations. Geophys J Int 180(1):88–100

Milne G, Mitrovica J (1996) Postglacial sea-level change on a rotating Earth: first results from a gravitationally self-consistent sea-level equation. Geophys J Int 126(3):F13–F20

Milne GA, Mitrovica JX (1998) Postglacial sea-level change on a rotating Earth. Geophys J Int 133(1):1–19

Milne GA, Mitrovica JX (2008) Searching for eustasy in deglacial sea-level histories. Quatern Sci Rev 27(25–26):2292–2302

Milne GA, Mitrovica JX, Davis JL (1999) Near-field hydro-isostasy: the implementation of a revised sea-level equation. Geophys J Int 139(2):464–482

Mitrovica JX, Peltier W (1991) On postglacial geoid subsidence over the equatorial oceans. J Geophys Res: Solid Earth 96(B12):20,053–20,071

Mitrovica J, Milne G (2002) On the origin of late Holocene sea-level highstands within equatorial ocean basins. Quatern Sci Rev 21(20):2179–2190

Mitrovica JX, Milne GA (2003) On post-glacial sea level: I. General theory. Geophys J Int 154(2):253–267

Mitrovica J, Davis J, Shapiro I (1994) A spectral formalism for computing three-dimensional deformations due to surface loads. J Geophys Res 99:7057–7073

Mitrovica J, Gomez N, Morrow E, Hay C, Latychev K, Tamisiea M (2011) On the robustness of predictions of sea level fingerprints. Geophys J Int 187(2):729–742

Peltier W (1999) Global sea level rise and glacial isostatic adjustment. Glob Planet Change 20 (2–3):93–123

Peltier W (2004) Global glacial isostasy and the surface of the ice-age Earth: the ICE-5G (VM2) model and GRACE. Annu Rev Earth Planet Sci 32:111–149

Pfeffer J, Allemand P (2016) The key role of vertical land motions in coastal sea level variations: a global synthesis of multisatellite altimetry, tide gauge data and GPS measurements. Earth Planet Sci Lett 439:39–47

Pfeffer J, Spada G, Mémin A, Boy J-P, Allemand P (2017) Decoding the origins of vertical land motions observed today at coasts. Geophys J Int 210(1):148–165

Plag H-P, Jüettner H-U (2001) Inversion of global tide gauge data for present-day ice load changes. Memoirs of National Institute of Polar Research. Special issue 54:301–317

Spada G (2017) Glacial isostatic adjustment and contemporary sea level rise: an overview. Surv Geophys 38(1):1–33

Spada G, Stocchi P (2007) SELEN: a Fortran 90 program for solving the "Sea Level Equation". Comput Geosci 33(4):538–562

Spada G, Melini D (2019) SELEN^4 (SELEN version 4.0): a Fortran program for solving the gravitationally and topographically self-consistent sea-level equation in glacial isostatic adjustment modeling. Geosci Model Dev 12(12):5055–5075

Spada G, Melini D, Galassi G, Colleoni F (2012) Modeling sea level changes and geodetic variations by glacial isostasy: the improved SELEN code. arXiv:1212.5061

Spada G, Stocchi P (2006) The sea level equation, theory and numerical examples. Aracne Editrice

Stocchi P, Spada G (2007) Glacio and hydro-isostasy in the Mediterranean sea: Clark zones and role of remote ice sheets. Ann Geophys 50(6):741–761

Suess E (1906) Face of the Earth. Clarendon Press, Oxford

Tamisiea ME (2011) Ongoing glacial isostatic contributions to observations of sea level change. Geophys J Int 186(3):1036–1044

Tushingham A, Peltier W (1991) ICE-3G—a new global model of late Pleistocene deglaciation based upon geophysical predictions of post-glacial relative sea level change. J Geophys Res 96(B3):4497–4523

Walcott R (1972) Past sea levels, eustasy and deformation of the Earth. Quatern Res 2(1):1–14

Wolf D, Hagedoorn J, Martinec Z (2002) A new time-domain method of implementing the sea-level equation in glacial-isostatic adjustment. In: AGU fall meeting abstracts, vol 2002, pp G12A–1059

Woodward R (1888) On the form and position of mean sea level. USGS Bull 48:87–170

Wöppelmann G, Marcos M (2016) Vertical land motion as a key to understanding sea level change and variability. Rev Geophys 54(1):64–92

Wu P (2004) Using commercial finite element packages for the study of earth deformations, sea levels and the state of stress. Geophys J Int 158(2):401–408

Chapter 6
Solving the Sea Level Equation

Hark, do you hear the sea?

(King Lear: Act IV)

6.1 Introduction

In this chapter we present an example of numerical solution of the Sea Level Equation, obtained by means of the open-source code `SELEN`[4] (Spada and Melini 2019b) that implements the SLE theories described throughout the book.[1,2] The solution is based upon the GIA model ICE-7G_NA(VM7) by Roy and Peltier (2015, 2017), which represents the latest iteration of the ICE-X suite of models developed by William Richard Peltier and collaborators at the University of Toronto. However, it is important to note that our solution does not exactly duplicate that of Roy and Peltier (2015, 2017) since our approach differs for the details of the spatiotemporal discretisation of the surface load and of the field variables, possibly for the rotational theory adopted and for the rheological setup of the model (in particular, we are using an incompressible Earth). For these reasons, it would be more correct to say that our solution is a *realisation* of Peltier's solution.[3] The exact origin of possible discrepancies between our results and Peltier's can only be established by a rigorous

---

[1] `SELEN`[4] can be downloaded from the page https://github.org/geodynamics/selen.

[2] Other SLE solvers like `ISSM-SESAW v1.0` of Adhikari et al. (2016) do not consider the viscous behaviour of the Earth and are more appropriate to describe the Earth's response on times scales not exceeding one century.

[3] See the WR Peltier's data set page at https://www.atmosp.physics.utoronto.ca/~peltier/data.php [last accessed on September 30, 2024].

G. Spada and D. Melini, *Glacial Isostatic Adjustment*, Springer Geophysics,
https://doi.org/10.1007/978-3-031-90538-4_6

benchmark test, following the example of Spada et al. (2011a) in the context of Post Glacial Rebound and of Martinec et al. (2018) in the framework of GIA.

The chapter is organised as follows. In Sect. 6.2 the numerical solution of the SLE is set up by defining the spatial and temporal grids on which all the input fields shall be discretised. Then, we describe the three basic inputs of the SLE, namely *(i)* the rheological profile employed in the Earth's mantle and lithosphere, *(ii)* the spatio-temporal evolution of the ice thickness and *(iii)* the present-day topography that, along with the ice distribution, defines the current ocean function. These inputs allow to obtain a numerical solution of the SLE adopting the iterative scheme described in Appendix E, whose convergence is studied in Sect. 6.3. After the numerical solution of the SLE is attained to a satisfactory degree of precision, in Sect. 6.4 predictions for a number of quantities of geophysical interest are presented and discussed in detail, including past sea-level variations, paleo-topography, the present day GIA fingerprints and the sea-level variations at tide gauges, polar motion, and the time variations of the Stokes coefficients of gravitational potential. The whole presentation is quite informal. At some points we shall give insight about the way we have implemented in `SELEN`[4] some of the fundamental GIA variables.

6.2 Setting up the Numerical Solution

In the following sections we shall first discuss the discretisation scheme for the numerical solution of the SLE. Then, we shall describe its basic input quantities. These are: *(i)* the Earth rheological profile, whose characteristics determine quantitatively the viscoelastic Love numbers as explained in Chap. 2; *(ii)* the spatial and temporal evolution of the global ice sheets, which is related to the *surface load* discussed in Chap. 3, and *(iii)* a model of present-day topography, which is used to reconstruct paleo-topography according to the PT Eq. (1.105). While the latter is usually prescribed according to a global relief model such as ETOPO1 (Amante and Eakins 2009), the rheological profile and the history of the ice sheets are generally inferred by optimising the fit of the GIA model to observational data including, for instance, proxies of palaeo-sea level or geodetic observations from ground-based or space-borne instrumentations (e.g., Roy and Peltier 2018).

6.2.1 Spatiotemporal Discretisation Grids

The numerical scheme for the solution of the SLE discussed in detail in Appendix E requires the discretisation of all the fields over suitable spatial and temporal grids. Hence, as a first step in setting up a numerical solution of the SLE for a particular GIA model, the parameters of the computational grids shall be defined coherently with the chosen model, and the desired maximum degree $n_{\max}$ used for the spherical harmonic expansions of spatial fields shall be set. The choice of these parameters represents a tradeoff between the spatial resolution and its computational cost, so that different experiments may be necessary to identify the optimal setup.

In the following, we shall employ a spatial grid with resolution parameter $R = 44$ (see Sect. E.2), which according to our experience represents a good compromise between the computational speed and the quality of the solution. The layout of the grid is shown in Fig. E.2 and, according to Table E.1, it covers the Earth's surface with $75,692$ pixels, with a horizontal spacing between adjacent pixels of about $0.84°$, corresponding to ~92 km on the Earth's surface. This spacing is close to the $1°$ spacing of the uniform longitude-latitude cartesian geographical grid on which the ICE-7G model is made available by Prof. WR Peltier from his data sets page, with $360 \times 180 = 64,800$ rectangular elements.

In the following, spectral quantities will be expanded up to a maximum harmonic degree $n_{max} = 256$. According to Tegmark (1996), a numerical quadrature on the $R = 44$ grid would be accurate for harmonic functions of degree up to $n_{max} \sim 476$ (see Table E.1), hence the chosen resolution largely ensures a numerically stable integration. A maximum harmonic degree $n_{max} = 512$, which is quite often adopted in GIA modelling, would however require a higher spatial resolution (at least $R = 60$, among the cases considered in Table E.1).

The choice of the time increment used for temporal discretisation is directly related to the temporal resolution of the ice thickness model that shall be supplied as an input to the SLE. In its original form, the ICE-7G_NA GIA model that we will consider in this chapter has a time resolution of 1 kyr before 20 kyr BP and 0.5 kyr BP afterwards.[4] Since our solution scheme assumes a constant time step (see Sect. E.1 for details), we shall employ a uniform time step of 0.5 kyr which, since the time-history of ICE-7G_NA model extends back to 26 kyr BP, results into a number of $N = 52$ discrete time steps, each 500 years long.

6.2.2 *Rheological Profile*

As discussed in detail in Chap. 2, the time-dependent response of the Earth to surface loads is described by *viscoelastic Love numbers* (see, in particular, Sect. 2.3). Those have been used in Chap. 3 to build the *surface response functions*, which in turn are essential ingredients in the formulation of the SLE. Coherently with the ICE-7G_NA GIA model, we shall assume the VM7 Earth viscosity profile of Roy and Peltier (2017), which is reproduced in Table 6.1. The VM7 profile has a 10-layer structure, including a 75 km thick elastic lithosphere, a high-viscosity ($\eta = 10^{22}$ Pa s) lithospheric mantle introduced by Peltier and Drummond (2008) to improve the fit with geodetic observations of horizontal post-glacial velocities, an upper mantle and transition zone with viscosity of 4.5×10^{20} Pa s and a five-layer lower mantle with viscosity that progressively increases with depth, reaching the value of $1.2 \times$

[4] 1 kyr is equivalent to $1,000$ years.

Table 6.1 Radial structure and parameters of the 1-D model VM7, employed in the ICE-7G_NA model of Roy and Peltier (2015, 2017). The model has ten layers: an elastic lithosphere (LT), a high-viscosity lithospheric mantle (LTM), the shallow upper mantle (SUM), the mantle transition zone (TZ), five lower mantle layers (LMx) and the fluid homogeneous core (C). At a given radius, the unperturbed gravity acceleration $g_0(r)$ is computed according to Eq. (2.6). The viscosity is expressed in Haskell units, with 1 Hu = 10^{21} Pa·s. The layer structure and viscosity profile are from Table 2 of Roy and Peltier (2017), while the values of ρ and μ are obtained by volume-averaging the elastic constants of the Preliminary Reference Earth Model (PREM, see Dziewonski and Anderson 1981)

Layer	$r_{\rm bot}$ km	$r_{\rm top}$ km	$r_{\rm top} - r_{\rm bot}$ km	ρ kg/m^3	$\mu \times 10^{11}$Pa	η Hu
LT	6296	6371	75	3161	0.5653	∞
LTM	6256	6296	40	3373	0.6725	10.00
SUM	5951	6256	305	3460	0.7430	0.45
TZ	5701	5951	250	3895	1.1092	0.45
LM1	5391	5701	310	4474	1.7422	0.90
LM2	4901	5391	490	4704	2.0060	1.80
LM3	4421	4901	480	4966	2.2905	6.00
LM4	3971	4421	450	5205	2.5572	10.00
LM5	3485	3971	485	5437	2.8256	12.00
C	0	3485	3485	10961	0	0

10^{22} Pa s at the core-mantle boundary. The core is homogeneous and inviscid (i.e., of vanishing viscosity).

In Table 6.1, the radii of the boundaries between layers and the layers viscosities are taken from Table 2 of Roy and Peltier (2017), while the density and rigidity for each layer have been obtained by volume-averaging the elastic structure of the Preliminary Reference Earth Model (PREM), by means of:

$$\rho = \frac{3}{r_{\rm top}^3 - r_{\rm bot}^3} \int_{r_{\rm bot}}^{r_{\rm top}} \rho_{\rm prem}(r) r^2 {\rm d}r \tag{6.1}$$

and

$$\mu = \frac{3}{r_{\rm top}^3 - r_{\rm bot}^3} \int_{r_{\rm bot}}^{r_{\rm top}} \rho_{\rm prem}(r) v_{S,{\rm prem}}^2(r)\, r^2 {\rm d}r\,, \tag{6.2}$$

where $\rho_{\rm prem}(r)$ and $v_{S,{\rm prem}}(r)$ are the density and S-wave velocity given by the functions in Table 1 of Dziewonski and Anderson (1981), respectively.

The spectrum of relaxation for model VM7 is shown in Fig. 6.1, where the relaxation times $\tau_{nj} = -1/s_{nj}$ $(j = 1, \ldots, M)$ are plotted as a function of the harmonic degree n, with $1 \le n \le n_{\rm max} = 256$. For each harmonic degree, there are $M = 4N_{\rm V} = 32$ relaxation modes, where $N_{\rm V} = 8$ is the number of viscoelastic layers in the model (see Sect. C.7.2). The structure of branches in Fig. 6.1 is similar to that for model B11 described in Sect. 2.4.3, with "transient", "core", "lithospheric"

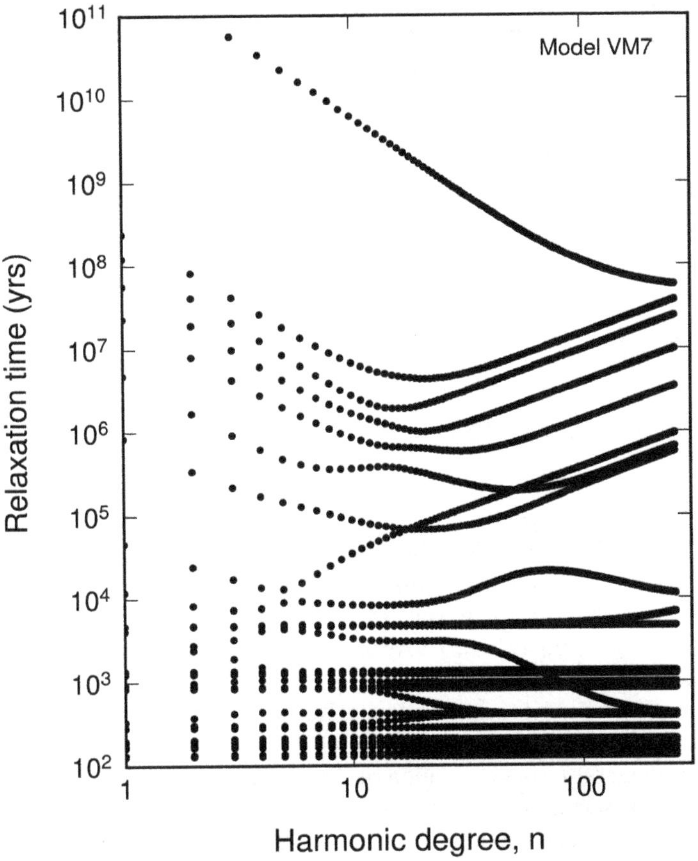

Fig. 6.1 Relaxation spectrum for the viscosity model VM7 of Roy and Peltier (2017), in the range of harmonic degrees $1 \leq n \leq 256$

and "mantle" modes. The number of modes and their detailed layout, however, is significantly different from that of model B11 as a consequence of the comparatively larger number of layers in model VM7. Remarkably, the relaxation times for VM7 span a much larger range of time scales than those for B11, with relaxation times ranging from a few centuries to values $\sim 10^{11}$ yr for the lowest harmonic degrees.[5]

Figure 6.2 shows the values of the elastic ($h_n^{L,e}$, $\ell_n^{L,e}$, $k_n^{L,e}$) and the fluid ($h_n^{L,f}$, $\ell_n^{L,f}$, $k_n^{L,f}$) LLNs for model VM7 as a function of harmonic degree n. As discussed in Chap. 2, those numbers represent the response immediately after loading ($t \to 0$) and after a long time (i.e., for $t \to \infty$), respectively. Since the elastic and fluid LLNs are not affected by the viscosity profile and depend only upon the elastic structure of the model, results for VM7 are almost coincident with those obtained for the

[5] For comparison, we note that the age of the Earth is estimated to be $\approx 4.5 \times 10^9$ yr.

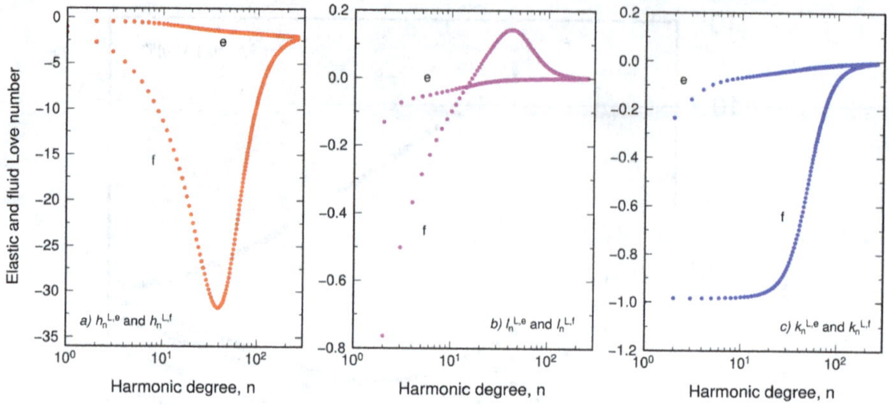

Fig. 6.2 Elastic (e) and fluid (f) LLNs h, ℓ and k for model VM7, in the range of harmonic degrees $1 \le n \le 256$

simpler model B11 (see Fig. 2.7), because the elastic structures of the two models are essentially two different discretisations of the same PREM 1-D profile.

Figure 6.3 shows the absolute values of the viscoelastic residues for VM7 which appear on the *r.h.s.* of Eq. (2.68). The residues, along with the relaxation spectrum in Fig. 6.1 and the elastic and fluid LLNs in Fig. 6.2, completely determine the response of the chosen Earth model to surface loads. Knowledge of those quantities allows to compute the $e_n^{s,u,g}$ constants and the functions $\beta_n^{s,u,g}(t)$ introduced in Chap. 3, which in turn are necessary for the evaluation of the *surface response functions* $\mathcal{R}^{\mathrm{sur}}(\gamma, t)$. From a practical standpoint, all quantities shown in Figs. 6.1, 6.2 and 6.3 can be computed using the open-source code TABOO (Spada 2003; Spada et al. 2011a),

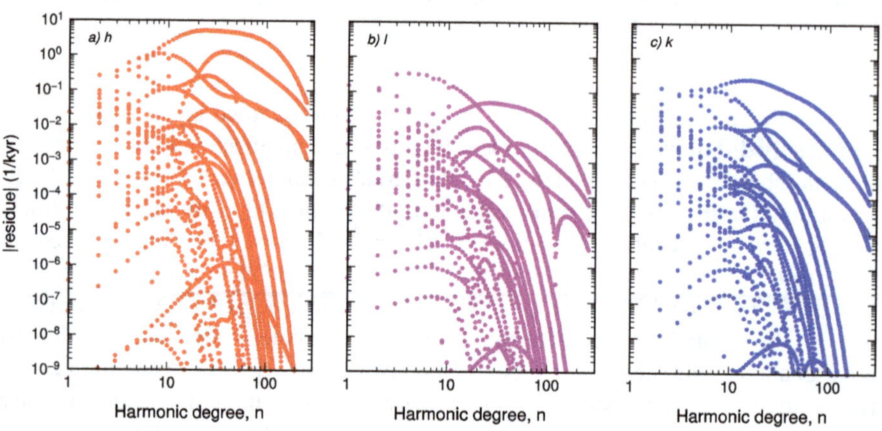

Fig. 6.3 Moduli of the viscous components of the h, ℓ and k LLNs for the 1-D model VM7, for harmonic degrees $1 \le n \le 256$

or with *ad hoc* numerical codes implementing the semi-analytical solution scheme described in Sect. C.7.

As discussed in Chap. 4, the *rotational response functions* $\mathcal{R}^{\text{rot}}(\gamma, t)$ require, in addition of the harmonic degree $n = 2$ viscoelastic LLNs, knowledge of the $n = 2$ tidal Love numbers. For model VM7, the elastic and fluid limits for the TLNs, as well as the viscoelastic residues, can be obtained following the same approach used to evaluate the corresponding quantities for the loading case, with the *proviso* that boundary conditions suitable for an external potential shall be prescribed in place of those for a surface load (see Sect. C.5). Tidal quantities needed to build the RRFs can also be evaluated with the open-source `TABOO` code. Since in this example we shall employ the *revised rotation theory* by Mitrovica et al. (2005) and Mitrovica and Wahr (2011), as discussed in Sect. 4.3.1.2, the fluid '*k*' TLN of an effectively fluid Earth is also needed. This quantity can be practically evaluated either by the `ALMA` Love numbers calculator (Spada 2008; Melini et al. 2022), by computing the long-timescale response of a variation of the VM7 model in which the lithosphere is assumed to be viscoelastic, or by explicitly building the solution for a layered fluid sphere through the fluid propagators discussed in Sect. C.4. For the VM7 model structure given in Table 6.1, we obtain $k_2^{\text{T,f}}(LT = 0) = 0.9427$ which, assuming an excess ellipticity correction $k^{\text{ee}} = 0.0082$ (Nakiboglu 1982; Mitrovica and Wahr 2011), yields a secular Love number $k^{\text{s}} = 0.9509$ (see Sect. 4.3.1.2).

6.2.3 *Ice Sheet Model*

The second essential ingredient for the solution of the SLE is a description of the spatio-temporal evolution of the ice sheets. This is provided by the $I(\gamma, t)$ function introduced in Sect. 1.4, corresponding to the ice thickness at a given point γ and a given time t. We remark that the solution of the SLE depends only upon the time history of the *variation* of the ice load $\mathcal{I}(\gamma, t) = I(\gamma, t) - I(\gamma, t_0)$, where t_0 denotes a reference equilibrium state (e.g., the LGM). However, since the actual ice thickness $I(\gamma, t_N)$ is relevant for the definition of the ocean function (see Sect. 1.4.1.1), it is convenient to prescribe an ice thickness evolution that at $t = t_N$ coincides with the present-day ice sheet thickness. In this example, we shall employ the ICE-7G_NA model by Roy and Peltier (2015, 2017), which has been developed on the basis of the ICE-6G(VM5a) GIA model (Argus et al. 2014; Peltier et al. 2015) with the aim of improving the agreement with geodetic observations in North America, but has been also shown to be broadly consistent with regional evidence of past variations of relative sea-level in the Mediterranean Sea (see Roy and Peltier 2018).

The spatio-temporal evolution of ice thickness according to ICE-7G_NA is made available on the home page of WR Peltier[6] as a set of global grids with 1° resolution, over a time range starting at 26 kyr BP and with a time step of 1 kyr until 20 kyr BP and 0.5 kyr afterwards. The total ice thickness variation between 26 kyr BP

[6] https://www.atmosp.physics.utoronto.ca/~peltier/data.php [last accessed on September 13, 2024].

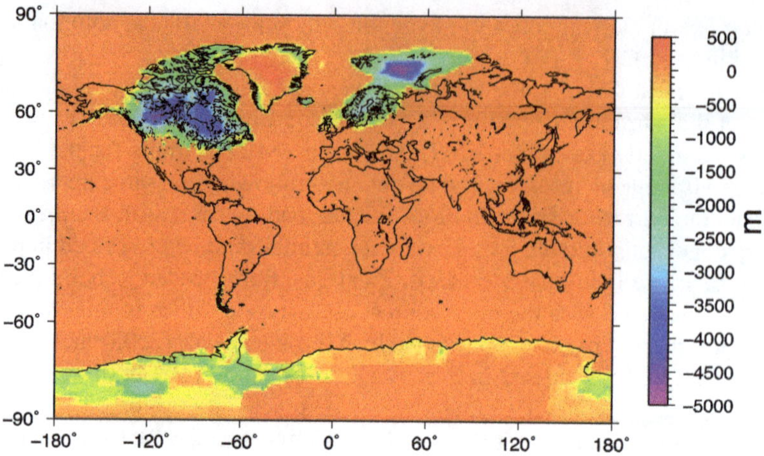

Fig. 6.4 Total ice thickness variation $\mathcal{I}_N = I(\gamma, t_N) - I(\gamma, t_0)$ for the ICE-7G GIA model of Roy and Peltier (2015, 2017)

and present time, namely $\mathcal{I}_N(\gamma) = I(\gamma, t_N) - I(\gamma, t_0)$, is shown in Fig. 6.4. It is apparent that most of the ice mass loss is distributed over North America, due to the melting of the Laurentide ice sheet which, at the Last Glacial Maximum, reached a thickness of about 4 km over a vast portion of Canada. The Fennoscandian ice sheet gives a substantially smaller contribution, with ice thickness in the range between 1 and 2 km across most of the formerly glaciated region. A minor contribution comes from Antarctica, where ice loss is concentrated across the Western Antarctica Ice Sheet (WAIS) and Antarctic Peninsula (AP). Remarkably, across internal regions of the Eastern Antarctica Ice Sheet (EAIS) and of the Greenland Ice Sheet (GRIS), ICE_7G assumes a small mass *gain* as denoted by the increase in ice thickness.

To better appreciate the time evolution of the melting of ice sheets it is useful to consider, at given time t, the total grounded ice mass $M^{\mathrm{i,gr}}(t)$. By a suitable normalization, this quantity is customarily expressed in terms of *equivalent sea level*:

$$\mathrm{ESL}(t) = \frac{M^{\mathrm{i,gr}}(t)}{\rho^{\mathrm{w}} A^{\mathrm{op}}}, \tag{6.3}$$

where ρ^{w} and A^{op} are the water density and the area of present-day oceans, respectively. When analysing a melting curve, it is convenient to consider the difference between the grounded ice mass at a given past epoch and that at present time, which is given by

$$\Delta\mathrm{ESL}(t) = \mathrm{ESL}(t) - \mathrm{ESL}(t_N), \tag{6.4}$$

and where, by recalling definitions (1.140) and (5.49), we have

$$\begin{aligned}\Delta \mathrm{ESL}(t) =& \frac{M^{\mathrm{i,gr}}(t) - M^{\mathrm{i,gr}}(t_N)}{\rho^{\mathrm{w}} A^{\mathrm{op}}} \\ =& \frac{M^{\mathrm{i,gr}}(t) - M^{\mathrm{i,gr}}(t_0)}{\rho^{\mathrm{w}} A^{\mathrm{op}}} - \frac{M^{\mathrm{i,gr}}(t_N) - M^{\mathrm{i,gr}}(t_0)}{\rho^{\mathrm{w}} A^{\mathrm{op}}} \\ =& \frac{\mu(t)}{\rho^{\mathrm{w}} A^{\mathrm{op}}} - \frac{\mu(t_N)}{\rho^{\mathrm{w}} A^{\mathrm{op}}} \\ =& \mathcal{S}^{\mathrm{eus}}(t_N) - \mathcal{S}^{\mathrm{eus}}(t)\,, \end{aligned} \tag{6.5}$$

where $\mathcal{S}^{\mathrm{eus}}(t)$ is the eustatic sea-level change defined in (5.49).

Figure 6.5 shows the grounded ice mass in the ICE-7G model in terms of equivalent sea level relative to present-day, with individual contributions from ice sheets in North America, Northern Europe, Greenland and Southern Hemisphere shown separately. The largest meltwater contribution to the total ESL of $\sim$130 m comes from the Laurentide ice sheet, with an ESL of $\sim$90 m, while the Fennoscandian ice sheet contributes with $\sim$25 m and the ice sheets in the southern hemisphere with $\sim$15 m. The smallest contribution (a few meters of ESL) comes from the Greenland ice sheet. Figure 6.5 shows that the melting rate alternates slow phases and rapid accelerations dubbed Meltwater Pulses (MWPs, Fairbanks 1989; Fairbanks et al. 1992), which occur at about 14 kyr BP (MWP-1a) and 11 kyr BP (MWP-1b), with most of the melting process ending at about 6 kyr BP. For a detailed geophysical analysis of the

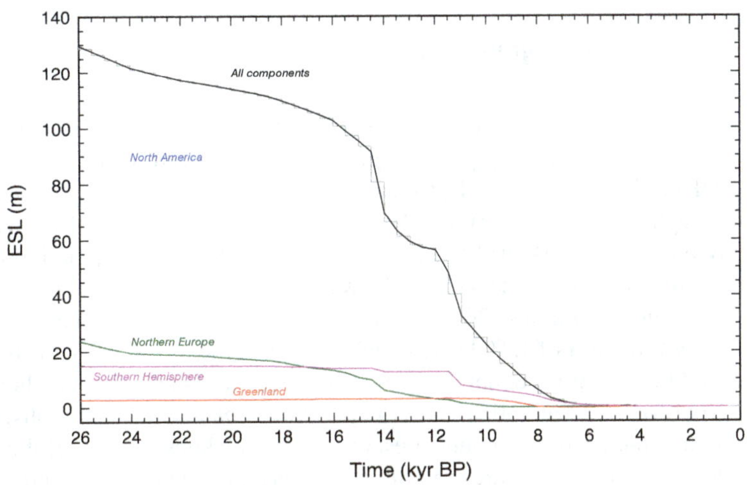

Fig. 6.5 Time-history of the Equivalent Sea Level (ESL) in the ICE-7G GIA model of Roy and Peltier (2015, 2017), since 26 kyrs BP. Individual contributions for the Southern Hemisphere (Antarctica, New Zealand and Patagonia), North America, Northern Europe and Greenland (also including Iceland), are shown in colours. The thin grey curve corresponds to the piecewise constant (PC) approximation employed in SELEN^4, with 500-year-long steps. Note that $\frac{d}{dt}\mathrm{ESL} = 0$ after time $\approx$ 6 kyrs BP, implying that the mass of the grounded ice remained constant thereafter

spatiotemporal evolution of ice sheets in the ICE-7G model, the readers are referred to the works of Roy and Peltier (2015, 2017, 2018).

To use the ICE-7G GIA model in the numerical solution scheme described in Appendix E, the ice thickness $I(\gamma, t)$ shall be discretised on the Tegmark grid pixels γ_p ($p = 1, \ldots, P$ and on a set of time steps $t_k = k\Delta t$ ($k = 0, \ldots, N$). Since the function $I(\gamma, t)$ is originally given on a uniform (but *not equal-area*) geographic grid with 1° resolution, its spatial discretisation on the points γ_p of the Tegmark grid can be accomplished through spherical interpolation, for instance using the `sphinterpolate` function of the Generic Mapping Tools (Wessel and Smith 1998; Wessel et al. 2013), which implements the algorithms by Renka (1997a, b). On the other hand, time discretisation requires the conversion from the original variable-step piecewise linear fuction into a piecewise constant function with uniform stepping. We choose a uniform time step of $\Delta t = 0.5$ kyr, corresponding to the smallest time step of the original ice model, and the resulting ESL curve is shown in gray in Fig. 6.5. Of course, it is possible to choose a smaller time discretisation step Δt, which will result in a better approximation of the original piecewise-linear time-history, but that will come at an increased computational cost. In any case, we remark that our spatio-temporal discretisation scheme shall always differ somewhat from that employed in the original ICE-7G model, hence, once again, we emphasise that we are solving the SLE for a *realisation* of the chosen GIA model on our computational grid.

6.2.4 Present-Day Topography

The third and last ingredient for the numerical solution of the SLE is a model of the present-day topography $T_N(\gamma) = T(\gamma, t_N)$, which is used to define the ocean function (OF) according to Eq. (1.96) and to obtain, by an iterative approach, the paleo-topography through the PT Eq. (1.105). A discretised model of present-day topography on the Tegmark grid can be obtained by spherical interpolation from a global relief model; in the present example, we use the ETOPO1 digital topographic model by Amante and Eakins (2009), which is shown in Fig. 6.6. In particular, we use the *bedrock* version of ETOPO1, which describes the topography of the solid Earth without taking into account the present-day ice sheets and glaciers; hence, in the figure, portions of Antarctica and Greenland have *negative* topography. However, when also taking into account present-day ice thickness $I(\gamma, t_N)$, the ocean function defined in (1.96) ensures that those regions are actually considered "dry" (i.e., $O(\gamma_p, t_N) = 0$; see Fig. 6.7). We remark that, to improve the quality of the SLE solution, it is important to employ a present-day topographic data set that accurately describes topography and bathymetry in Antarctica, particularly in regions where modern ice is not grounded (Purcell et al. 2016; Peltier et al. 2018). In this respect, it would be convenient to merge the ETOPO1 global relief model with an high-resolution elevation model of Antarctica such as Bedmap2 (Fretwell et al. 2013).

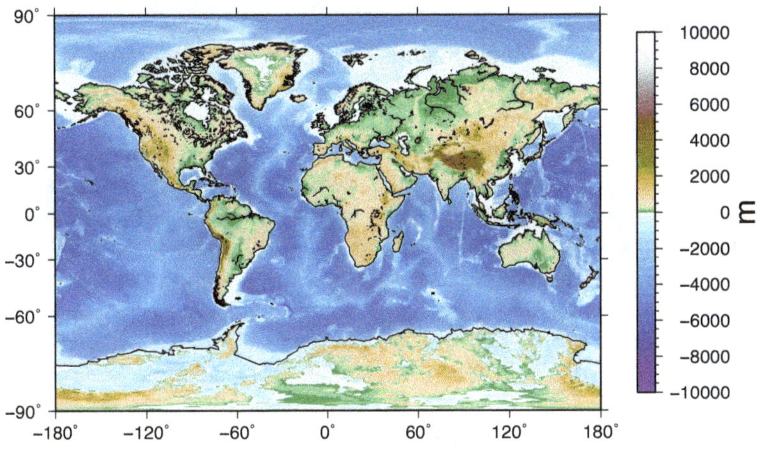

Fig. 6.6 Present-day bedrock topography according to the ETOPO1 global relief model (bedrock version) by Amante and Eakins (2009). The original topographic grid, given with 1 arc-minute resolution, has been resampled here on a 1° grid for the ease of plotting

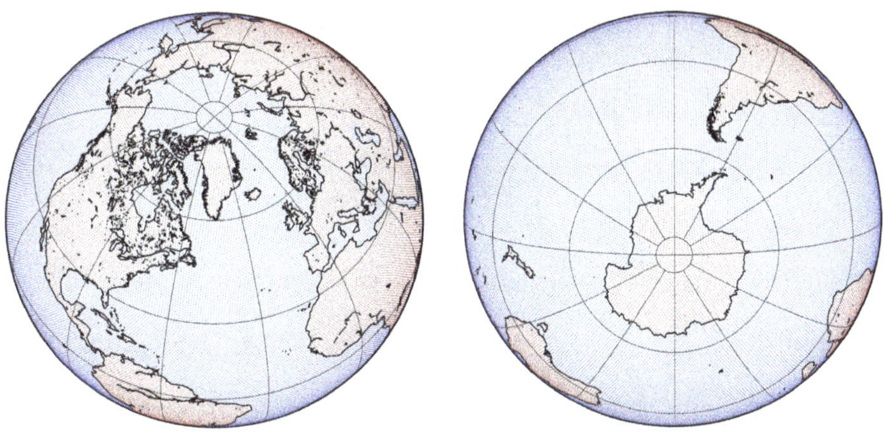

Fig. 6.7 Discretised present-day ocean function $O_{p,N}$ for a grid with resolution $R = 44$. In blue and brown pixels, the ocean function is $O_{p,N} = 1$ and $O_{p,N} = 0$, respectively

6.3 Convergence of the Numerical Solution

In the previous section we have described the Earth rheological profile and the spatio-temporal evolution model of ice sheets that we will use to obtain the numerical solution of the SLE discussed in this chapter. These represent the two basic inputs for the SLE and, once they are defined, the solution is completely determined. However, as we have shown in Chap. 5, due the implicit nature of the SLE it is not possible to obtain an exact solution, except if some simplifying assumptions are invoked.

For more realistic cases, such as the one described here, an approximate solution of the SLE shall be obtained through an iterative approach. It is therefore of particular importance to assess the convergence of the iterative solution scheme and to define a stopping criterion corresponding to the desired numerical accuracy of the solution.

As we have discussed in detail in Chap. 5, the numerical solution of the SLE involves two nested iterations. In the inner (or *internal*) iteration, an estimate of sea level change $\mathcal{S}(\gamma, t)$ is obtained by starting from an initial *ansatz*, using it to evaluate the right-hand side of Eqs. (5.28) or (5.38) to obtain an updated estimate, and iterating the process until some stopping criterion is satisfied. During the process, the ocean function $O(\gamma, t)$ is kept constant. In the outer (or *external*) iteration, once an acceptable solution for $\mathcal{S}$ is obtained, topography $T(\gamma, t)$ is updated according to the PT equation (1.105), a new estimate of the ocean function is obtained from (1.96), and a new internal iteration is executed with the updated ocean function. As initial conditions for the iterative scheme, it customarily assumed $\mathcal{S}(\gamma, t) \equiv \mathcal{S}^{\text{ave}}(t)$ and $O(\gamma, t) \equiv O(\gamma, t_N)$.

We start by studying the convergence of the internal iterations, a process during which the unknown field $\mathcal{S}(\gamma, t)$ is obtained for a given estimate of the ocean function. If $\mathcal{S}^{(n)}(\gamma, t)$ is the estimate of $\mathcal{S}$ after n internal iterations, with $\mathcal{S}^{(0)}(\gamma, t) = \mathcal{S}^{\text{ave}}(t)$ being the initial estimate, in analogy with (5.55) we define an average relative difference between internal iterations as

$$\Delta_n^{\text{int}} = \frac{\displaystyle\int_0^{t_N} \int_{\text{e}} \left|\mathcal{S}^{(n)}(\gamma, t) - \mathcal{S}^{(n-1)}(\gamma, t)\right| \mathrm{d}\gamma \, \mathrm{d}t}{\displaystyle\int_0^{t_N} \int_{\text{e}} \left|\mathcal{S}^{(n)}(\gamma, t)\right| \mathrm{d}\gamma \, \mathrm{d}t}, \tag{6.6}$$

whose discretised version, according to the scheme discussed in Appendix E, reads

$$\Delta_n^{\text{int}} = \frac{\displaystyle\sum_{p,k} \left|\mathcal{S}_{p,k}^{(n)} - \mathcal{S}_{p,k}^{(n-1)}\right|}{\displaystyle\sum_{p,k} \left|\mathcal{S}_{p,k}^{(n)}\right|}, \tag{6.7}$$

where $\mathcal{S}_{p,k}^{(n)} = \mathcal{S}^{(n)}(\gamma_p, t_k)$ is the discretised solution of the SLE after n internal iterations.

Figure 6.8a shows Δ_n^{int} as a function of n for the SLE setup described in Sect. 6.2 above. Internal convergence turns out to be quite regular, with Δ_n^{int} decreasing well below the 10^{-2} level after three internal iterations and reaching the 10^{-4} level at $n = 5$. The results shown in Fig. 6.8a have been obtained performing only one external iteration, i.e. assuming no horizontal migration of shorelines; however, the internal convergence remains essentially identical when more than one external iteration is performed. For the GIA model that we are discussing in this chapter, we will stop the

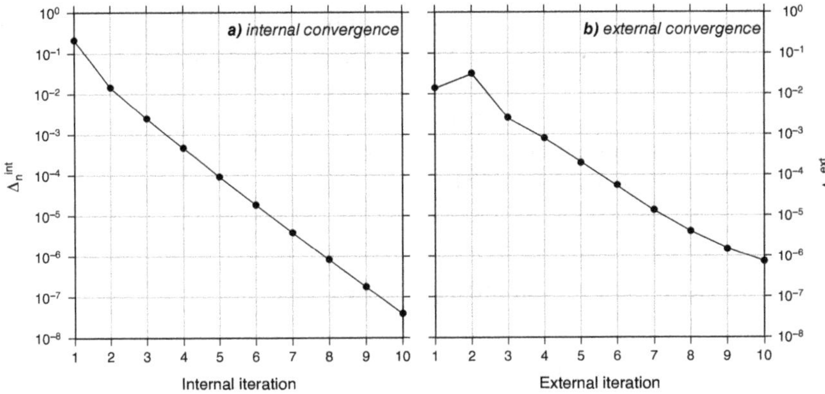

Fig. 6.8 Numerical convergence of the SLE for the example discussed in this chapter. Frame **a** shows $\Delta_n^{\rm int}$ defined in Eq. (6.6) as a function of the number of internal iterations, assuming that $O(\gamma, t)$ is constant in time and equal to the present-day OF. Frame **b** shows $\Delta_n^{\rm ext}$ defined in Eq. (6.8) as a function of the number of external iterations, using ten internal iterations to solve the SLE

solution at $n = 5$ internal iterations, which turns out to represent a good compromise between computational speed and convergence of the solution for most practical applications.

We now discuss the convergence of the external iterations, in which the ocean function $O(\gamma, t)$ is progressively refined on the basis of the solution of the SLE and of the PT equation (1.105). In close analogy with (6.6), we quantify the convergence of external iterations by means of

$$\Delta_n^{\rm ext} = \frac{\int_0^{t_N} \int_{\rm e} \left| O^{(n)}(\gamma, t) - O^{(n-1)}(\gamma, t) \right| {\rm d}\gamma \, {\rm d}t}{\int_0^{t_N} \int_{\rm e} \left| O^{(n)}(\gamma, t) \right| {\rm d}\gamma \, {\rm d}t}, \tag{6.8}$$

where $O^{(n)}(\gamma, t)$ is the estimate of the ocean function after n external iterations, and $O^{(0)}(\gamma, t) = O(\gamma, t_N)$ is its initial estimate, in which the ocean function is assumed to be constant in time and equal to that determined by present-day topography. The discretised version of (6.8) is

$$\Delta_n^{\rm ext} = \frac{\sum_{p,k} \left| O_{p,k}^{(n)} - O_{p,k}^{(n-1)} \right|}{\sum_{p,k} \left| O_{p,k}^{(n)} \right|}, \tag{6.9}$$

where $O_{p,k}^{(n)} = O^{(n)}(\gamma_p, t_k)$ is the discretised ocean function estimated after n external iterations.

Figure 6.8b shows Δ_n^{ext} as a function of the number of external iterations. After the first two iterations, the external convergence becomes regular, with Δ_n^{ext} decreasing below the 10^{-2} level for $n = 3$ and reaching about $\sim 2 \times 10^{-4}$ for $n = 5$. For the numerical example discussed in this chapter, we have chosen to stop the iterative scheme after 5 external iterations, so that according to Fig. 6.8 both the internal and external convergences will be such that $\Delta_n^{\text{int}} \sim \Delta_n^{\text{ext}} \sim 10^{-4}$.

6.4 Predictions of Observable GIA Quantities

In the previous sections we have discussed the discretisation parameters, the input quantities (the Earth's rheology and the ice thickness spatio-temporal evolution) and the numerical convergence for our example solution of the SLE, which shall be referred to as I7G-R44 to remark that it represents a *realisation* of the ICE-7G_NA model on a $R = 44$ Tegmark grid. In what follows, we shall discuss some quantities of geophysical interest that can be obtained from the numerical solution of the SLE. All the results have been computed with the `SELEN`[4] open-source code available from https://github.org/geodynamics/selen. The readers can configure the code with the input data and parameters summarised in Table 6.2 to reproduce the results described in the rest of the chapter or to further explore their sensitivity to the various parameters.

Table 6.2 Summary of input data, parameters and sources employed in the I7G-R44 solution of the SLE discussed in the present chapter

Ice model	ICE-7G_NA (Roy and Peltier 2015, 2017)
Viscosity profile	VM7 (see Table 2 of Roy and Peltier 2017)
Density and rigidity profiles	Volume-averaged PREM (Dziewonski and Anderson 1981)
Mantle layers	$N_V = 8$
Other layers	Elastic lithosphere and inviscid core
Present-day topography	ETOPO1 (Amante and Eakins 2009)
Tegmark resolution	$R = 44$ (75, 692 pixels)
Maximum harmonic degree	$n_{\text{max}} = 256$
External iterations	5
Internal iterations	5
Rotational feedback	Revised rotational theory (Mitrovica et al. 2005; Mitrovica and Wahr 2011)

6.4.1 Past Sea-Level Variations

The unknown in the Sea Level Equation is the sea-level change $\mathcal{S}(\gamma, t)$ with respect to the equilibrium state at $t \leq t_0$. Past sea levels at a given location can be determined on the basis of geological, biological or archaeological evidence (Pirazzoli 1991, 1993), hence it is possible to directly compare a solution of the SLE with observations to assess how a given GIA model agrees with available data or to tune its input parameters (rheological model and spatio-temporal evolution of ice-sheets) in order to improve the fit with observational constraints. While, in our formulation of the SLE, sea-level change is expressed with respect to the initial equilibrium state before the onset of deglaciation, observations of past sea levels are usually expressed as *relative sea level* (RSL), assuming present-day ($t = t_N$) as the reference state. Relative sea levels can be easily obtained by

$$\mathrm{RSL}(\gamma, t) = \mathcal{S}(\gamma, t) - \mathcal{S}(\gamma, t_N)\,, \tag{6.10}$$

where $\mathcal{S}(\gamma, t)$ is sea-level change, i.e., the solution of the SLE.

Figure 6.9 shows the locations of a set of 12 sites from the Tushingham and Peltier (1993) RSL database,[7] which we have chosen to be representative of formerly glaciated regions of North America (blue) and Northern Europe (red), of Antarctica (orange) and of the regions in the far-field of the late-Pleistocene ice complexes (green). For each site, predicted RSL curves obtained from the I7G-R44 solution

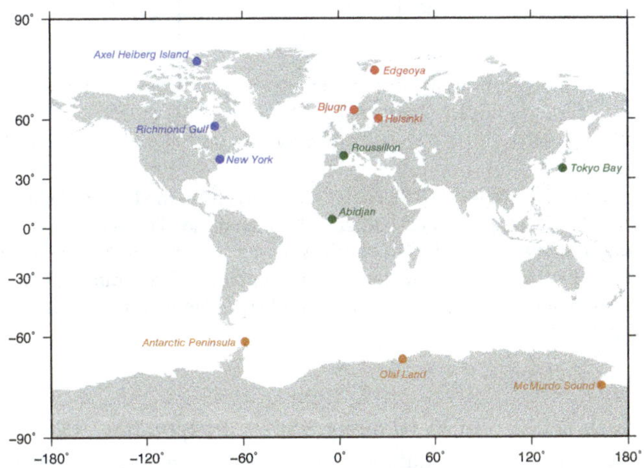

Fig. 6.9 Locations of twelve sites from the Tushingham and Peltier (1993) database for which the synthetic relative sea level (RSL) curves shown in Fig. 6.10 have been computed

---

[7] The data are included in the SELEN[4] package. The database is also available from the ftp site ftp://ftp.ncdc.noaa.gov/pub/data/paleo/paleocean/relative_sea_level/sealevel.dat [last accessed on December 29, 2024].

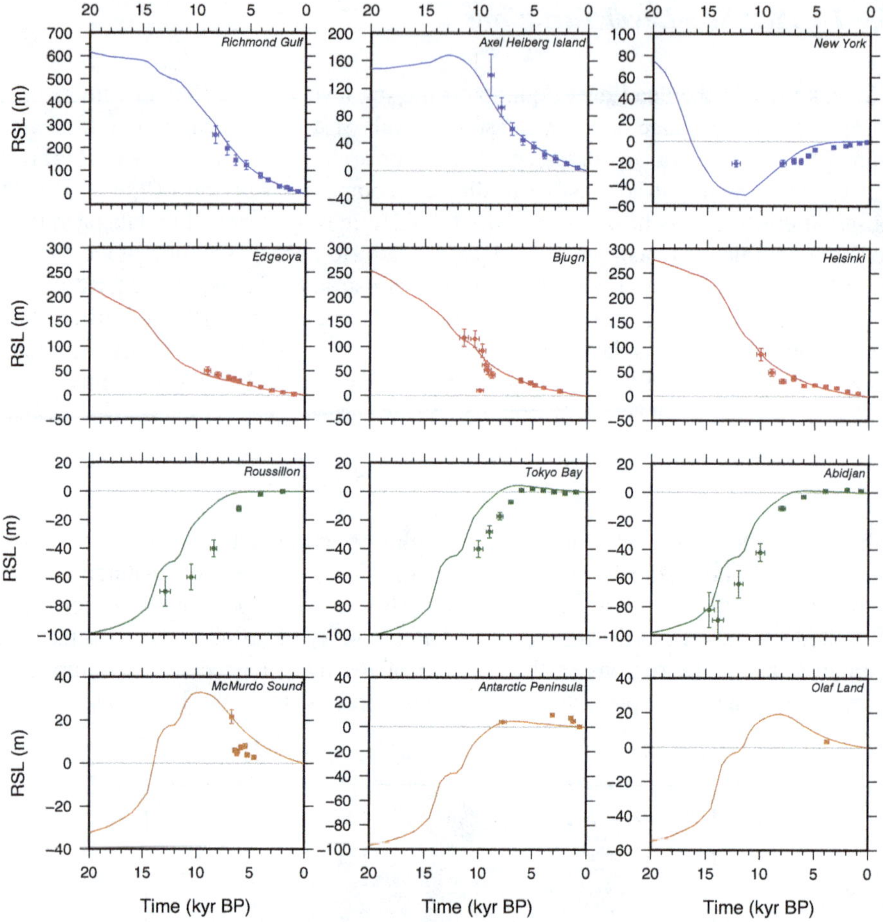

Fig. 6.10 Synthetic relative sea level (RSL) curves obtained from the I7G-R44 solution of the SLE at twelve sites from the Tushingham and Peltier (1993) database. The sites are shown in Fig. 6.9. From top to bottom, the rows show sites in North America (in blue), in Northern Europe (red), in the far field of formerly glaciated regions (green), and in Antarctica (orange). For each site, RSL data and the associated uncertainties are also shown. Please note that, for the ease of viewing, the panels have different vertical scales

of the SLE are shown in Fig. 6.10 along with corresponding RSL data from the Tushingham and Peltier (1993) database. It is important to note that the predicted RSL curves are obtained assuming a Maxwell rheology for the mantle; a discussion of possible effects of the transient Burgers rheology on the history of sea level is presented by Simon et al. (2022) and Lau (2023).

From a visual inspection of the RSL curves in Fig. 6.10, remarkable differences between sites are evident, since the amplitude and temporal dependence of each curve is related to the specific *RSL zone* in which the site is located (see Sect. 5.2).

Sites in the formerly glaciated regions of North America and Northern Europe are characterised by a sea level fall, due to the combined effects of solid Earth uplift in response to the melting of the ice load and of the loss of gravitational attraction of the oceans by the ice mass. Sites located in the peripheral regions in proximity of the former ice sheets, such as New York, show a first phase of emergence followed by sea-level rise due to the collapse of the lateral bulges (see Sect. 3.4). In the far field of the formerly glaciated regions, the three sites show considerably similar RSL curves characterised by sea-level rise, since in that case the sea-level change is mostly sensitive to the meltwater input and only marginally affected by the local hydro-isostatic response or by self-gravitation effects. For the RSL sites of Tokyo Bay and Abidjan, the curves reach a maximum at $\sim 6 \div 7$ kyr BP, after which RSL slowly falls to zero. Those sea-level *highstands* are related to hydro-isostatic effects such as "continental levering" (Mitrovica and Milne 2002). For sites in Antarctica, the RSL curves show a first phase of sea-level rise similar to that observed for far-field sites, since the ICE-7G_NA GIA model assumes almost no change in the Antarctica ice sheet until $\sim 14 \div 10$ kyr BP (see Fig. 6.5). After the onset of ice sheet melting in the Southern Hemisphere, RSL curves turn into a phase of emergence associated with the Earth's response to unloading.

Figure 6.10 shows that for some sites there is a quite good agreement between synthetic RSL curves and observations of past sea levels, while at other sites there are remarkable differences between modelled and observed sea levels. While a complete discussion of the fit of the ICE-7G_NA(VM7) GIA model with the Tushingham and Peltier (1993) database would be beyond the scope of present book, those differences may be due to geological processes (such as tectonic deformations or sediment loading) contributing to local vertical movements, or to the effects of lateral heterogeneities in the Earth structure that are not accounted for by the spherically symmetric rheological model adopted here. We also remark that the ICE-7G_NA(VM7) has been tuned to fit a different set of paleo-sealevel observations, so there is no *a priori* requirement that the model should fit the specific Tushingham and Peltier (1993) RSL database.

The spatial distribution of the different sea-level signatures discussed above is shown in Fig. 6.11. In the map, the colours correspond to four possible time evolutions of RSL curves from 20 kyr BP to present: continuous (i.e., monotonic) *emergence* or *submergence*, a phase of emergence followed by submergence, and *vice versa*. In regions located beneath the centres of the former ice complexes in North America and Northern Europe the SLE predicts a continuous sea-level fall, while regions located in the periphery of formerly glaciated areas show alternating phases of emergence and submergence related to the migration and collapse of the lateral bulges. Most (but not all) coastal regions are characterised by a sea-level highstand, which is also present in a wide oceanic region whose shape suggests a contribution from rotational feedback. In regions where sea-level change is not monotonic, the timing and relative importance of the emergence and submergence phases is controlled by various factors, including the distance from the former ice sheets and the melting time history of the different complexes. Hence, a comprehensive interpretation of patterns shown in Fig. 6.11 would demand a detailed analysis. In general, the four

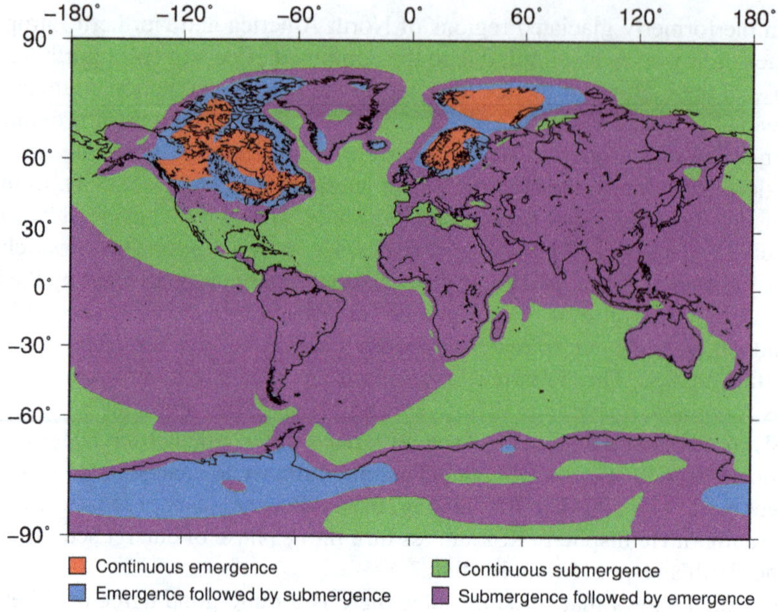

Fig. 6.11 Spatial distribution of different signatures of sea-level change predicted by the I7G-R44 solution of the SLE from 20 kyr BP to present. The four zones evidenced in this map are related to the *Clark's zones* introduced in Clark et al. (1978) and discussed in Sect. 5.2. Note that, as a consequence of (1.98), sea-level change is defined on both ocean and land regions

zones in Fig. 6.11 can be related to the *Clark's zones* discussed in Sect. 5.2 and shown in Fig. 5 of Clark et al. (1978). However, it should be taken into account that those results were obtained assuming a significantly different spatio-temporal evolution for the ice sheets (in which, for instance, there is no contribution from Antarctica) and without including the rotational feedback on sea level.

6.4.2 *Ice Age Paleo-Topography*

According to Peltier (1994), knowledge of the history of sea-level change $\mathcal{S}(\gamma, t)$ allows to reconstruct the *paleo-topography* at a given epoch in the past by virtue of the so-called PT Eq. (1.105). We note that, taking into account the definition of relative sea level (6.10), the PT equation can also be cast in the form

$$T(\gamma, t) = T(\gamma, t_N) - \mathrm{RSL}(\gamma, t)\,, \tag{6.11}$$

where $\mathrm{RSL}(\gamma, t)$ is given by Eq. (6.10).

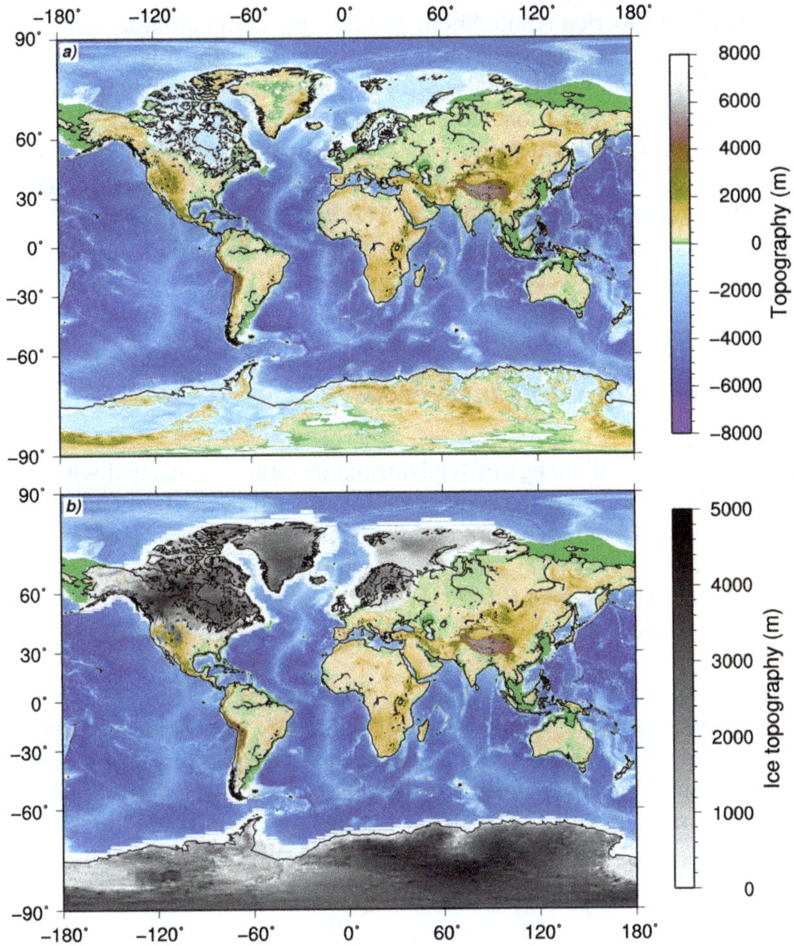

Fig. 6.12 Paleo-topography at 20 kyr BP according to the I7G-R44 solution of the SLE. Panel **a** shows the bedrock paleo-topography $T(\gamma, t)$ while **b** shows true paleo-topography, i.e., $T(\gamma, t) + I(\gamma, t)$. The color scale represents elevation of ice-free regions or regions with floating ice, while the grey scale represents elevation of regions with grounded ice (above or below sea level)

Figure 6.12a shows paleo-topography $T(\gamma, t)$ at 20 kyr BP obtained by the PT equation from our GIA solution I7G-R44. Since T represents elevation of the solid Earth (i.e., the *bedrock topography*), wide regions in North America, Northern Europe and Antarctica turn out to be below sea level ($T < 0$) as a consequence of the deformation due to the load exerted by ice sheets. Figure 6.12b shows *true paleo-topography*, defined as $T(\gamma, t) + I(\gamma, t)$, with the elevation of ice-covered regions represented by a grey scale.

The maps in Fig. 6.12 show the major land regions that were emerged at the Last Glacial Maximum. These include Beringia, Sundaland, Patagonia, Sahul, Doggerland, and the East Siberian Arctic Shelf. The time evolution of these regions during the last deglaciation, and their relevance in the context of the migrations of human populations, have been widely analysed in literature. For a detailed discussion, we refer the readers to e.g., Dobson and Dobson (1999); Dobson (2014); Dobson et al. (2020), for some general considerations to Hopkins et al. (1967); Jakobsson et al. (2017) and to Lodolo et al. (2020); Dobson et al. (2021) for a focus on some specific regions. The importance of spatially variable effects associated with the physical processes embedded in the SLE has been emphasised, in particular, by Spada and Galassi (2017) and more recently by Borreggine et al. (2022).

6.4.3 Present-Day GIA Fingerprints

The present-day rates of change of fundamental quantities associated with the SLE have a particular relevance in the context of GIA, since they can be directly observed through ground-based or space-borne geodetic instruments. These present-day rates are referred to as *GIA fingerprints*, in analogy with the signals dubbed as *finger-prints functions* by Plag and Jüettner (2001) that manifest contemporary variations in the state of the cryosphere driven by global warming (see also Spada 2017; Spada and Melini 2019a).

The GIA fingerprints considered here are the present-day rates of relative sea-level change ($\dot{\mathcal{S}}$), of vertical land motion (VLM, $\dot{\mathcal{U}}$), of absolute sea level ($\dot{\mathcal{N}}$), and of geoid height change ($\dot{\mathcal{G}}$). From a practical standpoint, they can be computed numerically by introducing a future time-step $t_{N+1} = t_N + \Delta t$ in the discretised time-history of the ice load, with $I(\gamma, t_{N+1}) = I(\gamma, t_N)$, and approximating the present-day rates as

$$\dot{\mathcal{F}}(\gamma, t_N) \approx \frac{\mathcal{F}(\gamma, t_{N+1}) - \mathcal{F}(\gamma, t_{N-1})}{2\,\Delta t}, \tag{6.12}$$

where symbol $\mathcal{F}$ stands for any of the scalar fields $\mathcal{S}$, $\mathcal{U}$, $\mathcal{N}$, and $\mathcal{G}$, and Δt is the discretisation time step.

Figure 6.13 shows the GIA fingerprints for the I7G-R44 SLE solution, evaluated according to Eq. (6.12). The spatial variability of the fingerprints is a consequence of the physical interactions between the ice sheets, the solid Earth and the oceans, including deformation, gravitational attraction and rotational feedbacks. The most evident feature is a marked sea-level fall in formerly glaciated regions in North America end Northern Europe, surrounded by swaths of sea-level rise. Those patterns are mostly due to significant rates of vertical uplift of the solid Earth under the former ice loads (up to several mm/yr), accompanied by the subsidence in the peripheral regions associated with the collapse of forebulges (see Sect. 3.4.3), and manifest a state of isostatic disequilibrium due to the still-ongoing readjustment after the melting of the ice loads. The imprint of GIA on vertical land motions is evident in observed rates of deformation from global GNSS networks (Fig. 6.14a), with isostatic readjustment representing the major contribution to present-day vertical movements

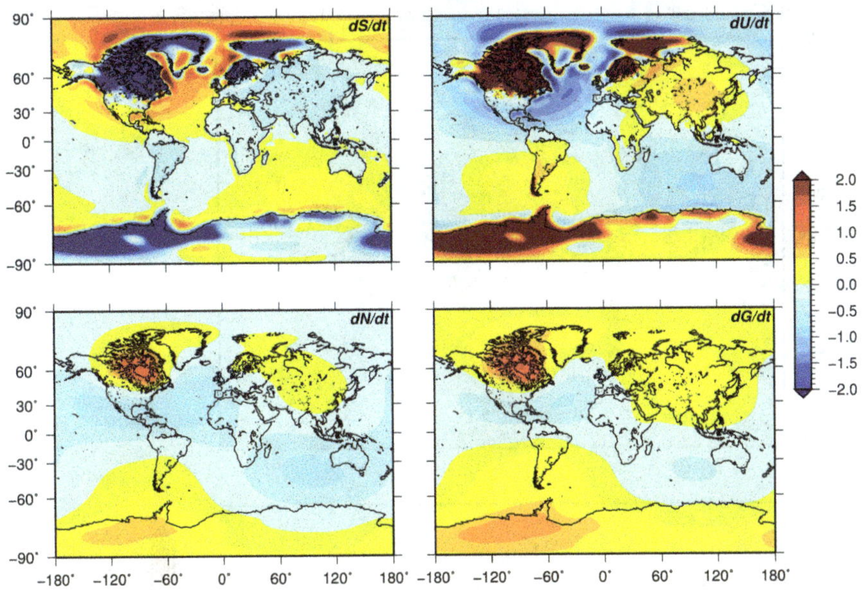

Fig. 6.13 Present-day GIA fingerprints for sea-level change $\dot{\mathcal{S}}(t_N)$, VLM $\dot{\mathcal{U}}(t_N)$, sea surface height $\dot{\mathcal{N}}(t_N)$, and geoid height $\dot{\mathcal{G}}(t_N)$ according to the I7G-R44 solution of the SLE. For the ease of viewing, the vertical scale is saturated; the full range of variability for each fingerprint is -12.7 to 3.1 mm/yr for $\dot{\mathcal{S}}$, -3.0 to 13.9 mm/yr for $\dot{\mathcal{U}}$, -0.8 to 1.4 mm/yr for $\dot{\mathcal{N}}$, and -0.5 to 1.6 mm/yr for $\dot{\mathcal{G}}$

in formerly glaciated regions (see, e.g., Kreemer et al. 2018). Fingerprints for $\dot{\mathcal{N}}$ and $\dot{\mathcal{G}}$ have considerably smaller amplitudes than $\dot{\mathcal{S}}$ and $\dot{\mathcal{U}}$, with the most prominent feature being a vertical movement of the sea surface height over North America at a rate of ≈ 1 mm/yr. The four-lobed pattern displayed by all the fingerprints manifests the effect of the rotational contribution, which is characterised by a strong symmetry of harmonic degree $n = 2$ and order $m = 1$.

We note that, by virtue of the SLE in its basic form (5.7), the four fingerprints shown in Fig. 6.13 are not independent of one another since the relation

$$\dot{\mathcal{S}}(\gamma, t) = \dot{\mathcal{N}} - \dot{\mathcal{U}} \tag{6.13}$$

holds true at all points $\gamma = (\theta, \lambda)$ and at any epoch t. In regions close to the former ice sheets, where the amplitudes of $\dot{\mathcal{S}}$ and $\dot{\mathcal{U}}$ are considerably larger than that of $\dot{\mathcal{N}}$, (6.13) leads to the crude approximation $\dot{\mathcal{S}} \approx -\dot{\mathcal{U}}$. However, in principle the SLE makes possible the apparently paradoxical condition of a sea-level rise in uplifting regions (or, *vice versa*, of sea-level fall in subsiding regions), if values of $\dot{\mathcal{N}}$ are such that (6.13) is valid. In addition, as a consequence of (5.8), fingerprints $\dot{\mathcal{N}}$ and $\dot{\mathcal{G}}$ are related by

$$\dot{\mathcal{N}}(\gamma, t) = \dot{\mathcal{G}} + \dot{c}\,, \tag{6.14}$$

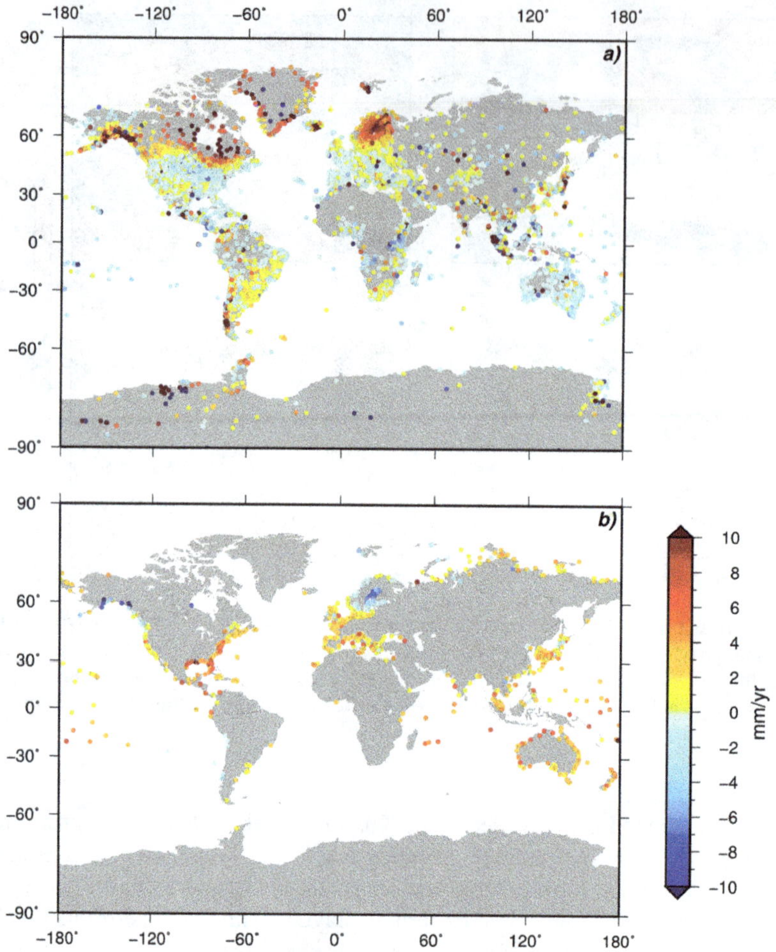

Fig. 6.14 Observations of present-day rates of vertical land motion (VLM) from global GNSS networks estimated by the Nevada Geodetic Laboratory (Blewitt et al. 2018) in the IGS14 reference frame **(a)** and rates of sea-level change estimated by the Permanent Service for Mean Sea Level **(b)**. Rates of VLM and of sea-level change have been downloaded respectively from the NGL website http://geodesy.unr.edu/vlm.php and from the PSMSL website https://psmsl.org/products/trends/ [both accessed on October 2, 2024]

where we recall that c is the FC76 constant that ensures mass conservation in the (oceans+ice) system. Hence, fingerprints $\dot{\mathcal{N}}$ and $\dot{\mathcal{G}}$ shown in Fig. 6.12 differ only by the spatially invariant term $\dot{c}$.

Of particular interest are the global whole-Earth and ocean averages of the GIA fingerprints, considered in Table 6.3. Since these have been discussed in detail by Spada and Melini (2019a), here we only have a cursory look at these values. We observe that $\langle\dot{\mathcal{U}}\rangle^e$ and $\langle\dot{\mathcal{G}}\rangle^e$ vanish because of mass conservation. We also note that,

Table 6.3 Whole Earth and ocean averages of GIA fingerprints $\dot{\mathcal{S}}$, $\dot{\mathcal{U}}$, $\dot{\mathcal{N}}$ and $\dot{\mathcal{G}}$ obtained from the I7G-R44 SLE solution. By value +0.000 we denote small positive numerical results that are zero within the displayed precision

Fingerprint $\dot{\mathcal{F}}$	Earth average $\langle\dot{\mathcal{F}}\rangle^{\mathrm{e}}$ (mm/yr)	Ocean average $\langle\dot{\mathcal{F}}\rangle^{\mathrm{o}}$ (mm/yr)
$\dot{\mathcal{S}}$	−0.227	−0.001
$\dot{\mathcal{U}}$	+0.000	−0.280
$\dot{\mathcal{N}}$	−0.227	−0.281
$\dot{\mathcal{G}}$	+0.000	−0.054

because of Eq. (5.8), the time derivative of the FC76 constant is $\dot{c} = \langle\dot{\mathcal{N}}\rangle^{\mathrm{e}} = -0.227$ mm/yr for the GIA model employed here. Furthermore, $\langle\dot{\mathcal{S}}\rangle^{\mathrm{e}} = \langle\dot{\mathcal{N}}\rangle^{\mathrm{e}}$ by virtue of (6.13). It is also interesting that $\langle\dot{\mathcal{S}}\rangle^{\mathrm{o}}$ is numerically found to be *very* close to the value $\langle\dot{\mathcal{S}}\rangle^{\mathrm{o}} \equiv 0$ that we would have obtained analytically adopting the FC76 theory, i.e. neglecting the time variations of the ocean function and the rotational effects (see Sect. 5.3.4); as a consequence, $\langle\dot{\mathcal{U}}\rangle^{\mathrm{o}}$ matches $\langle\dot{\mathcal{N}}\rangle^{\mathrm{o}}$ very closely. Taking the time derivative of Eq. (5.16) provides an explicit form for $\langle\dot{\mathcal{S}}\rangle^{\mathrm{o}} \equiv \dot{\mathcal{S}}_{ave}$; hence, the small (but non-zero) present-day value of $\langle\dot{\mathcal{S}}\rangle^{\mathrm{o}}$ in Table 6.3 may reflect minor variations of OF the associated with the variation of the area of the oceans, or small changes of the grounded ice mass $\mu(t)$, or both. Finally, we note that $\langle\dot{\mathcal{N}}\rangle^{\mathrm{o}}$ is the value that is customarily adopted to correct ocean-averaged sea-level altimetry data for the effects of GIA, generally close to −0.3 mm/yr for the GIA models available in the literature (see e.g., Tamisiea 2011; Spada and Melini 2019a).

6.4.4 Sea-Level Change at Tide Gauges

The evaluation of the $\dot{\mathcal{S}}$ GIA fingerprint at the locations of tide-gauge (TG) sites is of particular importance since it contributed to the interpretation of the instrumentally observed sea-level variations (Douglas 1991, 1997; Jevrejeva et al. 2014). Time series recorded by TG networks such that of the Permanent Service for Mean Sea Level (PSMSL, https://psmsl.org/) are commonly used to evaluate secular trends of present-day sea-level rise at global or regional scales. Those trends, displayed in Fig. 6.14b, show a general pattern of sea-level rise due to present-day increase in global temperatures (Douglas 1991, 1992, 1997). However, a sea-level fall is evident on TGs in Scandinavia and on the few observations available in the Hudson Bay, which is interpreted as the effect of GIA. It is therefore of paramount importance, for a correct estimation of the climate-driven global sea-level rise, to quantify the contribution of GIA on the secular trends recorded at TGs.

Table 6.4 lists the 23 TG sites which have been considered by Douglas (1997, D97 hereinafter) to estimate the global mean sea-level rise (GMSLR). Those sites have been selected according to a set of rigorous criteria, including the length and com-

Table 6.4 *In situ* rates of present-day sea-level change at the 23 TGs considered by Douglas (1997) for his estimate of global mean sea-level rise, as redetermined by Spada and Galassi (2012), and corresponding GIA rates of sea-level change according to our I7G-R44 SLE solution. Note that with +0.0 and −0.0 we denote small positive and negative numerical results, respectively, which are zero within the displayed precision

Region	n.	TG site	Observed rate (mm/yr)	GIA rate (mm/yr)
English channel	1	Newlyn	1.8 ± 0.1	+0.3
	2	Brest	1.4 ± 0.1	+0.3
Atlantic	3	Cascais	1.3 ± 0.1	+0.2
	4	Lagos	1.4 ± 0.2	−0.2
	5	S. C. Tenerife	1.6 ± 0.2	+0.3
Mediterranean sea	6	Marseille	1.2 ± 0.1	−0.0
	7	Genova	1.2 ± 0.1	−0.2
	8	Trieste	1.3 ± 0.1	−0.4
New Zealand	9	Auckland	1.3 ± 0.1	−0.2
	10	Dunedin	1.2 ± 0.1	−0.1
	11	Wellington	2.0 ± 0.3	−0.2
Pacific	12	Honolulu	1.5 ± 0.1	+0.0
North American West Coast	13	San Francisco	1.6 ± 0.1	+0.2
	14	Santa Monica	1.4 ± 0.2	+0.4
	15	La Jolla	2.1 ± 0.1	+0.4
	16	San Diego	2.1 ± 0.1	+0.4
Central America	17	Balboa	1.5 ± 0.1	+0.0
	18	Cristobal	1.4 ± 0.1	+0.1
	19	Quequen	0.9 ± 0.2	−0.5
	20	Buenos Aires	1.6 ± 0.2	−0.5
South East North America	21	Pensacola	2.2 ± 0.2	+0.4
	22	Key west	2.3 ± 0.1	+0.4
	23	Fernandina	2.0 ± 0.1	+0.3
Average			1.6 ± 0.1	+0.1

pleteness of the time series, the absence of nearby collisional tectonic boundaries, and the distance from previously ice-covered regions. Figure 6.15 shows observed rates of sea level change at the D97 sites, as they have been redetermined by Spada and Galassi (2012), as well as modeled values of $\dot{S}$ according to our I7G-R44 SLE solution; numerical values for both observed and modeled rates are listed in Table 6.4. The current rates of sea-level change due to GIA are found in the range between −0.5 and +0.4 mm/yr, which at some sites represent a significant fraction of the observed secular trend (over 50% at Quequen, on the Atlantic coast of South America). However, the average contribution to GIA at the 23 D97 sites is very close to zero, as we would expect for an ideally regular distribution of TGs across the oceans (see Table 6.3). This indicates that GIA only marginally affects the GMSLR estimate of

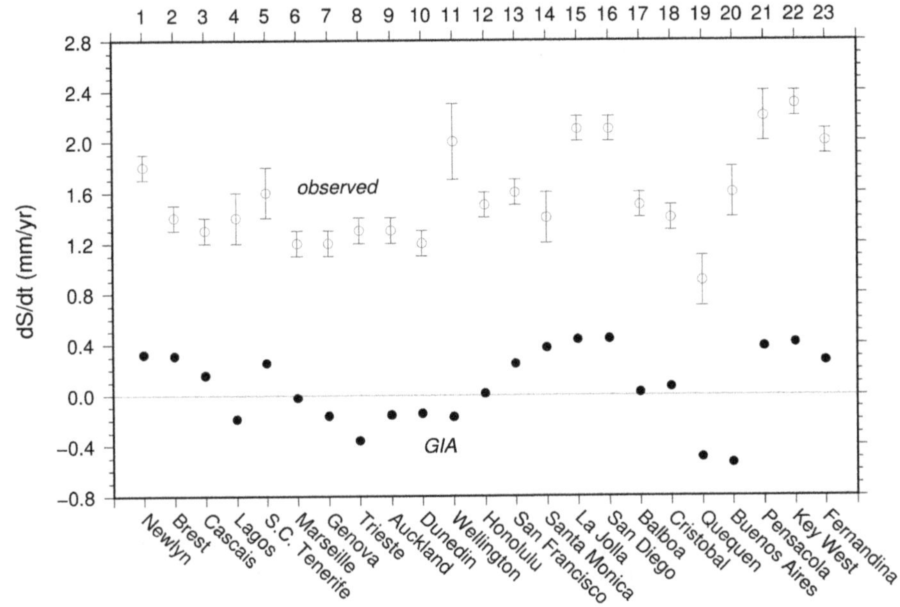

Fig. 6.15 Present-day rate of sea-level change $\dot{S}$ due to GIA at the 23 TG sites considered by Douglas (1997) according to the I7G-R44 GIA solution of the SLE (black circles). Also shown are observed rates of sea-level change with their associated uncertainties (open circles), redetermined by Spada and Galassi (2012), also reported in Table 6.4

1.6 ± 0.1 mm/yr, despite its importance on regional scales, demonstrated by the rich spatial variability of the fingerprints shown in Fig. 6.13 (Melini and Spada 2019).

6.4.5 Polar Motion[8]

The relation between changes in the rotational state of the Earth and geophysical processes such as the melting of late Pleistocene ice sheet has been widely discussed in the classical literature (for a historical account, see Munk and MacDonald 1960, and references herein). In the context of GIA, the feedback of changes in the polar motion has been introduced in the context of viscoelastic Earth models starting from the 80s (see, e.g., Sabadini and Peltier 1981; Sabadini et al. 1982, 1984; Ricard et al. 1993; Vermeersen et al. 1996b), and more recently a renewed interest has been motivated by applications to the problem of global climate change (Mitrovica and Wahr 2011).

[8] In this Section, we largely follow the text of the supplementary materials of Spada and Melini (2019b).

The fundamental solution of the Liouville equations for polar motion is given by Eq. (4.118). According to the general discretisation scheme outlined in Sect. E.3 and implemented in the `SELEN`[4] code, this solution can be evaluated numerically as

$$\mathbf{m}_k \equiv \mathbf{m}(t = t_k) = \sum_{k'=0}^{k-1} \psi_{k'}^{\mathrm{rig}} \left[A'^{\mathrm{e}} + A'^{\mathrm{s}} (k - k')\Delta t + \sum_{i=1}^{M'} \frac{A'_i}{a_i} \left(\mathrm{e}^{a_i (k-k')\Delta t} - 1 \right) \right], \tag{6.15}$$

where we recall that $\mathbf{m}$ is a complex variable, hence $\mathbf{m}_k = m_{1k} + \mathrm{i}\, m_{2k}$, Δt is the time discretisation step and $\mathbf{m}_k$ is expressed in the natural units of *radians*. For the rate of polar motion, we discretise the derivative of the fundamental solution of Liouville equations, given by (4.119), obtaining

$$\dot{\mathbf{m}}_k \equiv \dot{\mathbf{m}}(t = t_k) = \sum_{k'=0}^{k-1} \psi_{k'}^{\mathrm{rig}} \left[A'^{\mathrm{s}} + \sum_{i=1}^{M'} A'_i \mathrm{e}^{a_i (k-k')\Delta t} \right]. \tag{6.16}$$

To characterise the history of polar motion, it is useful to define some additional quantities. These are the *displacement of the pole of rotation* over the Earth's surface:

$$p_k = \sqrt{m_{1k}^2 + m_{2k}^2}, \tag{6.17}$$

the *rate of polar displacement*

$$\dot{p}_k = \frac{m_{1k}\dot{m}_{1k} + m_{2k}\dot{m}_{2k}}{\sqrt{m_{1k}^2 + m_{2k}^2}}, \tag{6.18}$$

and the *direction of polar motion* on the (x, y) plane

$$\phi_k = \tan^{-1}\left(\frac{m_{2k}}{m_{1k}}\right), \tag{6.19}$$

where $k = 1, \ldots N$ and where, by convention, the polar motion components $m_1 \equiv m_x$ and $m_2 \equiv m_y$ are measured along the Greenwich meridian and along the direction 90°E, respectively. With the above definitions, p_k and ϕ_k re given in units of *radians*, while $\dot{p}_k$ is expressed in *radians per unit time*, hence they shall be rescaled by suitable factors to obtain quantities in the more commonly employed units of deg and deg/Myr.

Figure 6.16 shows the evolution in time of the rotational pole position $\mathbf{m}$ and of its rate of change $\dot{\mathbf{m}}$ in response to GIA, obtained by Eqs. (6.15)–(6.16) above, for our I7G-R44 SLE solution. Figure 6.16a shows that, since the beginning of deglaciation, the pole has moved by about $p_N \approx 0.14°$ (corresponding to $\approx$ 15 km on the Earth's surface), roughly in the direction of the Hudson Bay ($\phi_N \approx -78°$E), consistently with seminal theoretical results by Sabadini and Peltier (1981) and Sabadini et al.

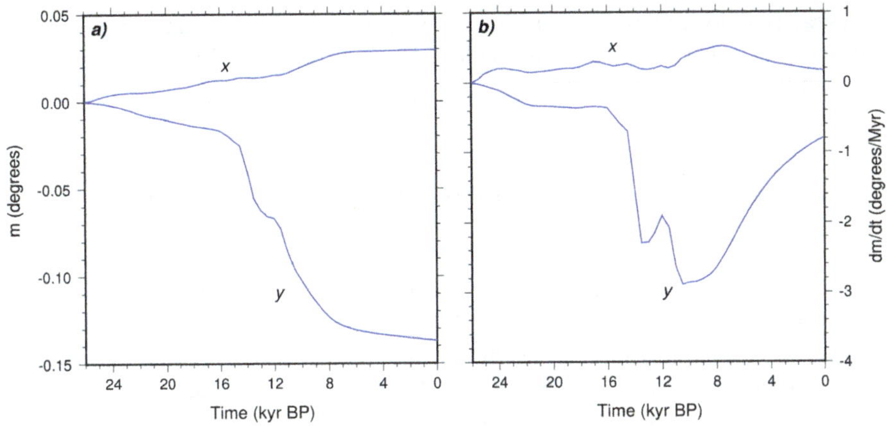

Fig. 6.16 Cartesian components (x, y) of the position vector **m** of the instantaneous pole of rotation **(a)** and their time derivatives **(b)** as a function of time according to the I7G-R44 solution of the SLE. The x axis points towards the prime meridian while the y axis is directed towards 90°E

(1982) and with polar motion observations (Dickman 1977). The rate of change of the position of the pole (Fig. 6.16b) shows an abrupt acceleration in the m_2 component at about 14 kyr BP, corresponding to the timing of the rapid melting episode referred to as Meltwater Pulse 1A (MWP-1A, between 14.3 and 12.8 kyr BP; see, e.g., Blanchon 2011). Since the onset of MWP-1A is not accompanied by any significant change in the m_1 component, the acceleration in the pole path has been roughly in the direction 90°E, with rates up to $\dot{p} \approx 3°/\text{Myr}$ that largely exceed the rate of polar motion drift observed since the year 1900, of about 1°/Myr (~10 cm/yr on the Earth's surface; see Dickman 1977).

6.4.6 *Stokes Coefficients*[9]

As a consequence of mass redistribution within the Earth system, GIA induces a perturbation of the gravity field, hence a GIA *fingerprint* is expected to contribute to the total field measured by space missions such as the Gravity Recovery and Climate Experiment (GRACE, see Bettadpur 2018) or the Gravity field and Ocean Circulation Explorer (GOCE, see e.g. van der Meijde et al. 2015). Modeling the GIA imprints on space-borne determinations of the Earth's gravitational potential is therefore of great importance, since the GIA contribution shall be properly accounted for a correct interpretation of the time-varying field in terms of hydrological or cryospheric processes.

[9] In this Section, we largely follow the text of the supplementary materials of Spada and Melini (2019b) concerning the Stokes coefficients.

In gravity field studies, the Earth's external gravitational potential is represented using a spherical harmonics expansion like that given by Eq. (1.27), which is reported here for the readers convenience:

$$W^{g}(r,\gamma,t) = \frac{GM}{r}\sum_{\substack{nm \\ m\geq 0}}\left(\frac{a}{r}\right)^{n}\left(c_{nm}(t)\cos m\lambda + s_{nm}(t)\sin m\lambda\right)\, P_{nm}(\cos\theta)\,, \tag{6.20}$$

where G is the universal gravitational constant, M is the mass of the Earth, a is its mean radius, $P_{nm}(\cos\theta)$ are the associated Legendre polynomials of harmonic degree n and order m, and the (non-dimensional) real-valued quantities c_{nm} and s_{nm} are the un-normalised cosine and sine Stokes coefficients (Heiskanen and Moritz 1981), whose explicit definition is given in Eq. (1.28). We note that due to mass exchanges between the components of the Earth system, c_{nm} and s_{nm} (and therefore W^{g}) are time-dependent (see e.g., Bettadpur 2018). By choosing the origin of the reference frame in the Earth's CM, and adopting fully normalised Legendre polynomials (FNALPs, see Sect. A.2), we can write:

$$W^{g}(r,\gamma,t) = \frac{GM}{r}\left[1+\sum_{n=2}^{n_{\max}}\left(\frac{a}{r}\right)^{n}\sum_{m=0}^{n}\left(\overline{c}_{nm}(t)\cos m\lambda + \overline{s}_{nm}(t)\sin m\lambda\right)\overline{P}_{nm}(\cos\theta)\right], \tag{6.21}$$

where $\overline{c}_{nm}$ and $\overline{s}_{nm}$ are the fully normalised Stokes coefficients and, by virtue of (A.20), it is easily shown that

$$\begin{pmatrix}\overline{c}_{nm}\\ \overline{s}_{nm}\end{pmatrix}(t) = \sqrt{\frac{1}{(2-\delta_{0m})(2n+1)}\cdot\frac{(n+m)!}{(n-m)!}}\begin{pmatrix}c_{nm}\\ s_{nm}\end{pmatrix}(t)\,. \tag{6.22}$$

By the Bruns formula (Hofmann-Wellenhof and Moritz 2006), the geopotential variation ΔW^{g} induced by GIA at the Earth's surface is related to the geoid response function through $\mathcal{G}(\gamma,t) = \Delta W^{g}/g_0$, where g_0 is the surface gravity acceleration. It is useful to recall that $\mathcal{G} = \mathcal{G}^{\rm sur} + \mathcal{G}^{\rm rot}$, where $\mathcal{G}^{\rm sur}$ and $\mathcal{G}^{\rm rot}$ are associated with changes in the surface loads (see Eq. 3.37) and changes of the rotational potential (Eq. 4.133), respectively. Furthermore, we recall that both terms contain a *direct effect* and an *indirect effect*, stemming from the '$\delta(t)$' and the '$k(t)$' terms in Eqs. (3.39) and (4.124), respectively. Henceforth, we assume that both direct and indirect contributions are included in $\mathcal{G}$, hence in the Stokes coefficients variations through ΔW^{g}. Note, however, that measurements of the gravitational field by satellite missions are not sensible to the direct centrifugal effect on gravity potential, since the instrument is not tied to the rotating Earth.[10] Hence, a comparison between the GRACE-derived

[10] About this point, it is interesting to read a personal communication of John Wahr to William Richard Peltier in 2011, reported in the supplementary Text S3 of Peltier et al. (2012).

rates of gravity change of harmonic degree 2 and order 1 and the GIA prediction ΔW^{g} may be misleading (see Chambers et al. 2010; Peltier et al. 2012; Chambers et al. 2012).

In close analogy with Eq. (6.21), we write

$$\Delta W^{g}(\gamma,t)=\frac{GM}{a}\sum_{n=2}^{n_{\max}}\sum_{m=0}^{n}\left(\overline{\delta c}_{nm}(t)\cos m\lambda+\overline{\delta s}_{nm}(t)\sin m\lambda\right)\overline{P}_{nm}(\cos\theta)\,, \tag{6.23}$$

where $\overline{\delta c}_{nm}(t)$ and $\overline{\delta s}_{nm}(t)$ are the normalised Stokes coefficients that describe variations of the geopotential induced by GIA. Recalling that $g_0=GM/a^2$, it follows that $\mathcal{G}(\gamma,t)=\Delta W^{g}/g_0$ can be also expressed in series of spherical harmonics

$$\mathcal{G}(\gamma,t)=a\sum_{n=2}^{n_{\max}}\sum_{m=0}^{n}\left(\overline{\delta c}_{nm}(t)\cos m\lambda+\overline{\delta s}_{nm}(t)\sin m\lambda\right)\overline{P}_{nm}(\cos\theta)\,, \tag{6.24}$$

where no terms of degree $n=0$ appear since *(i)* we are only considering mass-conserving loads for which $\mathcal{G}^{\mathrm{sur}}_{00}=0$ (see Eqs. 3.39 and 1.137) and *(ii)* $\mathcal{G}^{\mathrm{rot}}$ is a pure degree $n=2$ order $m=1$ quantity (see Sect. D.1). Furthermore, harmonic degree $n=1$ terms are also missing from (6.23)–(6.24), since we are conventionally using a reference frame with origin in the CM, so that there are no such terms in the surface GF for the geoid.

The present-day ($t=t_N$), GIA-induced rates of change of the Stokes coefficients in Eq. (6.24) are related with the harmonic coefficients in the CSH expansion of the $\dot{\mathcal{G}}$ fingerprint. In fact, taking the time-derivative and recalling Eq. (A.52), we obtain

$$\dot{\overline{\delta c}}_{nm}+\mathrm{i}\,\dot{\overline{\delta s}}_{nm}=a^{-1}\sqrt{2-\delta_{0m}}\;\dot{\mathcal{G}}^{*}_{nm}(t_N),\quad n\geq 2,\ m=0,\ldots n, \tag{6.25}$$

where the asterisk denotes complex conjugation (these expressions are consistent with those used by Melini and Spada 2019). We note that since the sea surface variation is defined as $\mathcal{N}=\mathcal{G}+c$ (see Eq. 5.8), where c is the so called "FC76 constant", for harmonic degrees $n\geq 2$ we have $\dot{\mathcal{G}}_{nm}(t)=\dot{\mathcal{N}}_{nm}(t)$, since $\dot{c}$ is harmonic of degree $n=0$. We observe that in the context of GRACE the spherical harmonics normalization convention does *not* include the *Condon-Shortley phase factor* $(-1)^m$ (see Bettadpur 2018, page 6). Hence, the right hand side of Eq. (6.25) should be multiplied by this factor in order to be compliant with the GRACE conventions for spherical harmonics.

At a given harmonic degree n, the *power spectrum* of the Stokes coefficients is

$$S^2(n)=\sum_{m=0}^{n}\left[\left(\dot{\overline{\delta c}}_{nm}\right)^2+\left(\dot{\overline{\delta s}}_{nm}\right)^2\right], \tag{6.26}$$

where the present day rates of change of the Stokes coefficients are given by Eq. (6.25). The spectrum provides information on how the "energy" of the GIA-induced variations of the geopotential is partitioned across each harmonic component (see e.g., Spada and Galassi 2015).

Performing a numerical differentiation in time by a symmetric difference quotient, the present-day rates of change of the Stokes coefficients can be computed, within the discretisation scheme adopted to solve the SLE, according to:

$$\dot{\overline{\delta c}}_{nm} = +a^{-1}(-1)^m\sqrt{2-\delta_{0m}}\ \mathrm{Re}\left(\frac{\mathcal{N}_{nm,N+1}-\mathcal{N}_{nm,N-1}}{2\Delta t}\right) \tag{6.27}$$

$$\dot{\overline{\delta s}}_{nm} = -a^{-1}(-1)^m\sqrt{2-\delta_{0m}}\ \mathrm{Im}\left(\frac{\mathcal{N}_{nm,N+1}-\mathcal{N}_{nm,N-1}}{2\Delta t}\right), \tag{6.28}$$

where Δt is the time step and the GRACE convention on the FNALPs is used.

Figure 6.17a shows the GIA-induced rates of change of Stokes coefficients $\dot{\overline{\delta c}}_{nm}$ and $\dot{\overline{\delta s}}_{nm}$ for $2 \le n \le 6$, according to the I7G-R44 SLE solution. The largest amplitudes are found for the $n = 2$ coefficients, reflecting the strongly zonal pattern of $\dot{\mathcal{G}}$ associated with rotational feedback, evident from Fig. 6.13. For $n \ge 2$ the amplitudes of the GIA-induced rates of Stokes coefficients decay rapidly with the harmonic degree, with the only coefficients exceeding the 10^{-11} yr^{-1} level being $\dot{\overline{\delta c}}_{20} = 2.17$, $\dot{\overline{\delta s}}_{21} = 2.06$, $\dot{\overline{\delta s}}_{31} = -1.41$ and $\dot{\overline{\delta s}}_{41} = -1.07$ (all values are given in units of 10^{-11} yr^{-1}).

The spectral properties of the GIA imprints on the gravitational field are better visualised in Fig. 6.17b, which shows the power spectrum defined in (6.26) for $2 \le n \le 256$. The spectrum is *red* and shows an almost regular and rapid decay with n, following the power law $S^2(n) \approx n^{-4.5}$ for $n \ge 10$. For longer wavelengths

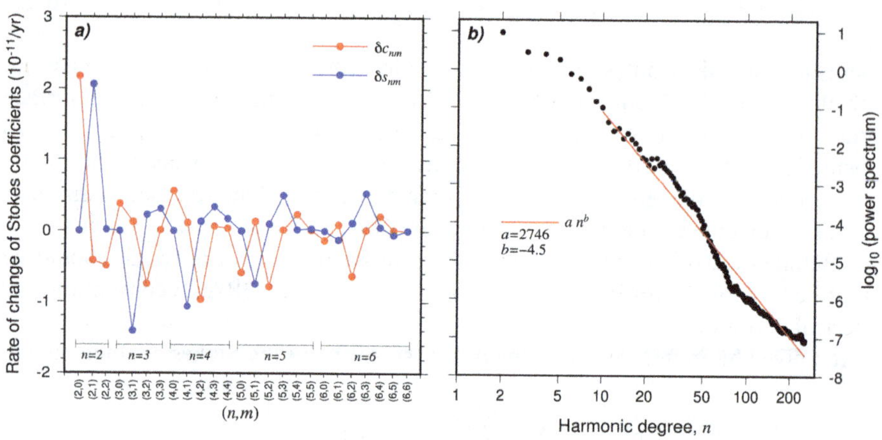

Fig. 6.17 Present-day GIA-induced rate of change of the low harmonic degree ($2 \le n \le 6$) Stokes coefficients $\dot{\overline{\delta c}}_{nm}$ and $\dot{\overline{\delta s}}_{nm}$ according to the I7G-R44 SLE solution (**a**) and their power spectrum up to degree $n = 256$. The solid red line in panel (**b**) represents the power law of the form $a\,n^b$ that best fits, in the least squares sense, the spectrum for $n \ge 10$

($n < 10$), where most of the energy is concentrated, the spectrum decays more slowly with increasing n. As noted by Spada and Galassi (2015), the power spectrum of the absolute sea-level variation observed during the *altimetry era* (1992-today) is also red but it drops with increasing harmonic degree n much less quickly, following a power law with exponent ~ -1.8. This is a clear indication of the comparatively smoother pattern (and of the lower energy) of the GIA component of present-day sea-level change.

References

Adhikari S, Ivins ER, Larour E (2016) ISSM-SESAW v1. 0: mesh-based computation of gravitationally consistent sea-level and geodetic signatures caused by cryosphere and climate driven mass change. Geosci Model Dev 9(3)

Amante C, Eakins B (2009) ETOPO1 arc-minute global relief model: procedures, data source and analysis. Technical Memorandum NESDIS NGDC; vol 24, National Oceanic and Atmospheric Administration

Argus DF, Peltier W, Drummond R, Moore AW (2014) The Antarctica component of postglacial rebound model ICE-6G_C (VM5a) based on GPS positioning, exposure age dating of ice thicknesses, and relative sea level histories. Geophys J Int 198(1):537–563

Bettadpur S (2018) Level-2 gravity field product user handbook. The GRACE Project (Jet Propulsion Laboratory). Pasadena, CA, p 2003

Blanchon P (2011) Meltwater pulses. In: Encyclopedia of modern coral reefs: structure, form and process. Encyclopedia of Earth Science Series, Springer, Berlin, pp 683–690. http://unam.academia.edu/PaulBlanchon/Papers

Blewitt G, Hammond WC, Kreemer C (2018) Harnessing the GPS data explosion for interdisciplinary science. Eos, Transactions American Geophysical Union, vol 99

Borreggine M, Powell E, Pico T, Mitrovica JX, Meadow R, Tryon C (2022) Not a bathtub: a consideration of sea-level physics for archaeological models of human migration. J Archaeol Sci 137:105507

Chambers DP, Wahr J, Tamisiea ME, Nerem RS (2012) Reply to comment by Peltier WR et al on "Ocean mass from GRACE and glacial isostatic adjustment". J Geophys Res: Solid Earth (1978–2012) 117(B11)

Chambers D, Wahr J, Tamisiea M, Nerem R (2010) Ocean mass from GRACE and glacial isostatic adjustment. J Geophys Res 115:B11415

Clark JA, Farrell WE, Peltier WR (1978) Global changes in postglacial sea level: a numerical calculation. Quatern Res 9(3):265–287

Dickman S (1977) Secular trend of the Earth's rotation pole: consideration of motion of the latitude observatories. Geophys J Roy Astron Soc 51(1):229–244

Dobson JE (2014) Aquaterra incognita: lost land beneath the sea. Geogr Rev 104(2):123–138

Dobson JE, Dobson JR (1999) Explore Aquaterra-Lost land beneath the sea. Geo World 12:30–31

Dobson JE, Spada G, Galassi G (2020) Global choke points may link sea level and human settlement at the last glacial maximum. Geogr Rev 110(4):595–620

Dobson JE, Spada G, Galassi G (2021) The Bering transitory archipelago: stepping stones for the first Americans. Comptes Rendus Géoscience 353(1):55–65

Douglas B (1992) Global sea level acceleration. J Geophys Res 97(C8):12,699–12,706

Douglas B (1991) Global sea level rise. J Geophys Res 96(C4):6981–6992

Douglas B (1997) Global sea level rise: a redetermination. Surv Geophys 18:279–292

Dziewonski AM, Anderson DL (1981) Preliminary reference Earth model. Phys Earth Planet Inter 25(4):297–356

Fairbanks RG, Charles CD, Wright JD (1992) Origin of global meltwater pulses. In: Radiocarbon after four decades. Springer, Berlin, pp 473–500

Fairbanks RG (1989) A 17,000-year glacio-eustatic sea level record: influence of glacial melting rates on the Younger Dryas event and deep-ocean circulation. Nature 342(6250):637–642

Fretwell P, Pritchard HD, Vaughan DG, Bamber JL, Barrand NE, Bell R, Bianchi C, Bingham R, Blankenship DD, Casassa G et al (2013) Bedmap2: improved ice bed, surface and thickness datasets for Antarctica. Cryosphere 7(1):375–393

Heiskanen WA, Moritz H (1981) Physical geodesy. Institute of Physical Geodesy, Technical University

Hofmann-Wellenhof B, Moritz H (2006) Physical geodesy. Springer Science & Business Media

Hopkins DM et al (1967) The Bering land bridge. Stanford University Press

Jakobsson M, Pearce C, Cronin TM, Backman J, Anderson LG, Barrientos N, Björk G, Coxall H, De Boer A, Mayer LA et al (2017) Post-glacial flooding of the Bering Land bridge dated to 11 cal ka BP based on new geophysical and sediment records. Clim Past 13(8):991

Jevrejeva S, Moore J, Grinsted A, Matthews A, Spada G (2014) Trends and acceleration in global and regional sea levels since 1807. Glob Planet Change 113:11–22

Kreemer C, Hammond WC, Blewitt G (2018) A robust estimation of the 3-D intraplate deformation of the North American plate from GPS. J Geophys Res: Solid Earth 123(5):4388–4412

Lau HC (2023) Transient rheology in sea level change: Implications for Meltwater Pulse 1A. Earth Planet Sci Lett 609:118106

Lodolo E, Galassi G, Spada G, Zecchin M, Civile D, Bressoux M (2020) Post-LGM coastline evolution of the NW Sicilian channel: comparing high-resolution geophysical data with glacial isostatic adjustment modeling. PLoS ONE 15(2):1–22

Martinec Z, Klemann V, van der Wal W, Riva R, Spada G, Sun Y, Melini D, Kachuck S, Barletta V, Simon K (2018) A benchmark study of numerical implementations of the sea level equation in GIA modelling. Geophys J Int 215(1):389–414

Melini D, Spada G (2019) Some remarks on glacial isostatic adjustment modelling uncertainties. Geophys J Int 218(1):401–413

Melini D, Saliby C, Spada G (2022) On computing viscoelastic Love numbers for general planetary models: the ALMA^3 code. Geophys J Int 231(3):1502–1517

Mitrovica J, Milne G (2002) On the origin of late Holocene sea-level high stands within equatorial ocean basins. Quatern Sci Rev 21(20):2179–2190

Mitrovica JX, Wahr J (2011) Ice Age Earth rotation. Annu Rev Earth Planet Sci 39:577–616

Mitrovica JX, Wahr J, Matsuyama I, Paulson A (2005) The rotational stability of an ice-age earth. Geophys J Int 161(2):491–506

Munk WH, MacDonald GJ (1960) The rotation of the earth: a geophysical discussion. Cambridge University Press, New York

Nakiboglu S (1982) Hydrostatic theory of the Earth and its mechanical implications. Phys Earth Planet Inter 28(4):302–311

Peltier WR (1994) Ice age paleotopography. Science 265(5169):195–201

Peltier W, Argus D, Drummond R (2015) Space geodesy constrains ice age terminal deglaciation: the global ICE-6G_C (VM5a) model. J Geophys Res: Solid Earth 120(1):450–487

Peltier W, Argus DF, R Drummond Comment on "An assessment of the ICE-6G_C (VM5a) Glacial Isostatic Adjustment Model" by Purcell et al (2018) J Geophys Res: Solid Earth 123(2):2019–2028

Peltier W, Drummond R (2008) Rheological stratification of the lithosphere: a direct inference based upon the geodetically observed pattern of the glacial isostatic adjustment of the North American continent. Geophys Res Lett 35(16)

Peltier W, Drummond R, Roy K (2012) Comment on "Ocean mass from GRACE and glacial isostatic adjustment" by Chambers DP et al. J Geophys Res: Solid Earth (1978–2012) 117(B11)

Pirazzoli PA (1991) World atlas of Holocene sea-level changes. Elsevier oceanography series, vol 58

Pirazzoli PA (1993) Global sea-level changes and their measurement. Glob Planet Change 8(3):135–148
Plag H-P, Jüettner H-U (2001) Inversion of global tide gauge data for present-day ice load changes. Memoirs of National Institute of Polar Research. Special issue 54:301–317
Purcell A, Tregoning P, Dehecq A (2016) An assessment of the ICE6G_C (VM5a) glacial isostatic adjustment model. J Geophys Res: Solid Earth 121(5):3939–3950
Renka RJ (1997) Algorithm 772: STRIPACK: Delaunay triangulation and Voronoi diagram on the surface of a sphere. ACM Trans Math Softw 23(3):416–434
Renka RJ (1997) Algorithm 773: SSRFPACK: interpolation of scattered data on the surface of a sphere with a surface under tension. ACM Trans Math Softw 23(3):435–442
Ricard Y, Spada G, Sabadini R (1993) Polar wandering of a dynamic Earth. Geophys J Int 113(2):284–298
Roy K, Peltier W (2017) Space-geodetic and water level gauge constraints on continental uplift and tilting over North America: regional convergence of the ICE-6G_C (VM5a/VM6) models. Geophys J Int 210(2):1115–1142
Roy K, Peltier W (2018) Relative sea level in the Western Mediterranean basin: a regional test of the ICE-7G_NA (VM7) model and a constraint on late Holocene Antarctic deglaciation. Quatern Sci Rev 183:76–87
Roy K, Peltier W (2015) Glacial isostatic adjustment, relative sea level history and mantle viscosity: reconciling relative sea level model predictions for the U.S. East coast with geological constraints. Geophys J Int 201(2):1156–1181
Sabadini R, Peltier W (1981) Pleistocene deglaciation and the Earth's rotation: implications for mantle viscosity. Geophys J Int 66(3):553–578
Sabadini R, Yuen DA, Boschi E (1982) Polar wandering and the forced responses of a rotating, multilayered, viscoelastic planet. J Geophys Res: Solid Earth 87(B4):2885–2903
Sabadini R, Yuen DA, Boschi E (1984) A comparison of the complete and truncated versions of the polar wander equations. J Geophys Res: Solid Earth 89(B9):7609–7620
Simon K, Riva R, Broerse T (2022) Identifying geographical patterns of transient deformation in the geological sea level record. J Geophys Res: Solid Earth 127(7). Cited by: 10; All Open Access, Green Open Access, Hybrid Gold Open Access
Spada G (2003) The theory behind TABOO. Samizdat Press. https://samizdat.mines.edu/home-old/
Spada G (2008) ALMA, a Fortran program for computing the viscoelastic Love numbers of a spherically symmetric planet. Comput Geosci 34(6):667–687
Spada G (2017) Glacial isostatic adjustment and contemporary sea level rise: an overview. Surv Geophys 38(1):1–33
Spada G, Galassi G (2012) New estimates of secular sea level rise from tide gauge data and GIA modelling. Geophys J Int 191(3):1067–1094
Spada G, Galassi G (2015) Spectral analysis of sea level during the altimetry era, and evidence for GIA and glacial melting fingerprints. Glob Planet Change 143:34–49
Spada G, Galassi G (2017) Extent and dynamic evolution of the lost land Aquaterra since the last glacial maximum. CR Geosci 349(4):151–158
Spada G, Barletta VR, Klemann V, Riva R, Martinec Z, Gasperini P, Lund B, Wolf D, Vermeersen L, King M (2011) A benchmark study for glacial isostatic adjustment codes. Geophys J Int 185(1):106–132
Spada G, Melini D (2019) On some properties of the glacial isostatic adjustment fingerprints. Water 11(9):1844
Spada G, Melini D (2019b) SELEN^4 (SELEN version 4.0): a Fortran program for solving the gravitationally and topographically self-consistent sea-level equation in glacial isostatic adjustment modeling. Geosci Model Dev 12(12):5055–5075
Tamisiea ME (2011) Ongoing glacial isostatic contributions to observations of sea level change. Geophys J Int 186(3):1036–1044

Tegmark M (1996) An icosahedron-based method for pixelizing the celestial sphere. Astrophys J 470:L81

Tushingham A, Peltier W (1993) Relative sea level database. Technical report 93–106, NOAA/NGDC Paleoclimatology Program, Boulder CO, USA

van der Meijde M, Pail R, Bingham R, Floberghagen R (2015) GOCE data, models, and applications: A review. Int J Appl Earth Obs Geoinformation 35:4–15. GOCE earth science applications and models(Based on the ESA GOCE solid earth workshop, 16–17 Oct 2012)

Vermeersen LLA, Sabadini R, Spada G (1996) Compressible rotational deformation. Geophys J Int 126:735–761

Wessel P, Smith WHF (1998) New, improved version of Generic Mapping Tools released. EOS Trans Am Geophys Union 79(47):579

Wessel P, Smith WH, Scharroo R, Luis J, Wobbe F (2013) Generic Mapping Tools: improved version released. EOS Trans Am Geophys Union 94(45):409–410

Appendix A
Spherical Harmonics

Spherical harmonic functions—or simply *spherical harmonics*—play an important role in the framework of the Sea Level Equation (SLE), since the equation is solved by a pseudo-spectral method in which all the relevant scalar fields are decomposed in series of spherical harmonics. But before that, the spherical harmonics (SHs) constitute a natural tool to describe any variable defined on the surface of a sphere, such as the Earth's gravity field and its time variations.

As discussed by Kaula (1967) and Brett (1988), SHs arise in a physical context as solutions of Laplace's equation in spherical coordinates. Their important property of *orthogonality*, however, suggests that SHs constitute a natural spectral representation for any function that varies over the surface of the sphere, regardless of whether the function is related to Laplace's equation. In this respect, SHs may be regarded as a tool for describing a spatial field defined over the surface of the sphere, in close analogy to how a Fourier series may describe a function that varies in time.

In this appendix we introduce the Legendre polynomials, the associated Legendre polynomials, and the complex spherical harmonics. The expansion of a scalar function in series of spherical harmonics is accomplished using both their *complex* and *real forms*. Here we only provide the basic definitions and properties; for a more in-depth presentation the readers are referred to the classical textbooks (e.g., Ferrers 1877; Laden 1938; MacRobert 1927; Sternberg and Smith 1944; Kaula 1967) and to more recent contributions (e.g., Atkinson and Han 2012; Copley 2015; Wieczorek and Meschede 2018) or online resources.[1]

[1] See, for instance: https://en.wikipedia.org/wiki/Spherical_harmonics and http://mathworld.wolfram.com/SphericalHarmonic.html [last accessed on September 27, 2024].

G. Spada and D. Melini, *Glacial Isostatic Adjustment*, Springer Geophysics,
https://doi.org/10.1007/978-3-031-90538-4

A.1 Legendre Polynomials

The Legendre polynomials $P_n(x)$ are[2] mutually orthogonal polynomials of *degree* n, defined in the interval $-1 \leq x \leq +1$. The LPs are solutions of the Legendre differential equation

$$\frac{\mathrm{d}}{\mathrm{d}x}\left((1-x^2)\,\frac{\mathrm{d}P_n(x)}{\mathrm{d}x}\right) + n(n+1)P_n(x) = 0\,, \tag{A.1}$$

and they are explicitly given by Rodrigues' formula

$$P_n(x) = \frac{1}{2^n n!}\frac{\mathrm{d}^n}{\mathrm{d}x^n}(x^2-1)^n\,. \tag{A.2}$$

Some *low degree* LPs are listed in Table A.1.

A.1.1 A Few Properties of Legendre Polynomials

Some properties of the LPs are worth to mention since they are employed several times throughout the book. For these and other properties, see e.g., Abramowitz et al. (1988).

1. Orthogonality.

$$\int_{-1}^{1} P_n(x)P_{n'}(x)\mathrm{d}x = \int_0^{\pi} P_n(\cos\theta)P_{n'}(\cos\theta)\sin\theta\mathrm{d}\theta = \frac{2\delta_{nn'}}{2n+1}\,. \tag{A.3}$$

2. Parity.

$$P_n(\pm 1) = (\pm 1)^n\,. \tag{A.4}$$

3. Shifted derivatives.

$$P_n(x) = \frac{P'_{n+1}(x) - P'_{n-1}(x)}{2n+1}\,, \quad n \geq 1\,. \tag{A.5}$$

4. Link with the Čebyšëv polynomials of 1st kind $T_n(z) = \cos(nz)$.

$$\int_z^1 \frac{P_n(x)\mathrm{d}x}{\sqrt{x-z}} = \frac{T_n(z) - T_{n+1}(z)}{(n+\frac{1}{2})\sqrt{1-z}}\,. \tag{A.6}$$

---

[2] See also http://mathworld.wolfram.com/LegendrePolynomial.html [last accessed on December 29, 2024].

5. Generating function.

$$\sum_{n=0}^{\infty} z^n P_n(\cos\beta) = \frac{1}{\sqrt{1-2z\cos\beta+z^2}}, \quad |z| \le 1. \tag{A.7}$$

6. Legendre sums. Certain infinite sums involving Legendre polynomials are of interest because they help in the numerical construction of the Green's functions (see Sect. 2.2.1). These are:

$$\sum_{n=0}^{\infty} P_n(\cos\theta) = \frac{1}{2\sin\frac{\theta}{2}}, \tag{A.8}$$

$$\sum_{n=0}^{\infty} n P_n(\cos\theta) = -\frac{1}{4\sin\frac{\theta}{2}}, \tag{A.9}$$

$$\sum_{n=1}^{\infty} \frac{1}{n} P_n(\cos\theta) = -\ln\left(\sin\frac{\theta}{2}\right) - \ln\left(1+\sin\frac{\theta}{2}\right), \tag{A.10}$$

$$\sum_{n=1}^{\infty} \frac{\partial P_n(\cos\theta)}{\partial\theta} = -\frac{\cos\frac{\theta}{2}}{4\sin^2\frac{\theta}{2}}, \tag{A.11}$$

$$\sum_{n=1}^{\infty} \frac{1}{n}\frac{\partial P_n(\cos\theta)}{\partial\theta} = -\frac{\cos\frac{\theta}{2}\left(1+2\sin\frac{\theta}{2}\right)}{2\sin\frac{\theta}{2}\left(1+\sin\frac{\theta}{2}\right)}. \tag{A.12}$$

For more Legendre sums and some discussion, see Appendix 1 of Farrell (1972), §8.926 of Gradshteyn et al. (1988), Appendix A of Sun and Sjöberg (1999) and, lastly, Fowler et al. (2019).

A.1.2 Expansion in Series of Legendre Polynomials

It is often useful to expand in series of LPs real-valued scalar functions $f = f(\theta)$, only depending upon co-latitude; hence these functions are symmetric *w.r.t.* the z axis. Their expansion reads:

$$f(\theta) = \sum_{n=0}^{\infty} f_n P_n(\cos\theta), \tag{A.13}$$

with real-valued coefficients f_n. By multiplying both sides by $P_{n'}(\cos\theta)$, integrating over the whole interval of co-latitudes ($0 \leq \theta \leq \pi$) and using the orthogonality of the LPs (A.3), the coefficients of the expansion are easily obtained:

$$f_n = \frac{2n+1}{2} \int_0^{\pi} f(\theta) P_n(\cos\theta) \sin\theta d\theta \,. \tag{A.14}$$

A.2 Associated Legendre Polynomials

The associated Legendre polynomials[3] $P_{nm}(x)$ (ALPs) are defined as the canonical solution to the general Legendre differential equation:

$$(1-x^2)\frac{d^2 P_{nm}(x)}{dx^2} - 2x\frac{dP_{nm}(x)}{dx} + \left(n(n+1) - \frac{m^2}{1-x^2}\right) P_{nm}(x) = 0 \,, \tag{A.15}$$

where integers n ($n = 0, 1, \ldots$) and m ($m = 0, 1, \ldots n$) are the *degree* and the *order* of the ALP, respectively.[4] The ALPs can be obtained in terms of m-order derivatives of ordinary Legendre polynomials, with:

$$P_{nm}(x) = (-1)^m (1-x^2)^{m/2} \frac{d^m}{dx^m} P_n(x) \,, \tag{A.16}$$

where factor $(-1)^m$ is known as the *Condon-Shortley phase*. The P_{nm}'s defined by Eq. (A.16) are also referred to as *un-normalised* ALPs. It is easily shown that, for $m = 0$, the general Legendre differential equation (A.15) coincides with (A.1), hence

$$P_{n0}(x) \equiv P_n(x) \,. \tag{A.17}$$

The ALPs are particularly useful when their argument is expressed in terms of angles, with $x = \cos\theta$. Some of low degree ALPs are listed in Table A.1.

We note that for the ALPs, the following orthogonality condition holds

$$\int_{-1}^{+1} P_{nm}(x) P_{n'm}(x) dx = \frac{2(n+m)!}{(2n+1)(n-m)!} \delta_{nn'} \,. \tag{A.18}$$

It is also worth to recall the special values of the ALPs:

$$P_{nm}(\pm 1) = (\pm 1)^n \delta_{m0} \,. \tag{A.19}$$

[3] A nice introduction is found at the page https://en.wikipedia.org/wiki/Associated_Legendre_polynomials [last accessed on December 29, 2024].

[4] Solutions of Eq. (A.15) corresponding to arbitrary real or complex values of n and m are referred to as associated Legendre *functions*. Note that for m odd, the ALPs are not actually polynomials, being dependent upon $\sin\theta$.

Table A.1 LPs P_n in terms of $x = \cos\theta$, and ALPs $P_{nm}(\cos\theta)$ of harmonic degree $0 \le n \le 2$ and order $0 \le m \le n$. We note that $P_{n0}(x) = P_n(x)$

Degree, n	Order, m	$P_n(x)$	$P_{nm}(\cos\theta)$
0	0	1	1
1	0	x	$\cos\theta$
1	1		$-\sin\theta$
2	0	$\frac{1}{2}(3x^2 - 1)$	$\frac{1}{2}\left(3\cos^2\theta - 1\right)$
2	1		$-3\sin\theta\cos\theta$
2	2		$3\sin^2\theta$

A.2.1 *Fully Normalised Associated Legendre Polynomials*

Along with the un-normalised P_{nm}'s, various *normalised* forms of the ALPs exist, which have been categorised by Wieczorek and Meschede (2018). One of these forms is of interest here, since it is usually employed in the field of geodesy and spectral analysis, and referred to as the *fully normalised* form of the ALP (or FNALP).

The FNALPs of degree n and order m are defined as

$$\overline{P}_{nm}(x) = \sqrt{(2 - \delta_{0m})(2n + 1)\frac{(n - m)!}{(n + m)!}}\, P_{nm}(x)\,, \tag{A.20}$$

where $P_{nm}(x)$ is the usual un-normalised form given by Eq. (A.16). Note that by virtue of definition (A.16), the FNALPs *include* the Condon-Shortley phase factor $(-1)^m$. Using (A.20) and (A.18) it is easily verified that for the FNALPs the following orthogonality condition holds true

$$\int_{-1}^{+1} \overline{P}_{nm}(x)\overline{P}_{n'm}(x)\mathrm{d}x = 2(2 - \delta_{0m})\,\delta_{nn'}\,, \tag{A.21}$$

in agreement with Wieczorek and Meschede (2018).

A.3 Complex Spherical Harmonics

The complex spherical harmonics (CSHs) are mutually orthogonal, complex-valued functions of co-latitude θ and longitude λ, defined over the surface of the unit sphere. In this work, we use conventionally the so-called 4π-normalised form of the CSHs.

Table A.2 4π-normalised complex spherical harmonics (CSHs) of degree $n \leq 2$ and order $0 \leq m \leq n$, for argument $\gamma = (\theta, \lambda)$, where θ is colatitude and λ is longitude. In the third and in the fourth columns, the CSHs are expressed in terms of $P_{nm}(x)$ and written explicitly in terms of trigonometric functions, respectively. Note that, for negative orders, $Y_{n\,-m}(\gamma) = (-1)^m Y^*_{nm}(\gamma)$

Degree, n	Order, m	$Y_{nm}(\gamma)$	$Y_{nm}(\gamma)$
0	0	1	1
1	0	$\sqrt{3}P_{10}(\cos\theta)$	$\sqrt{3}\cos\theta$
1	1	$\sqrt{\frac{3}{2}}P_{11}(\cos\theta)e^{i\lambda}$	$-\sqrt{\frac{3}{2}}\sin\theta e^{i\lambda}$
2	0	$\sqrt{5}P_{20}(\cos\theta)$	$\frac{\sqrt{5}}{2}(3\cos^2\theta - 1)$
2	1	$\sqrt{\frac{5}{6}}P_{21}(\cos\theta)e^{i\lambda}$	$-3\sqrt{\frac{5}{6}}\sin\theta\cos\theta e^{i\lambda}$
2	2	$\sqrt{\frac{5}{24}}P_{22}(\cos\theta)e^{2i\lambda}$	$3\sqrt{\frac{5}{24}}\sin^2\theta e^{2i\lambda}$

Other choices, characterised by different normalisations,[5] are however possible, as explained in Wieczorek and Meschede (2018).

The 4π-normalised CSHs are defined as

$$Y_{nm}(\gamma) = \mu_{nm}\, P_{nm}(\cos\theta)\, e^{im\lambda} \quad m \geq 0\,, \tag{A.22}$$

with the normalisation factor

$$\mu_{nm} \equiv \sqrt{(2n+1)\frac{(n-m)!}{(n+m)!}}\,, \tag{A.23}$$

where integers n ($n = 0, 1, \ldots$) and m ($|m| \leq n$) are the harmonic *degree* and *order* of the CSH, respectively, and the P_{nm}'s are given by (A.16).

The CSHs with negative order are defined as:

$$Y_{n\,-m}(\gamma) = (-1)^m Y^*_{nm}(\gamma)\,, \tag{A.24}$$

where the asterisk denotes complex conjugation (c.c.).

We observe that, using Eq. (A.20) and definition (A.22), the CSHs can be expressed in terms of FNALPs as follows:

$$Y_{nm}(\gamma) = \frac{1}{\sqrt{2-\delta_{0m}}}\,\overline{P}_{nm}(\cos\theta)e^{im\lambda}\,. \tag{A.25}$$

[5] Indeed, in a previous work about the SLE (Spada and Stocchi 2006), "quantum mechanics" (or "1-normalised") complex spherical harmonics have been adopted.

A.3.1 A Few Properties of the Complex Spherical Harmonics

We only give a few essential properties, needed for the development of the SLE theory in the body of the book.

1. Orthogonality. The CSHs are orthogonal over the unit sphere. In particular, due to the particular choice of the normalisation factor μ_{lm} (A.23), the following *orthogonality condition* holds true:

$$\int_{\gamma} Y_{nm}(\gamma) Y^{*}_{n'm'}(\gamma)\, \mathrm{d}\gamma = 4\pi\, \delta_{nn'} \delta_{mm'}, \tag{A.26}$$

which motivates the CSHs attribute "4π-normalised".

2. Degree 0 CSH. By definition of CSHs (A.22), since $P_{00} = 1$ (see Table A.1), we have:

$$Y_{00}(\gamma) = 1\,, \tag{A.27}$$

with other low-degree CSHs listed in Table A.1.

3. Addition theorem. In Fig. A.1, we consider two points P and S on the unit sphere, with spherical coordinates $\gamma = (\theta, \lambda)$ and $\gamma' = (\theta', \lambda')$, respectively, and we denote with α the angular distance between γ and γ' along the great circle connecting them.

The *addition theorem* for CSHs states that

$$P_n(\cos\alpha) = \frac{1}{2n+1} \sum_{m=-n}^{n} Y^{*}_{nm}(\gamma') Y_{nm}(\gamma)\,, \tag{A.28}$$

where $P_n(x)$ is the degree n LP defined by Eq. (A.2) and where by the law of cosines (or "cosine rule for sides") in spherical trigonometry, we have

$$\cos\alpha = \cos\theta \cos\theta' + \sin\theta \sin\theta' \cos(\lambda - \lambda')\,. \tag{A.29}$$

A proof of the addition theorem, based on induction, is given by Coster and Hart (1991).

4. Derivatives of the addition theorem. It is sometimes useful to consider the *derivatives* of the addition theorem. The subject has been presented in detail by Winch and Roberts (1995). Here, we are particularly interested into the partial derivatives of the addition theorem given by Eq. (A.28) with respect to variables θ (co-latitude) and λ (longitude), which represent the geographical coordinates of the "observer" P in Fig. A.1.

The derivative of (A.28) *w.r.t.* θ is

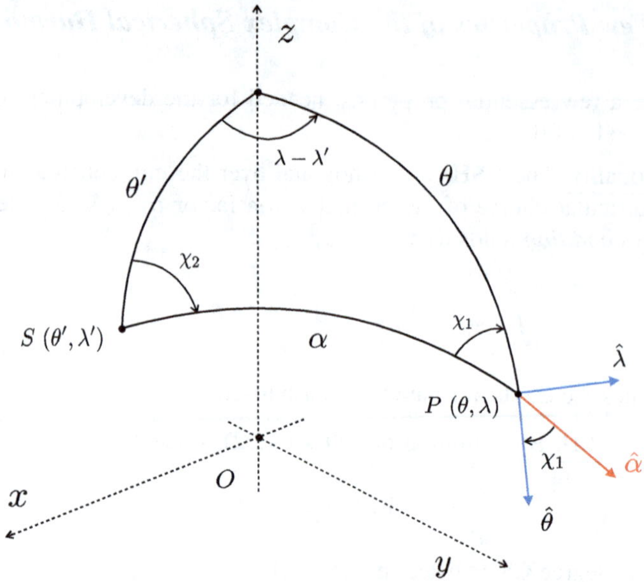

Fig. A.1 Spherical triangle with sides θ, θ' and α, used to illustrate the addition theorem for spherical harmonics. The two points S and P, located on the surface of the sphere, are the "source" and the "observer" of coordinates $\gamma' = (\theta', \lambda')$ and $\gamma = (\theta, \lambda)$, respectively. Unit vectors in the direction of increasing source-observer angular distance $\hat{\alpha}$, of increasing co-latitude $\hat{\theta}$ and of increasing longitude $\hat{\lambda}$ are also shown. The unit vectors are tangent to the Earth's surface

$$\frac{\partial P_n(\cos\alpha)}{\partial\alpha}\,\hat{\alpha}\cdot\hat{\theta} = \frac{1}{2n+1}\sum_{m=-n}^{n} Y^*_{nm}(\gamma')\,\nabla_\theta Y_{nm}(\gamma)\,, \tag{A.30}$$

where $\hat{\alpha}$ and $\hat{\theta}$ are the unit vectors in the direction of increasing α and θ, respectively (see Fig. A.1), $\nabla_\theta = \frac{\partial}{\partial\theta}$ is the $\hat{\theta}$ component of the horizontal gradient and according to the first of Eqs. (7) of Winch and Roberts (1995) we have used $\frac{\partial\alpha}{\partial\theta} = \cos\chi_1$.

In a similar fashion, the derivative of (A.28) *w.r.t.* λ is

$$\frac{\partial P_n(\cos\alpha)}{\partial\alpha}\,\hat{\alpha}\cdot\hat{\lambda} = \frac{1}{2n+1}\sum_{m=-n}^{n} Y^*_{nm}(\gamma')\,\nabla_\lambda Y_{nm}(\gamma)\,, \tag{A.31}$$

where $\hat{\lambda}$ is the unit vector in the direction of increasing longitude λ (see Fig. A.1), $\nabla_\lambda = \frac{1}{\sin\theta}\frac{\partial}{\partial\lambda}$ is the $\hat{\lambda}$ component of the horizontal gradient, and we have used the first of Eqs. (8) of Winch and Roberts (1995), i.e., $\frac{\partial\alpha}{\partial\lambda} = \sin\theta\sin\chi_1$.

4. Average of the CSHs. All the CSHs have a vanishing average over the sphere, except $Y_{00}(\gamma)$. This can be seen using the definition of whole-Earth average (1.136), the orthogonality condition (A.26) and taking (A.27) into consideration:

$$\begin{aligned}\langle Y_{nm}(\gamma)\rangle^{\mathrm{e}} \equiv& \frac{1}{4\pi a^2}\int_{\mathrm{e}} Y_{nm}(\gamma)\mathrm{d}A = \frac{1}{4\pi a^2}\int_{\mathrm{e}} Y_{nm}(\gamma)a^2\mathrm{d}\gamma \\ =& \frac{1}{4\pi}\int_{\gamma} Y_{nm}(\gamma)Y_{00}^{*}(\gamma)\mathrm{d}\gamma \;=\; \delta_{n0}\delta_{m0}\,. \end{aligned} \tag{A.32}$$

5. Legendre equation for the CSHs. As a consequence of Legendre equation (A.15), for the CSHs the following relation holds:

$$\nabla_{\theta}^2 Y + \cot\theta\;\nabla_{\theta}Y + \nabla_{\lambda}^2 Y = -n(n+1)Y\,, \tag{A.33}$$

where $\nabla_{\theta} = \frac{\partial}{\partial\theta}$ and $\nabla_{\lambda} = \frac{1}{\sin\theta}\frac{\partial}{\partial\lambda}$ are the $\hat{\theta}$ and $\hat{\lambda}$ components of the horizontal gradient.

6. "Other harmonics". As discussed in Spada and Galassi (2015), and references therein, mutually orthogonal spherical harmonics can be also defined on subsets of the unit sphere.

A.3.2 *Expansion in Series of CSHs*

It is often useful to expand a scalar, square-integrable, real-valued, time-dependent function as an infinite series of CSHs, with

$$F(\gamma, t) = \sum_{nm} F_{nm}(t)Y_{nm}(\gamma)\,, \tag{A.34}$$

where $F_{nm}(t)$ are the (complex) *coefficients of the CSH expansion* and we have used the notation

$$\sum_{nm} \equiv \sum_{n=0}^{\infty}\sum_{m=-n}^{n}\,. \tag{A.35}$$

By multiplying both sides of Eq. (A.34) by $Y_{n'm'}^{*}(\gamma)$, integrating over the unit sphere and using the orthogonality condition (A.26) we obtain:

$$\begin{aligned} F(\gamma, t)Y_{n'm'}^{*}(\gamma) =& \sum_{nm} F_{nm}(t)Y_{nm}(\gamma)Y_{n'm'}^{*}(\gamma) \\ \int_{\gamma} F(\gamma, t)Y_{n'm'}^{*}(\gamma)\mathrm{d}\gamma =& \sum_{nm} F_{nm}(t)\int_{\gamma} Y_{nm}(\gamma)Y_{n'm'}^{*}(\gamma)\mathrm{d}\gamma, \\ \int_{\gamma} F(\gamma, t)Y_{n'm'}^{*}(\gamma)\mathrm{d}\gamma =& 4\pi\sum_{nm} F_{nm}(t)\delta_{nn'}\delta_{mm'} \;=\; 4\pi\, F_{n'm'}(t)\,. \end{aligned}$$

Hence, for an arbitrary degree n and order m, the coefficients of the CSH expansion are:

$$F_{nm}(t) = \frac{1}{4\pi} \int_\gamma F(\gamma, t) Y^*_{nm}(\gamma) \mathrm{d}\gamma \,, \tag{A.36}$$

so that taking (A.24) into account, we have the following condition on the coefficients

$$F_{n-m}(t) = (-1)^m F^*_{nm}(t) \,, \tag{A.37}$$

where F^*_{nm} is the c.c. of F_{nm}.

It can be easily verified that the degree $n = 0$ coefficient in the CSH expansion (A.34) simply represents the spatial average of $F(\gamma, t)$ over the sphere. In fact, using $n = m = 0$ and Eq. (A.27) into (A.36) we see that

$$F_{00}(t) = \frac{1}{4\pi} \int_\gamma F(\gamma, t) \mathrm{d}\gamma = \frac{1}{4\pi a^2} \int_\mathrm{e} F(\gamma, t) \mathrm{d}A \;=\; \langle F(\gamma, t) \rangle^\mathrm{e}, \tag{A.38}$$

where $\langle F(\gamma, t) \rangle^\mathrm{e}$ represents the *spatial average* of $F(\gamma, t)$ over a sphere of radius a (see Eq. 1.136).

In closing, we note that an expansion in series of Legendre polynomials can be formally considered as an expansion in series of CSH. In fact, starting from (A.13), we have:

$$\begin{aligned} f(\theta) &\equiv \sum_{n=0}^{\infty} f_n \, P_n(\cos\theta) \\ &= \sum_{nm} f_n \, P_{nm}(\cos\theta) \, \mathrm{e}^{\mathrm{i}m\lambda} \, \delta_{m0} \\ &= \sum_{nm} \frac{f_n \, \delta_{m0}}{\mu_{nm}} \, Y_{nm}(\theta, \lambda) \\ &= \sum_{nm} f_{nm} \, Y_{nm}(\theta, \lambda) \,, \end{aligned} \tag{A.39}$$

with coefficients:

$$f_{nm} = \frac{f_n \, \delta_{m0}}{\mu_{nm}} \,, \tag{A.40}$$

where, because of (A.23), we note that

$$f_{n0} = \frac{f_n}{\sqrt{2n+1}} \,. \tag{A.41}$$

A.4 Real Spherical Harmonics

In geodesy and geodynamics, spherical harmonic functions in "real form" are generally preferred to complex harmonics (see e.g., Heiskanen and Moritz 1981). The advantage of CSHs is that they are relatively easy to manipulate in theoretical computations, also thanks to the properties of the complex exponent function e^{ix} (see the definition of CSH A.22). However, complex harmonics are in a certain sense numerically redundant, having both a real and an imaginary part and involving both positive and *negative* orders. Real spherical harmonics (RSHs) have no imaginary parts, are restricted to positive orders and allow for a simpler visualization, being associated with a single scalar field.

It is clear that, for any function $F(\theta, \lambda)$ defined on the surface of a unit sphere, the CSH and the RSH representations are totally equivalent, in principle. Here we define the RSHs and we determine the equivalence between CSH and RSH expansions of (real) scalar fields, either involving the ALPs (associated Legendre polynomials) or the FNALPs (fully normalised associated Legendre polynomials).

A real (unnormalised) RSH of degree n and order m has the form

$$S_{nm}(\theta, \lambda) = (c_{nm} \cos m\lambda + s_{nm} \sin m\lambda)\, P_{nm}(\cos\theta)\,, \tag{A.42}$$

where c_{nm} and s_{nm} are real-valued constants, the P_{nm}'s are the ALPs defined by Eq. (A.16) and for a given harmonic degree n the order m is $0 \leq m \leq n$. A similar expression can be also established using the FNALPs $\overline{P}_{nm}(\cos\theta)$, with appropriate (real) constants $\overline{c}_{nm}$ and $\overline{s}_{nm}$.

The RSHs have nice symmetries that are determined by the values of degree n and order m (for a few examples, see Fig. A.2). The *zonal harmonics* are those with $m = 0$, for which the surface of the sphere is subdivided into zones bounded by n parallels, whose latitudes are determined by the zeroes of the Legendre polynomial $P_n(\cos\theta)$. For *tesseral harmonics*, $0 < m < n$ and the sphere is divided into quadrangles bounded by $2m$ meridians and $n - m$ parallels. Finally, *sectorial* (or *sectoral*) *harmonics* are those with $m = n$; they divide the sphere into sectors bounded by $2m$ meridians. With increasing n and m the nodal lines (those parallels and meridians where $S_{nm}(\theta, \lambda) = 0$) become closer and closer, and the areas of the zones, quadrangles and sectors become smaller and smaller.

A.4.1 Expansion in Series of ALPs

Here we show that a general CSH expansion like

$$F(\gamma) = \sum_{nm} F_{nm} Y_{nm}(\gamma)\,, \tag{A.43}$$

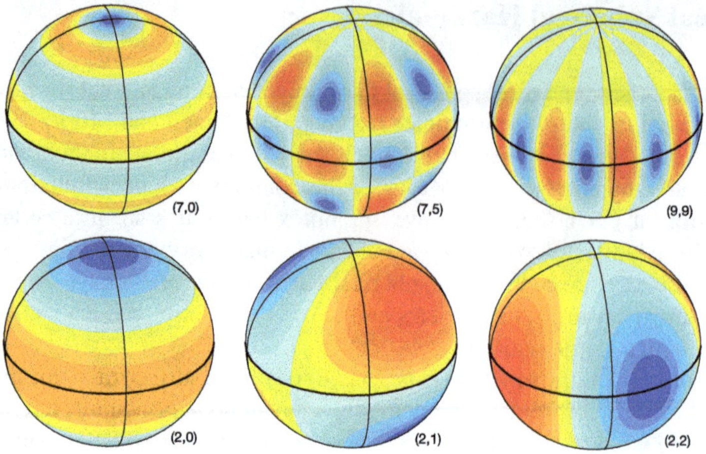

Fig. A.2 A few examples of RSHs, defined by (A.42), adopting certain values of constants c_{nm} and s_{nm}, for various symmetries according to the values of harmonic degree and order (n, m) shown. The bottom row shows all the three possible harmonic degree $n = 2$ RSHs. Reddish and bluish hues denote regions in which the RSHs take opposite signs, separated by nodal lines

where we have dropped the time-dependence of F, is equivalent to the RSH expansion:

$$F(\gamma) = \sum_{\substack{nm \\ m \geq 0}} (c_{nm} \cos m\lambda + s_{nm} \sin m\lambda)\, P_{nm}(\cos\theta)\,, \tag{A.44}$$

where the cosine and sine real-valued coefficients c_{nm} and s_{nm} are univocally determined by the complex-valued coefficients F_{nm}. The demonstration is straightforward, and is based on condition (A.37), on definition (A.22), and some algebra:

$$\begin{aligned} F(\gamma) &= \sum_{nm} F_{nm}\, Y_{nm}(\gamma) \\ &= \sum_{\substack{nm \\ m \geq 0}} (2 - \delta_{0m})\, (\mathrm{Re}(F_{nm})\mathrm{Re}(Y_{nm}) - \mathrm{Im}(F_{nm})\mathrm{Im}(Y_{nm})) \\ &= \sum_{\substack{nm \\ m \geq 0}} (2 - \delta_{0m})\, \mu_{nm}\, (\mathrm{Re}(F_{nm}) \cos m\lambda - \mathrm{Im}(F_{nm}) \sin m\lambda)\;\, P_{nm}(\cos\theta)\,, \end{aligned}$$

which has the form (A.44) provided that

$$c_{nm} = +\,(2 - \delta_{0m})\;\, \mu_{nm}\; \mathrm{Re}\,(F_{nm}) \tag{A.45}$$

$$s_{nm} = -\,(2 - \delta_{0m})\;\, \mu_{nm}\; \mathrm{Im}\,(F_{nm})\;, \tag{A.46}$$

hence:

$$c_{nm} + \mathrm{i}\, s_{nm} = (2 - \delta_{0m})\ \mu_{nm}\ F^*_{nm}\,, \tag{A.47}$$

where F^*_{nm} is the c.c. (complex conjugate) of F_{nm}.

A.4.2 Expansion in Series of FNALPs

Here we show that a CSH expansion like

$$F(\gamma) = \sum_{nm} F_{nm} Y_{nm}(\gamma), \tag{A.48}$$

is equivalent to:

$$F(\gamma) = \sum_{\substack{nm \\ m \geq 0}} (\overline{c}_{nm} \cos m\lambda + \overline{s}_{nm} \sin m\lambda)\, \overline{P}_{nm}(\cos\theta)\,, \tag{A.49}$$

where the $\overline{P}_{nm}(\cos\theta)$'s are the FNALPs and the cosine and sine real-valued coefficients $\overline{c}_{nm}$ and $\overline{s}_{nm}$ are univocally determined by F_{nm}. The demonstration, based on condition (A.37) and Eq. (A.25), is straightforward and is left to the readers. The result is:

$$\overline{c}_{nm} = +\sqrt{2 - \delta_{0m}}\ \mathrm{Re}\,(F_{nm}) \tag{A.50}$$

$$\overline{s}_{nm} = -\sqrt{2 - \delta_{0m}}\ \mathrm{Im}\,(F_{nm})\,, \tag{A.51}$$

hence

$$\overline{c}_{nm} + \mathrm{i}\,\overline{s}_{nm} = \sqrt{2 - \delta_{0m}}\ F^*_{nm}\,, \tag{A.52}$$

where F^*_{nm} is the c.c. of F_{nm}.

A.5 Vector Spherical Harmonics

The vector spherical harmonics (VSHs) constitute a natural extension of CSHs aimed at conveniently representing a vector function in a spherical coordinate system. Here we will give only a brief account of VSHs; for a detailed discussion and a comprehensive list of their properties we refer the readers to Barrera et al. (1985) and references therein.

In analogy with the scalar expansion (A.34), it can be shown that any vector function $\vec{V}$ can be expanded as

$$\vec{V}(r,\gamma) = \sum_{nm} \left(v_{nm}\vec{Y}_{nm}(\gamma) + w_{nm}\vec{\Psi}_{nm}(\gamma) + z_{nm}\vec{\Phi}_{nm}(\gamma) \right) , \tag{A.53}$$

where the expansion coefficients v_{nm}, w_{nm} and z_{nm} are radial functions (i.e., they do not depend upon angular coordinates) and we have introduced the *vector spherical harmonics*

$$\vec{Y}_{nm} = Y_{nm}\hat{r} , \tag{A.54}$$

$$\vec{\Psi}_{nm} = r\nabla Y_{nm} , \tag{A.55}$$

$$\vec{\Phi}_{nm} = \vec{r} \times \nabla Y_{nm} . \tag{A.56}$$

As a consequence of the definition above, and taking the CSHs orthogonality properties into account (see A.26), it can be shown that

$$\int_\gamma \vec{Y}_{nm} \cdot \vec{Y}^*_{n'm'} \,\mathrm{d}\gamma = 4\pi\delta_{nn'}\delta_{mm'} , \tag{A.57}$$

$$\int_\gamma \vec{\Psi}_{nm} \cdot \vec{\Psi}^*_{n'm'} \,\mathrm{d}\gamma = 4\pi n(n+1)\delta_{nn'}\delta_{mm'} , \tag{A.58}$$

$$\int_\gamma \vec{\Phi}_{nm} \cdot \vec{\Phi}^*_{n'm'} \,\mathrm{d}\gamma = 4\pi n(n+1)\delta_{nn'}\delta_{mm'} , \tag{A.59}$$

$$\int_\gamma \vec{Y}_{nm} \cdot \vec{\Psi}^*_{n'm'} \,\mathrm{d}\gamma = \int_\gamma \vec{Y}_{nm} \cdot \vec{\Phi}^*_{n'm'} \,\mathrm{d}\gamma = \int_\gamma \vec{\Phi}_{nm} \cdot \vec{\Psi}^*_{n'm'} \,\mathrm{d}\gamma = 0 , \tag{A.60}$$

hence, for any vector function $\vec{V}$ the coefficients of the VSH expansion are given by

$$v_{nm} = \frac{1}{4\pi} \int_\gamma \vec{V} \cdot \vec{Y}^*_{nm} \,\mathrm{d}\gamma , \tag{A.61}$$

$$w_{nm} = \frac{1}{4\pi n(n+1)} \int_\gamma \vec{V} \cdot \vec{\Psi}^*_{nm} \,\mathrm{d}\gamma , \tag{A.62}$$

$$z_{nm} = \frac{1}{4\pi n(n+1)} \int_\gamma \vec{V} \cdot \vec{\Phi}^*_{nm} \,\mathrm{d}\gamma . \tag{A.63}$$

The general VSH expansion (A.53) greatly simplifies in the case of a vector function with axial symmetry, which is of particular relevance for the description of the response of a spherically symmetric Earth to surface loading or tidal forcing. In that case, *i)* $\vec{V}$ is only a function of radius r and of colatitude θ and *ii)* its horizontal component has the direction $\hat{\theta}$. Since there is no dependence upon λ, only the $m = 0$ terms contribute to the expansion (A.53), hence

$$\vec{V}(r,\theta) = \sum_n \left[v_{n0}Y_{n0}\hat{r} + w_{n0}\left(\frac{\partial}{\partial\theta}Y_{n0}\right)\hat{\theta} + z_{n0}\hat{r} \times \left(\frac{\partial}{\partial\theta}Y_{n0}\hat{\theta}\right) \right] , \tag{A.64}$$

where we have used the expression (B.4) for the gradient operator in spherical coordinates. Since the third term in the expansion (A.64) is orthogonal to both $\hat{r}$ and $\hat{\theta}$, it immediately follows that for an axisymmetric vector it must be $z_{n0} = 0$. By recalling the definition (A.22) of CSHs, we can thus rewrite (A.64) as

$$\vec{V}(r, \theta) = \sum_n \left[a_n P_n(\cos\theta)\hat{r} + b_n \frac{\partial}{\partial\theta} P_n(\cos\theta)\hat{\theta} \right], \tag{A.65}$$

where we have defined the new expansion coefficients $a_n = \mu_{n0} v_{n0}$ and $b_n = \mu_{n0} w_{n0}$.

Appendix B
Vector Calculus and Transforms

In the first section of this appendix, the differential operators gradient, curl, divergence, and laplacian *and all that* are presented in a rather informal way, mostly following Schey (2005), and also giving some important theorems. In the second section, we deal with the general toroidal-poloidal decomposition of solenoidal fields, important for the derivation of the equations that govern the equilibrium of a layered incompressible Earth. Last, we provide some basic definitions and properties concerning the Laplace transforms, which are often used throughout the book.

B.1 Differential Operators

B.1.1 Gradient

In a three-dimensional cartesian coordinates system, the *gradient* of the scalar field

$$f = f(x_1, x_2, x_3) \tag{B.1}$$

is vector

$$\nabla f = \frac{\partial f}{\partial x_1}\hat{x}_1 + \frac{\partial f}{\partial x_2}\hat{x}_2 + \frac{\partial f}{\partial x_3}\hat{x}_3\,, \tag{B.2}$$

where symbol ∇ ("*nabla*" or "*del*"), denotes the *vector differential operator* and $\hat{x}_i$ ($i = 1, 2, 3$) are the unit vectors along the three axes (see Fig. B.1a). We note that sometimes the gradient of f is indicated by grad f. In *index notation*, the gradient of a scalar f reads $f_{,i}$. The concept of gradient can be extended to higher rank *tensors*. For example, the gradient of the first rank tensor w_i (a vector) is the second rank tensor $w_{i,k}$.

G. Spada and D. Melini, *Glacial Isostatic Adjustment*, Springer Geophysics,
https://doi.org/10.1007/978-3-031-90538-4

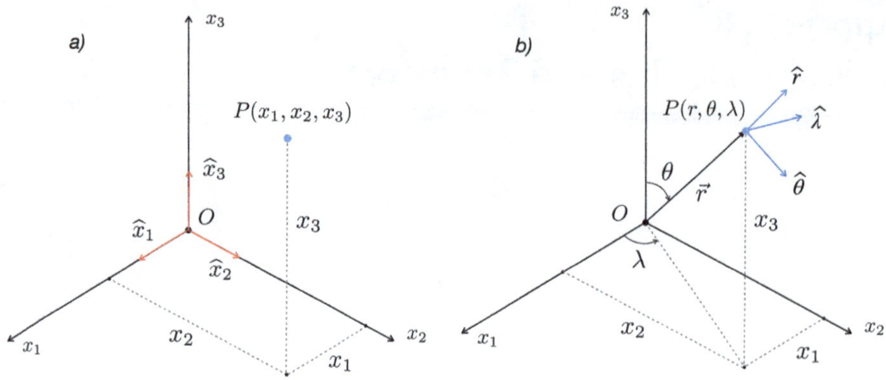

Fig. B.1 Cartesian (x_1, x_2, x_3) and spherical coordinates (r, θ, λ) of a point P in space. The cartesian unit vectors $(\hat{x}_1, \hat{x}_2, \hat{x}_3)$ and the spherical ones $(\hat{r}, \hat{\theta}, \hat{\lambda})$ are also shown. The relationship between cartesian and spherical coordinates, and *vice versa*, are given by Eqs. (1.1) and (1.3) in the body of the text, respectively

The physical interpretation of ∇f is intuitive. In fact, we note that the variation of f is corresponding to an elementary displacement $d\vec{x}$ is

$$df = \frac{\partial f}{\partial x_i} \mathrm{d}x_i = \nabla f \cdot \mathrm{d}\vec{x} = ||\nabla f||\, ||\mathrm{d}\vec{x}|| \cos\alpha \,, \tag{B.3}$$

where α is the angle that ∇f forms with $d\vec{x}$. Hence, the gradient ∇f defines the direction of the fastest increase of f.

In spherical coordinates, the gradient is expressed as follows:

$$\nabla f = \frac{\partial f}{\partial r} \hat{r} + \frac{1}{r} \frac{\partial f}{\partial \theta} \hat{\theta} + \frac{1}{r \sin\theta} \frac{\partial f}{\partial \lambda} \hat{\lambda} \,, \tag{B.4}$$

where the mutually orthogonal unit vectors $(\hat{r}, \hat{\theta}, \hat{\lambda})$ are shown in Fig. B.1b.

Sometimes it is useful to write the gradient as

$$\nabla f = \frac{\partial f}{\partial r} \hat{r} + \frac{1}{r} \nabla_h f \,, \tag{B.5}$$

where the *horizontal gradient* operator is

$$\nabla_h = \hat{\theta}\, \nabla_\theta + \hat{\lambda}\, \nabla_\lambda \,, \tag{B.6}$$

with

$$\nabla_\theta = \frac{\partial}{\partial \theta} \tag{B.7}$$

and

$$\nabla_\lambda = \frac{1}{\sin\theta}\frac{\partial}{\partial\lambda}\,, \tag{B.8}$$

respectively.

For a vector field $\vec{v} = (v_1, v_2, v_3)$, the gradient is replaced with the second rank tensor

$$J_{ij} = \frac{\partial v_i}{\partial x_j}\,, \tag{B.9}$$

which is called the *Jacobian matrix*[6] (or the *Jacobian*) of $\vec{v}$. In index notation, the Jacobian matrix reads $v_{i,j}$.

B.1.2 Curl

The *curl* of vector field

$$\vec{v} = (v_1, v_2, v_3) \tag{B.10}$$

is a vector, defined as

$$\nabla\times\vec{v} = \left(\frac{\partial v_3}{\partial x_2} - \frac{\partial v_2}{\partial x_3}\right)\hat{x}_1 + \left(\frac{\partial v_1}{\partial x_3} - \frac{\partial v_3}{\partial x_1}\right)\hat{x}_2 + \left(\frac{\partial v_2}{\partial x_1} - \frac{\partial v_1}{\partial x_2}\right)\hat{x}_3\,, \tag{B.11}$$

which can be interpreted as the vector (cross) product ($\times$) between the gradient operator ∇ and vector $\vec{v}$. The concept of curl is tightly connected to the concept of *vorticity* of a vector field (see e.g., Schey 2005). In some textbooks, the curl of $\vec{v}$ is denoted by rot $\vec{v}$. In index notation, the curl is expressed as $\epsilon_{ijk}v_{j,k}$ where ϵ_{ijk} is the Levi-Civita alternating symbol.[7]

It is sometimes useful to express the curl in spherical coordinates (see Fig. B.1b). Denoting as usual with (r, θ, λ) the radius, co-latitude and longitude, the components of $\nabla\times\vec{v}$ in spherical coordinates are:

---

[6] We note that the Jacobian of $\vec{v}$ can also written as the *dyadic product* $\nabla\vec{v}$, with the dyadic product $\vec{a}\vec{b}$ between vectors $\vec{a}$ and $\vec{b}$ being the rank two tensor $(\vec{a}\vec{b})_{ij} \equiv a_i b_j$.

[7] We recall that the Levi-Civita alternating symbol, also referred to as *Ricci symbol* after GR Curbastro (see https://en.wikipedia.org/wiki/Levi-Civita_symbol [last accessed on December 29, 2024]), is a constant tensor defined as $\epsilon_{ijk} = +1$ when (i, j, k) is an even permutation of the fundamental triplet $(1, 2, 3)$, $\epsilon_{ijk} = -1$ when (i, j, k) is an odd permutation, and $\epsilon_{ijk} = 0$ if one index is repeated.

$$(\nabla \times \vec{v})_r = \frac{1}{r \sin\theta}\left(\frac{\partial}{\partial\theta}(v_\lambda \sin\theta) - \frac{\partial v_\theta}{\partial\lambda}\right) \tag{B.12}$$

$$(\nabla \times \vec{v})_\theta = \frac{1}{r}\left(\frac{1}{\sin\theta}\frac{\partial v_r}{\partial\lambda} - \frac{\partial}{\partial r}(r v_\lambda)\right) \tag{B.13}$$

$$(\nabla \times \vec{v})_\lambda = \frac{1}{r}\left(\frac{\partial}{\partial r}(r v_\theta) - \frac{\partial v_r}{\partial\theta}\right), \tag{B.14}$$

or equivalently:

$$(\nabla \times \vec{v})_r = \frac{1}{r}\frac{\partial v_\lambda}{\partial\theta} + \frac{\cot\theta}{r} v_\lambda - \frac{1}{r\sin\theta}\frac{\partial v_\theta}{\partial\lambda} \tag{B.15}$$

$$(\nabla \times \vec{v})_\theta = \frac{1}{r\sin\theta}\frac{\partial v_r}{\partial\lambda} - \frac{\partial v_\lambda}{\partial r} - \frac{v_\lambda}{r} \tag{B.16}$$

$$(\nabla \times \vec{v})_\lambda = \frac{\partial v_\theta}{\partial r} + \frac{v_\theta}{r} - \frac{1}{r}\frac{\partial v_r}{\partial\theta}, \tag{B.17}$$

where $(v_r, v_\theta, v_\lambda)$ are the spherical components of $\vec{v}$ (e.g., Schey 2005).

Irrotational (or *conservative*) vector fields are denoted by the property

$$\nabla \times \vec{v} = 0, \tag{B.18}$$

i.e., their curl vanishes everywhere. Since for any scalar ϕ we have

$$\nabla \times (\nabla\phi) = 0, \tag{B.19}$$

that is easily verified using the definitions of curl and gradient, irrotational fields can be associated to a *scalar potential* ϕ such that

$$\vec{v} = \nabla\phi. \tag{B.20}$$

A fundamental example of conservative field is the *gravitational field*, usually denoted by $\vec{g}$, for which a scalar potential W called *gravitational potential* exists, such[8] that $\vec{g} = \nabla W$ (or $g_i = W_{,i}$; see e.g., Hofmann-Wellenhof and Moritz 2006).

B.1.3 Divergence

The *divergence* of a vector field $\vec{v}$ with cartesian components (v_1, v_2, v_3), is

$$\nabla \cdot \vec{v} = \frac{\partial v_1}{\partial x_1} + \frac{\partial v_2}{\partial x_2} + \frac{\partial v_3}{\partial x_3}, \tag{B.21}$$

[8] This is the choice commonly adopted in Geodesy. In the textbooks of physics the choice $\vec{g} = -\nabla\phi$ is more often encountered (see page 137 of Schey 2005, for a discussion).

which can be interpreted as the scalar (dot) product $(\cdot)$ between the gradient operator ∇ and vector $\vec{v}$. The divergence is tightly connected with the concept of *flux* (see e.g., Schey 2005) and sometimes is denoted by div $\vec{v}$. In index notation, the divergence is $v_{k,k}$, where we have used the Einstein summation convention. The concept of divergence can be extended to higher order quantities (tensors). For example, the divergence of the *second rank* tensor Γ_{ij} is the *first rank* tensor (vector) $\Gamma_{ij,j}$.

In spherical coordinates, the divergence reads:

$$\nabla \cdot \vec{v} = \frac{1}{r^2}\frac{\partial}{\partial r}(r^2 v_r) + \frac{1}{r\sin\theta}\frac{\partial}{\partial\theta}(v_\theta \sin\theta) + \frac{1}{r\sin\theta}\frac{\partial v_\lambda}{\partial\lambda}, \tag{B.22}$$

or

$$\nabla \cdot \vec{v} = \frac{\partial v_r}{\partial r} + \frac{2v_r}{r} + \frac{1}{r}\left(\cot\theta\, v_\theta + \frac{\partial v_\theta}{\partial\theta} + \frac{1}{\sin\theta}\frac{\partial v_\lambda}{\partial\lambda}\right), \tag{B.23}$$

where $\vec{v} = (v_r, v_\theta, v_\lambda)$.

Solenoidal fields are those fields characterised by the condition

$$\nabla \cdot \vec{v} = 0\,, \tag{B.24}$$

i.e., their divergence vanishes identically (in index notation $v_{k,k} = 0$). In view of the importance of solenoidal fields in the study of the displacements induced by surface loads, they are considered in the dedicated Sect. B.2. Since

$$\nabla \cdot (\nabla \times \vec{v}) = 0 \tag{B.25}$$

in consequence of the definition of divergence and of curl, solenoidal fields can be associated to a *vector potential* $\vec{a}$ such that

$$\vec{v} = \nabla \times \vec{a}\,. \tag{B.26}$$

A well known example of solenoidal field is the *magnetic field*, usually denoted with $\vec{B}$, for which a *magnetic vector potential* $\vec{A}$ exists, with $\vec{B} = \nabla \times \vec{A}$ (see e.g., Jackson 1999).

B.1.4 Laplacian

The *laplacian* is a second order differential operator named after the french scholar Pierre-Simon Laplace (1749–1827). The laplacian of the scalar field $f = f(x_1, x_2, x_3)$ is the scalar

$$\nabla^2 f = \frac{\partial^2 f}{\partial x_1^2} + \frac{\partial^2 f}{\partial x_2^2} + \frac{\partial^2 f}{\partial x_3^2}, \tag{B.27}$$

which can be interpreted as $\nabla \cdot \nabla f$, i.e., the divergence of the gradient of f; indeed, sometimes the laplacian of f is written as div(grad f). Another common notation for the laplacian of f is Δf, whereas in index notation, the laplacian of f is $f_{,kk}$. The laplacian of a vector $\vec{v} = (v_1, v_2, v_3)$ is also a vector, whose components in a cartesian reference frame are the laplacians of the individual components, i.e., $\nabla^2 \vec{v} = \left(\nabla^2 v_1, \nabla^2 v_2, \nabla^2 v_3\right)$ or $v_{i,kk}$ in index notation. In general, the laplacian of a vector field can be defined through the identity

$$\nabla \times \nabla \times \vec{v} = \nabla(\nabla \cdot \vec{v}) - \nabla^2 \vec{v} \tag{B.28}$$

(e.g., Schey 2005), that in the case of *solenoidal fields* ($\nabla \cdot \vec{v} \equiv 0$) gives

$$\nabla^2 \vec{v} = -\nabla \times \nabla \times \vec{v}, \tag{B.29}$$

which shall be particularly useful in the development of the incompressible field equations later on.

In Appendix A, the properties of the spherical harmonic functions are studied in detail. They are solutions of Laplace's equation:

$$\nabla^2 f = 0, \tag{B.30}$$

where in spherical coordinates the laplacian reads

$$\nabla^2 f = \frac{1}{r^2}\frac{\partial}{\partial r}\left(r^2 \frac{\partial f}{\partial r}\right) + \frac{1}{r^2 \sin\theta}\frac{\partial}{\partial \theta}\left(\sin\theta \frac{\partial f}{\partial \theta}\right) + \frac{1}{r^2 \sin^2\theta}\frac{\partial^2 f}{\partial \lambda^2}, \tag{B.31}$$

or equivalently

$$\nabla^2 f = \frac{\partial^2 f}{\partial r^2} + \frac{2}{r}\frac{\partial f}{\partial r} + \frac{1}{r^2}\frac{\partial^2 f}{\partial \theta^2} + \frac{\cot\theta}{r^2}\frac{\partial f}{\partial \theta} + \frac{1}{r^2 \sin^2\theta}\frac{\partial^2 f}{\partial \lambda^2}. \tag{B.32}$$

The $\nabla^2 f$ term in the Laplace's equation (B.30) can be conveniently expanded in spherical harmonics. Indeed, if f is a scalar function allowing a CSH expansion of the form (A.34), i.e.

$$f = \sum_{nm} f_{nm} Y_{nm}, \tag{B.33}$$

with the f_{nm} coefficients depending only upon r, (B.32) gives

$$\begin{aligned}
\nabla^2 f &= \sum_{nm}\left(\frac{\partial^2 f_{nm}}{\partial r^2}+\frac{2}{r}\frac{\partial f_{nm}}{\partial r}\right)Y_{nm}+\frac{f_{nm}}{r^2}\underbrace{\left(\frac{\partial^2 Y_{nm}}{\partial\theta^2}+\cot\theta\frac{\partial Y_{nm}}{\partial\theta}+\frac{1}{\sin^2\theta}\frac{\partial^2 Y_{nm}}{\partial\lambda^2}\right)}_{-n(n+1)Y_{nm}} \\
&= \sum_{nm}\left(\frac{\partial^2 f_{nm}}{\partial r^2}+\frac{2}{r}\frac{\partial f_{nm}}{\partial r}-\frac{n(n+1)}{r^2}f_{nm}\right)Y_{nm} \\
&= \sum_{nm}\left(\nabla_r^2 f_{nm}\right)Y_{nm}\,, \qquad \text{(B.34)}
\end{aligned}$$

where we have made use of identity (A.33) and we have introduced the differential operator

$$\nabla_r^2 \equiv \frac{\partial^2}{\partial r^2}+\frac{2}{r}\frac{\partial}{\partial r}-\frac{n(n+1)}{r^2}\,, \qquad \text{(B.35)}$$

which is often referred to as the *radial component of the laplacian operator.*

B.1.5 Two Important Theorems

There are two important integral theorems that involve differential operators. The readers are referred to e.g. Schey (2005) for their demonstration and a physical discussion. Let us consider a vector field $\vec{F}$.

The *divergence theorem* states that:

$$\int_V \left(\nabla\cdot\vec{F}\right)\mathrm{d}V = \int_S \vec{F}\cdot\hat{n}\,\mathrm{d}S\,, \qquad \text{(B.36)}$$

where V is the volume of space bounded by closed surface S and $\hat{n}$ is the outer normal to the surface. The *r.h.s.* represents the *flux* $\Phi_S(\vec{F})$ of $\vec{F}$ across S. The divergence theorem implies that for *solenoidal fields* ($\nabla\cdot\vec{F}=0$), $\Phi_S(\vec{F})$ vanishes across any closed surface.

The *curl theorem* states that:

$$\int_S \left(\nabla\times\vec{F}\right)\cdot\hat{n}\,\mathrm{d}S = \int_C \vec{F}\cdot\mathrm{d}\vec{l}\,, \qquad \text{(B.37)}$$

where S is an open surface with boundary C and the *r.h.s.* represents the *circulation* $\Gamma_C(\vec{F})$ of $\vec{F}$ along C. The curl theorem implies that for *irrotational fields* ($\nabla\times\vec{F}=0$), the circulation vanishes along any closed path.

B.2 T/P Decomposition of a Solenoidal Field

Solenoidal fields are those vector fields $\vec{B}(r, \gamma)$ for which

$$\nabla \cdot \vec{B} = 0 . \tag{B.38}$$

They are often discussed in the treaties of geomegnetism, since the Earth's magnetic field is solenoidal (see e.g., Bloxham and Jackson 1991). However, they are also central in the study of mantle flow in global geodynamics (e.g., Vigny 1989; Peltier 1989) and in the modelling of elastic and viscoelastic deformations induced by earthquakes on a global scale (see e.g., Pollitz 1992; Piersanti et al. 1995). Furthermore, they have a number of applications in planetary sciences (see e.g., Toepfer et al. 2021).

For solenoidal fields,[9] the following unique decomposition holds:

$$\vec{B}(r, \gamma) = \vec{B}^{t} + \vec{B}^{p} , \tag{B.39}$$

where $\gamma = (\theta, \lambda)$ and $\vec{B}^{t}$ and $\vec{B}^{p}$ are the *toroidal* and *poloidal components* of $\vec{B}$, respectively. This is known as *Mie representation* of $\vec{B}$; sometimes, it is also referred to as *toroidal-poloidal (T/P) decomposition* (Backus 1986; Backus et al. 1996). As pointed out by Smylie (1965), the separation of a solenoidal field into toroidal and poloidal components has been invoked by many authors, one of the first being Lamb (1881a, b). For non-solenoidal fields, the Mie representation is not valid; in this case, a more general formalism is employed, which involves the *Helmoltz representation* of the tangent vector and vector spherical harmonics (see Sect. A.5 and, e.g., Barrera et al. 1985; Dahlen and Tromp 1998).

In Eq. (B.39), the toroidal and poloidal components of $\vec{B}$ are:

$$\vec{B}^{t}(r, \gamma) = \nabla \times (T \hat{r}) \tag{B.40}$$

and

$$\vec{B}^{p}(r, \gamma) = \nabla \times \nabla \times (P \hat{r}) , \tag{B.41}$$

respectively, where $T(r, \gamma)$ and $P(r, \gamma)$ are referred to as the *toroidal scalar* and the *poloidal scalar*, respectively. Note that according to Eqs. (B.39–B.41) a solenoidal field is completely and uniquely determined by $T(r, \gamma)$ and $P(r, \gamma)$. These scalars have a vanishing average over the unit sphere i.e., $\int_{\gamma} T d\gamma = \int_{\gamma} P d\gamma = 0$. Furthermore, based upon definitions above, $\vec{B}^{t} \perp \vec{B}^{p}$ and $\vec{B}^{t} \perp \hat{r}$. Hence, the toroidal field is tangential. In addition, the curl of a toroidal field is poloidal, and, in turn, the curl of a poloidal field is toroidal (see e.g., Dahlen and Tromp 1998).

---

[9] Note that, in general, all the fields considered in this section are time-dependent but this is left implicit to simplify notation.

With the aid of Eqs. (B.12–B.14), and after some manipulation, Eqs. (B.40) and (B.41) can be made explicit, obtaining

$$B_r^t(r,\gamma) = 0 \tag{B.42}$$

$$B_\theta^t(r,\gamma) = \frac{1}{r\sin\theta}\frac{\partial T}{\partial\lambda} \tag{B.43}$$

$$B_\lambda^t(r,\gamma) = -\frac{1}{r}\frac{\partial T}{\partial\theta}, \tag{B.44}$$

and

$$B_r^p(r,\gamma) = -\frac{1}{r^2}\left(\frac{\partial^2 P}{\partial\theta^2} + \cot\theta\,\frac{\partial P}{\partial\theta} + \frac{1}{\sin^2\theta}\frac{\partial^2 P}{\partial\lambda^2}\right) \tag{B.45}$$

$$B_\theta^p(r,\gamma) = \frac{1}{r}\frac{\partial^2 P}{\partial r\partial\theta} \tag{B.46}$$

$$B_\lambda^p(r,\gamma) = \frac{1}{r\sin\theta}\frac{\partial^2 P}{\partial r\partial\lambda}, \tag{B.47}$$

respectively.

To find suitable spectral representations for the toroidal and poloidal components of $\vec{B}$, it is now convenient to expand the toroidal and poloidal scalars in series, with:

$$T(r,\gamma) = \sum_{nm} t_{nm} Y_{nm} \tag{B.48}$$

$$P(r,\gamma) = \sum_{nm} p_{nm} Y_{nm}, \tag{B.49}$$

where $Y_{nm}(\gamma)$ is the 4π-normalised complex spherical harmonic (CSH) of degree and order (n, m) defined in Sect. A.3 while $t_{nm} = t_{nm}(r)$ and $p_{nm} = p_{nm}(r)$ denote appropriate toroidal and poloidal radial functions, respectively. By direct substitution, the toroidal components (B.42–B.44) become:

$$B_r^t(r,\gamma) = 0 \tag{B.50}$$

$$B_\theta^t(r,\gamma) = \frac{1}{\sin\theta}\sum_{nm}\frac{t_{nm}}{r}\frac{\partial Y_{nm}}{\partial\lambda} \tag{B.51}$$

$$B_\lambda^t(r,\gamma) = -\sum_{nm}\frac{t_{nm}}{r}\frac{\partial Y_{nm}}{\partial\theta}, \tag{B.52}$$

whereas the poloidal components (B.45–B.47) are:

$$B_r^P(r,\gamma) = \frac{1}{r^2}\sum_{nm} n(n+1)\,p_{nm}Y_{nm} \tag{B.53}$$

$$B_\theta^P(r,\gamma) = \sum_{nm} \frac{p'_{nm}}{r}\frac{\partial Y_{nm}}{\partial\theta} \tag{B.54}$$

$$B_\lambda^P(r,\gamma) = \frac{1}{\sin\theta}\sum_{nm} \frac{p'_{nm}}{r}\frac{\partial Y_{nm}}{\partial\lambda}\,, \tag{B.55}$$

where in (B.53) we have used property (A.33) of the CSHs, and the prime denotes the radial derivative.

Introducing the toroidal

$$\tau_{nm}^{(1)}(r) = \frac{t_{nm}}{r} \tag{B.56}$$

and the poloidal radial functions

$$\sigma_{nm}^{(1)}(r) = \frac{n(n+1)}{r^2}\,p_{nm} \tag{B.57}$$

$$\sigma_{nm}^{(2)}(r) = \frac{p'_{nm}}{r}\,, \tag{B.58}$$

and using (B.39), from (B.50–B.52) and (B.53–B.55), the sought spectral forms for the components of the solenoidal vector $\vec{B}$ read

$$B_r(r,\gamma) = \sum_{nm} \sigma_{nm}^{(1)}\,Y_{nm} \tag{B.59}$$

$$B_\theta(r,\gamma) = \sum_{nm}\left(+\sigma_{nm}^{(2)}\frac{\partial Y_{nm}}{\partial\theta} + \tau_{nm}^{(1)}\frac{1}{\sin\theta}\frac{\partial Y_{nm}}{\partial\lambda}\right) \tag{B.60}$$

$$B_\lambda(r,\gamma) = \sum_{nm}\left(-\tau_{nm}^{(1)}\frac{\partial Y_{nm}}{\partial\theta} + \sigma_{nm}^{(2)}\frac{1}{\sin\theta}\frac{\partial Y_{nm}}{\partial\lambda}\right), \tag{B.61}$$

where we observe that the radial component B_r is expanded in series of CSH as a "true scalar" (see Eq. A.34), while the horizontal components B_θ and B_λ are expanded in terms of the horizontal gradient of the CSHs, in agreement with Barrera et al. (1985). Note, however, that since $\sigma_{nm}^{(1)}(r)$ and $\sigma_{nm}^{(2)}(r)$ are not independent (see Eqs. B.57 and B.58), only the two scalars p_{nm} and t_{nm} are effectively sufficient in order to define the T/P decomposition of a solenoidal field.

As a final remark, we observe that for *purely poloidal fields*, we have $t_{nm} = 0$ hence $\tau^{(1)}_{nm}(r) = 0$, so that (B.59–B.61) reduce to

$$B_r(r, \gamma) = \sum_{nm} \sigma^{(1)}_{nm} Y_{nm} \tag{B.62}$$

$$B_\theta(r, \gamma) = \sum_{nm} \sigma^{(2)}_{nm} \frac{\partial Y_{nm}}{\partial \theta} \tag{B.63}$$

$$B_\lambda(r, \gamma) = \sum_{nm} \sigma^{(2)}_{nm} \frac{1}{\sin\theta} \frac{\partial Y_{nm}}{\partial \lambda} . \tag{B.64}$$

B.3 Laplace Transforms

The *Laplace transform*, named after Pierre-Simon Laplace (1749–1827), is an integral transform that converts a function of a real variable t (time) to a function of a complex variable s (complex frequency).

The Laplace transform of a function $f(t)$ is defined as

$$LT\{f(t)\} \equiv \int_{0^-}^{\infty} \mathrm{e}^{-st} f(t)\mathrm{d}t , \tag{B.65}$$

where it is assumed that the integral converges and we have complied with the definition given by Lundberg et al. (2007). In general, we shall also denote the transform of $f(t)$ with the compact notation $\tilde{f}(s)$. In (B.65), the notation 0^- for the lower integration boundary has the meaning

$$\int_{0^-}^{\infty} \mathrm{e}^{-st} f(t)\mathrm{d}t \equiv \lim_{\varepsilon \to 0^-} \int_{\varepsilon}^{\infty} \mathrm{e}^{-st} f(t)\mathrm{d}t . \tag{B.66}$$

From the definition (B.65) it is straightforward to prove through integration by parts the "rule for derivatives" for Laplace transforms:

$$LT\left\{f'(t)\right\} = s\tilde{f}(s) - f(0^-) , \tag{B.67}$$

while, by the definition of time convolution

$$f(t) * g(t) \equiv \int_{-\infty}^{+\infty} f(t - t')g(t')\mathrm{d}t' , \tag{B.68}$$

where $f(t)$ and $g(t)$ are any two given functions of time, it is easily shown that (B.65) leads to

$$LT\{(f * g)(t)\} = \tilde{f}(s)\,\tilde{g}(s) . \tag{B.69}$$

Table B.1 A few useful Laplace transforms

n.	Time domain: $f(t) = LT^{-1}\{f(s)\}$	Laplace domain: $\tilde{f}(s) = LT\{f(t)\}$
(1).	$f(t)$	$\tilde{f}(s) = \int_{0^-}^{\infty} \mathrm{e}^{-st} f(t)\mathrm{d}t$
(2).	$f'(t)$	$s\tilde{f}(s) - f(0^-)$
(3).	$\delta(t)$	1
(4).	$H(t)$	$\frac{1}{s}$
(5).	$H(t-\tau)$	$\frac{\mathrm{e}^{-s\tau}}{s}$
(6).	$H(t)\,\mathrm{e}^{at}$	$\frac{1}{s-a}$
(7).	$H(t-\tau)\int_0^{t-\tau} f(t')\mathrm{d}t'$	$\frac{\mathrm{e}^{-s\tau}}{s}\tilde{f}(s)$
(8).	$f(t) * g(t)$	$\tilde{f}(s)\tilde{g}(s)$
(9).	$\sum_{k=1}^{M} \frac{p(\alpha_k)}{q'(\alpha_k)}\mathrm{e}^{\alpha_k t}$	$\frac{p(s)}{q(s)}$

Table B.1 summarises the properties discussed above and lists the Laplace transforms of some functions of interest used throughout this book. For more exhaustive tables, the readers are referred to e.g. Oberhettinger and Badii (2012).

The *inverse Laplace transform*, or *Laplace anti-transform*, is the operator that maps a function $f(s)$ into another function $g(t)$ such that $LT\{g(t)\} = f(s)$. An integral formula for the inverse Laplace transform is the *Bromwich integral*

$$LT^{-1}\{f(s)\} = \frac{1}{2\pi \mathrm{i}} \lim_{R\to\infty} \int_{\gamma-\mathrm{i}R}^{\gamma+\mathrm{i}R} f(s)\mathrm{e}^{st}\mathrm{d}t\,, \tag{B.70}$$

where real number γ is chosen in such a way that line $s = \gamma$ lies to the right of all singularities of $f(s)$. Definition (B.70) provides a way for computing the inverse Laplace transform by the Cauchy residue theorem:

$$LT^{-1}\{f(s)\} = \sum_{k=1}^{M} \mathrm{Res}(f(s)\mathrm{e}^{st}, s_k)\,, \tag{B.71}$$

where $s_1, \ldots, s_M$ are the *poles* of $f(s)$ and, for a given complex-valued function $f(z)$, $\mathrm{Res}(f, z_0)$ is the *residue* of $f(z)$ at the singular point z_0. If z_0 is a *simple pole*, one has:

$$\mathrm{Res}(f, z_0) = \lim_{z \to z_0} (z - z_0) f(z) . \tag{B.72}$$

In the case where $f(s)$ is a *strictly proper rational function* of the form

$$f(s) = \frac{p(s)}{q(s)}, \tag{B.73}$$

where $p(s)$ and $q(s)$ are two polynomials such that $\deg(p) < \deg(q)$, Eqs. (B.71) and (B.72) lead to

$$LT^{-1}\left\{\frac{p(s)}{q(s)}\right\} = \sum_{k=1}^{M} \frac{p(s_k)}{q'(s_k)} \mathrm{e}^{s_k t} , \tag{B.74}$$

where constants $s_1, \ldots, s_M$ are the M (simple) roots of algebraic equation $q(s) = 0$.

Finally, two important relations exist between the limits of a function $f(t)$ and of its Laplace transform $\tilde{f}(s)$, namely:

$$\lim_{t \to \infty} f(t) = \lim_{s \to 0} s \tilde{f}(s) . \tag{B.75}$$

and

$$\lim_{t \to 0} f(t) = \lim_{s \to \infty} s \tilde{f}(s) . \tag{B.76}$$

which are referred to as the *final value theorem* and *the initial value theorem*, respectively.

Appendix C
Viscoelastic Self-gravitating Sphere

In this appendix we discuss in detail how the equilibrium equations for a self-gravitating, spherically symmetric, layered viscoelastic sphere subject to tidal or loading perturbations can be formulated and solved analytically. We will first introduce the *viscoelastic field equations* in their most general form, suitable for a compressible material, and discuss how they can be simplified in the incompressible limit. We will then obtain the exact analytical solution for the deformation of an incompressible homogeneous shell, and an approximated solution valid in the compressible case. Field equations for an inviscid fluid will also be formulated and solved, and boundary conditions valid at a the solid-fluid interface and at the surface of the Earth will be discussed. Finally, we will present solutions for a homogeneous sphere and outline how the results of this appendix can be employed to solve a multi-layered, spherically symmetric model.

The analytical solutions discussed in this appendix are based upon the reduction of the three-dimensional field equations into two decoupled sets of scalar ODEs, made possible by the spherical symmetry of the Earth structure, which represents one of the classical problems in global geodynamics (Backus 1967). Its solution turns out to be quite cumbersome but instructive, and therefore it shall be outlined in full detail.[10]

[10] A large part of the material presented here is a transposition of an unpublished document that GS has originally drafted in the late 1990s for the students of the *"Tectonophysics"* course at the University of Bologna (Italy), and contained in a green cover copybook that has been recently rediscovered in an archive.

G. Spada and D. Melini, *Glacial Isostatic Adjustment*, Springer Geophysics,
https://doi.org/10.1007/978-3-031-90538-4

C.1 Field Equations

Love numbers, introduced in Chap. 2, are ultimately obtained by solving the field equations for a self-gravitating, non-rotating, spherically symmetric Earth. Here we mainly follow the classical work of Wu and Peltier (1982), where the elastic solutions originally developed by Backus (1967) have been extended to the case of a linear Maxwell viscoelastic rheology. Such extension has set the foundations of all the subsequent investigations on Glacial Isostatic Adjustment (GIA) and has opened the route to the modelization of the phenomenon of global post-seismic deformations. Subsequently, the field equations have been generalised to more complex linear rheologies. For some important references, see Sect. 1.3.

C.1.1 General Form of the Field Equations

According to the theory of the elastic gravitational free oscillations (Backus 1967), the *momentum conservation* equation reads

$$\nabla \cdot \tilde{\tau} - \rho_0 \nabla \tilde{\phi}_1 - \nabla \left(\rho_0 g_0 \tilde{\mathbf{u}} \cdot \hat{r} \right) - \tilde{\rho}_1 g_0 \hat{r} = 0 \,, \quad \text{Momentum Equation} \qquad \text{(C.1)}$$

where the limit of vanishing frequency has been taken, since inertial forces can be neglected due to the long time scale characterizing the surface loading problem compared to the time scale of propagation of seismic waves (see also Rayleigh 1906; Smylie and Mansinha 1971a; Farrell 1972; Cathles 1975; Sabadini et al. 2016). Note that, based upon the correspondence principle of linear viscoelasticity (see Sect. 1.3.9), we are dealing with Laplace-transformed quantities that here and in the following equations shall be denoted by an overtilde. Essentially, Eq. (C.1) is a generalised form of the well known Navier-Cauchy equilibrium equation in continuum mechanics $\nabla \cdot \tau + \rho \, \mathbf{g} = 0$ where τ is the stress field, ρ is density and $\mathbf{g}$ is the gravitational acceleration (e.g., Fung 1965). However, while in the Navier-Cauchy equation density and gravity are assumed to be constants, (C.1) properly accounts for gravity field variations and density changes induced by deformation within the Earth.

The momentum equation (C.1) is linearised with respect to perturbations relative to a hydrostatic state of equilibrium in which $\nabla p_0 = -\rho_0 g_0 \hat{r}$ where $p_0 = p_0(r)$ is the pressure, and the gravitational potential $\phi_0 = \phi_0(r, \gamma)$ obeys Poisson's equation $\nabla^2 \phi_0 = 4\pi G \rho_0$, where $\rho_0 = \rho_0(r)$ and $g_0 = g_0(r)$ are the unperturbed density field of the Earth and gravitational acceleration, respectively. In Eq. (C.1), $\tilde{\phi}_1 = \tilde{\phi}_1(r, \gamma) \ll \phi_0(r, \gamma)$ is the gravitational potential variation, $\tilde{\rho}_1 = \tilde{\rho}_1(r, \gamma) \ll \rho_0(r)$ is the density variation, $\tilde{\mathbf{u}} = \tilde{\mathbf{u}}(r, \gamma)$ is the displacement vector, $\tilde{\tau} = \tilde{\tau}(r, \gamma)$ is the stress field, and we recall that $\gamma = (\theta, \lambda)$ where θ is co-latitude and λ is longitude. As emphasised by Wu and Peltier (1982), the *prestress* term $\nabla \left(\rho_0 g_0 \tilde{\mathbf{u}} \cdot \hat{r} \right)$ ensures that in the limit of long times the equilibrium equations correctly reproduce those valid for a viscous fluid.

The total gravitational potential variation $\tilde{\phi}_1(r, \gamma)$ in Eq. (C.1) accounts both for the *direct* potential variation due to the surface load and for the *indirect* contribution due to the internal redistribution of mass affected by the load-induced deformation (see Wu and Peltier 1982, page 438). The gravitational potential variation obeys Poisson's equation

$$\nabla^2 \tilde{\phi}_1 = 4\pi G \tilde{\rho}_1 \,, \quad \text{Poisson's Equation} \tag{C.2}$$

whereas following Backus (1967), Smylie and Mansinha (1971a) and Wu and Peltier (1982), the density perturbation $\tilde{\rho}_1$ is obtained by linearizing the continuity equation, which provides

$$\tilde{\rho}_1 = -\rho_0 \tilde{\Delta} - \tilde{\mathbf{u}} \cdot \partial_r \rho_0 \, \hat{r} \,, \quad \text{Continuity Equation} \tag{C.3}$$

where the *dilatation* is

$$\tilde{\Delta} = \nabla \cdot \tilde{\mathbf{u}} \,. \tag{C.4}$$

Note that Eqs. (C.2) and (C.3) form, along with the momentum equation (C.1), a coupled set of linear PDEs.

In the Laplace domain, the stress-strain constitutive equation reads

$$\tilde{\tau}_{ij} = \tilde{\lambda} \tilde{\epsilon}_{kk} \delta_{ij} + 2\tilde{\mu} \tilde{\epsilon}_{ij} \,, \quad \text{Constitutive Equation} \tag{C.5}$$

with the infinitesimal strain

$$\tilde{\epsilon}_{ij} = \frac{1}{2} \left(\tilde{u}_{i,j} + \tilde{u}_{j,i} \right) \,, \tag{C.6}$$

where $\tilde{u}_{i,j}$ denotes the Jacobian of the displacement field and $\tilde{u}_{j,i}$ its transposed. Note that

$$\tilde{\epsilon}_{kk} \equiv \tilde{u}_{k,k} = \nabla \cdot \tilde{\mathbf{u}} \equiv \tilde{\Delta} \tag{C.7}$$

represents the local relative variation of the volume of the continuum (Wu and Peltier 1982; Ranalli 1995).

In the particular case of the Maxwell rheology, in Sect. 1.3.9, we have obtained the complex shear modulus

$$\tilde{\mu} = \frac{\mu s}{s + \dfrac{\mu}{\eta}} \,, \tag{C.8}$$

where μ and η are the elastic shear modulus and newtonian viscosity, respectively, s is the complex Laplace variable, and the corresponding complex form of first Lamé constant is

$$\tilde{\lambda} = \frac{\lambda s + \dfrac{\mu\kappa}{\eta}}{s + \dfrac{\mu}{\eta}}, \tag{C.9}$$

where λ is the first Lamé constant and the (frequency-independent) bulk modulus is:

$$\kappa = \tilde{\lambda} + \frac{2}{3}\tilde{\mu}\,. \tag{C.10}$$

Equations (C.1), (C.2), (C.3) and (C.5), constitute the fundamental field equations for a self-gravitating, non-rotating, and spherically symmetric Earth. Provided that appropriate complex moduli $\tilde{\mu}$ and $\tilde{\lambda}$ are made explicit, they are valid for an arbitrary linear rheological law.

C.1.2 Field Equations for a Compressible Homogeneous *Shell*

In the case of an homogeneous shell bounded by radii r_1 and r_2, like the one depicted in Fig. 2.5c, the form of the field equations is greatly simplified. Since the shell is homogeneous, its density ρ_0 is a constant and (C.3) reduces to

$$\tilde{\rho}_1 = -\rho_0\tilde{\Delta}\,, \quad \text{Continuity Equation} \tag{C.11}$$

stating that the the variations of density are only due to local changes in the volume of the continuum. As a consequence, (C.2) can be written as

$$\nabla^2\tilde{\phi}_1 = -4\pi G\rho_0\tilde{\Delta}\,. \quad \text{Poisson's Equation} \tag{C.12}$$

Assuming that ρ_0 is constant and taking into account the continuity equation (C.11), the momentum conservation equation (C.1) for a homogeneous shell takes the form

$$\nabla\cdot\tilde{\tau} - \rho_0\nabla\left(\tilde{\phi}_1 + g_0\tilde{\mathbf{u}}\cdot\hat{r}\right) + \rho_0 g_0\tilde{\Delta}\,\hat{r} = 0\,. \tag{C.13}$$

From the constitutive equation (C.5), the first term in (C.13) is

$$\begin{aligned}\tilde{\tau}_{ij,j} =& (\tilde{\lambda}\tilde{\Delta}\,\delta_{ij})_{,j} + 2\,\tilde{\mu}\,\tilde{\epsilon}_{ij,j}\\ =& \tilde{\lambda}\tilde{\Delta}_{,i} + \tilde{\mu}\,(\tilde{u}_{i,j} + \tilde{u}_{j,i})_{,j}\\ =& \tilde{\lambda}\tilde{\Delta}_{,i} + \tilde{\mu}\,(\tilde{u}_{i,jj} + \tilde{u}_{j,ij})\\ =& (\tilde{\lambda} + \tilde{\mu})\tilde{\Delta}_{,i} + \tilde{\mu}\,\tilde{u}_{i,jj}\,,\end{aligned} \tag{C.14}$$

where we have used the definition of dilatation $\tilde{\Delta}$ (C.7). In vector form

$$\nabla \cdot \tilde{\tau} = (\tilde{\lambda} + \tilde{\mu})\nabla\tilde{\Delta} + \tilde{\mu}\nabla^2\tilde{\mathbf{u}}\,, \tag{C.15}$$

which, substituted into (C.13) leads to:

$$\nabla\left[(\tilde{\lambda} + 2\tilde{\mu})\tilde{\Delta} - \rho_0\tilde{\phi}_1 - \rho_0 g_0 \tilde{\mathbf{u}} \cdot \hat{r}\right] + \rho_0 g_0 \tilde{\Delta}\hat{r} - \tilde{\mu}\left(\nabla \times \nabla \times \tilde{\mathbf{u}}\right) = 0\,, \quad \text{Momentum Equation} \tag{C.16}$$

where we have also used identity (B.28). The PDEs (C.11), (C.12) and (C.16), along with the constitutive equation (C.5), represent the complete set of field equations for a *compressible* and *homogeneous* shell.

C.1.3 *Field Equations for an* Incompressible Homogeneous *Shell*

In the case of an *incompressible* and *homogeneous* shell, the form of the field equations is further simplified and an exact analytical solution is possible. This case is particularly important since *most* (but of course, *not all*) of the approaches to GIA are based on this approximation (see e.g., Vermeersen et al. 1996a; Vermeersen and Sabadini 1997). Similarly, in the study of the Earth's response to endogenous sources like earthquakes, the assumption of incompressibility has been often adopted (see e.g., Sabadini et al. 1995; Piersanti et al. 1995, 1997; Melini et al. 2008); notable exceptions however exist, which are summarised in the comprehensive treatise of Sabadini et al. (2016).

Since the shell is incompressible, its volume is not altered by deformation (the dilatation is $\tilde{\Delta} = 0$), so that

$$\nabla \cdot \tilde{\mathbf{u}} = 0\,, \quad \text{Continuity Equation} \tag{C.17}$$

or

$$\tilde{u}_{k,k} = 0\,, \tag{C.18}$$

implying that the displacement vector $\tilde{\mathbf{u}}$ is solenoidal (see Sect. B.2). Furthermore, since the shell is homogeneous, $\partial_r \rho_0 = 0$. Then, from (C.3), the incremental density field is

$$\tilde{\rho}_1 = 0\,, \tag{C.19}$$

so that from (C.2) we see that the gravitational potential variation is harmonic both inside and outside the Earth:

$$\nabla^2 \tilde{\phi}_1 = 0\,, \quad \text{Laplace's Equation} \tag{C.20}$$

or

$$\tilde{\phi}_{1,kk} = 0\,. \tag{C.21}$$

The total gravitational acceleration is

$$\tilde{\mathbf{g}} = -g_0 \hat{r} + \tilde{\mathbf{g}}_1\,, \tag{C.22}$$

where

$$\tilde{\mathbf{g}}_1 = -\nabla \tilde{\phi}_1 \tag{C.23}$$

is the contribution to gravity acceleration due to mass redistribution and deformation.

With $\tilde{\rho}_1 = 0$, the momentum equation (C.1) reduces to

$$\nabla \cdot \tilde{\tau} - \rho_0 \nabla \tilde{\phi}_1 - \nabla \left(\tilde{\mathbf{u}} \cdot \rho_0 g_0 \hat{r} \right) = 0\,, \tag{C.24}$$

or

$$\tilde{\tau}_{ij,j} - \rho_0 \tilde{\phi}_{1,i} - \left(\rho_0 g_0 \tilde{u}_r \right)_{,i} = 0\,. \tag{C.25}$$

Following Love (1911)—see, in particular, §154 of his book—the limit of incompressible materials is realised letting the first Lamé constant $\tilde{\lambda}$ to infinity and, simultaneously, letting the dilatation to zero, in such a way that their product remains finite:

$$\tilde{\Pi} = \lim_{\substack{\tilde{\lambda} \to \infty \\ \tilde{\Delta} \to 0}} \left(\tilde{\lambda} \tilde{\Delta} \right)\,. \tag{C.26}$$

As a consequence, recalling (C.7), the constitutive equation (C.5) takes the form

$$\tilde{\tau}_{ij} = \tilde{\Pi} \delta_{ij} + 2 \tilde{\mu} \tilde{\epsilon}_{ij}\,, \quad \text{Constitutive Equation} \tag{C.27}$$

where $\tilde{\mu}$ is the viscoelastic complex modulus that, in the particular case of the Maxwell rheology, is expressed by (C.8), and where the infinitesimal strain tensor $\tilde{\epsilon}_{ij}$ is given by (C.6). As to the physical meaning of $\tilde{\Pi}$, we note that by contracting the indices in (C.27) and imposing the constraint of incompressibility (C.17), one obtains

$$\tilde{\Pi} = \frac{1}{3} \tilde{\tau}_{kk}\,, \tag{C.28}$$

showing that $\tilde{\Pi}$ represents the *mean normal stress* acting within the continuum.

A convenient form of the equilibrium equation can be established observing that, from (C.27), we have

$$
\begin{aligned}
\tilde{\tau}_{ij,j} &= (\tilde{\Pi}\,\delta_{ij})_{,j} + 2\,\tilde{\mu}\,\tilde{\epsilon}_{ij,j} \\
&= \tilde{\Pi}_{,i} + \tilde{\mu}\,(\tilde{u}_{i,j} + \tilde{u}_{j,i})_{,j} \\
&= \tilde{\Pi}_{,i} + \tilde{\mu}\,(\tilde{u}_{i,jj} + \tilde{u}_{j,ij}) \\
&= \tilde{\Pi}_{,i} + \tilde{\mu}\,\tilde{u}_{i,jj}\,,
\end{aligned}
\tag{C.29}
$$

where $\tilde{u}_{j,ij} = \tilde{u}_{j,ji} = 0$ because the displacement field is solenoidal (see C.17). In vector form:

$$\nabla \cdot \tilde{\tau} = \nabla\tilde{\Pi} + \tilde{\mu}\nabla^2\tilde{\mathbf{u}} \tag{C.30}$$

which substituted into (C.24) taking the identity (B.29) into account yields:

$$\nabla\big(\tilde{\Pi} - \rho_0\tilde{\phi}_1 - \rho_0 g_0 \tilde{u}_r\big) - \tilde{\mu}\,(\nabla \times \nabla \times \tilde{\mathbf{u}}) = 0\,. \quad \text{Momentum Equation} \tag{C.31}$$

The four PDEs above (namely, C.17, C.20, C.27 and C.31) constitute the field equations for an *incompressible* and *homogeneous* shell that in the ensuing section are transformed into a linear set of first order ODEs.

C.2 Solution for the Incompressible Homogeneous Shell

In this section we will discuss in detail how, in the incompressible limit, the field equations for an homogeneous shell can be cast in the form of ODEs through a spherical harmonic expansion and analytically solved. The case of an incompressible shell is of particular relevance since, as we shall see in the following, in that case an *exact* analytical solution of the field equations is possible. We will obtain the most general solution of the field equations by considering both the poloidal and the toroidal components of the problem. However, if the external forcing is axisymmetric (as in the case of a surface point load appropriate for GIA), only the poloidal component is relevant.

C.2.1 Spectral Analysis of the Field Equations

In Sect. C.1.3 we have obtained the field equations for an *incompressible* and *homogeneous* shell. They constitute a set of PDEs where the unknowns are the three components of the displacement vector $\tilde{\mathbf{u}}$, the incremental gravitational potential $\tilde{\phi}_1$ and the mean normal stress $\tilde{\Pi}$. Our target here is to transform these PDEs into ODEs by a suitable expansion in spherical harmonics of the fields involved. The whole derivation, which is quite cumbersome, is split in steps for the readers convenience.

C.2.1.1 Step 0: Notation

In this and in the ensuing sections, we shall adopt a simplified notation for the CSHs and for their partial derivatives, namely:

$$Y \equiv Y_{nm} \tag{C.32}$$

$$Y_\theta \equiv \frac{\partial Y_{nm}}{\partial \theta} \tag{C.33}$$

$$Y_\lambda \equiv \frac{1}{\sin\theta}\frac{\partial Y_{nm}}{\partial \lambda}\,, \tag{C.34}$$

where subscripts *nm* (i.e., the harmonic degree and order) and the γ-dependence are implicit on the left hand sides. For the mixed second order derivatives we adopt the notation

$$Y_{\lambda\theta} \equiv (Y_\theta)_\lambda \equiv \frac{1}{\sin\theta}\frac{\partial^2 Y_{nm}}{\partial\theta\,\partial\lambda} \tag{C.35}$$

and

$$Y_{\theta\lambda} \equiv (Y_\lambda)_\theta \equiv -\frac{\cos\theta}{\sin^2\theta}\frac{\partial Y_{nm}}{\partial\lambda} + \frac{1}{\sin\theta}\frac{\partial^2 Y_{nm}}{\partial\theta\,\partial\lambda}\,. \tag{C.36}$$

Accordingly, it is easily verified that:

$$Y_{\theta\lambda} + \cot\theta\, Y_\lambda = Y_{\lambda\theta}\,, \tag{C.37}$$

and

$$Y_{\theta\theta} + \cot\theta\, Y_\theta + Y_{\lambda\lambda} = -n(n+1)Y\,, \tag{C.38}$$

where the first is just a consequence of definitions above, while the second derives from the fundamental property of the CSH expressed by the Legendre equation (A.33).

C.2.1.2 Step 1: CSH Expansion of the Unknown Fields

☐ <u>*Displacement* $(\tilde{u}_r,\ \tilde{u}_\theta,\ \tilde{u}_\lambda)$</u>. Since the displacement field is solenoidal (see C.17), we can take advantage of the Mie representation. Hence, coherently with the notation adopted here, Eqs. (B.59–B.61) take the form:

$$\tilde{u}_r(r,\gamma) = \sum_{nm} \tilde{U}_1\, Y \tag{C.39}$$

$$\tilde{u}_\theta(r,\gamma) = \sum_{nm} \left(+\tilde{U}_2 Y_\theta + \tilde{V}_1 Y_\lambda\right) \tag{C.40}$$

$$\tilde{u}_\lambda(r,\gamma) = \sum_{nm} \left(-\tilde{V}_1 Y_\theta + \tilde{U}_2 Y_\lambda\right), \tag{C.41}$$

where $\tilde{U}_1$, $\tilde{U}_2$ and $\tilde{V}_1$ are poloidal and toroidal radial functions, respectively, implicitly dependent upon degree and order nm and radius r.

☐ *Gravitational potential* $\tilde{\phi}_1$. Since the variation of gravitational potential $\tilde{\phi}_1$ is a true scalar, in analogy with (C.39) we assume the usual CSH expansion

$$\tilde{\phi}_1(r,\gamma) = \sum_{nm} \tilde{U}_5\, Y\,, \tag{C.42}$$

where $\tilde{U}_5(r)$ are suitable radial coefficients.

☐ *Mean normal stress* $\tilde{\Pi}$. Similarly to the CSH expansion of $\tilde{u}_r$ and $\tilde{\phi}_1$, for the mean normal stress we write

$$\tilde{\Pi}(r,\gamma) = \sum_{nm} \tilde{U}_8\, Y\,, \tag{C.43}$$

where $\tilde{U}_8(r)$ are radial coefficients. The rationale behind the choice of these particular names for radial coefficients of $\tilde{\phi}_1$ and $\tilde{\Pi}$ will become more clear further on.

Table C.1 Explicit expressions for the six independent components of the infinitesimal strain tensor $\epsilon_{ij} = \frac{1}{2}(u_{i,j} + u_{j,i})$ in spherical coordinates (see e.g., Fung 1965)

$$\epsilon_{rr} = \frac{\partial u_r}{\partial r}$$

$$\epsilon_{\theta\theta} = \frac{1}{r}\left(\frac{\partial u_\theta}{\partial \theta} + u_r\right)$$

$$\epsilon_{\lambda\lambda} = \frac{1}{r}\left(\frac{1}{\sin\theta}\frac{\partial u_\lambda}{\partial \lambda} + u_r + u_\theta \cot\theta\right)$$

$$\epsilon_{r\theta} = \epsilon_{\theta r} = \frac{1}{2}\left(\frac{\partial u_\theta}{\partial r} - \frac{u_\theta}{r} + \frac{1}{r}\frac{\partial u_r}{\partial \theta}\right)$$

$$\epsilon_{\theta\lambda} = \epsilon_{\lambda\theta} = \frac{1}{2r}\left(\frac{1}{\sin\theta}\frac{\partial u_\theta}{\partial \lambda} + \frac{\partial u_\lambda}{\partial \theta} - u_\lambda \cot\theta\right)$$

$$\epsilon_{r\lambda} = \epsilon_{\lambda r} = \frac{1}{2}\left(\frac{1}{r\sin\theta}\frac{\partial u_r}{\partial \lambda} + \frac{\partial u_\lambda}{\partial r} - \frac{u_\lambda}{r}\right)$$

☐ *Stress component* $\tilde{\tau}_{rr}$. By virtue of the definition (C.27) of the stress tensor $\tilde{\tau}_{ij}$, and making also explicit the infinitesimal strain tensor $\tilde{\epsilon}_{ij}$ in spherical coordinates (see Table C.1), we have

$$
\begin{aligned}
\tilde{\tau}_{rr}(r,\gamma) &= \tilde{\Pi} + 2\,\tilde{\mu}\,\tilde{\epsilon}_{rr} \\
&= \tilde{\Pi} + 2\,\tilde{\mu}\,\frac{\partial \tilde{u}_r}{\partial r} \\
&= \sum_{nm} \tilde{U}_8\, Y + 2\tilde{\mu}\,\frac{\partial}{\partial r}\sum_{nm} \tilde{U}_1\, Y \\
&= \sum_{nm}\left(\tilde{U}_8 + \tilde{\mu}\,\tilde{U}_3\right) Y\,,
\end{aligned}
\tag{C.44}
$$

where we have introduced the auxiliary poloidal radial function

$$
\tilde{U}_3(r) \equiv 2\,\tilde{U}_1'\,. \tag{C.45}
$$

☐ *Stress component* $\tilde{\tau}_{r\theta} = \tilde{\tau}_{\theta r}$.

$$
\begin{aligned}
\tilde{\tau}_{r\theta}(r,\gamma) &= 2\,\tilde{\mu}\,\tilde{\epsilon}_{r\theta} \\
&= \tilde{\mu}\left(\frac{\partial \tilde{u}_\theta}{\partial r} - \frac{\tilde{u}_\theta}{r} + \frac{1}{r}\frac{\partial \tilde{u}_r}{\partial \theta}\right) \\
&= \tilde{\mu}\left(\frac{\partial}{\partial r}\sum_{nm}\left(\tilde{U}_2 Y_\theta + \tilde{V}_1 Y_\lambda\right) - \frac{1}{r}\sum_{nm}\left(\tilde{U}_2 Y_\theta + \tilde{V}_1 Y_\lambda\right) + \frac{1}{r}\frac{\partial}{\partial \theta}\sum_{nm}\tilde{U}_1\, Y\right) \\
&= \tilde{\mu}\sum_{nm}\left(\tilde{U}_2'\, Y_\theta + \tilde{V}_1'\, Y_\lambda - \frac{\tilde{U}_2}{r} Y_\theta - \frac{\tilde{V}_1}{r} Y_\lambda + \frac{\tilde{U}_1}{r} Y_\theta\right) \\
&= \tilde{\mu}\sum_{nm}\left(\tilde{U}_4\, Y_\theta + \tilde{V}_2\, Y_\lambda\right),
\end{aligned}
\tag{C.46}
$$

with auxiliary poloidal and toroidal radial functions:

$$
\tilde{U}_4(r) \equiv \tilde{U}_2' - \frac{\tilde{U}_2}{r} + \frac{\tilde{U}_1}{r} \tag{C.47}
$$

$$
\tilde{V}_2(r) \equiv \tilde{V}_1' - \frac{\tilde{V}_1}{r}\,, \tag{C.48}
$$

respectively.

☐ *Stress component* $\tilde{\tau}_{r\lambda} = \tilde{\tau}_{\lambda r}$.

$$
\begin{aligned}
\tilde{\tau}_{r\lambda}(r,\gamma) &= 2\,\tilde{\mu}\,\tilde{\epsilon}_{r\lambda} \\
&= \tilde{\mu}\left(\frac{1}{r\sin\theta}\frac{\partial\tilde{u}_r}{\partial\lambda} + \frac{\partial\tilde{u}_\lambda}{\partial r} - \frac{\tilde{u}_\lambda}{r}\right) \\
&= \tilde{\mu}\left(\frac{1}{r\sin\theta}\frac{\partial}{\partial\lambda}\sum_{nm}\tilde{U}_1\,Y + \frac{\partial}{\partial r}\sum_{nm}\left(-\tilde{V}_1\,Y_\theta + \tilde{U}_1\,Y_\lambda\right) + \right. \\
&\quad \left. - \frac{1}{r}\sum_{nm}\left(-\tilde{V}_1\,Y_\theta + \tilde{U}_1\,Y_\lambda\right)\right) \\
&= \tilde{\mu}\sum_{nm}\left(\frac{\tilde{U}_1}{r}Y_\lambda - \tilde{V}_1'\,Y_\theta + \tilde{U}_2'\,Y_\lambda + \frac{\tilde{V}_1}{r}Y_\theta - \frac{\tilde{U}_2}{r}Y_\lambda\right) \\
&= \tilde{\mu}\sum_{nm}\left(\tilde{U}_4\,Y_\lambda - \tilde{V}_2\,Y_\theta\right), \qquad \text{(C.49)}
\end{aligned}
$$

where we note that $\tilde{\tau}_{r\lambda}$ is expressed by means of the same radial functions involved in the expression of $\tilde{\tau}_{r\theta}$ (C.46) and defined by (C.47) and (C.48), respectively.

☐ *Stress component* $\tilde{\tau}_{\theta\theta}$. For the remaining three components of the stress tensor (namely, $\tilde{\tau}_{\theta\theta}$, $\tilde{\tau}_{\theta\lambda}$ and $\tilde{\tau}_{\lambda\lambda}$) we only provide the final expressions, leaving the derivation as an exercise for the readers.

$$
\begin{aligned}
\tilde{\tau}_{\theta\theta}(r,\gamma) &= \tilde{\Pi} + 2\,\tilde{\mu}\,\tilde{\epsilon}_{\theta\theta} \\
&= \sum_{nm}\tilde{U}_8\,Y + \frac{2\,\tilde{\mu}}{r}\sum_{nm}\left(\tilde{U}_2\,Y_{\theta\theta} + \tilde{V}_1\,Y_{\theta\lambda} + \tilde{U}_1 Y\right). \qquad \text{(C.50)}
\end{aligned}
$$

☐ *Stress component* $\tilde{\tau}_{\theta\lambda} = \tilde{\tau}_{\lambda\theta}$.

$$
\begin{aligned}
\tilde{\tau}_{\theta\lambda}(r,\gamma) &= 2\,\tilde{\mu}\,\tilde{\epsilon}_{\theta\lambda} \qquad \text{(C.51)} \\
&= \frac{\tilde{\mu}}{r}\sum_{nm}\left(\tilde{V}_1\,(-Y_{\theta\theta} + \cot\theta\;Y_\theta + Y_{\lambda\lambda}) + \tilde{U}_2\,(Y_{\theta\lambda} + Y_{\lambda\theta} - \cot\theta\;Y_\lambda)\right).
\end{aligned}
$$

☐ *Stress Component* $\tilde{\tau}_{\lambda\lambda}$.

$$
\begin{aligned}
\tilde{\tau}_{\lambda\lambda}(r,\gamma) &= \tilde{\Pi} + 2\,\tilde{\mu}\,\tilde{\epsilon}_{\lambda\lambda} \\
&= \sum_{nm}\tilde{U}_8 Y + \qquad \text{(C.52)} \\
&\quad + \frac{2\tilde{\mu}}{r}\sum_{nm}\left(\tilde{U}_2\,(Y_{\lambda\lambda} + \cot\theta\;Y_\theta) + \tilde{V}_1\,(\cot\theta\;Y_\lambda - Y_{\lambda\theta}) + \tilde{U}_1 Y\right).
\end{aligned}
$$

☐ Trace of the stress tensor. In closing, we note that using (C.44), (C.50) and (C.52) along with identities (C.37–C.38), the continuity equation (C.43) and (C.59), the trace of the stress field is:

$$
\begin{aligned}
\mathrm{Tr}(\tilde{\tau}_{ij}) = & \, 3\sum_{nm}\tilde{U}_8 Y + \tilde{\mu}\sum_{nm}\tilde{U}_3 Y + \\
& + \frac{2\tilde{\mu}}{r}\sum_{nm}\Big(2\tilde{U}_1 Y + \tilde{U}_2 \underbrace{(Y_{\theta\theta} + \cot\theta + Y_{\lambda\lambda})}_{=-n(n+1)Y} + \tilde{V}_1 \underbrace{(Y_{\theta\lambda} + \cot\theta\, Y_\lambda - Y_{\lambda\theta})}_{\equiv 0}\Big) \\
= & \, 3\sum_{nm}\tilde{U}_8 Y + 2\tilde{\mu}\underbrace{\sum_{nm}\left(\tilde{U}_1' + \frac{2\tilde{U}_1}{r} - \frac{n(n+1)}{r}\tilde{U}_2\right)}_{=0} Y \;=\; 3\,\tilde{\Pi}\,, \qquad \text{(C.53)}
\end{aligned}
$$

which agrees with (C.28).

C.2.1.3 Step 2: Continuity Equation

According to (B.23), the continuity equation

$$
\nabla \cdot \tilde{\mathbf{u}} = 0 \qquad \text{(C.54)}
$$

in spherical coordinates reads

$$
\frac{\partial \tilde{u}_r}{\partial r} + \frac{2\tilde{u}_r}{r} + \frac{1}{r}\left(\cot\theta\,\tilde{u}_\theta + \frac{\partial \tilde{u}_\theta}{\partial \theta} + \frac{1}{\sin\theta}\frac{\partial \tilde{u}_\lambda}{\partial \lambda}\right) = 0\,, \qquad \text{(C.55)}
$$

which can be easily transformed into an ODE using the Mie representation for the individual components of the displacement field (see Eqs. C.39–C.41). By direct substitution, we obtain

$$
\begin{aligned}
& \sum_{nm}\left(\tilde{U}_1' + \frac{2}{r}\tilde{U}_1\right) Y + \\
& + \frac{1}{r}\Big[\cot\theta \sum_{nm}\left(\tilde{U}_2 Y_\theta + \tilde{V}_1 Y_\lambda\right) + \sum_{nm}\left(\tilde{U}_2 Y_{\theta\theta} + \tilde{V}_1 Y_{\theta\lambda}\right) + \\
& + \sum_{nm}\left(-\tilde{V}_1 Y_{\lambda\theta} + \tilde{U}_2 Y_{\lambda\lambda}\right)\Big] = 0, \qquad \text{(C.56)}
\end{aligned}
$$

or

$$\sum_{nm}\left(\tilde{U}_1' + \frac{2}{r}\tilde{U}_1\right)Y + \tag{C.57}$$

$$+\frac{1}{r}\sum_{nm}\tilde{U}_2\underbrace{(Y_{\theta\theta}+\cot\theta\ Y_\theta + Y_{\lambda\lambda})}_{=-n(n+1)Y} + \frac{1}{r}\sum_{nm}\tilde{V}_1\underbrace{(Y_{\theta\lambda}+\cot\theta\ Y_\lambda - Y_{\lambda\theta})}_{=0} = 0\,,$$

which by virtue of identities (C.37) and (C.38) becomes

$$\sum_{nm}\left(\tilde{U}_1' + \frac{2}{r}\tilde{U}_1 - \frac{n(n+1)}{r}\tilde{U}_2\right)Y = 0\,, \tag{C.58}$$

so that the mass conservation principle expressed by the continuity equation yields the ODE:

$$\tilde{U}_1' + \frac{2}{r}\tilde{U}_1 - \frac{n(n+1)}{r}\tilde{U}_2 = 0\,. \tag{C.59}$$

C.2.1.4 Step 3: Laplace's Equation

According to (B.32), the Laplace's equation for the incremental gravitational potential

$$\nabla^2\tilde{\phi}_1 = 0 \tag{C.60}$$

in spherical coordinates reads

$$\frac{\partial^2\tilde{\phi}_1}{\partial r^2} + \frac{2}{r}\frac{\partial\tilde{\phi}_1}{\partial r} + \frac{1}{r^2}\left(\frac{\partial^2\tilde{\phi}_1}{\partial\theta^2} + \cot\theta\frac{\partial\tilde{\phi}_1}{\partial\theta} + \frac{1}{\sin^2\theta}\frac{\partial^2\tilde{\phi}_1}{\partial\lambda^2}\right) = 0\,. \tag{C.61}$$

Using (C.42), by direct substitution we obtain

$$\sum_{nm}\left(\tilde{U}_5'' + \frac{2\tilde{U}_5'}{r}\right)Y + \frac{1}{r^2}\sum_{nm}\tilde{U}_5\underbrace{(Y_{\theta\theta}+\cot\theta\,Y_\theta + Y_{\lambda\lambda})}_{=-n(n+1)Y} = 0\,, \tag{C.62}$$

providing the second order ODE:

$$\tilde{U}_5'' + \frac{2\tilde{U}_5'}{r} - \frac{n(n+1)}{r^2}\tilde{U}_5 = 0\,. \tag{C.63}$$

The general solution of this equation can be established in terms of powers of r. Assuming $\tilde{U}_5(r) \sim r^p$ and substituting gives the algebraic equation $p^2 + p - n(n+1) = 0$ which has integer roots $p_1 = n$ and $p_2 = -(n+1)$, respectively. Hence, by linearity, the general solution of (C.63) inside the shell is the linear combination

$$\tilde{U}_5(r) = \tilde{c}_3 r^n + \tilde{c}_3^* r^{-(n+1)}\,, \tag{C.64}$$

where $\tilde{c}_3$ and $\tilde{c}_3^*$ are n-dependent constants. Note that the two terms on the *r.h.s.* are regular and singular for $r \to 0$, respectively (here the asterisk *does not* indicate the complex conjugate). Hence, inside the shell they are both admissible, whereas outside the Earth the first term is not. For the homogeneous Earth model (the so called *Kelvin model* discussed in Sect. C.6), only the first term is admissible inside the sphere since the second is singular in the origin.

Following a long tradition, it is convenient to reduce (C.63) to first order by introducing a new radial function $\tilde{U}_6$, related to the potential gradient, since this simplifies the application of the boundary conditions corresponding either to a surface and to a tidal load (see Sect. C.5). According to Farrell (1972, page 772) and Wu and Peltier (1982, page 439), variable $\tilde{U}_6$ is defined[11] as:

$$\tilde{U}_6(r) = \tilde{U}_5' + \frac{n+1}{r}\tilde{U}_5 + 4\pi G\rho_0\tilde{U}_1\,, \tag{C.65}$$

which itself can be cast in the form of a linear first order ODE:

$$\tilde{U}_5' = -4\pi G\rho_0\tilde{U}_1 - \frac{n+1}{r}\tilde{U}_5 + \tilde{U}_6\,. \tag{C.66}$$

Taking the radial derivative of (C.65) and using (C.66) with Laplace's equation (C.63) and continuity equation (C.59), after some algebra one obtains a further first order ODE that reads:

$$\tilde{U}_6' = -\frac{4\pi G\rho_0(n+1)}{r}\tilde{U}_1 + \frac{4\pi G\rho_0 n(n+1)}{r}\tilde{U}_2 + \frac{n-1}{r}\tilde{U}_6\,. \tag{C.67}$$

C.2.1.5 Step 4: Momentum Equation

Recalling (C.31), the momentum equation stems from two vector terms

$$\underbrace{\nabla\left(\tilde{\Pi} - \rho_0\tilde{\phi}_1 - \rho_0 g_0\tilde{u}_r\right)}_{1^{\text{st}}\text{ term}} + \tilde{\mu}\underbrace{\left(-\nabla\times\nabla\times\tilde{\mathbf{u}}\right)}_{2^{\text{nd}}\text{ term}} = 0 \tag{C.68}$$

that are handled separately below.

□ <u>1st term</u>. The derivation is a bit involved for the $\hat{r}$ component but straightforward for the two horizontals. We use the definition of gradient in spherical coordinates (B.4), the Mie representation of the displacement field (C.39–C.41), the scalar CSH expansions for $\tilde{\phi}_1$ (C.42) and $\tilde{\Pi}$ (C.43), the definition of potential stress (C.66),

---

[11] In the literature, this auxiliary variable is normally denoted by Q or q and sometimes referred to as *potential stress* (see e.g., Sabadini et al. 2016). However, note that different authors have used different definitions for this variable (see for example Longman 1962, page 847). This may lead to some confusion in the application of the appropriate surface boundary conditions (see Sect. C.5). Here we stick on definition (C.65), which agrees with Sabadini et al. (2016); see, in particular, their Eq. (1.93).

the expression of the unperturbed gravity gradient (2.7) and the scalar form of the continuity equation (C.59), obtaining:

$$\begin{aligned}[1^{\mathrm{st}}\ \text{term}]_r &= \frac{\partial}{\partial r}(\tilde{\Pi} - \rho_0\tilde{\phi}_1 - \rho_0 g_0 \tilde{u}_r) \\ &= \sum_{nm}\left(\tilde{U}_8' - \rho_0\tilde{U}_5' - \rho_0(g_0\tilde{U}_1)'\right)Y \\ &= \sum_{nm}\left(\tilde{U}_8' + \frac{4\rho_0 g_0}{r}\tilde{U}_1 - \frac{n(n+1)\rho_0 g_0}{r}\tilde{U}_2 + \right. \\ &\quad \left. + \frac{n+1}{r}\rho_0\tilde{U}_5 - \rho_0\tilde{U}_6\right)Y\,. \end{aligned} \tag{C.69}$$

$$\begin{aligned}[1^{\mathrm{st}}\ \text{term}]_\theta &= \frac{1}{r}\frac{\partial}{\partial\theta}(\tilde{\Pi} - \rho_0\tilde{\phi}_1 - \rho_0 g_0 \tilde{u}_r) \\ &= \frac{1}{r}\sum_{nm}\left(\tilde{U}_8 - \rho_0\tilde{U}_5 - \rho_0 g_0\tilde{U}_1\right)Y_\theta\,. \end{aligned} \tag{C.70}$$

$$\begin{aligned}[1^{\mathrm{st}}\ \text{term}]_\lambda &= \frac{1}{r\sin\theta}\frac{\partial}{\partial\lambda}(\tilde{\Pi} - \rho_0\tilde{\phi}_1 - \rho_0 g_0 \tilde{u}_r) \\ &= \frac{1}{r}\sum_{nm}\left(\tilde{U}_8 - \rho_0\tilde{U}_5 - \rho_0 g_0\tilde{U}_1\right)Y_\lambda\,. \end{aligned} \tag{C.71}$$

□ <u>2nd term</u>. This can be easily handled into two steps.

First, recalling the expression of the curl in spherical coordinates (B.15–B.17), we obtain:

$$(\nabla\times\tilde{\mathbf{u}})_r = \sum_{nm} L\frac{\tilde{V}_1}{r}Y \tag{C.72}$$

$$(\nabla\times\tilde{\mathbf{u}})_\theta = \sum_{nm}\left(-\tilde{U}_7 Y_\lambda + \tilde{V}_7 Y_\theta\right) \tag{C.73}$$

$$(\nabla\times\tilde{\mathbf{u}})_\lambda = \sum_{nm}\left(+\tilde{U}_7 Y_\lambda + \tilde{V}_7 Y_\theta\right), \tag{C.74}$$

where we have set

$$L \equiv n(n+1)\,, \tag{C.75}$$

and we have defined two new radial functions:

$$\tilde{U}_7(r) = \tilde{U}_2' + \frac{\tilde{U}_2}{r} - \frac{\tilde{U}_1}{r} \tag{C.76}$$

$$\tilde{V}_7(r) = \tilde{V}_1' + \frac{\tilde{V}_1}{r}\,. \tag{C.77}$$

Second, by applying the curl operator once more, after some algebra we obtain:

$$[2^{\text{nd}}\text{ term}]_r = \sum_{nm} L\frac{\tilde{U}_7}{r}Y\,. \tag{C.78}$$

$$[2^{\text{nd}}\text{ term}]_\theta = \sum_{nm}\left(\tilde{U}_7' + \frac{\tilde{U}_7}{r}\right)Y_\theta - \sum_{nm}\left(\frac{L}{r^2}\tilde{V}_1 - \tilde{V}_7' - \frac{\tilde{V}_7}{r}\right)Y_\lambda\,. \tag{C.79}$$

$$[2^{\text{nd}}\text{ term}]_\lambda = \sum_{nm}\left(\frac{L}{r^2}\tilde{V}_1 - \tilde{V}_7' - \frac{\tilde{V}_7}{r}\right)Y_\theta + \sum_{nm}\left(\tilde{U}_7' + \frac{\tilde{U}_7}{r}\right)Y_\lambda\,. \tag{C.80}$$

C.2.1.6 Step 5: $\hat{r}$ Component of the Momentum Equation

The $\hat{r}$ (*i.e.*, upward) component of the momentum equation leads to a poloidal ODE that is made explicit here. Using (C.68), (C.69), and (C.78), the radial equation reads:

$$\tilde{U}_8' + \frac{4\rho_0 g_0}{r}\tilde{U}_1 - \frac{L\rho_0 g_0}{r}\tilde{U}_2 + \frac{n+1}{r}\rho_0\tilde{U}_5 - \rho_0\tilde{U}_6 + \tilde{\mu}\frac{L}{r}\tilde{U}_7 = 0\,, \tag{C.81}$$

which recalling the definition of $\tilde{U}_7$ (C.76) and $\tilde{U}_4$ (C.47) can be also written as

$$\begin{aligned}\left(\tilde{U}_8 + \tilde{\mu}\tilde{U}_3\right)' - \tilde{\mu}\tilde{U}_3' + \frac{4\rho_0 g_0}{r}\tilde{U}_1 - \frac{L\rho_0 g_0}{r}\tilde{U}_2 + \\ \frac{n+1}{r}\rho_0\tilde{U}_5 - \rho_0\tilde{U}_6 + \tilde{\mu}\frac{L}{r}\left(\tilde{U}_4 + \frac{2\tilde{U}_2}{r} - \frac{2\tilde{U}_1}{r}\right) = 0\,,\end{aligned} \tag{C.82}$$

where we note that the radial function $(\tilde{U}_8 + \tilde{\mu}\tilde{U}_3)$ represents the coefficient of the expansion of the stress component $\tilde{\tau}_{rr}$ in series of CSH (see Eq. C.44). Furthermore, after some manipulations, it is possible to write $\tilde{U}_3'$ in terms of $\tilde{U}_1$, $\tilde{U}_2'$ and $\tilde{U}_4$. Indeed, using the definition $\tilde{U}_3 = 2\tilde{U}_1'$ (C.45) and recalling the continuity equation (C.59), one obtains the identity $\tilde{U}_3 = 2L\tilde{U}_2/r - 4\tilde{U}_1/r$. Evaluating the radial derivative of $\tilde{U}_3$ gives:

$$\begin{aligned}\tilde{U}_3'(r) &= 2L(\tilde{U}_2'/r - \tilde{U}_2/r^2) - 4(\tilde{U}_1'/r - \tilde{U}_1/r^2)\\ &= 2L\tilde{U}_2'/r - 2L\tilde{U}_2/r^2 - 4\tilde{U}_1'/r + 4\tilde{U}_1/r^2\\ &= (2L/r)(\tilde{U}_4 + \tilde{U}_2/r - \tilde{U}_1/r) - 2L\tilde{U}_2/r^2 - (4/r)(L\tilde{U}_2/r - 2\tilde{U}_1/r) + 4\tilde{U}_1/r^2\\ &= 2L\tilde{U}_4/r + (\tilde{U}_2/r)(\cancel{2L/r} - \cancel{2L/r} - 4L/r) + (\tilde{U}_1/r)(-2L/r + 8/r + 4/r)\\ &= 2L\tilde{U}_4/r - 4L\tilde{U}_2/r^2 + 2(6-L)\tilde{U}_1/r^2\,,\end{aligned} \tag{C.83}$$

so that by substituting into (C.82) we obtain:

$$\left(\tilde{U}_8 + \tilde{\mu}\tilde{U}_3\right)' - \frac{2\,L\tilde{\mu}}{r}\tilde{U}_4 + \frac{4\,L\tilde{\mu}}{r^2}\tilde{U}_2 - 2\tilde{\mu}(6-L)\frac{\tilde{U}_1}{r^2} + \frac{4\rho_0 g_0}{r}\tilde{U}_1 - \frac{L\rho_0 g_0}{r}\tilde{U}_2 + \\ + \frac{n+1}{r}\rho_0\tilde{U}_5 - \rho_0\tilde{U}_6 + \frac{\tilde{\mu}L}{r}\tilde{U}_4 + \frac{2\tilde{\mu}L}{r^2}\tilde{U}_2 - \frac{2\tilde{\mu}L}{r^2}\tilde{U}_1 = 0\,, \quad \text{(C.84)}$$

or

$$\left(\tilde{U}_8 + \tilde{\mu}\tilde{U}_3\right)' - \frac{12\tilde{\mu}}{r^2}\tilde{U}_1 + \frac{4\rho_0 g_0}{r}\tilde{U}_1 + \frac{6\,L\tilde{\mu}}{r^2}\tilde{U}_2 - \frac{L\rho_0 g_0}{r}\tilde{U}_2 - \frac{L\tilde{\mu}}{r}\tilde{U}_4 + \\ + \frac{n+1}{r}\rho_0\tilde{U}_5 - \rho_0\tilde{U}_6 = 0\,, \quad \text{(C.85)}$$

which is equivalent to the new poloidal ODE:

$$\left(\tilde{U}_8 + \tilde{\mu}\tilde{U}_3\right)' = \frac{4}{r}\left(\frac{3\tilde{\mu}}{r} - \beta_0\right)\tilde{U}_1 - \frac{L}{r}\left(\frac{6\tilde{\mu}}{r} - \beta_0\right)\tilde{U}_2 + \frac{L\tilde{\mu}}{r}\tilde{U}_4 - \frac{n+1}{r}\rho_0\tilde{U}_5 + \rho_0\tilde{U}_6\,, \quad \text{(C.86)}$$

where we have defined the radial function:

$$\beta_0(r) \equiv \rho_0 g_0\,. \quad \text{(C.87)}$$

C.2.1.7 Step 6: $\hat{\theta}$ Component of the Momentum Equation

The $\hat{\theta}$ (southward) component of the momentum equation leads to a couple of ODEs. These are obtained by setting to zero, for any given n and m, the terms multiplying Y_λ and Y_θ in (C.79) and (C.70), respectively.

□ *The first* of the two equations only involves toroidal scalars and reads:

$$\frac{L}{r^2}\tilde{V}_1 - \frac{1}{r}\left(r\tilde{V}_7' + \tilde{V}_7\right) = 0\,, \quad \text{(C.88)}$$

where note that the auxiliary variable $\tilde{V}_7$ appears, defined as $\tilde{V}_7 = \tilde{V}_1' + \tilde{V}_1/r$ (see C.77). It is convenient to eliminate $\tilde{V}_7$ so to write the equation in terms of variable $\tilde{V}_2 = \tilde{V}_1' - \tilde{V}_1/r$ instead. The motivation is that $\tilde{V}_2$ enters the T/P decomposition of the tangential stress $\tilde{\tau}_{r\theta}$ (C.48) that, in turn, is directly involved in the expression of the surface boundary conditions (see Sect. C.5.1). The transformation of (C.88) is only a matter of algebra. The term in parentheses can be manipulated as follows:

$$
\begin{aligned}
r\tilde{V}_7' + \tilde{V}_7 &= (r\tilde{V}_7)' \\
&= (r\tilde{V}_1' + \tilde{V}_1)' \\
&= \tilde{V}_1'' + 2\tilde{V}_1' \\
&= r(\tilde{V}_2' + (\tilde{V}_1/r)') + 2\tilde{V}_2 + 2\tilde{V}_1/r \\
&= r\tilde{V}_2' + r(\tilde{V}_1'/r - \tilde{V}_1/r^2) + 2\tilde{V}_2 + 2\tilde{V}_1/r \\
&= r\tilde{V}_2' + \tilde{V}_1' - \tilde{V}_1/r + 2\tilde{V}_2 + 2\tilde{V}_1/r \\
&= r\tilde{V}_2' + \tilde{V}_2 + \cancel{\tilde{V}_1/r} - \cancel{\tilde{V}_1/r} + 2\tilde{V}_2 + 2\tilde{V}_1/r \\
&= r\tilde{V}_2' + 3\tilde{V}_2 + 2\tilde{V}_1/r\,,
\end{aligned}
$$

so that (C.88) becomes

$$
\frac{L}{r^2}\tilde{V}_1 - \tilde{V}_2' - 3\frac{\tilde{V}_2}{r} - 2\frac{\tilde{V}_1}{r^2} = 0\,, \tag{C.89}
$$

which is equivalent to the toroidal ODE:

$$
\tilde{V}_2'(r) = (L-2)\,\frac{\tilde{V}_1}{r^2} - 3\frac{\tilde{V}_2}{r}\,. \tag{C.90}
$$

□ *The second* of the two equations, which only involves poloidal scalars, reads

$$
\left(\tilde{U}_8 - \rho_0\tilde{U}_5 - \beta_0\tilde{U}_1\right) + \tilde{\mu}\left(r\tilde{U}_7' + \tilde{U}_7\right) = 0\,. \tag{C.91}
$$

Using the definition of $\tilde{U}_7$ given by Eq. (C.76), that of $\tilde{U}_4$ (C.47) and continuity equation (C.59), the term multiplying $\tilde{\mu}$ can be easily transformed, giving:

$$
\begin{aligned}
r\tilde{U}_7' + \tilde{U}_7 &= (r\tilde{U}_7)' \\
&= (r\tilde{U}_2' + \tilde{U}_2 - \tilde{U}_1)' \\
&= (r\tilde{U}_4 + 2\tilde{U}_2 - 2\tilde{U}_1)' \\
&= r\tilde{U}_4' + \tilde{U}_4 + 2\tilde{U}_2' - 2\tilde{U}_1' \\
&= r\tilde{U}_4' + \tilde{U}_4 + 2(\tilde{U}_4 + \tilde{U}_2/r - \tilde{U}_1/r) - 4\tilde{U}_1' + 2\tilde{U}_1' \\
&= r\tilde{U}_4' + 3\tilde{U}_4 - 2\tilde{U}_1/r + 2\tilde{U}_2/r - 4(-2\tilde{U}_1/r + L\tilde{U}_2/r) + \tilde{U}_3 \\
&= r\tilde{U}_4' + 3\tilde{U}_4 + 6\tilde{U}_1/r + (2-4L)\tilde{U}_2/r + \tilde{U}_3\,,
\end{aligned}
$$

so that (C.91) becomes

$$
(\tilde{U}_8 + \tilde{\mu}\tilde{U}_3) - \rho_0\tilde{U}_5 - \beta_0\tilde{U}_1 + \tilde{\mu}\left(r\tilde{U}_4' + 3\tilde{U}_4 + 6\tilde{U}_1/r + (2-4L)\tilde{U}_2/r\right) = 0\,, \tag{C.92}
$$

which provides the poloidal ODE:

$$\tilde{\mu}\tilde{U}_4' = -\frac{1}{r}\left(\frac{6\tilde{\mu}}{r} - \beta_0\right)\tilde{U}_1 - \frac{2\tilde{\mu}}{r^2}(1-2L)\tilde{U}_2 - \frac{1}{r}\left(\tilde{U}_8 + \tilde{\mu}\tilde{U}_3\right) - \frac{3\tilde{\mu}}{r}\tilde{U}_4 + \frac{\rho_0}{r}\tilde{U}_5 \,. \tag{C.93}$$

C.2.1.8 Step 7: $\hat{\lambda}$ Component of the Momentum Equation

It is easily verified that considering the $\hat{\lambda}$ (*i.e.*, eastward) component of the momentum equation (see C.71 and C.80) would lead to the same couple of scalar equations obtained by considering the $\hat{\theta}$ component (see previous step). Hence, it will not be taken into consideration further.

C.2.2 Solution of the Poloidal ODEs

In previous sections, we have encountered six mutually independent equations in the poloidal radial variables $\tilde{U}_i$ $(i = 1, \ldots 6)$, which are summarised here and organised as a system of ODEs. The toroidal radial variables $\tilde{V}_1$ and $\tilde{V}_2$ shall be considered in Sect. C.2.3 below. The first four poloidal ODEs are the expression of each of the fundamental field equations introduced in Sect. C.1, the fifth introduces an auxiliary variable (the *potential stress*) useful to formulate the surface boundary conditions in Sect. C.5 while the sixth is a consequence of Laplace's equation for the incremental gravitational potential.

C.2.2.1 Poloidal ODEs

Altogether, the six *linear* poloidal equations form a set of first-order ODEs that have been the subject of a vast number of analyses and have set the foundations of all subsequent GIA studies based upon the assumption of a laterally homogeneous, incompressible Earth. Despite the linear nature of the problem, the solution of the ODEs is not straightforward, since the system matrix is r-dependent.

The six poloidal ODEs are:

☐ 1. *Continuity equation* (see C.59):

$$\tilde{U}_1' = -\frac{2}{r}\tilde{U}_1 + \frac{L}{r}\tilde{U}_2 \,. \tag{C.94}$$

☐ 2. *Definition of auxiliary variable $\tilde{U}_4$* (C.47):

$$\tilde{U}_2' = -\frac{\tilde{U}_1}{r} + \frac{\tilde{U}_2}{r} + \tilde{U}_4 \,. \tag{C.95}$$

☐ 3. *Radial component of the Momentum Equation* (C.86):

$$\left(\tilde{U}_8 + \tilde{\mu}\tilde{U}_3\right)' = \frac{4}{r}\left(\frac{3\tilde{\mu}}{r} - \beta_0\right)\tilde{U}_1 - \frac{L}{r}\left(\frac{6\tilde{\mu}}{r} - \beta_0\right)\tilde{U}_2 + \frac{L\tilde{\mu}}{r}\tilde{U}_4 - \frac{n+1}{r}\rho_0\tilde{U}_5 + \rho_0\tilde{U}_6 \,. \tag{C.96}$$

☐ 4. *Horizontal component of the Momentum Equation* (C.93):

$$\tilde{\mu}\,\tilde{U}_4' = -\frac{1}{r}\left(\frac{6\tilde{\mu}}{r} - \beta_0\right)\tilde{U}_1 - \frac{2\tilde{\mu}}{r^2}(1-2L)\tilde{U}_2 - \frac{1}{r}\left(\tilde{U}_8 + \tilde{\mu}\tilde{U}_3\right) - \frac{3\tilde{\mu}}{r}\tilde{U}_4 + \frac{\rho_0}{r}\tilde{U}_5 \,. \tag{C.97}$$

☐ 5. *Definition of potential stress* (C.66):

$$\tilde{U}_5' = -4\pi G\rho_0\tilde{U}_1 - \frac{n+1}{r}\tilde{U}_5 + \tilde{U}_6 \,. \tag{C.98}$$

☐ 6. *Laplace's equation* (see C.67):

$$\tilde{U}_6' = -\frac{4\pi G\rho_0(n+1)}{r}\tilde{U}_1 + \frac{4\pi G\rho_0 L}{r}\tilde{U}_2 + \frac{n-1}{r}\tilde{U}_6 \,. \tag{C.99}$$

At this point, following a longstanding tradition (see e.g., Wu and Peltier 1982), it is customary to define six new radial variables (often referred to as the "$\tilde{y}$-variables"):

$$\begin{cases} \tilde{y}_1 = \tilde{U}_1 \\ \tilde{y}_2 = \tilde{U}_2 \\ \tilde{y}_3 = \tilde{U}_8 + \tilde{\mu}\tilde{U}_3 \\ \tilde{y}_4 = \tilde{\mu}\tilde{U}_4 \\ \tilde{y}_5 = \tilde{U}_5 \\ \tilde{y}_6 = \tilde{U}_6 \,, \end{cases} \tag{C.100}$$

in such a way that, at a given harmonic degree and order (n, m), the six poloidal equations above can be cast in the form of a homogeneous set of ODEs that in compact form reads:

$$\tilde{\mathbf{y}}'(r) = \tilde{\mathbf{S}}\,\tilde{\mathbf{y}} \tag{C.101}$$

where $\tilde{\mathbf{y}}$ is the column vector containing the six $\tilde{y}$-variables (C.100) and the 6×6 r- and n-dependent (but not m-dependent) poloidal system matrix $\tilde{\mathbf{S}}$ is:

$$\tilde{\mathbf{S}}(r) = \begin{pmatrix} -\frac{2}{r} & \frac{L}{r} & 0 & 0 & 0 & 0 \\ -\frac{1}{r} & \frac{1}{r} & 0 & \frac{1}{\tilde{\mu}} & 0 & 0 \\ \frac{4}{r}\left(\frac{3\tilde{\mu}}{r} - \beta_0\right) & -\frac{L}{r}\left(\frac{6\tilde{\mu}}{r} - \beta_0\right) & 0 & \frac{L}{r} & -\frac{\rho_0(n+1)}{r} & \rho_0 \\ -\frac{1}{r}\left(\frac{6\tilde{\mu}}{r} - \beta_0\right) & -\frac{2}{r^2}(1-2L)\,\tilde{\mu} & -\frac{1}{r} & -\frac{3}{r} & \frac{\rho_0}{r} & 0 \\ -4\pi G\rho_0 & 0 & 0 & 0 & -\frac{n+1}{r} & 1 \\ -\frac{4\pi G\rho_0(n+1)}{r} & \frac{4\pi G\rho_0 L}{r} & 0 & 0 & 0 & \frac{n-1}{r} \end{pmatrix} . \tag{C.102}$$

According to Lambeck (1988), the form (C.101) of the equilibrium equations was first given by Alterman et al. (1959) in their fundamental paper *on the oscillations of the Earth* (the readers are also referred to the work of Alterman et al. 1974). According to these authors, the y_i's are introduced in order to obviate the need to differentiate the empirically determined quantities ρ_0, μ and λ.

It should also be noted that the $\tilde{\mathbf{S}}$ matrix is often denoted by symbol $\tilde{\mathbf{A}}$ in the literature. It is possible to cast in the form (C.101) also the field equations for a compressible homogeneous shell, as will be shown in detail in Sect. C.3 below.

C.2.2.2 Solutions of the Poloidal ODEs

In the system (C.101) the four mechanical variables (i.e., displacements and stresses) are coupled with the two gravitational variables (i.e., the incremental potential and the potential stress), and the separation into two independent subsystems is not possible. Furthermore, since matrix $\tilde{\mathbf{S}}$ is r-dependent and shows no apparent symmetries, obtaining a fundamental matrix of the poloidal set (C.101) is not a straightforward exercise. We go back to the key radial function $\tilde{y}_5(r)$ associated with the (harmonic) total incremental gravitational potential $\tilde{\phi}_1(r, \gamma)$, from which all the others are obtained in sequence. Our approach only partly overlaps the one adopted by Sabadini et al. (2016), but it yields the same final results.

C.2.2.3 Radial Function $\tilde{y}_5$

This function is associated with Laplace's equation for the gravitational potential. Recalling (C.64) we have immediately

$$\tilde{y}_5(r) \equiv \tilde{U}_5(r) = \tilde{c}_3\, r^n + \tilde{c}_3^*\, r^{-(n+1)}\,, \tag{C.103}$$

where $\tilde{c}_3$ and $\tilde{c}_3^*$ are undetermined (and a priori n-dependent) constants. Note that here and in the following the asterisk is not denoting the complex conjugate but the fact that the corresponding term is singular for $r \to 0$, while the unstarred c constants weigh the regular powers of r.

C.2.2.4 Radial Function $\tilde{y}_1$

To solve for variable $\tilde{y}_1 \equiv \tilde{U}_1$, we start from the momentum equation (C.68), which we write in a slightly different way:

$$\nabla\left(\rho_0\tilde{\Phi}_1 + \beta_0\tilde{u}_r - \tilde{\Pi}\right) + \tilde{\mu}(\nabla \times \nabla \times \tilde{\mathbf{u}}) = 0\,. \tag{C.104}$$

Evaluating the divergence, also taking into account that $\tilde{\mu}$ is a constant, shows that $(\rho_0\tilde{\Phi}_1 + \beta_0\tilde{u}_r - \tilde{\Pi})$ is harmonic. Therefore, in close analogy with (C.103), for the corresponding radial coefficient we may write:

$$\left(\rho_0\tilde{U}_5 + \beta_0\tilde{U}_1 - \tilde{U}_8\right) = \tilde{\mu}\left(\tilde{c}_1\, r^n + \tilde{c}_1^*\, r^{-(n+1)}\right)\,, \tag{C.105}$$

where we have used the CSH expansions (C.39), (C.42) and (C.43), $\tilde{c}_1$ and $\tilde{c}_1^*$ are new integration constants and $\tilde{\mu}$ has been factorised for the sake of convenience.

We now note that the $\hat{r}$ component of the momentum equation (C.104) reads

$$\frac{\partial}{\partial r}\left(\rho_0\tilde{\Phi}_1 + \beta_0\tilde{u}_r - \tilde{\Pi}\right) + \tilde{\mu}(\nabla \times \nabla \times \tilde{\mathbf{u}})_r = 0\,, \tag{C.106}$$

that using (C.78) gives:

$$\left(\rho_0\tilde{U}_5' + (\beta_0\tilde{U}_1)' - \tilde{U}_8'\right) = \tilde{\mu}L\frac{\tilde{U}_7}{r}\,, \tag{C.107}$$

which compared with the radial derivative of (C.105) provides:

$$L\left(\tilde{U}_2' + \frac{\tilde{U}_2}{r} - \frac{\tilde{U}_1}{r}\right) = n\,\tilde{c}_1\, r^n - (n+1)\,\tilde{c}_1^*\, r^{-(n+1)}\,, \tag{C.108}$$

where we have also used the definition of $\tilde{U}_7$ (C.76). It is noteworthy that with the aid of the continuity equation (C.94) variable $\tilde{U}_2$ can be eliminated, obtaining a non-homogeneous second order linear ODEs in the unknown $\tilde{U}_1$ that reads

$$r^2\tilde{U}_1'' + 4r\tilde{U}_1' + (2-L)\tilde{U}_1 = n\,\tilde{c}_1\,r^{(n+1)} - (n+1)\tilde{c}_1^*\,r^{-n}\,. \tag{C.109}$$

The solution of (C.109) is straightforward. First, we seek the general solution of the associated homogeneous equation in the form: $\tilde{U}_1^{hom}(r) \sim r^p$, where p is unknown. By substitution, we find two possible integer solutions for p, giving: $\tilde{U}_1^{hom}(r) = \tilde{c}_2 r^{n-1} + \tilde{c}_2^* r^{-(n+2)}$, where $\tilde{c}_2$ and $\tilde{c}_2^*$ are new indeterminate constants. Second, it is convenient to search for a particular solution of the non-homogeneous equation in the form: $\tilde{U}_1^{par}(r) = \tilde{b}_1\,r^{n+1} + \tilde{b}_1^*\,r^{-n}$ which by direct substitution provides $\tilde{b}_1 = n/(2(2n+3))\,\tilde{c}_1$ and $\tilde{b}_1^* = (n+1)/(2(2n-1))\,\tilde{c}_1^*$. Hence:

$$\tilde{y}_1(r) \equiv \tilde{U}_1(r) = \frac{n}{2(2n+3)}\,\tilde{c}_1\,r^{n+1} + \tilde{c}_2\,r^{n-1} + \frac{n+1}{2(2n-1)}\,\tilde{c}_1^*\,r^{-n} + \tilde{c}_2^*\,r^{-(n+2)}\,. \tag{C.110}$$

C.2.2.5 Radial Function $\tilde{y}_6$

By its own definition, this variable is linked to $\tilde{U}_5$ and $\tilde{U}_1$ by (C.98). Using (C.110) and (C.103), after some algebra we obtain:

$$\begin{aligned}\tilde{y}_6(r) \equiv \tilde{U}_6(r) =& 2\pi G\rho_0 \frac{n}{2n+3}\,\tilde{c}_1\,r^{n+1} + 4\pi G\rho_0\,\tilde{c}_2\,r^{n-1} + (2n+1)\,\tilde{c}_3\,r^{n-1} + \\ &+ 2\pi G\rho_0\,\frac{n+1}{2n-1}\,\tilde{c}_1^*\,r^{-n} + 4\pi G\rho_0\,\tilde{c}_2^*\,r^{-(n+2)}\,.\end{aligned} \tag{C.111}$$

C.2.2.6 Radial Function $\tilde{y}_2$

Once $\tilde{U}_1$ is determined from (C.110), variable $\tilde{y}_2$ follows immediately from the continuity equation (C.94). After some manipulation, we obtain:

$$\begin{aligned}\tilde{y}_2(r) \equiv \tilde{U}_2(r) =& \frac{n+3}{2(n+1)(2n+3)}\,\tilde{c}_1\,r^{n+1} + \frac{1}{n}\,\tilde{c}_2\,r^{n-1} + \\ &+ \frac{2-n}{2n(2n-1)}\,\tilde{c}_1^*\,r^{-n} - \frac{1}{n+1}\,\tilde{c}_2^*\,r^{-(n+2)}\,.\end{aligned} \tag{C.112}$$

C.2.2.7 Radial Function $\tilde{y}_4$

This radial function describes the poloidal part of the stress tensor component $\tilde{\tau}_{r\theta}$ (see C.46). According to (C.47), it is defined in terms of $\tilde{U}_1$ and $\tilde{U}_2$. Thus, using

(C.110) and (C.112), after some cumbersome (but straightforward) algebra, for this variable we obtain:

$$\tilde{y}_4(r) \equiv \tilde{\mu}\,\tilde{U}_4(r) = \frac{n(n+2)}{(2n+1)(2n+3)}\,\tilde{\mu}\,\tilde{c}_1\,r^n + \frac{2(n-1)}{n}\,\tilde{\mu}\,\tilde{c}_2 r^{n-2} + \\ + \frac{n^2-1}{n(2n-1)}\,\tilde{\mu}\,\tilde{c}_1^*\,r^{-(n+1)} + \frac{2(n+2)}{n+1}\,\tilde{\mu}\,\tilde{c}_2^*\,r^{-(n+3)}\,. \quad \text{(C.113)}$$

C.2.2.8 Radial Function $\tilde{y}_3$

Function $\tilde{y}_3$ describes the poloidal part of the stress tensor component $\tilde{\tau}_{rr}$ (see C.44). Accordingly, it is defined in terms of scalars $\tilde{U}_3$ and $\tilde{U}_8$, which here are considered separately below.

Using definition (C.45) and taking the radial derivative of (C.110), for $\tilde{U}_3$ we have:

$$\tilde{U}_3(r) = \frac{n(n+1)}{2n+3}\,\tilde{c}_1\,r^n + 2(n-1)\,\tilde{c}_2\,r^{n-2} + \\ - \frac{n(n+1)}{2n-1}\,\tilde{c}_1^*\,r^{-(n+1)} - 2(n+2)\,\tilde{c}_2^*\,r^{-(n+3)}\,. \quad \text{(C.114)}$$

From (C.105), it is possible to obtain $\tilde{U}_8$ in terms of the known solutions $\tilde{U}_1$ (C.110) and $\tilde{U}_5$ (C.103). After some algebra we arrive to:

$$\tilde{U}_8(r) = \left(\frac{n}{2(2n+3)}\beta_0 r - \tilde{\mu}\right)\tilde{c}_1\,r^n + \beta_0\,\tilde{c}_2\,r^{n-1} + \rho_0\,\tilde{c}_3\,r^n + \quad \text{(C.115)} \\ + \left(\frac{n+1}{2(2n-1)}\beta_0 r - \tilde{\mu}\right)\tilde{c}_1^*\,r^{-(n+1)} + \beta_0\,\tilde{c}_2^*\,r^{-(n+2)} + \rho_0\,\tilde{c}_3^*\,r^{-(n+1)}\,.$$

Thus:

$$\tilde{y}_3(r) \equiv \tilde{U}_8(r) + \tilde{\mu}\tilde{U}_3(r) =$$

$$= \left(\frac{n\beta_0 r + 2(n^2-n-3)\tilde{\mu}}{2(2n+3)}\right)\tilde{c}_1 r^n + (\beta_0 r + 2(n-1)\tilde{\mu})\,\tilde{c}_2 r^{n-2} + \rho_0\tilde{c}_3 r^n + \\ + \left(\frac{(n+1)\beta_0 r - 2(n^2+3n-1)\tilde{\mu}}{2(2n-1)}\right)\tilde{c}_1^* r^{-(n+1)} + \\ + (\beta_0 r - 2(n+2)\tilde{\mu})\,\tilde{c}_2^* r^{-(n+3)} + \rho_0\tilde{c}_3^* r^{-(n+1)}\,. \quad \text{(C.116)}$$

C.2.2.9 Poloidal Fundamental Solution for the Shell

The results obtained in this section can be summarised as:

$$\tilde{\mathbf{y}}(r) = \tilde{\mathbf{Y}}(r)\,\tilde{\mathbf{C}}\,, \tag{C.117}$$

where $\tilde{\mathbf{y}}$ represents the $\tilde{y}_i$ 6-elements column vector, $\tilde{\mathbf{Y}}(r)$ is the 6×6 *fundamental matrix* of the poloidal set of ODEs, whose lines contain the set of linearly independent solutions that we have determined, and

$$\tilde{\mathbf{C}} = (\tilde{c}_1, \tilde{c}_2, \tilde{c}_3, \tilde{c}_1^*, \tilde{c}_2^*, \tilde{c}_3^*)^t \tag{C.118}$$

is a vector of arbitrary constants.

It is important to remark that the general solution (C.117) is valid for any type of forcing, not necessarily characterised by any kind of symmetry. So, in general the solution vector $\tilde{\mathbf{y}}(r) = \tilde{\mathbf{y}}_{nm}(r)$ shall be dependent upon degree and order (n, m) and the same holds true for the constant vector $\tilde{\mathbf{C}} = \tilde{\mathbf{C}}_{nm}$, whereas it is clear from the above derivation that the fundamental matrix $\tilde{\mathbf{Y}}(r) = \tilde{\mathbf{Y}}_n(r)$ shall be only n- (and r-) dependent. However, in some relevant geophysical problems there is a zonal symmetry in the forcing. This is the case of a point mass acting along the z-axis or of a tidal attraction exerted along the same direction. In such particular geometry, the components of the solution vector $\tilde{\mathbf{y}}(r)$ shall be characterised by a zonal symmetry, since the Earth models considered in this book are spherically symmetric. As a consequence, $\tilde{\mathbf{y}}(r)$=$\tilde{\mathbf{y}}_n(r)$ and $\tilde{\mathbf{C}} = \tilde{\mathbf{C}}_n$ shall be only n-dependent, in this case.[12]

In order to simplify notation, following e.g., Sabadini et al. (1982) it is convenient to introduce a block structure

$$\tilde{\mathbf{Y}}(r) = \left(\tilde{\mathbf{N}}|\tilde{\mathbf{N}}^*\right), \tag{C.119}$$

where $\tilde{\mathbf{N}}$ and $\tilde{\mathbf{N}}^*$ are 6×3 arrays:

$$\tilde{\mathbf{N}}(r) = \begin{pmatrix} \frac{n}{2(2n+3)} r^{n+1} & r^{n-1} & 0 \\ \frac{n+3}{2(n+1)(2n+3)} r^{n+1} & \frac{1}{n} r^{n-1} & 0 \\ \left(\frac{n\beta_0 r + 2(n^2-n-3)\tilde{\mu}}{2(2n+3)}\right) r^n & (\beta_0 r + 2(n-1)\tilde{\mu})\, r^{n-2} & \rho_0 r^n \\ \frac{n(n+2)}{(2n+1)(2n+3)}\, \tilde{\mu}\, r^n & \frac{2(n-1)}{n}\, \tilde{\mu}\, r^{n-2} & 0 \\ 0 & 0 & r^n \\ 2\pi G\rho_0\, \frac{n}{2n+3}\, r^{n+1} & 4\pi G\rho_0\, r^{n-1} & (2n+1)\, r^{n-1} \end{pmatrix} \tag{C.120}$$

[12] This is not true, for instance, in the case of an internal double-couple representing a seismic dislocation, for which the forcing also includes $m = \pm 1$ and $m = \pm 2$ terms (see, e.g., Piersanti et al. 1995).

and

$$\tilde{\mathbf{N}}^*(r) = \begin{pmatrix} \frac{n+1}{2(2n-1)}\, r^{-n} & r^{-(n+2)} & 0 \\ \frac{2-n}{2n(2n-1)}\, r^{-n} & -\frac{1}{n+1}\, r^{-(n+2)} & 0 \\ \left(\frac{(n+1)\beta_0 r - 2(n^2+3n-1)\tilde{\mu}}{2(2n-1)}\right) r^{-(n+1)} & (\beta_0 r - 2(n+2)\tilde{\mu})\, r^{-(n+3)} & \rho_0 r^{-(n+1)} \\ \frac{n^2-1}{n(2n-1)}\, \tilde{\mu}\, r^{-(n+1)} & \frac{2(n+2)}{n+1}\, \tilde{\mu}\, r^{-(n+3)} & 0 \\ 0 & 0 & r^{-(n+1)} \\ 2\pi G \rho_0\, \frac{n+1}{2n-1}\, r^{-n} & 4\pi G \rho_0\, r^{-(n+2)} & 0 \end{pmatrix}, \tag{C.121}$$

containing regular and non-regular powers of radius r for $r \to 0$, respectively.

C.2.3 *Solution of the Toroidal ODEs*

Until now, we have encountered two mutually independent ODEs in the toroidal radial variables $\tilde{V}_1$ and $\tilde{V}_2$, respectively. The first is just a useful definition associated with the horizontal stress $\tilde{\tau}_{r\theta}$, while the second derives from the expression of the horizontal component of the momentum equation.

C.2.3.1 Toroidal ODEs

The two toroidal ODEs are:

☐ 1. *Definition of auxiliary variable* $\tilde{V}_2$ (see C.48):

$$\tilde{V}_1' = \frac{\tilde{V}_1}{r} + \tilde{V}_2 \,. \tag{C.122}$$

☐ 2. *Horizontal component of the momentum equation* (C.90):

$$\tilde{V}_2' = (L-2)\,\frac{\tilde{V}_1}{r^2} - 3\frac{\tilde{V}_2}{r} \,. \tag{C.123}$$

It is now convenient to introduce the two $\tilde{z}$-variables:

$$\begin{cases} \tilde{z}_1 = \tilde{V}_1 \\ \tilde{z}_2 = \tilde{\mu}\tilde{V}_2 \,, \end{cases} \tag{C.124}$$

so that the toroidal ODEs above can be cast, for any harmonic degree and order (n, m), in the form of a homogeneous set of two ODEs:

$$\tilde{\mathbf{z}}'(r) = \tilde{\mathbf{T}}\,\tilde{\mathbf{z}} \,, \tag{C.125}$$

where $\tilde{\mathbf{z}}$ is the column vector containing the two $\tilde{z}$-variables (C.124) and the 2×2 r- and n-dependent (but not m-dependent) toroidal system matrix $\tilde{\mathbf{T}}$ is:

$$\tilde{\mathbf{T}}(r) = \begin{pmatrix} \dfrac{1}{r} & \dfrac{1}{\tilde{\mu}} \\ \dfrac{\tilde{\mu}(L-2)}{r^2} & -\dfrac{3}{r} \end{pmatrix} . \tag{C.126}$$

It is important to remark, at this point, that the toroidal system (C.125) is fully decoupled from the poloidal one (C.101). As we shall discuss in Sect. C.5 below, boundary conditions associated with surface loads and external potentials can be expressed only in terms of poloidal variables, hence in these cases toroidal modes are not excited.

C.2.3.2 Toroidal Fundamental Solutions for the Shell

The solution of the toroidal system (C.125) requires less work compared to the poloidal system (C.101).

Solving (C.122) for variable $\tilde{V}_2$ and substituting into (C.123) yields, after some algebra, an homogeneous second order equation for $\tilde{V}_1$ that reads:

$$r^2\, \tilde{V}_1'' + 2r\, \tilde{V}_1' - L\, \tilde{V}_1 = 0\,, \tag{C.127}$$

for which we search a test solution of the type $\tilde{V}_1 \sim r^q$, where q is a constant. By substitution, equation $q^2 + q - L = 0$ is obtained, which is satisfied for (integer) values $q = n$ and $q = -(n+1)$. Hence we obtain the toroidal fundamental solutions:

$$\tilde{V}_1(r) = \tilde{d}_1\, r^n + \tilde{d}_1^*\, r^{-(n+1)}\,, \tag{C.128}$$

and

$$\tilde{V}_2(r) = \tilde{d}_1\, (n-1)\, r^{n-1} - \tilde{d}_1^*\, (n+2) r^{-(n+2)}\,, \tag{C.129}$$

where $\tilde{d}_1$ and $\tilde{d}_1^*$ are (possibly n-dependent) constants. Recalling the second of (C.124), these two latter equations can be arranged as follows:

$$\tilde{\mathbf{z}}(r) = \tilde{\mathbf{Z}}\, \tilde{\mathbf{d}} \tag{C.130}$$

with the array

$$\tilde{\mathbf{Z}}(r) = \begin{pmatrix} r^n & r^{-(n+1)} \\ \tilde{\mu}(n-1)r^{n-1} & -\tilde{\mu}(n+2)r^{-(n+2)} \end{pmatrix} \tag{C.131}$$

and the vector of arbitrary constants:

$$\tilde{\mathbf{d}} = (\tilde{d}_1, \tilde{d}_1^*)^t \,, \tag{C.132}$$

in accord with Piersanti et al. (1995, see their Eq. 30).

C.3 Solution for the Compressible Homogeneous Shell

In the previous section we have discussed in full detail how the field equations can be cast in spectral form and analytically solved for the case of an incompressible homogeneous shell. We will now discuss how an analytical solution can be obtained for a compressible shell (i.e., if the incompressibility approximation is relaxed). We shall mostly follow Wu and Peltier (1982) and Vermeersen et al. (1996a), which in turn are based upon the solutions of the elasto-gravitational field equations obtained by Alterman et al. (1959) and Gilbert and Backus (1968) in the context of the free oscillations of the Earth.

C.3.1 Spectral Analysis of the Field Equations

In Sect. C.1 we have obtained the field equations for a *compressible* and *homogeneous* shell, which constitute a set of coupled PDEs where the unknowns are the three components of the displacement vector $\tilde{\mathbf{u}}$, the incremental gravitational potential $\tilde{\phi}_1$ and the dilatation $\tilde{\Delta} \equiv \nabla \cdot \tilde{\mathbf{u}}$. We will now reduce these PDEs to a set of ODEs by means of a suitable spectral expansion of the fields involved. We will closely follow the procedure described in Sect. C.2.1 for the incompressible case but with some notable differences.

C.3.1.1 Step 0: Notation

We first observe that, differently from the case of the incompressible shell, we cannot adopt the Mie representation for the displacement field since $\tilde{\mathbf{u}}$ is not solenoidal (i.e., $\nabla \cdot \tilde{\mathbf{u}} \equiv \tilde{\Delta} \neq 0$). However, due to the spherical symmetry of the Earth model and to the axial symmetry of the boundary conditions (see Sect. C.5), we can assume that $\tilde{\mathbf{u}}$ is an axially symmetric vector which according to (A.64) can be written as

$$\tilde{\mathbf{u}} = \sum_n \tilde{u} P \,\hat{r} + \tilde{v} P_\theta \,\hat{\theta} \,, \tag{C.133}$$

where, adopting the same simplified convention as in (C.32) for the angular derivatives,

$$P \equiv P_n(\cos\theta) \tag{C.134}$$

$$P_\theta \equiv \frac{\partial}{\partial\theta} P_n(\cos\theta) \tag{C.135}$$

are the associated Legendre functions and their derivatives *w.r.t.* colatitude θ, the harmonic degree n has been left implicit and $(\tilde{u}, \tilde{v})$ are suitable radial functions. We finally observe that, from identity (C.38) with $m = 0$, it follows that

$$P_{\theta\theta} + \cot\theta\, P_\theta = -n(n+1)P\,, \tag{C.136}$$

where $P_{\theta\theta} \equiv \partial^2 P_n(\cos\theta)/\partial\theta^2$.

We note that it would be also possible to expand the above fields in full vector spherical harmonics (see Sect. A.5), without making any a priori assumption about the symmetry of the problem. This approach, as shown by Gilbert and Backus (1968), leads to the same equations that we will obtain in the following. Similarly to the incompressible case, the degeneracy in harmonic order m would only be removed by non-axisymmetric boundary conditions, which are not of interest here.

C.3.1.2 Step 1: CSH Expansion of the Unknown Fields

☐ *Displacement* $(\tilde{u}_r, \tilde{u}_\theta, \tilde{u}_\phi)$. According to the expansion (C.133) for the displacement vector $\tilde{\mathbf{u}}$ it immediately follows that

$$\tilde{u}_r \equiv \sum_n \tilde{u} P\,, \tag{C.137}$$

$$\tilde{u}_\theta \equiv \sum_n \tilde{v} P_\theta\,, \tag{C.138}$$

$$\tilde{u}_\lambda \equiv 0\,. \tag{C.139}$$

☐ *Gravitational potential* $\tilde{\phi}_1$. The variation of gravitational potential $\tilde{\phi}_1$ is a scalar with axial symmetry, hence it can be expanded in Legendre polynomials as

$$\tilde{\phi}_1 = \sum_n \tilde{\phi}\, P\,, \tag{C.140}$$

where $\tilde{\phi}$ are suitable radial coefficients.

☐ *Dilatation* $\tilde{\Delta}$. Similarly to the variation of gravitational potential $\tilde{\phi}_1$, the dilatation $\tilde{\Delta}$ is an axisymmetric scalar field and can be written as

$$\tilde{\Delta} = \sum_n \tilde{\chi} P\,, \tag{C.141}$$

with $\tilde{\chi}$ being appropriate radial coefficients.

□ <u>*Stress component* $\tilde{\tau}_{rr}$.</u> The $\tilde{\tau}_{rr}$ component of the stress tensor can be readily obtained from the constitutive equation (C.5), taking into account (C.133) and (C.141), and using the expressions for components of the infinitesimal strain listed in Table C.1:

$$\begin{aligned}
\tilde{\tau}_{rr} &= \tilde{\lambda}\tilde{\Delta} + 2\tilde{\mu}\tilde{\epsilon}_{rr} \\
&= \tilde{\lambda}\sum_n \tilde{\chi} P + 2\tilde{\mu}\frac{\partial}{\partial r}\sum_n \tilde{u} P \\
&= \sum_n \left(\tilde{\lambda}\tilde{\chi} + 2\tilde{\mu}\tilde{u}'\right) P \\
&= \sum_n \tilde{T}_r P\,,
\end{aligned} \tag{C.142}$$

where we have introduced the radial function

$$\tilde{T}_r \equiv \tilde{\lambda}\tilde{\chi} + 2\tilde{\mu}\tilde{u}'\,, \tag{C.143}$$

which in the incompressible limit is easily shown to be coincident with the harmonic coefficients in (C.44).

□ <u>*Stress component* $\tilde{\tau}_{r\theta}$.</u> In analogy with $\tilde{\tau}_{rr}$, the $\tilde{\tau}_{r\theta}$ component of the stress tensor can be written as

$$\begin{aligned}
\tilde{\tau}_{r\theta} &= 2\tilde{\mu}\tilde{\epsilon}_{r\theta} \\
&= \tilde{\mu}\left(\frac{\partial \tilde{u}_\theta}{\partial r} - \frac{\tilde{u}_\theta}{r} + \frac{1}{r}\frac{\partial \tilde{u}_r}{\partial \theta}\right) \\
&= \tilde{\mu}\left(\frac{\partial}{\partial r}\sum_n \tilde{v} P_\theta - \frac{1}{r}\sum_n \tilde{v} P_\theta + \frac{1}{r}\frac{\partial}{\partial \theta}\sum_n \tilde{u} P\right) \\
&= \tilde{\mu}\sum_n \left(\tilde{v}' - \frac{\tilde{v}}{r} + \frac{\tilde{u}}{r}\right) P_\theta \\
&= \sum_n \tilde{T}_\theta P_\theta\,,
\end{aligned} \tag{C.144}$$

where we have defined the radial function

$$\tilde{T}_\theta \equiv \tilde{\mu}\left(\tilde{v}' - \frac{\tilde{v}}{r} + \frac{\tilde{u}}{r}\right), \tag{C.145}$$

which is easily recognised as the counterpart of (C.47) for the compressible problem. The remaining four independent components of the stress tensor $\tilde{\tau}$ will not be needed in the following, hence their expansion in Legendre polynomials is left as an exercise for the readers.

C.3.1.3 Step 2: Continuity Equation

In spectral form, the continuity equation (C.11) for a homogeneous compressible shell simply reads

$$\tilde{\rho}_1 = -\rho_0 \tilde{\chi} \tag{C.146}$$

Using the expression (B.23) for the divergence in spherical coordinates with (C.133), we can write explicitly the dilatation $\nabla \cdot \tilde{\mathbf{u}}$ as

$$\begin{aligned} \tilde{\Delta} &= \sum_n \left(\tilde{u}' + \frac{2\tilde{u}}{r} \right) P + \frac{\tilde{v}}{r} \underbrace{(\cot\theta P_\theta + P_{\theta\theta})}_{-n(n+1)P} \\ &= \sum_n \left(\tilde{u}' + \frac{2}{r}\tilde{u} - \frac{L}{r}\tilde{v} \right) P , \end{aligned} \tag{C.147}$$

where we recall that $L = n(n+1)$. From (C.147) and (C.141) we immediately have

$$\tilde{\chi} = \tilde{u}' + \frac{2}{r}\tilde{u} - \frac{L}{r}\tilde{v} . \tag{C.148}$$

C.3.1.4 Step 3: Poisson's Equation

According to (B.32), in spectral form Poisson's equation (C.12) for the gravitational potential reads

$$\tilde{\phi}'' + \frac{2}{r}\tilde{\phi}' - \frac{L}{r^2}\tilde{\phi} = -4\pi G\rho_0 \tilde{\chi} . \tag{C.149}$$

In analogy with the incompressible case, we introduce the auxiliary radial function

$$\tilde{Q} = \tilde{\phi}' + \frac{n+1}{r}\tilde{\phi} + 4\pi G\rho_0 \tilde{u} , \tag{C.150}$$

which corresponds to the first order ODE for $\tilde{\phi}$

$$\tilde{\phi}' = -4\pi G\rho_0 \tilde{u} - \frac{n+1}{r}\tilde{\phi} + \tilde{Q} . \tag{C.151}$$

We now obtain a first order ODE for $\tilde{Q}$ by differentiating the definition (C.150) and using (C.149), (C.151) and (C.148):

$$
\begin{aligned}
\tilde{Q}' &= \tilde{\phi}'' + \frac{n+1}{r}\tilde{\phi}' - \frac{n+1}{r^2}\tilde{\phi} + 4\pi G\rho_0\tilde{u}' \\
&= -4\pi G\rho_0\tilde{\chi} - \frac{2}{r}\tilde{\phi}' + \frac{L}{r^2}\tilde{\phi} + \frac{n+1}{r}\tilde{\phi}' - \frac{n+1}{r^2}\tilde{\phi} + 4\pi G\rho_0\tilde{u}' \\
&= -4\pi G\rho_0\tilde{\chi} + \frac{n-1}{r}\tilde{\phi}' + \frac{(n+1)(n-1)}{r^2}\tilde{\phi} + 4\pi G\rho_0\tilde{u}' \\
&= -4\pi G\rho_0\tilde{\chi} - 4\pi G\rho_0\frac{n-1}{r}\tilde{u} + \frac{n-1}{r}\tilde{Q} + 4\pi G\rho_0\tilde{u}' \\
&= \cancel{-4\pi G\rho_0\tilde{u}'} - 8\pi G\rho_0\frac{\tilde{u}}{r} + 4\pi G\rho_0\frac{L}{r}\tilde{v} - 4\pi G\rho_0\frac{n-1}{r}\tilde{u} + \frac{n-1}{r}\tilde{Q} + \cancel{4\pi G\rho_0\tilde{u}'} \\
&= -4\pi G\rho_0\frac{n+1}{r}\tilde{u} + 4\pi G\rho_0\frac{L}{r}\tilde{v} + \frac{n-1}{r}\tilde{Q}\,. \qquad \text{(C.152)}
\end{aligned}
$$

C.3.1.5 Step 4: Momentum Equation

We recall that the momentum equation (C.1) for a compressible homogeneous shell takes the form

$$
\nabla\left[(\tilde{\lambda}+2\tilde{\mu})\tilde{\Delta} - \rho_0\tilde{\phi}_1 - \rho_0 g_0\tilde{\mathbf{u}}\cdot\hat{r}\right] + \rho_0 g_0\tilde{\Delta}\hat{r} - \tilde{\mu}\,(\nabla\times\nabla\times\tilde{\mathbf{u}}) = 0\,.
$$

While, using (C.133–C.141) and (B.4), the spectral expansion of the first two terms is straightforward, the third term requires some additional work. By using (B.12–B.14) with (C.133), we obtain

$$
\begin{aligned}
\nabla\times\tilde{\mathbf{u}} &= \sum_n \frac{1}{r}\left(\frac{\partial}{\partial r}(r\tilde{v}P_\theta) - \frac{\partial}{\partial\theta}(\tilde{u}P)\right)\hat{\lambda} \\
&= \sum_n \left(\tilde{v}' + \frac{\tilde{v}}{r} - \frac{\tilde{u}}{r}\right)P_\theta\hat{\lambda} \\
&= \sum_n \tilde{H}P_\theta\hat{\lambda}\,, \qquad \text{(C.153)}
\end{aligned}
$$

where we have defined

$$
\tilde{H} \equiv \tilde{v}' + \frac{\tilde{v}}{r} - \frac{\tilde{u}}{r}\,. \qquad \text{(C.154)}
$$

By using once more (B.12–B.14) with (C.153), we have

$$
\begin{aligned}
\nabla \times \nabla \times \tilde{\mathbf{u}} &= \nabla \times \sum_n \tilde{H} P_\theta \hat{\lambda} \\
&= \sum_n \frac{1}{r \sin\theta} \frac{\partial}{\partial \theta} \left(\tilde{H} P_\theta \sin\theta \right) \hat{r} - \frac{1}{r} \frac{\partial}{\partial r} \left(r \tilde{H} P_\theta \right) \hat{\theta} \\
&= \sum_n \frac{\tilde{H}}{r} \underbrace{(P_{\theta\theta} + \cot\theta \, P_\theta)}_{-n(n+1)P} \hat{r} - \frac{1}{r} (r \tilde{H})' P_\theta \hat{\theta} \\
&= \sum_n -\frac{L}{r} \tilde{H} P \hat{r} - \frac{1}{r} (r \tilde{H})' P_\theta \hat{\theta} \, .
\end{aligned}
\tag{C.155}
$$

C.3.1.6 Step 5: $\hat{r}$ Component of the Momentum Equation

The spectral form of the $\hat{r}$ component of the momentum equation (C.16) can now be made explicit using (C.133), (C.140), (C.141) and (C.155), leading to

$$
(\tilde{\lambda} + 2\tilde{\mu}) \tilde{\chi}' - \rho_0 \tilde{\phi}' - \rho_0 (g_0 \tilde{u})' + \rho_0 g_0 \tilde{\chi} + \tilde{\mu} \frac{L}{r} \tilde{H} = 0 \, . \tag{C.156}
$$

By taking the radial derivative of (C.148) and rearranging we obtain

$$
(\tilde{\lambda} + 2\tilde{\mu}) \tilde{\chi}' = \frac{\partial}{\partial r} \underbrace{\left(\tilde{\lambda} \tilde{\chi} + 2\tilde{\mu} \tilde{u}' \right)}_{\tilde{T}_r} + \frac{4\tilde{\mu}}{r} \left(\tilde{u}' - \frac{\tilde{u}}{r} \right) - \frac{2\tilde{\mu} L}{r} \left(\tilde{v}' - \frac{\tilde{v}}{r} \right) , \tag{C.157}
$$

where we have used the definition (C.143) of the radial function $\tilde{T}_r$ representing the harmonic coefficient of the spectral expansion of the $\tilde{\tau}_{rr}$ component of the stress tensor. By using (C.157) and (C.154) into (C.156), we obtain

$$
\begin{aligned}
0 &= \tilde{T}_r' - \rho_0 \tilde{\phi}' - \rho_0 (g_0 \tilde{u})' + \rho_0 g_0 \tilde{\chi} + \frac{4\tilde{\mu}}{r} \left(\tilde{u}' - \frac{\tilde{u}}{r} \right) - \frac{2\tilde{\mu} L}{r} \left(\tilde{v}' - \frac{\tilde{v}}{r} \right) \\
&\quad + \frac{\tilde{\mu} L}{r} \left(\tilde{v}' + \frac{\tilde{v}}{r} - \frac{\tilde{u}}{r} \right) \\
&= \tilde{T}_r' - \rho_0 \tilde{\phi}' - \rho_0 (g_0 \tilde{u})' + \rho_0 g_0 \tilde{\chi} + \frac{4\tilde{\mu}}{r} \left(\tilde{u}' - \frac{\tilde{u}}{r} \right) + \frac{\tilde{\mu} L}{r} \left(\frac{3\tilde{v}}{r} - \frac{\tilde{u}}{r} - \tilde{v}' \right) .
\end{aligned}
\tag{C.158}
$$

C.3.1.7 Step 6: $\hat{\theta}$ Component of the Momentum Equation

The spectral form of the $\hat{\theta}$ component of the momentum equation (C.16) can be readily obtained using the expression (B.4) for the gradient operator in spherical coordinates and the expansion (C.155) for the double curl of $\tilde{\mathbf{u}}$; after eliminating the common factor r^{-1}, it reads

$$(\tilde{\lambda} + 2\tilde{\mu})\tilde{\chi} - \rho_0\tilde{\phi} + \rho_0 g_0 \tilde{u} + \tilde{\mu}(r\tilde{H})' = 0\,. \tag{C.159}$$

Using (C.148) and (C.154), Eq. (C.159) can be expanded as

$$\begin{aligned} 0 &= \tilde{\lambda}\tilde{\chi} - \rho_0\tilde{\phi} + \rho_0 g_0 \tilde{u} + 2\tilde{\mu}\left(\tilde{u}' + \frac{2}{r}\tilde{u} - \frac{L}{r}\tilde{v}\right) + \tilde{\mu}\left(r\tilde{v}'' + 2\tilde{v}' - \tilde{u}'\right) \\ &= \tilde{\lambda}\tilde{\chi} - \rho_0\tilde{\phi} + \rho_0 g_0 \tilde{u} + \tilde{\mu}\left(r\tilde{v}'' + \tilde{u}' + 2\tilde{v}' + \frac{4}{r}\tilde{u} - \frac{2L}{r}\tilde{v}\right). \end{aligned} \tag{C.160}$$

By differentiating the definition (C.145) of the harmonic coefficients for the $\tilde{\tau}_{r\theta}$ component of the stress tensor, we now obtain

$$r\tilde{v}'' = \frac{1}{\tilde{\mu}}r\tilde{T}_\theta' - \tilde{u}' + \tilde{v}' + \frac{\tilde{u}}{r} - \frac{\tilde{v}}{r}\,, \tag{C.161}$$

which, upon substitution in (C.160) gives

$$\tilde{\lambda}\tilde{\chi} - \rho_0\tilde{\phi} + \rho_0 g_0 \tilde{u} + r\tilde{T}_\theta' + \tilde{\mu}\left(3\tilde{v}' + \frac{5}{r}\tilde{u} - \frac{2L+1}{r}\tilde{v}\right) = 0\,. \tag{C.162}$$

C.3.1.8 Spectral ODEs for the Compressible Shell

The set of coupled spectral ODEs obtained above can now be cast in the form of a system of first-order ODEs for the unknown fields ($\tilde{u}$, $\tilde{v}$, $\tilde{T}_r$, $\tilde{T}_\theta$, $\tilde{\phi}$, $\tilde{Q}$).

☐ *Radial function $\tilde{u}$*. From the definitions (C.148) and (C.143) of $\tilde{\chi}$ and $\tilde{T}_r$ we have

$$\tilde{T}_r = (\tilde{\lambda} + 2\tilde{\mu})\tilde{u}' + \frac{2\tilde{\lambda}}{r}\tilde{u} - \frac{\tilde{\lambda}L}{r}\tilde{v}\,, \tag{C.163}$$

which can be transformed into an ODE for $\tilde{u}$:

$$\tilde{u}' = -\frac{\tilde{\lambda}}{\tilde{\lambda} + 2\tilde{\mu}}\frac{2}{r}\tilde{u} + \frac{\tilde{\lambda}}{\tilde{\lambda} + 2\tilde{\mu}}\frac{L}{r}\tilde{v} + \frac{1}{\tilde{\lambda} + 2\tilde{\mu}}\tilde{T}_r\,. \tag{C.164}$$

☐ *Radial function $\tilde{v}$*. From the definition (C.145) we can immediately write an ODE for $\tilde{v}$ that reads

$$\tilde{v}' = -\frac{1}{r}\tilde{u} + \frac{1}{r}\tilde{v} + \frac{1}{\tilde{\mu}}\tilde{T}_\theta\,. \tag{C.165}$$

☐ *Radial function* $\tilde{\phi}$. An ODE for the radial function $\tilde{\phi}$ has been obtained already from the definition of $\tilde{Q}$ (see Eq. C.151) and it is reported here for convenience:

$$\tilde{\phi}' = -4\pi G\rho_0\tilde{u} - \frac{n+1}{r}\tilde{\phi} + \tilde{Q}\,. \tag{C.166}$$

☐ *Radial function* $\tilde{Q}$. An ODE for the radial function $\tilde{Q}$ has been obtained in (C.152) from Poisson's equation, and it reads

$$\tilde{Q}' = -4\pi G\rho_0\frac{n+1}{r}\tilde{u} + 4\pi G\rho_0\frac{L}{r}\tilde{v} + \frac{n-1}{r}\tilde{Q}\,. \tag{C.167}$$

☐ *Radial function* $\tilde{T}_r$. It can be obtained from (C.158) by using (C.164–C.166), (C.148) and the expression (2.7) for the radial derivative of the background gravity acceleration g_0. After some cumbersome but straightforward algebra, we obtain

$$\begin{aligned}\tilde{T}_r' = \frac{4}{r}\left(\frac{\tilde{\mu}}{r}\frac{3\tilde{\lambda}+2\tilde{\mu}}{\tilde{\lambda}+2\tilde{\mu}} - \beta_0\right)\tilde{u} + \frac{L}{r}\left(\beta_0 - \frac{2\tilde{\mu}}{r}\frac{3\tilde{\lambda}+2\tilde{\mu}}{\tilde{\lambda}+2\tilde{\mu}}\right)\tilde{v} +\\ - \frac{4}{r}\frac{\tilde{\mu}}{\tilde{\lambda}+2\tilde{\mu}}\tilde{T}_r + \frac{L}{r}\tilde{T}_\theta - \rho_0\frac{n+1}{r}\tilde{\phi} + \rho_0\tilde{Q}\,,\end{aligned} \tag{C.168}$$

where we recall that $\beta_0(r) \equiv \rho_0 g_0$ (see Eq. C.87 above).

☐ *Radial function* $\tilde{T}_\theta$. It is obtained by substituting into (C.162) the expressions (C.148), (C.164) and (C.165) for $\tilde{\chi}$, $\tilde{u}'$ and $\tilde{v}'$, respectively. After some algebra, we obtain

$$\begin{aligned}\tilde{T}_\theta' = \frac{1}{r}\left(\beta_0 - 2\frac{\tilde{\mu}}{r}\frac{3\tilde{\lambda}+2\tilde{\mu}}{\tilde{\lambda}+2\tilde{\mu}}\right)\tilde{u} + \frac{2\tilde{\mu}}{r^2}\left(2L\frac{\tilde{\lambda}+\tilde{\mu}}{\tilde{\lambda}+2\tilde{\mu}} - 1\right)\tilde{v} +\\ - \frac{1}{r}\frac{\tilde{\lambda}}{\tilde{\lambda}+2\tilde{\mu}}\tilde{T}_r - \frac{3}{r}\tilde{T}_\theta + \frac{\rho_0}{r}\tilde{\phi}\,.\end{aligned} \tag{C.169}$$

Along the exact same lines of the incompressible problem, the six ODEs derived above can be written in a compact form introducing the $\tilde{y}$-variables:

$$\begin{cases} \tilde{y}_1 = \tilde{u} \\ \tilde{y}_2 = \tilde{v} \\ \tilde{y}_3 = \tilde{T}_r \\ \tilde{y}_4 = \tilde{T}_\theta \\ \tilde{y}_5 = \tilde{\phi} \\ \tilde{y}_6 = \tilde{Q}\,, \end{cases} \tag{C.170}$$

in such a way that the unknown vector $\tilde{\mathbf{y}}$ is solution of the homogeneous set of ODEs

$$\tilde{\mathbf{y}}'(r) = \tilde{\mathbf{S}}\,\tilde{\mathbf{y}}\,, \tag{C.171}$$

with the 6×6 r- and n-dependent system matrix $\tilde{\mathbf{S}}$ for the compressible problem given by:

$$\tilde{\mathbf{S}}(r) = \begin{pmatrix} -\frac{\tilde{\lambda}}{\tilde{\lambda}+2\tilde{\mu}}\frac{2}{r} & \frac{\tilde{\lambda}}{\tilde{\lambda}+2\tilde{\mu}}\frac{L}{r} & \frac{1}{\tilde{\lambda}+2\tilde{\mu}} & 0 & 0 & 0 \\ -\frac{1}{r} & \frac{1}{r} & 0 & \frac{1}{\tilde{\mu}} & 0 & 0 \\ \frac{4}{r}\left(\frac{3\kappa}{r}\frac{\tilde{\mu}}{\tilde{\lambda}+2\tilde{\mu}} - \beta_0\right) & \frac{L}{r}\left(\beta_0 - \frac{6\kappa}{r}\frac{\tilde{\mu}}{\tilde{\lambda}+2\tilde{\mu}}\right) & -\frac{4}{r}\frac{\tilde{\mu}}{\tilde{\lambda}+2\tilde{\mu}} & \frac{L}{r} & -\rho_0\frac{n+1}{r} & \rho_0 \\ \frac{1}{r}\left(\beta_0 - \frac{6\kappa}{r}\frac{\tilde{\mu}}{\tilde{\lambda}+2\tilde{\mu}}\right) & \frac{2\tilde{\mu}}{r^2}\left(2L\frac{\tilde{\lambda}+\tilde{\mu}}{\tilde{\lambda}+2\tilde{\mu}} - 1\right) & -\frac{1}{r}\frac{\tilde{\lambda}}{\tilde{\lambda}+2\tilde{\mu}} & -\frac{3}{r} & \frac{\rho_0}{r} & 0 \\ -4\pi G\rho_0 & 0 & 0 & 0 & -\frac{n+1}{r} & 1 \\ -4\pi G\rho_0\frac{n+1}{r} & 4\pi G\rho_0\frac{L}{r} & 0 & 0 & 0 & \frac{n-1}{r} \end{pmatrix}, \tag{C.172}$$

where we have made use of the definition (C.10) for the bulk modulus κ. It is easily verified that matrix $\tilde{\mathbf{S}}(r)$ matches expression (1.95) of Sabadini et al. (2016), provided that their definition $\tilde{\beta} = \tilde{\lambda} + 2\tilde{\mu}$ and our identity $\beta_0 \equiv \rho_0 g_0$ are taken into consideration.

C.3.2 *Solution of the Spectral ODEs*

In the previous section, we have formulated a set of six first-order, coupled homogeneous ODEs for the compressible shell, for which we will now obtain an analytical solution. Contrary to the incompressible case, it is not possible to obtain an *exact* solution for the compressible system of ODEs (C.171), and some simplifying approximations about the radial dependence of g_0 shall be introduced. Nevertheless, analytical solutions for the compressible problem are of great relevance, since they provide invaluable insight about the effect of finite compressibility on the relaxation spectrum of the system.

C.3.2.1 Radial Function $\tilde{\chi}$

As a first step, we shall obtain an analytical solution for the radial function $\tilde{\chi}$ defined in (C.141). To this aim, we will need to recast the components of the momentum equation in a suitable form, as discussed in detail below.

First, we obtain a new vector equation by taking the curl of (C.16), which reads

$$\rho_0 \nabla \times (g_0 \tilde{\Delta} \hat{r}) - \tilde{\mu} \, (\nabla \times \nabla \times \tilde{\omega}) = 0 \, , \tag{C.173}$$

where we have defined the solenoidal vector

$$\tilde{\omega} = \nabla \times \tilde{\mathbf{u}} \, . \tag{C.174}$$

By virtue of identity (C.38), and recalling that $g_0 \hat{r} = \nabla \phi_0$, Eq. (C.173) can be rearranged as[13]

$$\begin{aligned} & \rho_0 \nabla \times (g_0 \tilde{\Delta} \hat{r}) - \tilde{\mu} \, (\nabla \times \nabla \times \tilde{\omega}) = \\ & = \rho_0 \tilde{\Delta} \nabla \times \left(g_0 \hat{r} \right) + \rho_0 (\nabla \tilde{\Delta}) \times (g_0 \hat{r}) + \tilde{\mu} \nabla^2 \tilde{\omega} \\ & = \cancel{\rho_0 \tilde{\Delta} \nabla \times (\nabla \phi_0)} + \rho_0 (\nabla \tilde{\Delta}) \times (g_0 \hat{r}) + \tilde{\mu} \nabla^2 \tilde{\omega} \\ & = -\rho_0 g_0 \hat{r} \times (\nabla \tilde{\Delta}) + \tilde{\mu} \nabla^2 \tilde{\omega} = 0 \, , \end{aligned} \tag{C.175}$$

hence we obtain

$$\frac{\tilde{\mu}}{\rho_0} \nabla^2 \tilde{\omega} = g_0 \hat{r} \times \nabla \tilde{\Delta} \, , \tag{C.176}$$

which matches Eq. (35) of Wu and Peltier (1982).

We now consider the divergence of the momentum conservation expressed by (C.16) which, taking also Poisson's equation (C.12) into account, leads to the scalar equation

$$\frac{\tilde{\lambda} + 2\tilde{\mu}}{\rho_0} \nabla^2 \tilde{\Delta} + 4\pi G \rho_0 \tilde{\Delta} - \nabla^2 (g_0 \tilde{\mathbf{u}} \cdot \hat{r}) + \nabla \cdot (g_0 \tilde{\Delta} \hat{r}) = 0 \, . \tag{C.177}$$

Following the seminal works of Alterman et al. (1959) and Gilbert and Backus (1968), we now assume that the ratio

$$\xi \equiv \frac{g_0}{r} \tag{C.178}$$

is approximately constant in each homogeneous layer (see also Vermeersen et al. 1996b). This, of course, would be strictly exact only in the case of a homogeneous Earth model, for which $g_0 = \frac{4}{3} \pi G \rho_0 r$ hence $\xi = \frac{4}{3} \pi G \rho_0$, while for a layered Earth

[13] Note that here g_0 is the magnitude of the gravity vector, i.e. $g_0 = |\vec{g}_0|$. Hence, taking into account that the unperturbed state is spherically symmetric (i.e., ϕ_0 is a function of r only) and that $\vec{g}_0$ is directed *downward* (i.e., it points in the opposite direction of $\hat{r}$), the equation $\vec{g}_0 = -\nabla \phi_0$ is equivalent to $g_0 \hat{r} = \nabla \phi_0$.

ξ can be assumed to be equal to its average value within the shell.[14] As noted by Gilbert and Backus (1968), ξ has the dimensions of frequency squared.

With $g_0 = \xi r$, Eq. (C.177) takes the form

$$\frac{\tilde{\lambda} + 2\tilde{\mu}}{\rho_0} \nabla^2 \tilde{\Delta} + 4\pi G \rho_0 \tilde{\Delta} - \xi \nabla^2 (\tilde{\mathbf{u}} \cdot \mathbf{r}) + \xi \nabla \cdot (\tilde{\Delta} \mathbf{r}) = 0 \,, \tag{C.182}$$

of which we consider the third and the fourth terms. Since they are scalars, we can evaluate them in any coordinate system of choice. Apart from the ξ factor, in a cartesian system the third term is:

$$\begin{aligned}
\nabla^2 (\tilde{\mathbf{u}} \cdot \mathbf{r}) &= (\tilde{u}_i r_i)_{,jj} \\
&= \tilde{u}_{i,jj} r_i + \cancel{\tilde{u}_i r_{i,jj}} + 2\tilde{u}_{i,j} r_{i,j} \\
&= \tilde{u}_{i,jj} r_i + 2\tilde{u}_{i,j} \delta_{i,j} \\
&= \tilde{u}_{i,jj} r_i + 2\tilde{u}_{i,i} \\
&= (\nabla^2 \tilde{\mathbf{u}}) \cdot \mathbf{r} + 2\tilde{\Delta} \,,
\end{aligned} \tag{C.183}$$

while the fourth term is:

$$\begin{aligned}
\nabla \cdot (\tilde{\Delta} \mathbf{r}) &= (\nabla \tilde{\Delta}) \cdot \mathbf{r} + \tilde{\Delta} (\nabla \cdot \mathbf{r}) \\
&= \nabla (\nabla \cdot \tilde{\mathbf{u}}) \cdot \mathbf{r} + 3\tilde{\Delta} \\
&= (\nabla \times \tilde{\omega}) \cdot \mathbf{r} + (\nabla^2 \tilde{\mathbf{u}}) \cdot \mathbf{r} + 3\tilde{\Delta} \,,
\end{aligned} \tag{C.184}$$

where we have used (B.28) and the identity $\nabla \cdot \mathbf{r} = 3$. Using (C.183) and (C.184) in (C.182), we finally obtain

---

[14] The average value of ξ in a homogeneous shell of density ρ_0 and bounded by radii r_0 and r_1 (with $r_0 < r_1$) can be easily evaluated analytically. Indeed, in a layered Earth model such as the one shown in Fig. 2.1, the mass $M_{\rm in}$ contained within radius r is:

$$M_{\rm in}(r) = \left(M_0 - \frac{4}{3} \pi r_0^3 \rho_0 \right) + \frac{4}{3} \pi r^3 \rho_0 = \Delta M_0 + \frac{4}{3} \pi r^3 \rho_0 \,, \tag{C.179}$$

where $M_0 = M_{\rm in}(r_0)$ is the mass contained inside the inner boundary of the shell and ΔM_0 represents the excess mass contained inside $r = r_0$ with respect to a homogeneous model of density ρ_0. It immediately follows that

$$\xi(r) = \frac{g_0(r)}{r} = \frac{G \Delta M_0}{r^3} + \frac{4}{3} \pi G \rho_0 \,, \tag{C.180}$$

hence for a homogeneous model (for which $\Delta M_0 = 0$), $\xi(r) = \frac{4}{3} \pi G \rho_0 =$ constant, as expected. The average of $\xi(r)$ in the shell is given by

$$\langle \xi \rangle = \frac{1}{r_1 - r_0} \int_{r_0}^{r_1} \xi(r) \mathrm{d}r = \frac{G \Delta M_0}{2} \frac{r_0 + r_1}{r_0^2 r_1^2} + \frac{4}{3} \pi G \rho_0 \,. \tag{C.181}$$

$$\frac{\tilde{\lambda}+2\tilde{\mu}}{\rho_0}\nabla^2\tilde{\Delta}+(4\pi G\rho_0+\xi)\tilde{\Delta}+\xi\mathbf{r}\cdot(\nabla\times\tilde{\omega})=0\,, \tag{C.185}$$

with matches Eq. (4) of Vermeersen et al. (1996b) and, for the case of a uniform sphere for which $\xi=\frac{4}{3}\pi G\rho_0$, also Eq. (36) of Wu and Peltier (1982).[15]

We now seek a spectral form of Eqs. (C.176) and (C.185). Taking into account (C.155), (C.38) and (B.12–B.14), the vector laplacian of $\tilde{\omega}$ on the left hand side of (C.176) can be written as

$$\begin{aligned}
\nabla^2\tilde{\omega} &= -\nabla\times(\nabla\times\nabla\times\tilde{\mathbf{u}})\\
&=\nabla\times\left(\sum_n\frac{L}{r}\tilde{H}P\hat{r}+\frac{1}{r}(r\tilde{H})'P_\theta\hat{\theta}\right)\\
&=\frac{1}{r}\sum_n\left[\frac{\partial}{\partial r}\left(r\cdot\frac{1}{r}(r\tilde{H})'P_\theta\right)-\frac{\partial}{\partial\theta}\left(\frac{L}{r}\tilde{H}P\right)\right]\hat{\lambda}\\
&=\frac{1}{r}\sum_n\left((r\tilde{H})''-\frac{L}{r}\tilde{H}\right)P_\theta\hat{\lambda}\\
&=\sum_n\left(\tilde{H}''+\frac{2\tilde{H}'}{r}-\frac{L}{r^2}\tilde{H}\right)P_\theta\hat{\lambda}\,,
\end{aligned} \tag{C.186}$$

while the vector product on the *r.h.s.* of (C.176) by virtue of (B.4) yields

$$\begin{aligned}
\hat{r}\times\nabla\tilde{\Delta} &=\hat{r}\times\nabla\left(\sum_n\tilde{\chi}_nP\right)\\
&=\hat{r}\times\left(\sum_n\tilde{\chi}'P\hat{r}+\frac{1}{r}\tilde{\chi}P_\theta\hat{\theta}\right)\\
&=\frac{1}{r}\sum_n\tilde{\chi}P_\theta\hat{\lambda}\,,
\end{aligned} \tag{C.187}$$

where we have used the identities $\hat{r}\times\hat{r}=0$ and $\hat{r}\times\hat{\theta}=\hat{\lambda}$. Hence, by using (C.186) and (C.187) into (C.176), we obtain

$$\nabla_r^2\tilde{H}=\frac{\xi}{\tilde{v}_s^2}\tilde{\chi}\,, \tag{C.188}$$

where ∇_r^2 is the radial part of the laplacian operator defined in (B.35) and

[15] Note that Eq. (36) of Wu and Peltier (1982) contains a misprint; in the second term on the right hand side, $\xi\mathbf{e}_r$ should read $g_0\mathbf{e}_r$.

$$\tilde{\upsilon}_s = \sqrt{\frac{\tilde{\mu}}{\rho_0}} \tag{C.189}$$

is the complex S-wave velocity.

Taking (C.155) into account, Eq. (C.185) is easily cast in spectral form as

$$\tilde{\upsilon}_P^2 \nabla_r^2 \tilde{\chi} + (4\pi G\rho_0 + \xi)\tilde{\chi} - \xi L \tilde{H} = 0\,, \tag{C.190}$$

where

$$\tilde{\upsilon}_P = \sqrt{\frac{\tilde{\lambda} + 2\tilde{\mu}}{\rho_0}} \tag{C.191}$$

is the complex P wave velocity. It is now convenient to define the quantity

$$\tilde{\alpha}^2 = \frac{4\pi G\rho_0 + \xi}{\tilde{\upsilon}_P^2}\,, \tag{C.192}$$

so that (C.190) takes the form

$$(\nabla_r^2 + \tilde{\alpha}^2)\tilde{\chi} = \frac{\xi L}{\tilde{\upsilon}_P^2}\tilde{H}\,. \tag{C.193}$$

Inspection of (C.188) and (C.193) shows that it is possible to eliminate $\tilde{H}$ and obtain an ODE in the only unknown $\tilde{\chi}$. Indeed, applying the ∇_r^2 operator to both sides of (C.188) and combining the result with (C.193), we obtain

$$\nabla_r^2 \tilde{H} = \frac{\tilde{\upsilon}_P^2}{L\xi}\nabla_r^2(\nabla_r^2 + \tilde{\alpha}^2)\tilde{\chi} = \xi\tilde{\upsilon}_S^2\tilde{\chi}\,, \tag{C.194}$$

which can be written as

$$\nabla_r^2(\nabla_r^2 + \tilde{\alpha}^2)\tilde{\chi} = \frac{1}{4}\tilde{\gamma}^2\,, \tag{C.195}$$

where

$$\tilde{\gamma}^2 = \frac{4L\xi^2}{\tilde{\upsilon}_P^2\tilde{\upsilon}_S^2}\,. \tag{C.196}$$

It is now easily verified that (C.195) can be cast in the symmetrical form

$$(\nabla_r^2 + \tilde{k}^2)(\nabla_r^2 + \tilde{q}^2)\tilde{\chi} = 0\,, \tag{C.197}$$

where we have defined the *complex wavenumbers*

$$\tilde{k}^2 = \frac{\tilde{\alpha}^2 + \sqrt{\tilde{\alpha}^4 + \tilde{\gamma}^2}}{2} \tag{C.198}$$

$$\tilde{q}^2 = \frac{\tilde{\alpha}^2 - \sqrt{\tilde{\alpha}^4 + \tilde{\gamma}^2}}{2}. \tag{C.199}$$

It is immediately seen that (C.197) allows the trivial solution $\tilde{\chi} = 0$, corresponding to an incompressible (divergence–free) mode. We note, however, that this solution is substantally different from the incompressible solutions obtained in Sect. C.2 since, by virtue of the finite value of $\tilde{\lambda}$, it is characterised by *vanishing mean normal stress*. In addition, solutions of equation

$$(\nabla_r^2 + \tilde{p}^2)F(r) = 0\,, \tag{C.200}$$

where $\tilde{p}$ can be any of the two wavenumbers $\tilde{k}$ and $\tilde{q}$, are also solutions of (C.197). Equation (C.200) is easily recognised as the radial part of the Helmoltz equation and, by making explicit the ∇_r^2 operator and defining the variable $z = \tilde{p}r$, it reads

$$z^2 \frac{\partial^2}{\partial z^2} F(z) + 2z \frac{\partial}{\partial z} F(z) + \left[z^2 - n(n+1)\right] F(z) = 0\,, \tag{C.201}$$

which represents the *spherical Bessel differential equation*, whose solutions are the *spherical Bessel functions* $F(z) = j_n(z)$ and $F(z) = y_n(z)$, with $j_n(z)$ and $y_n(z)$ being, respectively, regular and singular for $z \to 0$. Explicit forms of $j_n(z)$ and $y_n(z)$ for some values of n are listed in Table C.2. By linearity, the general solution of (C.197) reads

$$\tilde{\chi}(r) = \tilde{c}_1 j_n(\tilde{k}r) + \tilde{c}_2 j_n(\tilde{q}r) + \tilde{c}_1^* y_n(\tilde{k}r) + \tilde{c}_2^* y_n(\tilde{q}r)\,, \tag{C.202}$$

Table C.2 Explicit forms of spherical Bessel functions $j_n(z)$ and $y_n(z)$

n	$j_n(z)$	$y_n(z)$
0	$\frac{\sin z}{z}$	$-\frac{\cos z}{z}$
1	$\frac{\sin z}{z^2} - \frac{\cos z}{z}$	$-\frac{\cos z}{z^2} - \frac{\sin z}{z}$
2	$\left(\frac{3}{z^3} - \frac{1}{z}\right)\sin z - \frac{3}{z^2}\cos z$	$\left(\frac{1}{z} - \frac{3}{z^3}\right)\cos z - \frac{3}{z^2}\sin z$
3	$\left(\frac{15}{z^4} - \frac{6}{z^2}\right)\sin z + \left(\frac{1}{z} - \frac{15}{z^3}\right)\cos z$	$\left(\frac{6}{z^2} - \frac{15}{z^4}\right)\cos z + \left(\frac{1}{z} - \frac{15}{z^3}\right)\sin z$
4	$\left(\frac{1}{z} - \frac{45}{z^3} + \frac{105}{z^5}\right)\sin z + \left(\frac{10}{z^2} - \frac{105}{z^4}\right)\cos z$	$\left(-\frac{1}{z} + \frac{45}{z^3} - \frac{105}{z^5}\right)\cos z + \left(\frac{10}{z^2} - \frac{105}{z^4}\right)\sin z$

where $\tilde{c}_1$, $\tilde{c}_2$, $\tilde{c}_1^*$ and $\tilde{c}_2^*$ are integration constants and the star denotes the terms that are singular in the limit $r \to 0$ (note that, in this context, the star *does not stand for complex conjugation*).

We shall now obtain analytical solutions for the radial functions ($\tilde{u}$, $\tilde{v}$, $\tilde{\phi}$, $\tilde{Q}$, $\tilde{T}_r$, $\tilde{T}_\theta$), considering separately the cases $\tilde{\chi} \neq 0$ and $\tilde{\chi} = 0$.

C.3.2.2 Fundamental Solutions for $\tilde{\chi} \neq 0$

We will now obtain analytical solutions corresponding to $\tilde{\chi}$ in the form (C.202). Without loss of generality, we will assume

$$\tilde{\chi}(r) = \tilde{c}_1\, j_n(\tilde{k}r)\,, \tag{C.203}$$

and, by linearity, the other three fundamental solutions will be simply obtained by writing y_n or $\tilde{q}$ *in lieu* of j_n or $\tilde{k}$, and replacing $\tilde{c}_1$ with the corresponding integration constant in (C.202).

□ *Radial function $\tilde{\phi}$*. The perturbation to gravitational potential is a solution of Poisson's equation (C.12), which in spectral form reads

$$\nabla_r^2\tilde{\phi} = -4\pi G\rho_0\tilde{\chi} = -4\pi G\rho_0\tilde{c}_1\, j_n(\tilde{k}r)\,. \tag{C.204}$$

and for which we seek a solution of the form

$$\tilde{\phi} = \tilde{A}\, j_n(\tilde{k}r)\,. \tag{C.205}$$

By substituting (C.205) into (C.204), and using the spherical Bessel equation (C.201), we obtain

$$\begin{aligned}\nabla_r^2\tilde{\phi} &= \tilde{A}\left[\tilde{k}^2 j_n''(\tilde{k}r) + \frac{2\tilde{k}}{r} j_n'(\tilde{k}r) - \frac{L}{r^2} j_n(\tilde{k}r)\right] \\ &= \frac{\tilde{A}}{r^2}\underbrace{\left[(\tilde{k}r)^2 j_n''(\tilde{k}r) + 2(\tilde{k}r) j_n'(\tilde{k}r) - L j_n(\tilde{k}r)\right]}_{-(\tilde{k}r)^2 j_n(\tilde{k}r)} \\ &= -\tilde{A}\tilde{k}^2\,,\end{aligned} \tag{C.206}$$

and by comparing (C.206) with (C.204) we have $\tilde{A} = 4\pi G\rho_0\tilde{c}_1/\tilde{k}^2$, hence

$$\tilde{\phi} = \tilde{c}_1\frac{4\pi G\rho_0}{\tilde{k}^2} j_n(\tilde{k}r)\,. \tag{C.207}$$

□ <u>*Radial functions $\tilde{u}$ and $\tilde{v}$*</u>. We first consider the radial function $\tilde{H}$ which, by virtue of (C.188), is a solution of the spectral ODE

$$\nabla_r^2 \tilde{H} = \tilde{c}_1 \frac{\xi}{\tilde{v}_s^2} j_n(\tilde{k}r) , \tag{C.208}$$

which is formally identical to (C.204), hence in analogy with (C.207) we can immediately write its solution as

$$\tilde{H} = \tilde{c}_1 \, \tilde{C}_k j_n(\tilde{k}r) , \tag{C.209}$$

where

$$\tilde{C}_k = -\frac{\xi}{\tilde{v}_s^2 \tilde{k}^2} . \tag{C.210}$$

By using (C.203) and (C.209) in (C.148) and (C.154), we obtain the following system of ODEs for $\tilde{u}$ and $\tilde{v}$:

$$\tilde{u}' + \frac{2}{r}\tilde{u} - \frac{L}{r}\tilde{v} = \tilde{c}_1 \, j_n(\tilde{k}r) \tag{C.211}$$

$$\tilde{v}' + \frac{\tilde{v}}{r} - \frac{\tilde{u}}{r} = \tilde{c}_1 \, \tilde{C}_k j_n(\tilde{k}r) . \tag{C.212}$$

Equations (C.211–C.212) suggest a solution in terms of a combination of functions j_n and j_n'. If we assume a solution for $\tilde{u}$ and $\tilde{v}$ of the form

$$\tilde{u}(r) = \tilde{A}_u \frac{j_n(\tilde{k}r)}{r} + \tilde{B}_u j_n'(\tilde{k}r) \tag{C.213}$$

$$\tilde{v}(r) = \tilde{A}_v \frac{j_n(\tilde{k}r)}{r} + \tilde{B}_v j_n'(\tilde{k}r) , \tag{C.214}$$

by substituting (C.213–C.214) into (C.211–C.212) and eliminating the second derivatives j_n'' with the spherical Bessel equation (C.201), after some cumbersome algebra the constants $\tilde{A}_u$, $\tilde{B}_u$, $\tilde{A}_v$, $\tilde{B}_v$ can be determined by equating the coefficients of j_n and j_n', leading to

$$\tilde{u}(r) = -\frac{\tilde{c}_1}{\tilde{k}} \left(L\tilde{C}_k \frac{j_n(\tilde{k}r)}{\tilde{k}r} + j_n'(\tilde{k}r) \right) \tag{C.215}$$

$$\tilde{v}(r) = -\frac{\tilde{c}_1}{\tilde{k}} \left((\tilde{C}_k + 1) \frac{j_n(\tilde{k}r)}{\tilde{k}r} - \tilde{C}_k j_n'(\tilde{k}r) \right) . \tag{C.216}$$

□ <u>*Radial function $\tilde{T}_r$*</u>. The solution for radial function $\tilde{T}_r$ can be obtained from its definition (C.143), using (C.203) and (C.215). After some cumbersome algebra, we obtain

$$\tilde{T}_r = \tilde{c}_1 \left[\tilde{\lambda} j_n(\tilde{k}r) + 2\tilde{\mu} L \frac{\tilde{C}_k}{\tilde{k}r} \left(\frac{j_n(\tilde{k}r)}{\tilde{k}r} - j_n'(\tilde{k}r) \right) - 2\tilde{\mu} j_n''(\tilde{k}r) \right] . \tag{C.217}$$

☐ <u>*Radial function* $\tilde{T}_\theta$</u>. The solution for radial function $\tilde{T}_\theta$ follows from definition (C.145), using solutions (C.215–C.215) for $\tilde{u}$ and $\tilde{v}$; it reads

$$\tilde{T}_\theta = -\tilde{c}_1 \tilde{\mu} \left[\tilde{C}_k j_n(\tilde{k}r) - 2\frac{\tilde{C}_k + 1}{\tilde{k}r} \left(\frac{j_n(\tilde{k}r)}{\tilde{k}r} - j_n'(\tilde{k}r) \right) + 2\tilde{C}_k j''(\tilde{k}r) \right] . \tag{C.218}$$

☐ <u>*Radial function* $\tilde{Q}$</u>. The fundamental solution for $\tilde{Q}$ can be easily obtained from (C.150), using (C.207) and (C.215):

$$\begin{aligned} \tilde{Q} &= \tilde{\phi}' + \frac{n+1}{r} \tilde{\phi} + 4\pi G \rho \tilde{u} \\ &= \tilde{c}_1 \frac{4\pi G \rho_0}{\tilde{k}} j_n'(\tilde{k}r) + \tilde{c}_1 \frac{4\pi G \rho_0 (n+1)}{\tilde{k}^2 r} j_n(\tilde{k}r) \\ &\quad - \tilde{c}_1 \frac{4\pi G \rho_0 L \tilde{C}_k}{\tilde{k}^2 r} j_n(\tilde{k}r) - \tilde{c}_1 \frac{4\pi G \rho_0}{\tilde{k}} j_n'(\tilde{k}r) \\ &= \tilde{c}_1 \frac{4\pi G \rho_0 (n+1)(1 - n\tilde{C}_k)}{\tilde{k}^2 r} j_n(\tilde{k}r) . \end{aligned} \tag{C.219}$$

In closing, we note that the presence of $(\tilde{k}r)$ terms at the denominator in expressions for $\tilde{u}$, $\tilde{v}$, $\tilde{T}_r$, $\tilde{T}_\theta$ and $\tilde{Q}$ might suggest a singularity for $r \to 0$ also in regular solutions involving j_n. However, the existence of this singularity can be easily ruled out. Indeed, according e.g. to Eq. (10.1.4) of Abramowitz et al. (1988), for $z \to 0$ the spherical Bessel function $j_n(z)$ has the asymptotic limit

$$j_n(z) \sim \frac{z^n}{(2n+1)!!} , \tag{C.220}$$

where $(2n+1)!! = 1 \cdot 3 \cdot 5 \cdots (2n+1)$. Hence, in solutions of (C.202) containing j_n, the terms $j_n(\tilde{k}r)/(\tilde{k}r)$ appearing in (C.215), (C.216) and (C.219) are regular in $r = 0$ for harmonic degrees $n \geq 1$, since they have the asymptotic limit $(\tilde{k}r)^{n-1}/(2n+1)!!$. In a similar fashion, the solutions for radial functions $\tilde{T}_r$ and $\tilde{T}_\theta$ given in (C.217–C.218) contain terms

$$\frac{1}{\tilde{k}r} \left(\frac{j_n(\tilde{k}r)}{\tilde{k}r} - j_n'(\tilde{k}r) \right) \sim -\frac{(n-1)}{(2n+1)!!} (\tilde{k}r)^{n-2} \tag{C.221}$$

and

$$j_n''(\tilde{k}r) \sim \frac{n(n-1)}{(2n+1)!!}(\tilde{k}r)^{n-2}\,, \tag{C.222}$$

all of which are regular in $r = 0$ for $n \geq 1$. For $n = 0$ a different approach is needed for the solution of the field equations, hence the expressions obtained here would be not valid.

C.3.2.3 Fundamental Solutions for $\tilde{\chi} = 0$

We will now obtain the fundamental solutions corresponding to the trivial solution $\tilde{\chi} = 0$ of (C.197). Those solutions have zero dilatation ($\tilde{\Delta} = 0$), but are significantly different from the incompressible solutions obtained in Sect. C.2 since they have vanishing normal stress because of the finite value of $\tilde{\lambda}$. As a consequence, as we shall see in the following, there are only two linearly independent solutions associated with $\tilde{\chi} = 0$ which, in addition to the four solutions associated with (C.202), form the full set of six fundamental solutions for the compressibile shell.

- ☐ <u>*Radial function $\tilde{\phi}$*</u>. For $\tilde{\chi} = 0$, Poisson's equation (C.12) reduces to Laplace's equation

$$\nabla_r^2 \tilde{\phi} = 0\,, \tag{C.223}$$

whose solution is a linear combination of terms r^n and r^{-n-1}. For reasons that shall be clear in the following, we choose integration constants such that

$$\tilde{\phi} = \tilde{c}_3 \xi r^n + \tilde{c}_3^* \xi r^{-(n+1)}\,. \tag{C.224}$$

- ☐ <u>*Radial function $\tilde{u}$*</u>. To obtain the solution for the radial function $\tilde{u}$, we need a suitable form of the compressible momentum equation in the case of vanishing dilatation. For $\tilde{\Delta} = 0$, the momentum equation (C.16) takes the form

$$\rho_0 \nabla \left(\tilde{\phi}_1 + g_0 \tilde{\mathbf{u}} \cdot \hat{r}\right) + \tilde{\mu}\left(\nabla \times \nabla \times \tilde{\mathbf{u}}\right) = 0\,, \tag{C.225}$$

and its divergence and radial component are, respectively,

$$\nabla^2 \left(\tilde{\phi}_1 + g_0 \tilde{\mathbf{u}} \cdot \hat{r}\right) = 0 \tag{C.226}$$

and

$$\rho_0 \frac{\partial}{\partial r}\left(\tilde{\phi}_1 + g_0 \tilde{\mathbf{u}} \cdot \hat{r}\right) + \tilde{\mu}\left(\nabla \times \nabla \times \tilde{\mathbf{u}}\right) \cdot \hat{r} = 0. \tag{C.227}$$

Taking into account (C.155) and assuming $g_0 = \xi r$, the spectral forms of (C.226–C.227) are

$$\nabla_r^2 \left(\tilde{\phi} + \xi r \tilde{u}\right) = 0 \tag{C.228}$$

and

$$\rho_0\left(\tilde{\phi}+\xi r\tilde{u}\right)'-\tilde{\mu}\frac{L}{r}\tilde{H}=0\,. \tag{C.229}$$

From (C.148), for $\tilde{\chi}=0$ we obtain

$$\tilde{v}=\frac{1}{L}(r\tilde{u}'+2\tilde{u})\,, \tag{C.230}$$

hence (C.154) can be written in terms of $\tilde{u}$ as

$$\begin{aligned}\tilde{H}&=\frac{1}{L}\left(r\tilde{u}''+3\tilde{u}'\right)+\frac{1}{L}\left(\tilde{u}'+\frac{2\tilde{u}}{r}\right)-\frac{\tilde{u}}{r}\\&=\frac{1}{Lr}\left(r^2\tilde{u}''+4r\tilde{u}'+2\tilde{u}\right)-\frac{\tilde{u}}{r}\\&=\frac{1}{L}\frac{\left(r^2\tilde{u}\right)''}{r}-\frac{\tilde{u}}{r}\,,\end{aligned} \tag{C.231}$$

and by substituting (C.231) in (C.229) we obtain

$$\frac{\rho_0}{\tilde{\mu}}r^2\left(\tilde{\phi}+\xi r\tilde{u}\right)'-\left(r^2\tilde{u}\right)''+L\tilde{u}=0\,. \tag{C.232}$$

According to (C.228), function $(\tilde{\phi}+\xi r\tilde{u})$ is harmonic, so that

$$\tilde{\phi}+\xi r\tilde{u}=\tilde{c}_4\xi r^n+\tilde{c}_4^*\xi r^{-(n+1)} \tag{C.233}$$

or, equivalently,

$$\tilde{u}=(\tilde{c}_4-\tilde{c}_3)r^{n-1}+(\tilde{c}_4^*-\tilde{c}_3^*)r^{-(n+2)}\,. \tag{C.234}$$

The constants $\tilde{c}_4$ and $\tilde{c}_4^*$ shall now be determined by substituting (C.233) and (C.234) into (C.232). We have

$$\begin{aligned}0&=\frac{\rho_0}{\tilde{\mu}}\xi r^2\left(\tilde{c}_4r^n+\tilde{c}_4^*r^{-(n+1)}\right)'-\left[(\tilde{c}_4-\tilde{c}_3)r^{n+1}+(\tilde{c}_4^*-\tilde{c}_3^*)r^{-n}\right]''\\&\quad+n(n+1)\left[(\tilde{c}_4-\tilde{c}_3)r^{n-1}+(\tilde{c}_4^*-\tilde{c}_3^*)r^{-(n+2)}\right]\\&=\frac{\rho_0}{\tilde{\mu}}\xi r^2\left(\tilde{c}_4r^n+\tilde{c}_4^*r^{-(n+1)}\right)'\\&\quad-\left[(\tilde{c}_4-\tilde{c}_3)n(n+1)r^{n-1}+(\tilde{c}_4^*-\tilde{c}_3^*)n(n+1)r^{-(n+2)}\right]\\&\quad+n(n+1)\left[(\tilde{c}_4-\tilde{c}_3)r^{n-1}+(\tilde{c}_4^*-\tilde{c}_3^*)r^{-(n+2)}\right]\\&=\frac{\rho_0}{\tilde{\mu}}\xi\left(\tilde{c}_4nr^{n-1}-\tilde{c}_4^*(n+1)r^{-n}\right)\,,\end{aligned} \tag{C.235}$$

which implies $\tilde{c}_4 = \tilde{c}_4^* = 0$, hence the solution for $\tilde{u}$ can be written from (C.234) as

$$\tilde{u} = -\tilde{c}_3 r^{n-1} - \tilde{c}_3^* r^{-(n+2)} . \tag{C.236}$$

□ *Radial function* $\tilde{v}$. The solution for $\tilde{v}$ can be easily obtained from (C.230) and (C.236) as

$$\begin{aligned}
\tilde{v} =& -\frac{1}{n(n+1)}\left(\tilde{c}_3(n-1)r^{n-1} - \tilde{c}_3^*(n+2)r^{-(n+2)} + 2\tilde{c}_3 r^{n-1} + 2\tilde{c}_3^* r^{-(n+2)}\right) \\
=& -\frac{1}{n(n+1)}\left(\tilde{c}_3(n+1)r^{n-1} - n\tilde{c}_3^* r^{-(n+2)}\right) \\
=& -\tilde{c}_3\frac{r^{n-1}}{n} + \tilde{c}_3^*\frac{r^{-(n+2)}}{n+1} .
\end{aligned} \tag{C.237}$$

□ *Radial function* $\tilde{T}_r$. From its definition (C.143), taking into account that $\tilde{\chi} = 0$, we immediately have

$$\tilde{T}_r = 2\tilde{\mu}\tilde{u}' = -2\tilde{c}_3\tilde{\mu}(n-1)r^{n-2} + 2\tilde{c}_3^*\tilde{\mu}(n+2)r^{-(n+3)} . \tag{C.238}$$

□ *Radial function* $\tilde{T}_\theta$. It can be obtained using (C.236) and (C.237) in (C.145). We have

$$\begin{aligned}
\frac{1}{\tilde{\mu}}\tilde{T}_\theta =& \tilde{v}' - \frac{\tilde{v}}{r} + \frac{\tilde{u}}{r} \\
=& -\tilde{c}_3\frac{n-1}{n}r^{n-2} - \tilde{c}_3^*\frac{n+2}{n+1}r^{-(n+3)} + \tilde{c}_3\frac{1}{n}r^{n-2} - \tilde{c}_3^*\frac{1}{n+1}r^{-(n+3)} \\
& - \tilde{c}_3 r^{n-2} - \tilde{c}_3^* r^{-(n+3)} \\
=& -2\tilde{c}_3\frac{n-1}{n}r^{n-2} - 2\tilde{c}_3^*\frac{n+2}{n+1}r^{-(n+3)} ,
\end{aligned} \tag{C.239}$$

hence

$$\tilde{T}_\theta = -2\tilde{c}_3\tilde{\mu}\frac{n-1}{n}r^{n-2} - 2\tilde{c}_3^*\tilde{\mu}\frac{n+2}{n+1}r^{-(n+3)} . \tag{C.240}$$

□ *Radial function* $\tilde{Q}$. From (C.150), (C.224) and (C.236) we have

$$\begin{aligned}
\tilde{Q} =& \tilde{c}_3\xi n r^{n-1} - \cancel{\tilde{c}_3^*\xi(n+1)r^{-(n+2)}} + \tilde{c}_3\xi(n+1)r^{n-1} + \cancel{\tilde{c}_3^*\xi(n+1)r^{-(n+2)}} \\
& -4\pi G\rho_0\tilde{c}_3 r^{n-1} - 4\pi G\rho_0\tilde{c}_3^* r^{-(n+2)} \\
=& \tilde{c}_3\left((2n+1)\xi - 4\pi G\rho_0\right)r^{n-1} - 4\pi G\rho_0\tilde{c}_3^* r^{-(n+2)} .
\end{aligned} \tag{C.241}$$

C.3.2.4 Compressible Fundamental Solution for the Shell

Similarly to the incompressible case, the results obtained in this section can be written in compact form as:

$$\tilde{\mathbf{y}}(r) = \tilde{\mathbf{Y}}(r)\,\tilde{\mathbf{C}}\,, \tag{C.242}$$

where $\tilde{\mathbf{y}} = (\tilde{u}, \tilde{v}, \tilde{T}_r, \tilde{T}_\theta, \tilde{\phi}, \tilde{Q})^t$ is a 6-elements column vector containing the unknown radial functions, $\tilde{\mathbf{Y}}(r)$ is the 6×6 fundamental matrix of the compressible problem, whose lines contain the set of linearly independent solutions for each variable that we have determined above, and $\tilde{\mathbf{C}} = (\tilde{c}_1, \tilde{c}_2, \tilde{c}_3, \tilde{c}_1^*, \tilde{c}_2^*, \tilde{c}_3^*)^t$ is a vector of integration constants. Along the same lines of the incompressible case, we use the block structure (C.119), where $\tilde{\mathbf{N}}$ and $\tilde{\mathbf{N}}^*$ are 6×3 arrays containing the regular and irregular solutions, respectively. For the solutions which are regular in $r = 0$ we have

$$\tilde{\mathbf{N}}(r) = \left(\tilde{\mathbf{Y}}_1 | \tilde{\mathbf{Y}}_2 | \tilde{\mathbf{Y}}_3\right), \tag{C.243}$$

with:

$$\tilde{\mathbf{Y}}_1(r) = \begin{pmatrix} -\frac{1}{\tilde{k}}\left(n(n+1)\tilde{C}_k \frac{j_n(\tilde{k}r)}{\tilde{k}r} + j_n'(\tilde{k}r)\right) \\ -\frac{1}{\tilde{k}}\left((\tilde{C}_k+1)\frac{j_n(\tilde{k}r)}{\tilde{k}r} - \tilde{C}_k j_n'(\tilde{k}r)\right) \\ \tilde{\lambda} j_n(\tilde{k}r) + 2\tilde{\mu} n(n+1)\frac{\tilde{C}_k}{\tilde{k}r}\left(\frac{j_n(\tilde{k}r)}{\tilde{k}r} - j_n'(\tilde{k}r)\right) - 2\tilde{\mu} j_n''(\tilde{k}r) \\ -\tilde{\mu}\left[\tilde{C}_k j_n(\tilde{k}r) - 2\frac{\tilde{C}_k+1}{\tilde{k}r}\left(\frac{j_n(\tilde{k}r)}{\tilde{k}r} - j_n'(\tilde{k}r)\right) + 2\tilde{C}_k j''(\tilde{k}r)\right] \\ \frac{4\pi G\rho_0}{\tilde{k}^2} j_n(\tilde{k}r) \\ \frac{4\pi G\rho_0(n+1)(1-n\tilde{C}_k)}{\tilde{k}^2 r} j_n(\tilde{k}r) \end{pmatrix}, \tag{C.244}$$

$$\tilde{\mathbf{Y}}_2(r) = \begin{pmatrix} -\frac{1}{\tilde{q}}\left(n(n+1)\tilde{C}_q \frac{j_n(\tilde{q}r)}{\tilde{q}r} + j_n'(\tilde{q}r)\right) \\ -\frac{1}{\tilde{q}}\left((\tilde{C}_q+1)\frac{j_n(\tilde{q}r)}{\tilde{q}r} - \tilde{C}_q j_n'(\tilde{q}r)\right) \\ \tilde{\lambda} j_n(\tilde{q}r) + 2\tilde{\mu} n(n+1)\frac{\tilde{C}_q}{\tilde{q}r}\left(\frac{j_n(\tilde{q}r)}{\tilde{q}r} - j_n'(\tilde{q}r)\right) - 2\tilde{\mu} j_n''(\tilde{q}r) \\ -\tilde{\mu}\left[\tilde{C}_q j_n(\tilde{q}r) - 2\frac{\tilde{C}_q+1}{\tilde{q}r}\left(\frac{j_n(\tilde{q}r)}{\tilde{q}r} - j_n'(\tilde{q}r)\right) + 2\tilde{C}_q j''(\tilde{q}r)\right] \\ \frac{4\pi G\rho_0}{\tilde{q}^2} j_n(\tilde{q}r) \\ \frac{4\pi G\rho_0(n+1)(1-n\tilde{C}_q)}{\tilde{q}^2 r} j_n(\tilde{q}r) \end{pmatrix}, \tag{C.245}$$

and

$$\tilde{\mathbf{Y}}_3(r) = \begin{pmatrix} nr^{n-1} \\ r^{n-1} \\ 2\tilde{\mu} n(n-1)r^{n-2} \\ 2(n-1)r^{n-2} \\ -n\xi r^n \\ (4\pi G\rho_0 - (2n+1)\xi)nr^{n-1} \end{pmatrix}, \tag{C.246}$$

and where, for consistency with Wu and Peltier (1982) and Vermeersen et al. (1996a), $\tilde{\mathbf{Y}}_3$ has been multiplied by a factor $(-n)$, while the constant $\tilde{C}_q$ appearing in (C.245) is given by (C.210) with $\tilde{q}$ in the place of $\tilde{k}$, i.e.

$$\tilde{C}_q = -\frac{\xi}{\tilde{v}_s^2 \tilde{q}^2} \, . \tag{C.247}$$

Solutions (C.244–C.245) are consistent with Eq. (43) by Wu and Peltier (1982) and with Eqs. (27–28) by Vermeersen et al. (1996a); solution (C.246) is also consistent with Eq. (44) of Wu and Peltier (1982) and with Eq. (29) of Vermeersen et al. (1996a) if a uniform sphere is assumed, for which $4\pi G \rho_0 = 3\xi$.

For the solutions which are singular in $r = 0$ we have

$$\tilde{\mathbf{N}}^*(r) = \left(\tilde{\mathbf{Y}}_1^* | \tilde{\mathbf{Y}}_2^* | \tilde{\mathbf{Y}}_3^*\right), \tag{C.248}$$

where

$$\tilde{\mathbf{Y}}_1^*(r) = \begin{pmatrix} -\frac{1}{\tilde{k}}\left(n(n+1)\tilde{C}_k \frac{y_n(\tilde{k}r)}{\tilde{k}r} + y_n'(\tilde{k}r)\right) \\ -\frac{1}{\tilde{k}}\left((\tilde{C}_k + 1)\frac{y_n(\tilde{k}r)}{\tilde{k}r} - \tilde{C}_k y_n'(\tilde{k}r)\right) \\ \tilde{\lambda} y_n(\tilde{k}r) + 2\tilde{\mu} n(n+1)\frac{\tilde{C}_k}{\tilde{k}r}\left(\frac{y_n(\tilde{k}r)}{\tilde{k}r} - y_n'(\tilde{k}r)\right) - 2\tilde{\mu} y_n''(\tilde{k}r) \\ -\tilde{\mu}\left[\tilde{C}_k y_n(\tilde{k}r) - 2\frac{\tilde{C}_k+1}{\tilde{k}r}\left(\frac{y_n(\tilde{k}r)}{\tilde{k}r} - y_n'(\tilde{k}r)\right) + 2\tilde{C}_k y''(\tilde{k}r)\right] \\ \frac{4\pi G \rho_0}{\tilde{k}^2} y_n(\tilde{k}r) \\ \frac{4\pi G \rho_0 (n+1)(1-n\tilde{C}_k)}{\tilde{k}^2 r} y_n(\tilde{k}r) \end{pmatrix}, \tag{C.249}$$

$$\tilde{\mathbf{Y}}_2^*(r) = \begin{pmatrix} -\frac{1}{\tilde{q}}\left(n(n+1)\tilde{C}_q \frac{y_n(\tilde{q}r)}{\tilde{q}r} + y_n'(\tilde{q}r)\right) \\ -\frac{1}{\tilde{q}}\left((\tilde{C}_q + 1)\frac{y_n(\tilde{q}r)}{\tilde{q}r} - \tilde{C}_q y_n'(\tilde{q}r)\right) \\ \tilde{\lambda} y_n(\tilde{q}r) + 2\tilde{\mu} n(n+1)\frac{\tilde{C}_q}{\tilde{q}r}\left(\frac{y_n(\tilde{q}r)}{\tilde{q}r} - y_n'(\tilde{q}r)\right) - 2\tilde{\mu} y_n''(\tilde{q}r) \\ -\tilde{\mu}\left[\tilde{C}_q y_n(\tilde{q}r) - 2\frac{\tilde{C}_q+1}{\tilde{q}r}\left(\frac{y_n(\tilde{q}r)}{\tilde{q}r} - y_n'(\tilde{q}r)\right) + 2\tilde{C}_q y''(\tilde{q}r)\right] \\ \frac{4\pi G \rho_0}{\tilde{q}^2} y_n(\tilde{q}r) \\ \frac{4\pi G \rho_0 (n+1)(1-n\tilde{C}_q)}{\tilde{q}^2 r} y_n(\tilde{q}r) \end{pmatrix}, \tag{C.250}$$

and

$$\tilde{\mathbf{Y}}_3^*(r) = \begin{pmatrix} -(n+1)r^{-(n+2)} \\ r^{-(n+2)} \\ 2\tilde{\mu}(n+1)(n+2)r^{-(n+3)} \\ -2(n+2)r^{-(n+3)} \\ (n+1)\xi r^{-(n+1)} \\ -4\pi G \rho_0 (n+1) r^{-(n+2)} \end{pmatrix}, \tag{C.251}$$

and where, for consistency with Vermeersen et al. (1996a), we have multiplied $\tilde{\mathbf{Y}}_3^*$ by a factor $(n+1)$. Solutions (C.249–C.250) are consistent with Eqs. (30–31) of Vermeersen et al. (1996a), while solution (C.251) is consistent with their Eq. (32) only if $\xi = \frac{4}{3}\pi G\rho_0$ is assumed, which would be appropriate for a homogeneous sphere.[16]

C.4 Fluid Layers

It is of interest now to discuss how the field equations obtained in Sect. C.1 shall be modified when considering a perfectly fluid layer. The relevance of this problem is twofold. First, the Earth's outer core behaves as an inviscid (i.e., zero-viscosity) fluid, hence appropriate solutions for the field equations in the fluid case shall be assumed in the core. Second, the long term response of Maxwell viscoelastic layers approximates that of an inviscid fluid as the time since the onset of the forcing goes to infinity. Therefore, exact solutions for a fluid rheology are useful since they represent the limiting behaviour of a viscoelastic model.

C.4.1 Equilibrium Equations for a Fluid

The response of a self-gravitating fluid to static deformations has been first investigated by Longman (1963), who solved the field equations for the mantle discussed in Sect. C.1 in the limit of vanishing shear modulus, i.e., for a weak solid with $\mu \to 0$. This approach, however, leads to some inconsistencies dubbed as the "Longman paradox", which have been hotly debated in literature in the 1970s (see Smylie and Mansinha 1971a, b; Dahlen 1971; Longman 1975, and references therein). As pointed out by Chinnery (1975), the problem ultimately arises because the concept of *material element* cannot strictly be applied to a fluid substance. Hence in this section, following Chinnery (1975) and Crossley and Gubbins (1975), we shall obtain the equilibrium equations for a self-gravitating fluid from first principles, rather than deriving them from the $\mu \to 0$ limit of the solid mantle equations.

We assume that the fluid is in a condition of hydrostatic equilibrium both in the unperturbed and perturbed states. Hence

$$\nabla p = -\rho \nabla \phi\,, \tag{C.252}$$

[16] There is actually a mismatch between the 5th row of vector $\mathbf{Y}_6$ in Eq. (32) of Vermeersen et al. (1996a) and our solution $\tilde{\mathbf{Y}}_3^*$ in (C.251). We cannot assess if there is a misprint in Vermeersen et al. (1996a); however, we have validated numerically (in the elastic limit) our compressible fundamental solutions and found them to be consistent with independent results by Martens et al. (2019).

where p is the hydrostatic pressure, ρ is density and ϕ is the gravitational potential, which obeys Poisson's equation, namely:

$$\nabla^2\phi = 4\pi G\rho\,. \tag{C.253}$$

In the perturbed state, we write the pressure, density and gravitational potential as

$$p(r,\gamma) = p_0(r) + p_1(r,\gamma) \tag{C.254}$$
$$\rho(r,\gamma) = \rho_0(r) + \rho_1(r,\gamma) \tag{C.255}$$
$$\phi(r,\gamma) = \phi_0(r) + \phi_1(r,\gamma)\,, \tag{C.256}$$

where r is radius, $\gamma = (\theta, \lambda)$, θ is colatitude, λ is longitude, subscript 0 denotes quantities in the unperturbed state, in which all variables are functions of r only, and p_1, ρ_1 and ϕ_1 are small perturbations. By substituting these expressions into (C.252), recalling that $\nabla p_0 = -\rho_0\nabla\phi_0$ and neglecting second-order terms, we obtain

$$\nabla p_1 = -\rho_1\nabla\phi_0 - \rho_0\nabla\phi_1\,. \tag{C.257}$$

We now recall that, since the gravity acceleration is $\mathbf{g} = -\nabla\phi$, if we denote with g_0 the *magnitude of the downward-directed gravity vector* we have

$$g_0 = \frac{\partial\phi_0}{\partial r}\,. \tag{C.258}$$

Hence, the radial component of (C.257) is

$$\frac{\partial p_1}{\partial r} = -\rho_0\frac{\partial\phi_1}{\partial r} - \rho_1 g_0\,, \tag{C.259}$$

while the horizontal components, taking into account that ϕ_0 only depends upon r, are:

$$\frac{\partial p_1}{\partial\theta} = -\rho_0\frac{\partial\phi_1}{\partial\theta}\,, \tag{C.260}$$
$$\frac{\partial p_1}{\partial\lambda} = -\rho_0\frac{\partial\phi_1}{\partial\lambda}\,. \tag{C.261}$$

Since ρ_0 is also a function of r only, these equations imply

$$p_1 = -\rho_0\phi_1\,, \tag{C.262}$$

which, substituted into (C.259), leads to

$$\frac{\partial\rho_0}{\partial r}\phi_1 = \rho_1 g_0\,. \tag{C.263}$$

Since Poisson's equation (C.253) holds true both in the unperturbed and perturbed states, by virtue of (C.255) and (C.256) also the perturbed gravitational potential shall be a solution of Poisson's equation

$$\nabla^2 \phi_1 = 4\pi G \rho_1 \,, \tag{C.264}$$

which using (C.263) may be written as

$$\nabla^2 \phi_1 = \frac{4\pi G}{g_0} \rho_0' \phi_1 \,, \tag{C.265}$$

where the prime indicates the derivative *w.r.t.* r.

We now consider an equipotential surface passing through a given observation point in the unperturbed state, and defined by $\phi_0(r) =$ constant. In the deformed state, this surface will be displaced radially by a quantity $\epsilon_\phi(r, \gamma)$, such as

$$\phi(r + \epsilon_\phi) = \phi_0(r) \,, \tag{C.266}$$

which, for small perturbations and neglecting second-order terms, can be rearranged as

$$\phi(r) = \phi_0(r) - \epsilon_\phi(r, \gamma)\phi_0'(r) \,. \tag{C.267}$$

In a similar fashion, we define the radial displacement ϵ_ρ of an equal-density surface and the radial displacement ϵ_p of an equal pressure (*isobaric*) surface as

$$\rho(r) = \rho_0(r) - \epsilon_\rho(r, \gamma)\, \rho_0'(r) \,, \tag{C.268}$$

$$p(r) = p_0(r) - \epsilon_p(r, \gamma)\, p_0'(r) \,. \tag{C.269}$$

By the definition of the gravity vector $\mathbf{g} = -\nabla\phi$, it follows that equipotential surfaces are always orthogonal to the local direction of $\mathbf{g}$. By the condition of hydrostatic equilibrium (C.252), we have $\nabla p = \rho \mathbf{g}$, hence also isobaric surfaces are perpendicular to the direction of $\mathbf{g}$. Finally, we note that

$$\nabla \times (\rho \mathbf{g}) = \rho(\nabla \times \mathbf{g}) + (\nabla\rho) \times \mathbf{g} \,, \tag{C.270}$$

and, since both $\rho \mathbf{g} = \nabla p$ and $\mathbf{g} = -\nabla\phi$ are irrotational vectors (i.e., their curl vanishes), it follows that

$$(\nabla\rho) \times \mathbf{g} = 0 \,, \tag{C.271}$$

hence also surfaces of equal density are orthogonal to $\mathbf{g}$. Therefore, it is concluded that *equipotential surfaces, isobaric surfaces and surfaces of equal density are parallel both in the unperturbed and perturbed states.* Hence, equipotential surfaces shall be also surfaces of constant density and pressure.

We now consider, in the unperturbed state, an equipotential surface $\phi_0(r) =$ constant, which in the deformed state moves radially by $\epsilon_\phi(r, \gamma)$. The density and pressure on the deformed equipotential surface will be, to first order,

$$\rho^* = \rho(r) + \rho_0'(r)\epsilon_\rho(r, \gamma) = \rho_0(r) + \left[\epsilon_\phi(r, \gamma) - \epsilon_\rho(r, \gamma)\right] \rho_0'(r)\,, \qquad \text{(C.272)}$$
$$p^* = p(r) + p_0'(r)\epsilon_\phi(r, \gamma) = p_0(r) + \left[\epsilon_\phi(r, \gamma) - \epsilon_p(r, \gamma)\right] p_0'(r)\,, \qquad \text{(C.273)}$$

where we have used (C.268) and (C.269). Since density and pressure on the deformed equipotential surface shall be constant, ρ^* and p^* in the equations above cannot be a function of the angular variables γ. In general, this is possible only if

$$\epsilon_\phi = \epsilon_\rho = \epsilon_p \equiv \epsilon(r, \gamma)\,, \qquad \text{(C.274)}$$

hence *equipotential surfaces, isobaric surfaces and equal density surfaces deform radially by the same amount,*[17] which we denoted in (C.274) as $\epsilon(r, \gamma)$. In a stratified fluid layer in which $\rho_0(r)$ varies continuously, $\epsilon(r, \gamma)$ may be regarded as the radial deformation of an infinitely thin layer of a given density (a *material surface*). In this respect, this quantity is a close analogue of the radial deformation u_r of a solid layer. In a homogeneous layer this interpretation is not possible, but ϵ is still well defined in terms of equipotential and isobaric surfaces. An important consequence of (C.274) is that, by (C.272–C.273), *a fluid parcel will not experience changes neither in pressure nor in density between the unperturbed and perturbed states*, i.e. $d\rho = dp = 0$.

C.4.2 Spectral Form of the Equilibrium Equations

In this section we will obtain a set of ODEs for the static deformation of a fluid through a suitable expansion in spherical harmonics, along the same lines that have been followed in Sect. C.2.1.

We start by expanding in series of CSH the perturbation to gravitational potential ϕ_1 and the radial displacement ϵ of equipotential, isobaric or material surfaces. Since these are scalars, we have:

$$\phi_1(r, \gamma) = \sum_{nm} w_5(r)\, Y(\gamma)\,, \qquad \text{(C.275)}$$

and

[17] Actually, if the fields appearing in (C.272–C.273) are expanded in spherical harmonics, at degree $n = 0$ the *r.h.s.* will have no dependence upon γ even if the $n = 0$ coefficients of ϵ_ϕ, ϵ_ρ and ϵ_p are different, since $Y_{00} = 1$. It follows that, *only at harmonic degree zero*, equipotential, isobaric and material surfaces can be displaced by different amounts. However, in the context of GIA the degree $n = 0$ deformations are not taken into account, since the corresponding surface mass distributions would violate the mass conservation principle (see Sect. 1.4.4).

$$\epsilon(r, \gamma) = \sum_{nm} w_1(r)\, Y(\gamma)\,, \tag{C.276}$$

respectively, where w_1 and w_5 are appropriate radial coefficients (note that in order to simplify notation the *nm* dependence of these coefficients is left implicit). By substituting (C.275) into Poisson's equation (C.265) we obtain

$$w_5'' + \frac{2}{r} w_5' - \left[\frac{L}{r^2} w_5 + \frac{4\pi G}{g_0} \rho_0'\right] w_5 = 0\,, \tag{C.277}$$

where we recall that $L = n(n+1)$, while by substituting (C.275) and (C.276) into (C.267), we obtain:

$$w_1(r) = -\frac{w_5}{g_0}\,, \tag{C.278}$$

which is analogous to the well-known Bruns formula for geoidal undulations (e.g., Heiskanen and Moritz 1981).

We now consider the case of a *homogeneous fluid layer*, in which $\rho_0' = 0$. In that case, (C.277) becomes formally identical to (C.63), valid for the solid mantle, which manifests Laplace's equation:

$$w_5'' + \frac{2}{r}\, w_5' - \frac{L}{r^2}\, w_5 = 0\,. \tag{C.279}$$

In analogy with (C.65), it is convenient to introduce the w_6 auxiliary variable (the *potential stress*, see Sabadini et al. 2016):

$$w_6(r) \equiv w_5' + \frac{n+1}{r} w_5 + 4\pi G \rho_0 w_1\,, \tag{C.280}$$

which, by virtue of (C.278), leads to

$$w_5'(r) = \left(\frac{4\pi G \rho_0}{g_0} - \frac{n+1}{r}\right) w_5 + w_6\,, \tag{C.281}$$

while, taking the radial derivative of (C.280), using (C.281) and (C.279), after some algebra we obtain

$$w_6'(r) = \frac{8\pi G \rho_0}{g_0 r}(n-1) w_5 + \left(\frac{n-1}{r} - \frac{4\pi G \rho_0}{g_0}\right) w_6\,. \tag{C.282}$$

Equations (C.281) and (C.282) constitute a set of first order linear ODEs describing the static deformations in a fluid homogeneous layer. These can be written in a compact matrix form as

$$\frac{\mathrm{d}}{\mathrm{d}r}\begin{pmatrix} w_5 \\ w_6 \end{pmatrix} = \begin{pmatrix} \dfrac{4\pi G\rho_0}{g_0} - \dfrac{n+1}{r} & 1 \\ \dfrac{8\pi G\rho_0}{g_0 r}(n-1) & \dfrac{n-1}{r} - \dfrac{4\pi G\rho_0}{g_0} \end{pmatrix} \begin{pmatrix} w_5 \\ w_6 \end{pmatrix}, \tag{C.283}$$

in close analogy with (C.101) and (C.171), valid for a solid shell in the incompressible and compressible case, respectively.

C.4.3 Solution of the ODEs for a Homogeneous Fluid

A solution for the set of ODEs (C.283) in a fluid shell can be obtained in a straightforward way. Recalling (C.64), the solution of (C.279) can be written in the form

$$w_5(r) = c_\mathrm{f} r^n + c_\mathrm{f}^* r^{-(n+1)}, \tag{C.284}$$

where c_f and c_f^* are integration constants.[18] By its own definition (C.280), after some algebra the corresponding solution for w_6 can be written as

$$w_6(r) = c_\mathrm{f}\left(\frac{2n+1}{r} - \frac{4\pi G\rho_0}{g_0}\right) r^n - c_\mathrm{f}^* \frac{4\pi G\rho_0}{g_0} r^{-(n+1)}. \tag{C.285}$$

The results above can be cast in the same form of (C.117) and (C.242), with:

$$\mathbf{w}(r) = \mathbf{Y}_\mathrm{f}(r)\mathbf{C}_\mathrm{f}, \tag{C.286}$$

where $\mathbf{w} = (w_5, w_6)^t$, $\mathbf{C}_\mathrm{f} = (c_\mathrm{f}, c_\mathrm{f}^*)^t$ and $\mathbf{Y}_\mathrm{f}(r)$ is a 2×2 fundamental matrix for the fluid ODEs

$$\mathbf{Y}_\mathrm{f}(r) = \begin{pmatrix} r^n & r^{-(n+1)} \\ \left(\dfrac{2n+1}{r} - \dfrac{4\pi G\rho_0}{g_0}\right) r^n & -\dfrac{4\pi G\rho_0}{g_0} r^{-(n+1)} \end{pmatrix}. \tag{C.287}$$

C.4.4 Boundary Conditions at Solid-Fluid Interfaces

At the interface between a solid and a fluid layer, suitable boundary conditions shall be applied to prescribe how the solution in the liquid and solid layers transition into each other.

[18] As in Sect. C.2.2.2, here the asterisk is *not* denoting the complex conjugate. The unstarred and starred constants correspond to regular and singular terms for $r \to 0$, respectively.

We suppose that we have a fluid-solid interface at a radius $r = b$, with the layer at $r < b$ being fluid and that at $r > b$ being solid; this would be the case for the boundary between the Earth's outer core and the lower mantle. However, our derivation of the BCs will be independent of the order of the two layers and remains valid also for the case of a fluid layer encompassing a solid one, as in the case of the boundary between the inner and outer core.

For $r < b$, the state of the liquid layer is completely described by the vector $\mathbf{w} = (w_5, w_6)^t$, which is a solution of the (C.283) ODE. Conversely, for $r > b$ the state of the mantle is determined by the vector $\tilde{\mathbf{y}} = (\tilde{y}_1, \ldots, \tilde{y}_6)^t$ defined in (C.100), which is a solution of (C.101).[19] In what follows, we shall describe in detail how each component of the $\tilde{\mathbf{y}}(b^+)$ vector in the solid layer can be written in terms of components of the $\mathbf{w}(b^-)$ vector in the fluid layer, where with b^+ and b^- we denote two radii located in the solid and fluid layers, respectively, and infinitely close to the solid-fluid interface at $r = b$.

- ☐ <u>*Radial displacement*</u>. We have shown above that, in a fluid layer, equipotential surfaces, isobaric surfaces and material surfaces are displaced by the same amount $w_1 = -w_5/g_0$. In a solid layer this condition no longer holds true, hence the displacement of the material surface at $r = b^+$ is independent from the displacement of the equipotential surface $\phi(b) =$ constant. We will therefore assume that, for each harmonic degree, the radial functions w_1 and $\tilde{y}_1$ may differ by an arbitrary constant $\tilde{K}_2$[20]:

$$\tilde{y}_1(b^+) = -\frac{w_5(b^-)}{g_0(b)} + \tilde{K}_2 \, . \tag{C.288}$$

- ☐ <u>*Tangential displacement*</u>. Since a fluid cannot sustain shear stresses, the horizontal displacements of the solid layer are mechanically decoupled from the fluid layer and we assume *free slip* conditions at the interface. Hence,

$$\tilde{y}_2(b^+) = \tilde{K}_3 \, , \tag{C.289}$$

where, for each harmonic degree, $\tilde{K}_3$ is an arbitrary constant.

- ☐ <u>*Normal stress*</u>. We prescribe that normal stress shall be continuous across the fluid-solid interface. We recall that, in the fluid layer, isobaric surfaces are shifted radially by a quantity $\epsilon(r, \gamma)$ whose spherical harmonic coefficients are $w_1 = -w_5/g_0$. From (C.288), the deformed boundary between the fluid and solid layer is at a radial distance $\tilde{K}_2$ with respect to the equipotential (and isobaric)

---

[19] For a compressible mantle, the $\tilde{\mathbf{y}}$ vector is defined in (C.170) and it is a solution of (C.171).

[20] Note that, contrary to the solution vector $\tilde{\mathbf{y}}$ valid for the solid mantle, the solution $\mathbf{w}$ in the fluid is *not* explicitly dependent upon the Laplace variable s. However, the constant $\tilde{K}_2$ can be s-dependent. The same holds true for other constants that shall be introduced in the following to describe the boundary conditions between the liquid and solid layers.

displaced surface. Hence, in the perturbed state, the hydrostatic pressure just below the boundary surface can be written as

$$p(b^-) = p_0(b) + p_0'(b)\tilde{K}_2 = p_0(b) - \rho_0(b^-)g_0(b)\tilde{K}_2\,, \tag{C.290}$$

where we have used the condition of hydrostatic equilibrium $p_0' = -\rho_0 g_0$ and higher order terms have been neglected. Taking into account that, by convention, positive components of the stress tensor correspond to a traction, (C.290) leads to

$$\tilde{y}_3(b^+) = \rho_0(b^-)g_0(b)\tilde{K}_2\,. \tag{C.291}$$

- ☐ Shear stress. Since the fluid layer shall be in hydrostatic equilibrium both in the unperturbed and perturbed states, the shear stress at the solid-fluid interface vanishes:

$$\tilde{y}_4(b^+) = 0\,. \tag{C.292}$$

- ☐ Gravitational potential. The gravitational potential shall be continuous across the solid-fluid interface. Hence,

$$\tilde{y}_5(b^+) = w_5(b^-)\,. \tag{C.293}$$

- ☐ Potential stress. Obtaining boundary conditions for the *potential stress* requires some more work, since the definitions (C.65) and (C.280) are given in terms of variables $\tilde{y}_1$ and w_1, which have different physical meanings. We will follow the same approach by Chinnery (1975). In the unperturbed state, consider a point P in the solid layer, infinitely close to the solid-fluid interface, and a point Q in the fluid layer, infinitely close to P. By continuity of the gravitational potential and of its derivative across the interface, we have

$$\phi_0(P) = \phi_0(Q)\,, \tag{C.294}$$
$$\phi_0'(P) = \phi_0'(Q)\,. \tag{C.295}$$

In addition, the unperturbed gravitational potential ϕ_0 is a solution of Poisson's equation, hence

$$\phi_0''(P) + \frac{2}{r}\phi_0'(P) = 4\pi G\rho_0(P)\,, \tag{C.296}$$

$$\phi_0''(Q) + \frac{2}{r}\phi_0'(Q) = 4\pi G\rho_0(Q)\,. \tag{C.297}$$

In the deformed state, we assume that the point P has moved into a new position P'. If we consider a point Q' in the fluid layer, infinitely close to P', by continuity of the gravitational potential we have

$$\tilde{\phi}(P') = \phi(Q')\,, \tag{C.298}$$
$$\tilde{\phi}'(P') = \phi'(Q')\,. \tag{C.299}$$

It is important now to remark that, while the radial component of PP' is related to the $\tilde{y}_1$ variable, QQ' has no relation with w_1, since vertical displacement is not defined within the fluid. Keeping only first-order terms, we can write

$$\tilde{\phi}'(P') = \tilde{\phi}'(P) + \tilde{u}_r(P)\,\phi_0''(P) \tag{C.300}$$
$$\phi'(Q') = \phi'(Q) + \tilde{u}_r(P)\,\phi_0''(Q)\,, \tag{C.301}$$

where we have assumed that the radial components of both PP' and QQ' are equal to $\tilde{u}_r(P)$, since points Q and Q' have been chosen infinitely close to P and P', respectively. By using Eqs. (C.300–C.301), continuity conditions (C.294–C.295) and (C.298–C.299), and Poisson equations (C.296–C.297), we obtain

$$\begin{aligned}
\tilde{\phi}_1'(P) - \phi_1'(Q) &= [\tilde{\phi}'(P) - \phi_0'(P)] - [\phi'(Q) - \phi_0'(Q)] \\
&= \tilde{\phi}'(P) - \phi'(Q) \\
&= [\tilde{\phi}'(P') - \tilde{u}_r(P)\phi_0''(P)] - [\phi'(Q') - \tilde{u}_r(P)\phi_0''(Q)] \\
&= \tilde{u}_r(P)\left[\phi_0''(Q) - \phi_0''(P)\right] \\
&= 4\pi G\tilde{u}_r(P)\,[\rho_0(Q) - \rho_0(P)]\,,
\end{aligned} \tag{C.302}$$

which can be expanded in spherical harmonics, leading to:

$$\tilde{y}_5'(P) - w_5'(Q) = 4\pi G\tilde{y}_1(P)\,[\rho_0(Q) - \rho_0(P)]\;. \tag{C.303}$$

Hence, recalling the definitions (C.65) and (C.280), along with the continuity condition (C.293), we have

$$\begin{aligned}
\tilde{y}_6(P) - w_6(Q) &= \tilde{y}_5'(P) - w_6'(Q) + \frac{n+1}{b}[\tilde{y}_5(P) - w_5(Q)] + \\
&\quad + 4\pi G\rho_0(P)\tilde{y}_1(P) - 4\pi G\rho_0(Q)w_1(Q) \\
&= 4\pi G\tilde{y}_1(P)\rho_0(Q) - 4\pi G\tilde{y}_1(P)\rho_0(P) + \\
&\quad + 4\pi G\rho_0(P)\tilde{y}_1(P) - 4\pi G\rho_0(Q)w_1(Q) \\
&= 4\pi G\rho_0(Q)\left[\tilde{y}_1(P) - w_1(Q)\right]\,.
\end{aligned} \tag{C.304}$$

Equation (C.304), taking into account the boundary condition (C.288) for the radial displacement, leads to the following BC for the potential stress at the fluid-solid interface:

$$\tilde{y}_6(b^+) = w_6(b^-) + 4\pi G\rho_0(b^-)\tilde{K}_2\,. \tag{C.305}$$

We just conclude that the boundary conditions discussed above can be written in compact form as

$$\tilde{\mathbf{y}}(b^+) = \begin{pmatrix} -w_5(b^-)/g_0(b) \\ 0 \\ 0 \\ 0 \\ w_5(b^-) \\ w_6(b^-) \end{pmatrix} + \begin{pmatrix} 1 \\ 0 \\ \rho_0(b^-)g_0(b) \\ 0 \\ 0 \\ 4\pi G\rho_0(b^-) \end{pmatrix} \tilde{K}_2 + \begin{pmatrix} 0 \\ 1 \\ 0 \\ 0 \\ 0 \\ 0 \end{pmatrix} \tilde{K}_3 . \tag{C.306}$$

We now turn our attention to the case of a uniform, spherical liquid core surrounded by a solid mantle. This would correspond to an Earth model in which the inner solid core is not included, as it is often done in GIA studies (see e.g., Wu and Ni 1996; Spada et al. 2011a; Martinec et al. 2018). In this case, only the regular solution in (C.287) is valid; hence, taking into account that inside a uniform core the gravitational acceleration is $g_0(r) = \frac{4}{3}\pi G\rho_0 r$, the solution vector $\mathbf{w}$ can be written as

$$\mathbf{w}(r) = \tilde{K}_1 \begin{pmatrix} r^n \\ 2(n-1)r^{n-1} \end{pmatrix} , \tag{C.307}$$

where $\tilde{K}_1$ is another constant of integration. Using (C.307) in (C.306), and recalling that the y-variables are in general dependent upon the Laplace variable s, the solution vector at the base of the mantle takes the form

$$\tilde{\mathbf{y}}(b^+) = \mathbf{J}(b)\tilde{\mathbf{c}}_{\text{cmb}} , \tag{C.308}$$

where we have defined the array:

$$\mathbf{J}(b) = \begin{pmatrix} -b^n/g_0(b) & 1 & 0 \\ 0 & 0 & 1 \\ 0 & \rho_0(b^-)g_0(b) & 0 \\ 0 & 0 & 0 \\ b^n & 0 & 0 \\ 2(n-1)b^{n-1} & 4\pi G\rho_0(b^-) & 0 \end{pmatrix} , \tag{C.309}$$

and

$$\tilde{\mathbf{c}}_{\text{cmb}} = (\tilde{K}_1, \tilde{K}_2, \tilde{K}_3)^t \tag{C.310}$$

is a vector of integration constants that shall be determined by prescribing appropriate boundary conditions at the surface (see Sects. C.5 and C.7). We note that array (C.309) matches the expression given by Wu and Peltier (1982) and Sabadini et al. (1982).

C.5 Surface Boundary Conditions and Love Numbers

Here we describe how to obtain surface boundary conditions (BCs) for mass loading (Sect. C.5.1) and for tidal loading (Sect. C.5.2), mostly following a number of previous works in the literature. It is important to note that these conditions are the same for incompressible and compressible Earth models (Vermeersen et al. 1996b; Sabadini et al. 2016). We have made the effort of unifying the notation, deriving the BCs from first principles. The subject is *per se* quite burdensome.

C.5.1 Boundary Conditions for Mass Loading

These BCs are suitable when external masses (i.e., a layer of sediments, an ice sheet, a body of water or a volcanic edifice ...) are in physical contact with the Earth's surface. According to Longman (1962), for mass loading, at the deformed surface of the Earth the normal and tangential stress must be equal to the gravitational stress acting on the surface distribution of mass. This means that *(i)* the weight of the load is compensated by the vertical stress and *(ii)* the surface is free from tangential stress.

Condition *(i)*, involving vertical stress, reads:

$$\tilde{\tau}_{rr}(a) = \left(\tilde{\mathbf{g}} \cdot \hat{r}\right) \tilde{\mathcal{L}} \approx -g_0 \tilde{\mathcal{L}}, \tag{C.311}$$

where $\tilde{\mathbf{g}} = -g_0(r)\hat{r} + \tilde{\mathbf{g}}_1$ is gravity acceleration and $\tilde{\mathcal{L}}$ is the surface load variation, and in the second step we have neglected the term $\tilde{\mathbf{g}}_1 \cdot \hat{r}$ since $\tilde{\mathbf{g}}_1$ is a assumed to be a small perturbation in front of $g_0\hat{r}$. In terms of standard $\tilde{y}$-variables (see the third of Eqs. C.100 and C.44), the surface BC on vertical stress, which constitutes the *first poloidal BC* for mass loading, reads[21]

$$\tilde{y}_{3,nm}(a) = -g_0\tilde{\mathcal{L}}_{nm}\,, \tag{C.312}$$

where the $\tilde{\mathcal{L}}_{nm}$'s are the coefficients of the expansion of $\tilde{\mathcal{L}}$ in series of CSH.

Condition *(ii)*, involving tangential stress, is:

$$\tilde{\tau}_{r\theta}(a) = 0\,, \tag{C.313}$$

and according to (C.46) it implies, at the surface, the poloidal BC $\tilde{U}_4(a) = 0$ and the toroidal BC $\tilde{V}_2(a) = 0$. In terms of $\tilde{y}$ and $\tilde{z}$-variables, these translate into:

$$\tilde{y}_{4,nm}(a) = 0\,, \tag{C.314}$$

[21] BC (C.312) is equivalent to the first of Eqs. (16) of Longman (1962) and to the first of the three equations on page 774 of Farrell (1972). However, while (C.312) holds for a mass load of arbitrary geometry, these authors have adopted an axially symmetric mass distribution, which we shall consider as a special case in Sect. C.5.1. We note that Eq. (C.312) matches (1.109) of Sabadini et al. (2016), provided that their coefficients σ^L_{nm} are assimilated to our $\tilde{\mathcal{L}}_{nm}$'s.

which represents the *second poloidal BC* for mass loading, and into:

$$\tilde{z}_{2,nm}(a) = 0\,, \tag{C.315}$$

which constitutes the *toroidal BC* for mass loading, respectively.[22]

The third poloidal BC, which also involves gravitational variables, requires some more work. This condition can be established by applying Gauss's law for gravity, stating that *the gravitational flux through any closed surface is proportional to the enclosed mass.* Gauss's law is equivalent to Newton's law of universal gravitation. We note that in Longman (1962) and Wu and Peltier (1982) this BC is defined rather axiomatically, while in Cathles (1975), Lanzano (1982) and Sabadini et al. (2016) it has been properly motivated and discussed. Others, like James (1991), have applied Gauss's law without quoting it explicitly. We anticipate that Gauss's law shall also be useful in defining *third tidal loading BC* on Sect. C.5.2.

In the present context, Gauss's law reads:

$$\int_S \left(\tilde{\mathbf{g}}_1 \cdot \hat{n}\right) \,\mathrm{d}S = -4\pi G\, \Delta m\,, \tag{C.316}$$

where $\tilde{\mathbf{g}}_1$ is the total gravity field perturbation defined by Eq. (C.23), the integral represents its flux over the closed cylinder-shaped surface S of base area ΔA and width dh that contains a portion of the Earth's external layers (the "*Gaussian pillbox*" of Fig. C.1), and Δm is the mass variation inside S. This stems from two contributions, with $\Delta m = \Delta m^{load} + \Delta m^{disp}$, where $\Delta m^{load} = \tilde{\mathcal{L}}\,\Delta A$ is the intrinsic mass of the surface load and $\Delta m^{disp} = \rho_0 \tilde{u}_r\,\Delta A$ accounts for the deflection induced by the load, respectively.

Neglecting the flux across the lateral surface of the pillbox, Gauss's law (C.316) is equivalent to:

$$\left(\left(\tilde{\mathbf{g}}_1 \cdot \hat{r}\right)^{\text{ext}} - \left(\tilde{\mathbf{g}}_1 \cdot \hat{r}\right)^{\text{int}}\right) \Delta A = -4\pi G \left(\tilde{\mathcal{L}} + \rho_0 \tilde{u}_r\right) \Delta A\,, \tag{C.317}$$

where *ext* and *int* refer to variables evaluated at the external ($\hat{n} = \hat{r}$) and at the internal ($\hat{n} = -\hat{r}$) surface of the pillbox, respectively (see Fig. C.1). Since $\tilde{\mathbf{g}}_1 = -\nabla\tilde{\phi}_1$ (see Eq. C.23), from above we obtain

$$-\frac{\partial}{\partial r}\tilde{\phi}_1^{\text{ext}} + \frac{\partial}{\partial r}\tilde{\phi}_1^{\text{int}} = -4\pi G \left(\tilde{\mathcal{L}} + \rho_0 \tilde{u}_r\right)\,, \tag{C.318}$$

where we have used the identity $(\nabla\tilde{\phi}_1) \cdot \hat{r} = \partial\tilde{\phi}_1/\partial r$ (see B.4).

[22] Consideration of the tangential stress component $\tilde{\tau}_{r\lambda}$, given by (C.49), would not provide new BCs, but simply a duplication of (C.314) and (C.315). We note that Eq. (C.314) reproduces the second of Eq. (16) of Longman (1962) and the second of equations on page 774 of Farrell (1972). Equation (C.315) matches Eq. (105) of Lanzano (1982).

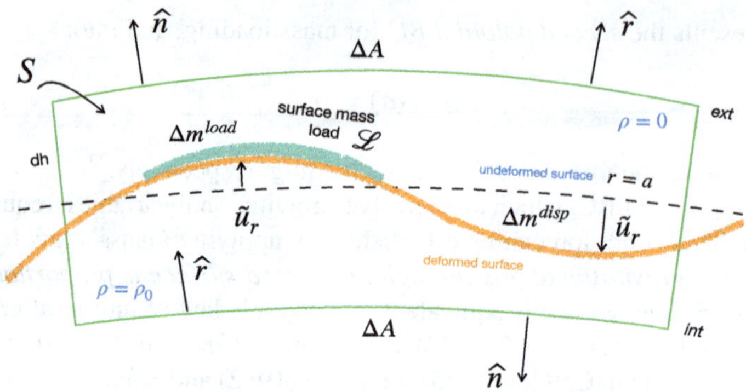

Fig. C.1 Geometry for the derivation of the third poloidal BC. For ease of visualisation, an enlarged Gaussian pillbox of base area ΔA and thickness $\ll a$ is shown in section. If the base area ΔA is infinitesimal, the displacement $\tilde{u}_r$ can be considered constant within the pillbox

Following e.g. Wu and Peltier (1982), it is convenient to write the total potential variation as the sum of two contributions, with:

$$\tilde{\phi}_1(r, \gamma) = \tilde{\phi}_2 + \tilde{\phi}_3 , \tag{C.319}$$

where $\tilde{\phi}_2(r, \gamma)$ is directly associated with the surface mass load,[23] whereas $\tilde{\phi}_3(r, \gamma)$ originates from the dynamical effects caused by mass

[23] The relationship between the direct contribution to gravitational potential $\tilde{\phi}_2$ and the surface load $\tilde{\mathcal{L}}$ can be established from first principles, assuming a rigid Earth. By Newton's law:

$$\tilde{\phi}_2(r, \gamma) = -G \int_{\mathrm{e}} \frac{\tilde{\mathcal{L}} \mathrm{d}A}{d} , \tag{C.320}$$

where the integration is over the Earth's surface e, we have used the definition of surface load (see Sect. 1.4.3.1), d is the distance that separates the observer at (r, γ) from the mass element $dM = \tilde{\mathcal{L}} dA$ at (a, γ'), and the element of area is $dA = a^2 \sin\theta' d\theta' d\lambda'$. Apart from the minus sign (due to a different convention), Eq. (C.320) is equivalent to Eq. (14) of Wu and Peltier (1982). By expanding $\tilde{\mathcal{L}}$ in series of CSHs with coefficients $\tilde{\mathcal{L}}_{nm}$ (see A.36) and using the generating function of Legendre polynomials (A.7) and the addition theorem (A.28), two distinct solutions for $\tilde{\phi}_2(r, \gamma)$ are obtained, according to the value of r, with the corresponding CSH coefficients:

$$\tilde{\phi}_{2,nm}(r) = -\frac{4\pi G a}{2n+1} \left(\frac{r}{a}\right)^n \tilde{\mathcal{L}}_{nm} , \quad \text{for } r \leq a , \tag{C.321}$$

and

$$\tilde{\phi}_{2,nm}(r) = -\frac{4\pi G a}{2n+1} \left(\frac{a}{r}\right)^{n+1} \tilde{\mathcal{L}}_{nm} , \quad \text{for } r \geq a , \tag{C.322}$$

implying that $\tilde{\phi}_{2,nm}(r)$ is continuous across the Earth's surface, with:

$$\tilde{\phi}_{2,nm}(a) = -\frac{4\pi G a}{2n+1} \tilde{\mathcal{L}}_{nm} . \tag{C.323}$$

redistribution.[24] It is important to note that the total gravitational potential perturbation $\tilde{\phi}_1(r,\gamma)$ is harmonic externally since $\tilde{\rho}_1 = 0$ outside the Earth (see C.2).

Expanding all the variables in (C.318) in series of CSHs and considering the individual radial coefficients gives

$$-\frac{\mathrm{d}}{\mathrm{d}r}\tilde{\phi}^{\mathrm{ext}}_{1,nm} + \frac{\mathrm{d}}{\mathrm{d}r}\tilde{\phi}^{\mathrm{int}}_{1,nm} = -4\pi G\left(\tilde{\mathcal{L}}_{nm} + \rho_0\,\tilde{U}_{1,nm}\right), \tag{C.324}$$

where we have used (C.39). But since $\tilde{\phi}^{\mathrm{ext}}_1(r,\gamma)$ is a solution of Laplace's equation, its harmonic coefficients have the form:

$$\tilde{\phi}^{\mathrm{ext}}_{1,nm}(r) = \tilde{c}_{nm}\left(\frac{a}{r}\right)^{n+1}, \tag{C.325}$$

where the $\tilde{c}_{nm}$'s are arbitrary constants and we have just considered regular terms for $r \to \infty$; thus, it is easily verified that

$$\frac{\mathrm{d}}{\mathrm{d}r}\tilde{\phi}^{\mathrm{ext}}_{1,nm} = -\left(\frac{n+1}{r}\right)\tilde{\phi}^{\mathrm{ext}}_{1,nm}, \tag{C.326}$$

that substituted into (C.324) gives

$$\left(\frac{n+1}{r}\right)\tilde{\phi}^{\mathrm{ext}}_{1,nm} + \frac{\mathrm{d}}{\mathrm{d}r}\tilde{\phi}^{\mathrm{int}}_{1,nm} = -4\pi G\left(\tilde{\mathcal{L}}_{nm} + \rho_0\,\tilde{U}_{1,nm}\right). \tag{C.327}$$

Adopting the continuity condition for the gravitational potential, we have:

$$\tilde{\phi}^{\mathrm{ext}}_{1,nm} = \tilde{\phi}^{\mathrm{int}}_{1,nm}, \tag{C.328}$$

which in fact constitutes the *fourth* poloidal BC. Upon rearranging, and recalling that $\tilde{\phi}^{\mathrm{ext}}_{1,nm} = \tilde{\phi}^{\mathrm{int}}_{1,nm} = \tilde{U}_{5,nm}$ (see C.42), we obtain:

$$\frac{\mathrm{d}}{\mathrm{d}r}\tilde{U}_{5,nm} + \left(\frac{n+1}{r}\right)\tilde{U}_{5,nm} + 4\pi G\rho_0\,\tilde{U}_{1,nm} = -4\pi G\tilde{\mathcal{L}}_{nm}, \tag{C.329}$$

where at the *l.h.s.* we recognise the expression of the *potential stress* (see Eq. C.65). Dropping superscript *int*, in terms of $\tilde{y}$-variables, this last equation gives the third poloidal BC[25] at the Earth's surface:

---

[24] Note that Farrell (1972) uses a different definition for the ϕ-variables, which may generate some confusion. Furthermore, Wu and Peltier (1982) denote our $\tilde{\phi}_1$ simply by letter ϕ.

[25] We remark that BC (C.330) is clearly a consequence of the particular definition of *potential stress* adopted here (see footnote on Sect. C.2.1.5). Al least another definition for $\tilde{y}_6$ has been employed in the literature, where the term $\left(\frac{n+1}{r}\right)\tilde{U}_{5,nm}$ is missing in Eq. (C.329), obviously leading to a different form of the third BC for surface loading. At page 84 of his PhD Thesis, James (1991) has commented upon this important aspect.

$$\tilde{y}_{6,nm}(a) = -4\pi G \tilde{\mathcal{L}}_{nm} \,. \tag{C.330}$$

By summarising the results obtained so far, the poloidal surface BCs for mass loading read:

$$\begin{cases} \tilde{y}_{3,nm}(a) = -g_0 \, \tilde{\mathcal{L}}_{nm} \\ \tilde{y}_{4,nm}(a) = 0 \\ \tilde{y}_{6,nm}(a) = -4\pi G \tilde{\mathcal{L}}_{nm} \,, \end{cases} \tag{C.331}$$

which are valid for a general surficial mass loading $\mathcal{L}(\gamma, t)$. By virtue of (C.323), it is useful to remark that these BCs can be cast in an equivalent form, only involving the direct potential variation at the Earth surface, namely:

$$\begin{cases} \tilde{y}_{3,nm}(a) = \rho_{\mathrm{e}} \left(\dfrac{2n+1}{3} \right) \tilde{\phi}_{2,nm} \\ \tilde{y}_{4,nm}(a) = 0 \\ \tilde{y}_{6,nm}(a) = \left(\dfrac{2n+1}{a} \right) \tilde{\phi}_{2,nm} \,, \end{cases} \tag{C.332}$$

where $\rho_{\mathrm{e}} = \frac{3m^{\mathrm{e}}}{4\pi a^3}$ is the average density of the Earth and we have used $g_0 \equiv g_0(a) = \frac{Gm^{\mathrm{e}}}{a^2}$.

The case of a *localised and impulsive unit mass* is of particular interest, since *loading Love numbers*—also referred to as *load-deformation* coefficients—are tightly associated with such a kind of forcing. Actually, in the seminal papers by Longman (1962), Farrell (1972) and Wu and Peltier (1982), the BCs for the general distributed load $\mathcal{L}(\gamma, t)$, given by Eq. (C.331), are not presented. Rather, these studies have considered the case of a *disc load* of half–amplitude α, and have obtained the poloidal BCs for a point load by taking the limit of an infinitesimal disc of angular radius $\alpha \to 0$. A slightly different approach shall be followed here, reaching the same conclusions.

As we have discussed in Sect. 2.2.2.1, for a localised and impulsive load placed at coordinates (θ', λ'), the surface load function reads:

$$\mathcal{L}^{\Delta}(\gamma, t) = \frac{m_{\mathrm{L}} \, \delta(t)}{a^2} \, \frac{\delta(\theta - \theta') \, \delta(\lambda - \lambda')}{\sin\theta} \,, \tag{C.333}$$

where m_{L} is the mass of the load and δ is Dirac's delta.[26] With the aid of (A.36), the Laplace-transformed harmonic coefficients of the expansion of $\mathcal{L}^{\Delta}(\gamma, t)$ in series of CSHs are

$$\tilde{\mathcal{L}}^{\Delta}_{nm} = \frac{m_{\mathrm{L}}}{4\pi a^2} \, \mu_{nm} \, P_{nm}(\cos\theta') \, \mathrm{e}^{-im\lambda'} \,, \tag{C.334}$$

[26] Note that $\mathcal{L}^{\Delta}(\gamma, t)$ has dimensions of a mass per unit area and per unit time, since the Dirac's delta has dimensions of the reciprocal of its argument.

where we have used the definition of CSH (A.22), we have taken advantage from the sifting property of Dirac's delta, we have noted that the Laplace transform of $\delta(t)$ is 1 (see Table B.1), and the normalization factor μ_{nm} is given by (A.23).

The case of a point mass located at the north pole (np), along the z-axis ($\theta' = 0$), is of particular importance. Since we consider a spherically symmetric Earth model,[27] we expect that such a special type of surface load—that hereby we shall denote with $\mathcal{L}^\delta(\gamma, t)$—shall be zonal in character, i.e., its harmonic coefficients only depend upon the harmonic degree n. By linearity, the Earth's response shall be purely zonal. Indeed, recalling that $P_{nm}(1) = \delta_{m0}$ (A.19) and using (C.334), the transform of the surface load coefficients shall be, in this case:

$$\tilde{\mathcal{L}}^\delta_{nm} = \frac{1}{4\pi a^2}\,\mu_{nm}\,\delta_{m0}\,, \tag{C.335}$$

which using (A.41) corresponds to Legendre polynomials expansion coefficients:

$$\tilde{\mathcal{L}}^\delta_{n} = \frac{2n+1}{4\pi a^2}\,, \tag{C.336}$$

where, without loss of generality, we have considered a unit mass letting $m_{\mathrm{L}} = 1$ kg.

By substituting (C.336) into Eq. (C.331), we find that in this particular case the BCs can be written as:

$$\begin{cases} \tilde{y}_{3,n}(a) = -\dfrac{(2n+1)\,g_0}{4\pi a^2} \\ \tilde{y}_{4,n}(a) = 0 \\ \tilde{y}_{6,n}(a) = -\dfrac{(2n+1)\,G}{a^2}\,, \end{cases} \quad \text{for mass loading} \tag{C.337}$$

where the $\tilde{y}_{i,n}$'s ($i = 3, 4, 6$) are coefficients of an expansion in series of Legendre polynomials, suitable to describe the response of the axially symmetric system considered here.[28]

We finally note that, using $\tilde{\mathcal{L}}_{nm} = \tilde{\mathcal{L}}^\delta_n$ into the general relationship given by (C.323), yields the Legendre coefficients of the direct gravitational potential in the form:

$$\tilde{\phi}^\delta_{2,n}(a) = -\frac{G}{a} \equiv -\frac{a\,g_0}{m^{\mathrm{e}}}\,, \tag{C.338}$$

that does not depend upon the degree n, i.e., its spectrum is white as it is characteristic of a point load. Equation (C.338) reproduces, apart from the sign, the result expressed by Eq. (16) of Wu and Peltier (1982).

---

[27] See Fig. 2.1 showing a general example of spherically symmetric layered Earth model.

[28] BCs (C.337) are consistent with previous studies employing the same formalism and conventions adopted here. For example, they agree with the three un-numbered equations at the beginning of page 774 of Farrell (1972) and with Eqs. (11a–c) of Wu and Peltier (1982), who have assumed a priori that the surface load is characterised by an azimutal symmetry. We also note that they match Eqs. (67a–c) of James (1991), keeping his definition (59) into account.

C.5.2 Boundary Conditions for Tidal Loading

When the perturbing masses are not in contact with the Earth's surface, as in the case of gravitational interactions, appropriate surface BCs are imposed, which are referred to as *tidal boundary conditions*. These BCs are poloidal in nature. The corresponding Love numbers, i.e., the *tidal (or "potential") Love numbers*, have a very broad range of applications in planetary sciences since they provide information upon the internal structure of celestial bodies. Here we follow various authors, although some details of the derivation are original. It is certain that in the literature there is some confusion about this topic, due to different conventions and notations adopted for the variables involved.

Since the external bodies are not in physical contact with the Earth, the vertical and horizontal stress vanish at the surface (*stress-free* BC). Similar to the mass loading case, the derivation of the necessary third BC is quite cumbersome. Actually, this BC is hiding a *fourth* BC, as we shall see. We invoke again Gauss's law (see C.316), keeping in mind that there is no direct contribution from the surface load whereas the contribution of the surface vertical displacement remains, being associated with the tidally-induced deformations. Then, letting $\tilde{\mathcal{L}} = 0$ in the expression of Gauss's law (C.318) yields:

$$\frac{\partial}{\partial r}\,\tilde{\phi}_1^{\mathrm{int}} - \frac{\partial}{\partial r}\,\tilde{\phi}_1^{\mathrm{ext}} + 4\pi G\rho_0\tilde{u}_r = 0\,, \qquad \text{(C.339)}$$

where we recall that superscripts *int* and *ext* refer to quantities evaluated at the external and at the internal surface of the Gaussian pillbox, respectively (see Fig. C.1).

As in the case of the surface mass loads, it is convenient to write the total potential variation as the sum of two contributions:

$$\tilde{\phi}_1(r,\gamma) = \tilde{\phi}_2 + \tilde{\phi}_3\,, \qquad \text{(C.340)}$$

where $\tilde{\phi}_2(r,\gamma)$ directly arises from the gravitational attraction of the tide-rising body, while $\tilde{\phi}_3(r,\gamma)$ is due to the tidally-induced deformations. This separation of $\tilde{\phi}_1(r,\gamma)$ is also invoked by Takeuchi (1950) and Takeuchi et al. (1962). Substituting (C.340) into (C.339) and expanding in series of CSHs yields:

$$\frac{\mathrm{d}}{\mathrm{d}r}\tilde{\phi}_{1,nm}^{\mathrm{int}} - \frac{\mathrm{d}}{\mathrm{d}r}\tilde{\phi}_{2,nm}^{\mathrm{ext}} - \frac{\mathrm{d}}{\mathrm{d}r}\tilde{\phi}_{3,nm}^{\mathrm{ext}} + 4\pi G\rho_0\tilde{U}_{1,nm} = 0\,, \qquad \text{(C.341)}$$

where we have used (C.39) and we have made the *nm* dependence explicit.

Following e.g., Agnew (2007), the external tidal potential—also referred to as the *tide rising potential*—is the harmonic function:

$$\phi_2(r,\gamma,t) = \sum_{\substack{nm\\ n\geq 2}} c_{nm}(t)\left(\frac{r}{a}\right)^n Y_{nm}(\gamma)\,, \qquad \text{(C.342)}$$

where coefficients $c_{nm}(t)$ are determined by the geometry of the two-body system.[29] Note that in (C.342), only degrees $n \geq 2$ are included, since *i)* a degree $n = 0$ term would produce a spatially uniform "tide" and *ii)* $\phi_2(r, \gamma, t)$ is expressed in a reference frame whose origin coincides with the centre of mass of the system (see e.g., Agnew 2007).

From (C.342), the harmonic coefficients of the tide rising potential, evaluated at the external surface of the pillbox, are:

$$\tilde{\phi}^{\text{ext}}_{2,nm}(r) = \tilde{c}_{nm} \left(\frac{r}{a}\right)^n , \tag{C.344}$$

hence

$$\frac{\text{d}}{\text{d}r}\tilde{\phi}^{\text{ext}}_{2,nm}(r) = \left(\frac{n}{r}\right) \tilde{\phi}^{\text{ext}}_{2,nm} , \tag{C.345}$$

which inserted into (C.341) gives:

$$\frac{\text{d}}{\text{d}r}\tilde{\phi}^{\text{int}}_{1,nm} - \left(\frac{n}{r}\right) \tilde{\phi}^{\text{ext}}_{2,nm} - \frac{\text{d}}{\text{d}r}\tilde{\phi}^{\text{ext}}_{3,nm} + 4\pi G\rho_0 \tilde{U}_{1,nm} = 0 . \tag{C.346}$$

Now we turn our attention on the $\tilde{\phi}^{\text{ext}}_{3,nm}$ term, associated with the tidally-induced deformation. We note that $\tilde{\phi}^{\text{ext}}_3(r, \gamma)$ is harmonic, since the tide rising potential $\tilde{\phi}^{\text{ext}}_2(r, \gamma)$ is harmonic and the total gravitational potential variation $\tilde{\phi}^{\text{ext}}_1(r, \gamma)$ is harmonic as well outside the Earth, regardless of the assumption of incompressibility (see Eq. C.2). Hence:

$$\tilde{\phi}^{\text{ext}}_{3,nm}(r) = \tilde{b}_{nm} \left(\frac{a}{r}\right)^{n+1} , \tag{C.347}$$

where in the solution of Laplace's equation for $\phi^{\text{ext}}_3(r, \gamma)$ we have only kept the term behaving regularly for $r \to \infty$, and the $\tilde{b}_{nm}$'s are arbitrary coefficients. Evaluating the radial derivative gives:

$$\frac{\text{d}}{\text{d}r}\tilde{\phi}^{\text{ext}}_{3,nm}(r) = -\left(\frac{n+1}{r}\right) \tilde{\phi}^{\text{ext}}_{3,nm} , \tag{C.348}$$

so that (C.346) becomes:

---

[29] Equation (C.342) is easily obtained from (2.5) of Agnew (2007) that gives the tide-rising potential. By invoking the addition theorem for spherical harmonics (A.28), after some straightforward manipulation it turns out that:

$$c_{nm}(t) = -\frac{GM_b}{a}\left(\frac{a}{R}\right)^{n+1} \frac{Y^*_{nm}(\gamma')}{2n+1} , \quad n \geq 2 , \tag{C.343}$$

where M_b is the mass of the attracting body, $R = R(t)$ is the distance that separates the centres of the two bodies, and $\gamma' = \gamma'(t) = (\theta', \lambda')$ where $\theta'(t)$ and $\lambda'(t)$ are the coordinates of the sub-body point. Note that the minus sign in Eq. (C.343) is coherent with our sign convention on potentials, which is contrary to Duncan Agnew's.

$$\frac{\mathrm{d}}{\mathrm{d}r}\tilde{\phi}^{\mathrm{int}}_{1,nm} - \left(\frac{n}{r}\right)\tilde{\phi}^{\mathrm{ext}}_{2,nm} + \left(\frac{n+1}{r}\right)\tilde{\phi}^{\mathrm{ext}}_{3,nm} + 4\pi G\rho_0 \tilde{U}_{1,nm} = 0\,. \tag{C.349}$$

We now rearrange this last equation conveniently, by adding and subtracting an *ad hoc* term, marked in braces, as follows:

$$\left(\frac{\mathrm{d}}{\mathrm{d}r}\tilde{\phi}^{\mathrm{int}}_{1,nm} + \underbrace{\left(\frac{n+1}{r}\right)\tilde{\phi}^{\mathrm{int}}_{1,nm}} + 4\pi G\rho_0 \tilde{U}_{1,nm}\right) - \underbrace{\left(\frac{n+1}{r}\right)\tilde{\phi}^{\mathrm{int}}_{1,nm}} + \\ - \left(\frac{n}{r}\right)\phi^{\mathrm{ext}}_{2,nm} + \left(\frac{n+1}{r}\right)\tilde{\phi}^{\mathrm{ext}}_{3,nm} = 0\,, \tag{C.350}$$

that using the definition of the *potential stress* $\tilde{Q}_{nm}$ (see Eq. C.65 and the footnote to Sect. C.2.1.5, recalling that $\tilde{Q}_{nm} \equiv \tilde{U}_{6,nm}$ and $\tilde{\phi}_{1,nm} \equiv \tilde{U}_{5,nm}$), yields:

$$\tilde{Q}^{\mathrm{int}}_{nm} - \left(\frac{n+1}{r}\right)\tilde{\phi}^{\mathrm{int}}_{1,nm} - \left(\frac{n}{r}\right)\phi^{\mathrm{ext}}_{2,nm} + \left(\frac{n+1}{r}\right)\tilde{\phi}^{\mathrm{ext}}_{3,nm} = 0\,, \tag{C.351}$$

which however is unsuitable as an explicit BC for $\tilde{Q}^{\mathrm{int}}_{nm}$, since it still contains the unknown $\tilde{\phi}^{\mathrm{int}}_{1,nm}$. Hence, a further condition is now required to eliminate $\tilde{\phi}^{\mathrm{int}}_{1,nm}$ from (C.351). Following e.g., Saito (1974), this condition (the *fourth tidal* BC) is simply the continuity of the total gravitational potential variation:

$$\phi^{\mathrm{int}}_{1,nm} = \phi^{\mathrm{ext}}_{1,nm}\,, \tag{C.352}$$

which using (C.340) provides

$$\tilde{Q}^{\mathrm{int}}_{nm} - \left(\frac{n+1}{r}\right)\left(\cancel{\tilde{\phi}^{\mathrm{ext}}_{3,nm}} + \phi^{\mathrm{ext}}_{2,nm}\right) - \left(\frac{n}{r}\right)\tilde{\phi}^{\mathrm{ext}}_{2,nm} + \left(\frac{n+1}{r}\right)\cancel{\tilde{\phi}^{\mathrm{ext}}_{3,nm}} = 0\,, \tag{C.353}$$

where we have emphasised that the two terms in $\phi^{\mathrm{ext}}_{3,nm}$ cancel out, so that after a little rearrangement we obtain:

$$\tilde{Q}^{\mathrm{int}}_{nm}(r) = \left(\frac{2n+1}{r}\right)\tilde{\phi}^{\mathrm{ext}}_{2,nm}\,. \tag{C.354}$$

As noted by Bos and Scherneck (2012), the remarkable property of this equation is that *it does not contain the unknown internal potential explicitly*.

Summarising the results so far, in terms of $\tilde{y}$-variables, the three surface tidal BCs read:

$$\begin{cases} \tilde{y}_{3,nm}(a) = 0 \\ \tilde{y}_{4,nm}(a) = 0 \\ \tilde{y}_{6,nm}(a) = \left(\dfrac{2n+1}{a}\right)\tilde{\phi}_{2,nm}, \end{cases} \tag{C.355}$$

and hold for a general external potential characterised by Laplace-transformed harmonic coefficients $\tilde{\phi}_{2,nm}(a)$. These BCs, of poloidal nature, constitute the tidal counterparts of Eq. (C.332), valid for a general mass load.[30]

The BCs (C.355) are valid for a generic point-like attracting body, of arbitrary time-varying coordinates (θ', λ') and distance $R(t)$ from the Earth's centre. In analogy with what we have done in Sect. C.5.1 for the mass load, it is convenient to consider the case of an impulsive (δ-like) tidal perturbation exerted by a point mass located along the z-axis, just above the north pole (*np*). This can be accomplished by letting $M_b = m_0\,\delta(t)$, $R(t) \to a$ and $\theta' = 0$ in (C.343), thus obtaining

$$\tilde{\phi}^{\delta}_{2,nm}(a) = \tilde{c}_{nm} = -\frac{G\,m_0}{a}\frac{\delta_{m0}}{\sqrt{2n+1}}, \tag{C.356}$$

where we taken advantage of the definition of CSH (A.22) and of the expression of μ_{nm} (A.23), also using $P_{nm}(1) = \delta_{m0}$ (see A.19). Considering, without loss of generality, a unit mass ($m_0 = 1$ kg), and recalling (A.41), we thus obtain:

$$\tilde{\phi}^{\delta}_{2,n}(a) = -\frac{G}{a} \equiv -\frac{a\,g_0}{m^{\mathrm{e}}}, \tag{C.357}$$

that represents the coefficients of the Legendre expansion of the tidal potential for the impulsive unit attracting point mass located just above the *np*. Thus, in this particular case, the tidal BCs are expressed, for $n \geq 2$, by:

$$\begin{cases} \tilde{y}_{3,n}(a) = 0 \\ \tilde{y}_{4,n}(a) = 0 & \text{for tidal loading} \\ \tilde{y}_{6,n}(a) = -\dfrac{(2n+1)\,G}{a^2}. \end{cases} \tag{C.358}$$

In closing, we note that these BCs reproduce their mass loading counterparts (C.337), except for the constraint on $\tilde{y}_{3,n}$. Hence, for a localised impulsive external mass, the tidal BCs can be just obtained from the mass loading BCs by switching off the condition on the vertical stress.

---

[30] These BCs are consistent with those given in Sabadini et al. (2016). See in particular their Eqs. (1.128) and (1.130), where their coefficients $\Phi^{T}_{nm}(a)$ correspond to our coefficients $\tilde{\phi}_{2,nm}(a)$.

C.5.3 Loading and Tidal Love Numbers

Love numbers, named after Augustus Edward Hough Love who first introduced them in his seminal works, are adimensional coefficients expressing the response of the Earth to surface loads or tidal forcings. In the context of tidal deformations of the Earth, Love (1911, see Sect. 59) assumed that the solid tidal deformation is proportional to the equilibrium tidal height as

$$U = h_2 \frac{W_2}{g_0}, \tag{C.359}$$

where W_2 is the *tide-raising potential*, U is the height of the solid tide, W_2/g_0 is, by Bruns formula, the height of the equilibrium tide, h_2 is an adimensional coefficient dubbed the *h tidal Love number of degree* 2. As a consequence of the deformation of the solid Earth, the gravitational potential will be further disturbed by a quantity W_2' which is assumed to be proportional to W_2 as

$$W_2' = k_2 W_2, \tag{C.360}$$

where k_2 is the *k tidal Love number of degree* 2. Shida (1912) introduced a third Love number ℓ_2 relating the horizontal deformation V associated to the solid tide to the equilibrium tidal height as

$$V = \ell_2 \frac{W_2}{g_0}. \tag{C.361}$$

According to our notation, at a given harmonic degree n the total perturbation to potential due to gravitational attraction of an external body is is given by $\tilde{\phi}_{1,n} = \tilde{\phi}_{2,n} + \tilde{\phi}_{3,n}$, where $\tilde{\phi}_{2,n}$ is the *direct* perturbation of the tide-raising body and $\tilde{\phi}_{3,n}$ is the *indirect* effect due to the deformation of the Earth. Hence, taking into account that the for tidal deformations, the definitions above can be generalised to an arbitrary harmonic degree n and written in compact form as[31]

$$\begin{pmatrix} \tilde{U}_{1,n}(a) \\ \tilde{U}_{2,n}(a) \\ \tilde{\phi}_{3,n}(a) \end{pmatrix} = \tilde{\phi}_2(a) \begin{pmatrix} -\tilde{h}_n^{\mathrm{T}}/g_0 \\ -\tilde{\ell}_n^{\mathrm{T}}/g_0 \\ \tilde{k}_n^{\mathrm{T}} \end{pmatrix}, \tag{C.362}$$

where $\tilde{U}_{1,n}$ and $\tilde{U}_{2,n}$ are, for harmonic degree n, the radial functions associated to vertical and horizontal deformation and defined in (C.39–C.41) and $\tilde{h}_n^{\mathrm{T}}, \tilde{\ell}_n^{\mathrm{T}}, \tilde{k}_n^{\mathrm{T}}$ are the *tidal Love numbers of degree n*. By using (C.357) and recalling that $\tilde{U}_{5,n} = \tilde{\phi}_2 + \tilde{\phi}_3$, where $\tilde{U}_{5,n}$ is the radial function defined in (C.42) at harmonic degree n, we can rewrite (C.362) as

[31] Note that definitions (C.359–C.361) are given in terms of the W_2 potential which has the opposite sign convention w.r.t. the $\tilde{\phi}_2$ potential.

$$\begin{pmatrix} \tilde{U}_{1,n}(a) \\ \tilde{U}_{2,n}(a) \\ \tilde{U}_{5,n}(a) \end{pmatrix} = \frac{a}{m^{\mathrm{e}}} \begin{pmatrix} \tilde{h}_n^{\mathrm{T}} \\ \tilde{\ell}_n^{\mathrm{T}} \\ -(1+\tilde{k}_n^{\mathrm{T}})g_0 \end{pmatrix}, \tag{C.363}$$

or, equivalently,

$$\begin{pmatrix} \tilde{h}_n^{\mathrm{T}} \\ \tilde{\ell}_n^{\mathrm{T}} \\ \tilde{k}_n^{\mathrm{T}} \end{pmatrix} = \begin{pmatrix} \frac{m^{\mathrm{e}}}{a}\tilde{U}_{1,n}(a) \\ \frac{m^{\mathrm{e}}}{a}\tilde{U}_{2,n}(a) \\ -1 - \frac{m^{\mathrm{e}}}{a g_0}\tilde{U}_{5,n}(a) \end{pmatrix}. \tag{C.364}$$

For surface loads it is possible to define *loading Love numbers*[32] $\tilde{h}_n^{\mathrm{L}}$, $\tilde{\ell}_n^{\mathrm{L}}$ and $\tilde{k}_n^{\mathrm{L}}$ in close analogy with (C.362). Since for a surface load the direct perturbation to potential $\tilde{\phi}_{2,n}$ in (C.338) has the same form as in the tidal case, the definition of loading Love numbers is identical to (C.363), with the radial functions on the *l.h.s.* computed with boundary conditions suitable for a surface load.

C.6 Homogeneous Earth Model

That of the deformation of a homogeneous, self-gravitating, elastic sphere is a classical problem in global geodynamics (see e.g., Jeffreys 1952). For this extremely simplified planetary model, often referred to as the *Kelvin sphere*, Love numbers are available in a fully analytical form both for surface and tidal forcings. This is certainly the main motivation of the success of this planetary model. The model is simple but at the same time it is rather realistic, since it accounts for the spherical geometry and self-gravitation, and it can provide a suitable solution in the case the internal structure of the planet is not considered of interest, or it is only known *on average*. For global problems such as the planetary response to tides, which only involves long-wavelength harmonics, the Kelvin sphere is certainly a very valid option (Melini et al. 2022).

Our purpose here is to show how to *solve the Kelvin sphere* taking advantage of the concepts illustrated in previous sections. The solution of the equilibrium equations for an incompressible elastic sphere has been first obtained by Gabriel Lamé in 1854 and, independently, by William Thomson (later knighted Lord Kelvin by Queen Victoria) in 1863. Those results have then been extended to the case of a compressible homogeneous sphere by A.E.H. Love (1911). Taking advantage of the correspondence principle, the solutions for an elastic homogeneous sphere can be extended to any linear viscoelastic rheology by substituting the elastic modulus μ and λ with appropriate complex moduli $\tilde{\mu}$ and $\tilde{\lambda}$ (see Sect. 1.3.9). In what follows, we will mostly focus on the case of an incompressible sphere, for which it is possible to obtain appealingly simple analytical expressions for the Love numbers.

[32] In some classical papers, for instance Longman (1963), the loading Love numbers are dubbed *load deformation coefficients*.

C.6.1 The Incompressible Homogeneous Sphere

The solution of the poloidal ODEs for an homogeneous incompressible shell has been discussed in Sect. C.2.2.2 and is expressed in the compact form (C.117). However, in a homogeneous sphere, only the solutions which are regular in $r = 0$ can be acceptable, hence in the present case we shall drop the singular part of (C.119). Therefore, the solution of the poloidal system in a homogeneous sphere takes the form

$$\tilde{\mathbf{y}}(r) = \tilde{\mathbf{N}}(r)\,\tilde{\mathbf{C}}\,, \tag{C.365}$$

where $\tilde{\mathbf{N}}$ is the regular matrix defined in (C.120) and $\tilde{\mathbf{C}} = (\tilde{c}_1, \tilde{c}_2, \tilde{c}_3)^t$ is a three-component vector that can be determined by prescribing suitable boundary conditions for $\tilde{\mathbf{y}}(a)$. This can be accomplished by solving for $\tilde{\mathbf{C}}$ the linear system

$$\tilde{\mathbf{b}}(a) = \tilde{\mathbf{R}}\,\tilde{\mathbf{C}}\,, \tag{C.366}$$

where vector $\tilde{\mathbf{b}}(a) = (\tilde{y}_3(a), \tilde{y}_4(a), \tilde{y}_6(a))^t$ is determined by the BCs (C.337) or (C.358) for loading or tidal excitations, respectively, while $\tilde{\mathbf{R}}$ is the 3×3 matrix containing the 3rd, 4th and 6th rows of $\tilde{\mathbf{N}}(a)$. After some cumbersome algebra that can be conveniently tackled with a symbolic manipulator, in agreement with Wu and Peltier (1982), for boundary conditions appropriate for a surface load we obtain:

$$\tilde{\mathbf{C}}^{\mathrm{L}} = \frac{1}{3a^n m^{\mathrm{e}} n(1+\lambda_n \tilde{\varepsilon})} \begin{pmatrix} 2(n^2-1)(2n+3) \\ -a^2 n^2 (n+2) \\ -3 a g_0 \tilde{\varepsilon}(2n^2+4n+3) \end{pmatrix}, \tag{C.367}$$

where m^{e} is the mass of the Earth and we have introduced the quantities

$$\tilde{\varepsilon}(s) = \frac{\tilde{\mu}}{\rho_0 g_0 a} \tag{C.368}$$

where ρ_0 is density and g_0 is surface gravity, and

$$\lambda_n = \frac{2n^2+4n+3}{n}\,, \tag{C.369}$$

where we note that

$$\lambda_2 = {}^{19}\!/_2\,. \tag{C.370}$$

Similarly, for boundary conditions corresponding to an external potential, we obtain

$$\tilde{\mathbf{C}}^{\mathrm{T}} = \frac{1}{2a^n m^{\mathrm{e}} n(n-1)(1+\lambda_n\tilde{\varepsilon})} \begin{pmatrix} -2(n^2-1)(2n+3) \\ a^2n^2(n+2) \\ -ag_0\left\{\left[2(2n^2+2n-1)\tilde{\varepsilon}+(2n+1)\right]n-6\tilde{\varepsilon}\right\} \end{pmatrix}. \tag{C.371}$$

Vectors (C.367) and (C.371) can now be used in (C.365) to obtain the full solution for the "$\tilde{y}$-*system*" for a surface load and a tidal potential, respectively. From (C.364), the corresponding Love numbers can then be obtained; after some manipulation, they can be cast in the forms

$$\begin{pmatrix} \tilde{h}_n^{\mathrm{L}} \\ \tilde{\ell}_n^{\mathrm{L}} \\ \tilde{k}_n^{\mathrm{L}} \end{pmatrix} = \frac{1}{1+\lambda_n\tilde{\varepsilon}} \begin{pmatrix} -\frac{2n+1}{3} \\ -\frac{1}{n} \\ -1 \end{pmatrix} \tag{C.372}$$

and

$$\begin{pmatrix} \tilde{h}_n^{\mathrm{T}} \\ \tilde{\ell}_n^{\mathrm{T}} \\ \tilde{k}_n^{\mathrm{T}} \end{pmatrix} = \frac{1}{1+\lambda_n\tilde{\varepsilon}} \begin{pmatrix} 1+\frac{3}{2(n-1)} \\ \frac{3}{2n(n-1)} \\ \frac{3}{2(n-1)} \end{pmatrix}, \tag{C.373}$$

where the quantities in the parentheses on the *r.h.s.* correspond to the *fluid limit* of the Love numbers, since $\tilde{\varepsilon} \to 0$ for $\tilde{\mu} \to 0$.

From (C.372–C.373), it can be easily seen that, for a given harmonic degree n, all six Love numbers can be written in the general form

$$\tilde{x}_n(s) = \frac{x_n^{\mathrm{f}}}{1+\lambda_n\tilde{\varepsilon}(s)}, \tag{C.374}$$

where $\tilde{x}_n$ is any of $\tilde{h}_n$, $\tilde{\ell}_n$ or $\tilde{k}_n$, x_n^{f} is the corresponding fluid limit and we have made explicit the dependence upon the Laplace variable s. It is now convenient to write $\tilde{\varepsilon}(s)$ as

$$\tilde{\varepsilon}(s) = \varepsilon_0 \frac{\tilde{\mu}(s)}{\mu}, \tag{C.375}$$

where the adimensional number

$$\varepsilon_0 = \frac{\mu}{\rho_0 g_0 a} \tag{C.376}$$

is a measure of the relative importance of elastic and gravitational forces, which for a homogeneous Earth model takes the value $\varepsilon_0 \simeq 0.42$ (see e.g., Consorzi et al. 2024; Francioni 2024). For a Maxwell rheology, using the the complex rigidity (C.8) in (C.375) yields

$$\tilde{\varepsilon}(s) = \varepsilon_0 \frac{s}{s+1/\tau_{\mathrm{M}}}, \tag{C.377}$$

where $\tau_\mathrm{M} = \eta/\mu$ is the *Maxwell characteristic time*. Using (C.377) in (C.375), after some algebra we obtain

$$\tilde{x}_n(s) = \frac{x_n^\mathrm{f}}{1+\lambda_n\varepsilon_0}\left[\frac{s+1/\tau_\mathrm{M}}{s+1/\tau_n}\right], \tag{C.378}$$

where

$$\tau_n = (1+\lambda_n\varepsilon_0)\tau_\mathrm{M} \tag{C.379}$$

is the *relaxation time* associated with harmonic degree n.

It is now useful to separate the elastic and viscous components of the Laplace-transformed Love numbers. As expected, the *elastic limit* of (C.378) is

$$x_n^\mathrm{e} = \lim_{s\to\infty}\tilde{x}_n(s) = \frac{x_n^f}{1+\lambda_n\varepsilon_0}, \tag{C.380}$$

while the non-elastic (or *viscous*) contribution is

$$\tilde{x}_n(s) - x_n^\mathrm{e} = x_n^\mathrm{e}\left[\frac{s+1/\tau_\mathrm{M}}{s+1/\tau_n} - 1\right] = x_n^\mathrm{e}\,\frac{1/\tau_\mathrm{M} - 1/\tau_n}{s+1/\tau_n}, \tag{C.381}$$

hence we can write (C.378) in the form

$$\tilde{x}_n(s) = x_n^\mathrm{e} + \frac{x_n^\mathrm{v}}{s+1/\tau_n}, \tag{C.382}$$

where we have introduced the *viscoelastic residue*

$$x_n^\mathrm{v} = x_n^\mathrm{e}\left(\frac{1}{\tau_\mathrm{M}} - \frac{1}{\tau_n}\right) = x_n^\mathrm{e}\,\frac{\lambda_n\varepsilon_0}{\tau_n}. \tag{C.383}$$

The Laplace-domain Love number in the form (C.382) is easily inverted using the expressions in Table B.1, yielding the time-domain Love number associated with an impulsive load, which reads

$$x_n^\delta(t) = x_n^\mathrm{e}\delta(t) + x_n^\mathrm{v}H(t)\mathrm{e}^{-t/\tau_n}. \tag{C.384}$$

As discussed also in Sect. 2.4.1, we note that the relaxation time τ_n always exceeds the Maxwell characteristic time, i.e. $\tau_n > \tau_\mathrm{M}$ for all harmonic degrees n. For large harmonic degrees, it is easily shown that the relaxation time follows the asymptotic relation

$$\tau_n \approx \frac{2\eta}{\rho_0 g_0 a}n = 2n\,\varepsilon_0\tau_\mathrm{M}, \tag{C.385}$$

which, as discussed by Wu and Peltier (1982), is in accordance with the relaxation expected for a viscous half-space (Haskell 1935), since large harmonic degrees correspond to small wavelengths at which deformation is insensitive to the sphericity of the Earth.

C.6.2 The Compressible Homogeneous Sphere

The compressible, elastic, homogeneous model has been first solved by Love in his classical textbooks and reconsidered by Hurford and Greenberg (2001) and Hurford et al. (2006) in the context of planetary instabilities caused by tidal forcing. The effects of compressibility are minor, when we deal with the Earth; however, compressibility can play a role for general celestial bodies. As discussed by Hurford et al. (2006), Love has applied his analytical solutions for the homogeneous sphere to two specific cases relevant for the study of Earth tides, and found an enhancement of h on the order of a few tens of percent due to effects of compressibility. However, depending on the radius and elastic parameters of the sphere, the enhancement can be much greater and may even become arbitrarily large.

For a compressible homogeneous sphere, loading and tidal Love numbers can be obtained with the same approach that we discussed for the incompressible sphere, provided that in (C.365) the regular part of the compressible fundamental matrix (C.243) is used in place of its incompressible counterpart. We remark that, for the homogeneous sphere, the compressible fundamental solutions (C.244–C.246) are exact, since the assumption $g_0/r =$ constant that has been made for their derivation holds true. However, explicit analytical expressions for compressible Love numbers are quite complicated and are therefore not reported here since they would not provide useful insights; for practical applications, they can be more conveniently obtained numerically, finding the $\tilde{\mathbf{C}}^{\mathrm{L}}$ or $\tilde{\mathbf{C}}^{\mathrm{T}}$ vectors by standard linear algebra methods (see Eqs. C.367 and C.371 above).

An important aspect of the compressible homogeneous model is its gravitational stability. As discussed by Vermeersen and Mitrovica (2000), this point has been first investigated in the context of an elastic model by Love in Chap. IX of his fundamental treatise (Love 1911). According to his findings, in an elastic self-gravitating sphere instabilities can arise when the resistance to compression is insufficient to balance the gravitational force, ultimately resulting in self-gravitational collapse. In particular, according to Love (1911, p. 111), when the gravitational energy lost exceeds the strain energy gained, the body is in an unstable state. The transition between stable and unstable states is represented by a critical condition in which, for a small deformation, the variation of strain energy is exactly equal to that of gravitational energy. In such a critical state, the equilibrium equations shall be satisfied by non-vanishing displacements under vanishing external forces, or, equivalently, the body must admit zero-frequency vibrations. The condition of gravitational stability originally formulated by Love (1911) has been revisited by Vermeersen and Mitrovica (2000), who obtained curves of marginal stability separating regions of the (λ, μ) plane in which an homogeneous elastic compressible model exhibits stable and unstable behaviours. Those curves correspond to the critical combinations of λ and μ for which the model allows non vanishing displacements in absence of external forcings.

By virtue of the correspondence principle, the Laplace-transformed response for a compressible homogeneous sphere with viscoelastic rheology can be obtained by substituting the explicit functional forms of the complex moduli $\tilde{\mu}(s)$ and $\tilde{\lambda}(s)$. As we shall discuss in more detail in Sect. C.7, this leads to a form of the Love numbers for an impulsive excitation that reads

$$x_n^{\delta}(t) = x_n^{\mathrm{e}}\delta(t) + \sum_{k=1}^{M} x_{n,k}^{\mathrm{v}} \mathrm{e}^{s_{n,k}t} , \tag{C.386}$$

which can be recognised as a generalization of the expression (C.383) obtained for the incompressible homogeneous sphere, and where $s_{n,k}$ are the M roots of the *secular equation* $\det(\tilde{\mathbf{R}}(s)) = 0$ at degree n, while the $x_{n,k}^{\mathrm{v}}$'s are dubbed *viscoelastic residues*. Hence, for $s = s_{n,k}$ the columns of $\tilde{\mathbf{R}}$ are linearly dependent, and the solution can satisfy null boundary conditions corresponding to $\tilde{\mathbf{b}} = 0$ in (C.366). In this respect, the M terms of the sum in (C.386) represent the spectrum of free relaxation modes of the model in absence of external forcings.

It is evident that, in (C.386), *negative* values of $s_{n,k}$ correspond to stable modes whose amplitudes decay with a characteristic time $\tau_{n,k} = -1/s_{n,k}$, while *positive* values of s_k correspond to unstable, exponentially growing modes. By comparing (C.386) with (C.384), it can be seen that in the incompressible case an homogeneous viscoelastic sphere with Maxwell rheology has a single stable mode with $s_n = -1/((1 + \lambda_n \varepsilon_0)\tau_{\mathrm{M}})$. This result is rather general, since Wolf (1994) has shown that no unstable modes exist in a homogeneous incompressible model with an arbitrary linear rheology.

For a compressible homogeneous sphere, determining the values of $s_{n,k}$ is rather difficult, since the secular equation involves spherical Bessel functions and therefore it cannot be solved analytically. Numerical experiments discussed by Wu and Peltier (1982) and Vermeersen et al. (1996a) have shown that a stable fundamental mantle mode exists, whose relaxation time deviates only slightly from the incompressible relaxation time (C.379). In addition, a set of *dilatation modes* (or D modes) appears, associated with roots of the secular equation that according to Vermeersen et al. (1996a) and Vermeersen and Mitrovica (2000) may be approximated by

$$s_{Dm} \approx -\frac{\Gamma_m \left(\lambda + \frac{2}{3}\mu\right) - 1}{\Gamma_m \left(\lambda + 2\mu\right) - 1} \frac{1}{\tau_{\mathrm{M}}} , \tag{C.387}$$

where Γ_m is a constant that, for $n = 2$, reads $\Gamma_m = (m\pi)^2/(4\pi\rho g a)$. It can be seen from (C.387) that, in some regions of the (λ, μ) plane, s_{Dm} may become positive and therefore the associated modes are unstable. It is important, however, to remark that the stability of those modes is not dependent upon viscosity but it is only determined by the elastic parameters of the model; in addition, Vermeersen and Mitrovica (2000) have shown that in a viscoelastic homogeneous sphere the D modes are always stable if its elastic parameters are such that the purely elastic response of the model is gravitationally stable.

A second class of unstable modes is represented by the Rayleigh-Taylor (RT) instabilities. Those have been first discussed in detail by Plag and Jüttner (1995) and are associated with density inversions, e.g. with regions where the density *decreases* with depth. RT modes are usually found in models where the density is a function of radius, if $\rho(r)$ is not a monotonically decreasing function of r (as in the case of the PREM model by Dziewonski and Anderson 1981); however, as shown by Hanyk et al. (1999), they can be excited also in a homogeneous model as a consequence of the density perturbations associated with finite compressibility. The importance of the contribution of RT modes to the sum in (C.386) and their associated timescales has been widely debated in literature with estimates of the latter ranging from a few kyrs to the order of the age of the Earth (Plag and Jüttner 1995; Hanyk et al. 1999). For the homogeneous sphere with compressible Maxwell rheology, Vermeersen and Mitrovica (2000) have found that unstable RT modes are associated with timescales at least one order of magnitude larger than the relaxation time of the stable M0 mode, with the fastest RT mode having a characteristic time of $\approx$ 10 kyr at $n = 2$ and $\approx$ 500 kyr at $n = 50$. In addition, for the h and k Love numbers, the viscoelastic residues (i.e., the $x^{\mathrm{v}}_{n,k}$'s in C.386) associated with RT modes are about one order of magnitude smaller than those of the stable M0 mode at low harmonic degrees, and two orders of magnitude smaller at $n = 50$. The situation is different for the ℓ Love number, with viscoelastic residues associated with RT modes being *larger* than those relative to the M0 mode, pointing to a significant instability of horizontal motions in a homogeneous compressible viscoelastic sphere.

C.7 The Layered Incompressible Sphere

In previous section we discussed the classical problem of the deformation of a homogeneous incompressible sphere, for which a fully analytical solution can be obtained. Such model, despite its ongoing relevance in global geodynamics, is too simple to accurately describe the response of the Earth to surface loads such those considered in the context of GIA. Hence, more realistic models of the Earth interior need to be introduced. Among these, the *layered sphere* has a particular relevance, since it provides a good trade-off between realism and computational complexity. In this model, the Earth rheological parameters (i.e., density, shear modulus and viscosity) are assumed to vary as a piece-wise constant function of radius; the model is therefore composed by a number of spherical shells for which solutions of the field equations are known (see Sect. C.2.2.2).

As opposed to the homogeneous sphere, for a layered sphere it is generally not possible to obtain the response in analytical form, except for some very simple two-layer models (Wu and Ni 1996). However, the solution for a complex layered model can still be written in a form that is easy to implement in a numerical code, by borrowing the *propagator method* developed in the context of free oscillations (Gilbert and Backus 1966). Several numerical codes for computing Love numbers of an incompressible layered sphere have developed within the GIA community, and there is generally a very good level of agreement between results independently obtained by different research groups (Spada et al. 2011a).

C.7.1 Impulse Response in the Laplace Domain

We consider a model including a uniform central spherical core with inviscid fluid rheology, surrounded by $N_{\rm L}$ spherical solid shells (or *layers*, see Fig. 2.1) with incompressible Maxwell viscoelastic rheology. The geometrical structure of the model is completely described by $N_{\rm L}+1$ radii $(r_0, r_1, \ldots, r_{N_{\rm L}})$, with r_0 being the radius of the core, r_i the outer radius of the i-th spherical layer, and $r_{N_{\rm L}} = a$ the Earth's radius. In the i-th layer we assume a density ρ_i and a rheology described by the complex shear modulus $\tilde{\mu}_i(s) = \mu_i s/(s + \mu_i/\eta_i)$, while the core is assumed to be fluid and inviscid, with uniform density ρ_0.

By (C.117), the Laplace-transformed solution vector in the i-th shell can be written as

$$\tilde{\mathbf{y}}_i(r) = \tilde{\mathbf{Y}}_i(r)\,\tilde{\mathbf{C}}_i\,, \tag{C.388}$$

where $\tilde{\mathbf{Y}}_i(r)$ is the fundamental matrix for the i-th layer, corresponding to a given harmonic degree n, obtained by using the rheological parameters $\rho = \rho_i$, $\tilde{\mu} = \tilde{\mu}_i$ in (C.120–C.121). From (C.388), it immediately follows that, if we consider two radii R_1 and R_2 entirely contained in the i-th layer (i.e., $r_{i-1} \le R_1 \le r_i$, $r_{i-1} \le R_2 \le r_i$), we have:

$$\tilde{\mathbf{y}}_i(R_2) = \left(\tilde{\mathbf{Y}}_i(R_2)\tilde{\mathbf{Y}}_i^{-1}(R_1)\right)\tilde{\mathbf{y}}_i(R_1) = \tilde{\mathbf{\Pi}}_i(R_1; R_2)\tilde{\mathbf{y}}_i(R_1)\,, \tag{C.389}$$

where we have defined the *propagator matrix*

$$\tilde{\mathbf{\Pi}}_i(R_1; R_2) = \tilde{\mathbf{Y}}_i(R_2)\tilde{\mathbf{Y}}_i^{-1}(R_1)\,. \tag{C.390}$$

Since *for an incompressible mantle* the inverse of the fundamental matrix $\tilde{\mathbf{Y}}^{-1}(r)$ has been determined in a closed form by Spada et al. (1990) (see also Spada et al. 1992b; Vermeersen and Sabadini 1997; Sabadini et al. 2016), the numerical evaluation of the propagator matrix is straightforward. In Sect. C.4 we have shown that at the core-mantle boundary $(r = r_0)$ the solution in the solid mantle takes the form (C.308):

$$\tilde{\mathbf{y}}_1(r_0) = \mathbf{J}(r_0)\tilde{\mathbf{c}}_{\rm cmb}\,, \tag{C.391}$$

where $\mathbf{J}(r)$ is the array defined in (C.309) and $\tilde{\mathbf{c}}_{\rm cmb}$ is a three-component vector to be determined through boundary conditions. By prescribing continuity of the solution across the internal interfaces, i.e. $\tilde{\mathbf{y}}_i(r_i) = \tilde{\mathbf{y}}_{i+1}(r_i)$ for $1 \le i \le N_\mathrm{L} - 1$, and iteratively applying (C.389) within each layer, we can propagate the solution from the solid side of the core-mantle boundary ($r = r_0^+$) up to the surface ($r = r_{N_\mathrm{L}} = a$) as[33]

$$\tilde{\mathbf{y}}_{N_\mathrm{L}}(a) = \tilde{\mathbf{\Pi}}_{N_\mathrm{L}}(a; r_{N_\mathrm{L}-1})\tilde{\mathbf{\Pi}}_{N_\mathrm{L}-1}(r_{N_\mathrm{L}-1}; r_{N_\mathrm{L}-2}) \cdots \\ \cdots \tilde{\mathbf{\Pi}}_2(r_2; r_1)\, \tilde{\mathbf{\Pi}}_1(r_1; r_0)\tilde{\mathbf{y}}_1(r_0)\,, \quad \text{(C.392)}$$

which, recalling (C.391), can be written as

$$\tilde{\mathbf{y}}_{N_\mathrm{L}}(a) = \tilde{\mathbf{\Lambda}}\mathbf{J}(r_0)\tilde{\mathbf{c}}_\mathrm{cmb}\,, \quad \text{(C.393)}$$

where we have defined the product

$$\tilde{\mathbf{\Lambda}} = \prod_{i=N_\mathrm{L}}^{1} \tilde{\mathbf{\Pi}}_i(r_i; r_{i-1})\,. \quad \text{(C.394)}$$

Equation (C.393) gives the solution at the surface in terms of the three unknown constants introduced to describe the transition of the solution from the fluid core into the viscoelastic mantle. These constants can be now determined by applying appropriate surface boundary conditions (BCs), which have been obtained in Sect. C.5. Surface BCs, both in case of surface loads and tidal potentials, can be expressed by prescribing appropriate values for the third, fourth and sixth components of the solution vector $\tilde{\mathbf{y}}(a)$. Hence, in a compact form they can be written as

$$\mathbf{P}_2\tilde{\mathbf{y}}(a) = \tilde{\mathbf{b}}\,, \quad \text{(C.395)}$$

where

$$\mathbf{P}_2 = \begin{pmatrix} 0\,0\,1\,0\,0\,0 \\ 0\,0\,0\,1\,0\,0 \\ 0\,0\,0\,0\,0\,1 \end{pmatrix} \quad \text{(C.396)}$$

is a projection operator whereas $\tilde{\mathbf{b}}$ is a three-component vector containing the surface BCs given by (C.337) for surface loads and (C.358) for an external potential. From (C.393) and (C.395), the unknown vector $\tilde{\mathbf{c}}_\mathrm{cmb}$ is therefore obtained as

$$\tilde{\mathbf{c}}_\mathrm{cmb} = \tilde{\mathbf{R}}^{-1}\tilde{\mathbf{b}}\,, \quad \text{(C.397)}$$

---

33 In the context of GIA, one or more layers in the outer part of the model are customarily assumed to be perfectly elastic, in order to model the lithosphere (see e.g., Spada et al. 2011a) or the crust (see e.g., Barletta et al. 2006). In that case, the structure of (C.392) is preserved, but the propagators corresponding to the elastic layers shall be built by assuming $\tilde{\mu} \equiv \mu$ in the fundamental matrix and in its inverse, where μ is the elastic shear modulus.

where we have defined

$$\tilde{\mathbf{R}}(s) = \mathbf{P}_2\,\tilde{\mathbf{\Lambda}}\,\mathbf{J}(r_0)\,, \tag{C.398}$$

and from (C.393) the solution at the surface takes the form

$$\tilde{\mathbf{y}}_{N_L}(a) = \tilde{\mathbf{\Lambda}}\mathbf{J}(r_0)\tilde{\mathbf{R}}^{-1}\tilde{\mathbf{b}}\,. \tag{C.399}$$

We now introduce a new projection operator selecting the first, second and fifth components of the solution vector, i.e.

$$\mathbf{P}_1 = \begin{pmatrix} 1\,0\,0\,0\,0\,0 \\ 0\,1\,0\,0\,0\,0 \\ 0\,0\,0\,0\,1\,0 \end{pmatrix}, \tag{C.400}$$

and define, in analogy with (C.398), the array

$$\tilde{\mathbf{Q}}(s) = \mathbf{P}_1\,\tilde{\mathbf{\Lambda}}\,\mathbf{J}(r_0)\,. \tag{C.401}$$

Then, the Laplace-transformed radial coefficients of the vertical and horizontal displacement, and of the perturbation of gravitational potential at the surface, corresponding to the 1st, 2nd and 5th elements of the solution vector in (C.399), can be expressed as[34]:

$$\tilde{\mathbf{x}}(s) = \tilde{\mathbf{Q}}\tilde{\mathbf{R}}^{-1}\tilde{\mathbf{b}}\,, \tag{C.402}$$

where

$$\tilde{\mathbf{x}}(s) = (\tilde{y}_1(a), \tilde{y}_2(a), \tilde{y}_5(a))^t\,. \tag{C.403}$$

Equation (C.402) provides a straightforward way to numerically evaluate the Laplace-transformed solution for the problem of the impulsive response of a given spherically symmetric planetary model to a mass or to tidal loading. We denote with $\mathbf{x}_\delta(t)$ the time-domain solution, which corresponds to the inverse Laplace transform of (C.402), i.e.

$$\mathbf{x}_\delta(t) = LT^{-1}\{\tilde{\mathbf{x}}(s)\}\,. \tag{C.404}$$

For a forcing with generic time-history $f(t)$, the response is given by a time convolution of $f(t)$ with the impulse response, i.e.,

$$\mathbf{x}_f(t) = f(t) * \mathbf{x}_\delta(t) = LT^{-1}\left\{\tilde{f}(s)\tilde{\mathbf{x}}(s)\right\}, \tag{C.405}$$

where $\tilde{f}(s)$ is the Laplace transform of $f(t)$ and we have used the identity $LT\{g(t) * h(t)\} = \tilde{g}(s)\tilde{h}(s)$. In the context of GIA, it is of particular relevance

[34] We note that (C.402) would be valid also for a compressible model, for which the fundamental matrix $\tilde{\mathbf{Y}}$ has been obtained in Sect. C.3.2.4. However, an analytical expression for its inverse is not available, so $\tilde{\mathbf{Y}}^{-1}$ would need to be determined through numerical methods.

the case of a stepwise time history, in which the external forcing is "switched on" at time $t = 0$ and kept constant. In that case, $f(t)$ coincides with the Heaviside step function $H(t)$ and, from Table B.1, we have

$$\mathbf{x}_H(t) = LT^{-1}\left\{\frac{1}{s}\tilde{\mathbf{x}}(s)\right\}. \tag{C.406}$$

By applying the initial and final value theorems (B.76) and (B.75) to (C.406), we obtain the important relations

$$\mathbf{x}^{\mathrm{e}} = \lim_{t\to 0} \mathbf{x}_H(t) = \lim_{s\to\infty} \tilde{\mathbf{x}}(s) \tag{C.407}$$

and

$$\mathbf{x}^{\mathrm{f}} = \lim_{t\to\infty} \mathbf{x}_H(t) = \lim_{s\to 0} \tilde{\mathbf{x}}(s)\,, \tag{C.408}$$

where $\mathbf{x}^{\mathrm{e}}$ and $\mathbf{x}^{\mathrm{f}}$ are the *elastic limit* and *fluid limit*, respectively. While $\mathbf{x}_e$ and $\mathbf{x}_f$ can in principle be computed respectively by solving the purely elastic problem and by using the solutions for fluid layers discussed in Sect. C.4, recovering the full time-dependence of the solution demands a numerical evaluation of the inverse Laplace transform in (C.405). This task can be accomplished by following different strategies, which are briefly outlined below.

C.7.2 Viscoelastic Normal Modes

The viscoelastic normal mode method, first introduced by Peltier (1974) and later refined by Wu and Peltier (1982), is based upon a semi-analytical Laplace inversion of (C.402), based on the Cauchy residue theorem. The main advantage of a normal mode analysis is that the time-domain response is written as a sum of exponential terms whose amplitudes and decay times have a direct physical meaning, providing a deep insight into the relaxation processes active in the system. In addition, the multi-exponential form allows to compute analytically the time convolution of $\mathbf{x}_\delta(t)$ with some load time-histories of relevance for GIA studies, providing a straightforward way to evaluate the response to load models of geophysical interest.

We start by rewriting the Laplace-transformed solution (C.402) in the form

$$\tilde{\mathbf{x}}(s) = \frac{\tilde{\mathbf{Q}}\tilde{\mathbf{R}}^{\dagger}\tilde{\mathbf{b}}}{N(s)}\,, \tag{C.409}$$

where $N(s) = \det(\tilde{\mathbf{R}})$, $\mathbf{A}^{\dagger}$ denotes the adjugate of matrix $\mathbf{A}$ (i.e., the transpose of its cofactor matrix) and we have used the relation $\mathbf{A}\mathbf{A}^{\dagger} = \det(\mathbf{A})\mathbf{I}$. Equation (C.409) is singular for the values of s which are solution of the *secular equation*

$$N(s) \equiv \det\left(\tilde{\mathbf{R}}(s)\right) = 0\,, \tag{C.410}$$

and, as discussed in Sect. C.6 above, we recall that the roots of (C.410) correspond to the values of s for which the homogeneous system obtained by prescribing null boundary conditions at the surface, i.e.,

$$\mathbf{P}_2\tilde{\mathbf{y}}(a) = 0\,, \tag{C.411}$$

allows non-trivial solutions. Hence, in close analogy with the problem of free oscillations, the values of s for which $N(s) = 0$ correspond to the *natural relaxation modes* of the system in absence of external forcing. It is now convenient to rewrite (C.409) in the form

$$\tilde{\mathbf{x}}(s) = \mathbf{x}^{\mathrm{e}} + \frac{\tilde{\mathbf{Q}}\tilde{\mathbf{R}}^{\dagger}\tilde{\mathbf{b}} - N(s)\mathbf{x}^{\mathrm{e}}}{N(s)}\,, \tag{C.412}$$

where $\mathbf{x}^{\mathrm{e}}$ is the elastic limit (C.407) and, by construction, the second term in (C.412) vanishes for $s \to \infty$, Hence, if the dependence of $\tilde{\mathbf{Q}}$ and $\tilde{\mathbf{R}}$ from s is polynomial, it follows that the second term in (C.412) is a *strictly proper rational function*, and in that case $N(s)$ is dubbed *secular polynomial*. Therefore, from the expressions listed in Table B.1, it is straightforward to write the inverse Laplace transform of (C.412) as

$$\mathbf{x}_{\delta}(t) = \mathbf{x}^{\mathrm{e}}\delta(t) + \sum_{k=1}^{M} \mathbf{r}_k \mathrm{e}^{s_k t}\,, \tag{C.413}$$

where $s_1, \ldots, s_M$ are the M roots of the secular Eq. C.410 and the constants

$$\mathbf{r}_k = \frac{\tilde{\mathbf{Q}}(s_k)\tilde{\mathbf{R}}^{\dagger}(s_k)\tilde{\mathbf{b}}}{N'(s_k)} \tag{C.414}$$

are the associated *viscoelastic residues*. We note that, by virtue of (C.364), also the viscoelastic Love numbers $h(t)$, $\ell(t)$ and $k(t)$ for an impulsive forcing would have a functional form similar to (C.413). From (C.412) it is evident that *negative roots* of the secular polynomial correspond to stable modes with a *decay time* $\tau_k = -1/s_k$, while *positive roots* would correspond to unstable modes associated with an exponentially growing response. As discussed in Sect. C.6 above, positive roots of $N(s)$ are related to gravitational instabilities that may arise in compressible rheologies, while for incompressible models with a stable density layering (i.e., in which density montonically increases with depth) it can be shown that all modes are stable (Wolf 1994; Plag and Jüttner 1995).

For a layered sphere with Maxwell rheology, it is possible to leverage the simple polynomial structure of matrix quantities appearing in (C.409) to write the secular polynomial $N(s)$ and the viscoelastic residuals $\mathbf{r}_k$ in a form amenable for numerical computations. By inspection of (C.120–C.121), it is easily realised that the fundamental matrix for a viscoelastic shell can be written as

$$\tilde{\mathbf{Y}} = \mathbf{M}_1 + \tilde{\mu}\, \mathbf{M}_2\,, \tag{C.415}$$

where $\tilde{\mu}$ is the complex modulus and $\mathbf{M}_1$ and $\mathbf{M}_2$ depend only upon density and radius. In a similar fashion, the inverse of the fundamental matrix, obtained by (Spada et al. 1992b) by algebraic manipulation, has the form

$$\tilde{\mathbf{Y}}^{-1} = \mathbf{N}_1 + \tilde{\mu}^{-1}\, \mathbf{N}_2\,, \tag{C.416}$$

where arrays $\mathbf{N}_1$ and $\mathbf{N}_2$ depend upon density and radius. By substituting into (C.415–C.416) the functional form (C.8) of $\tilde{\mu}(s)$ for a Maxwell body, it can be easily shown that

$$\tilde{\mathbf{Y}} = \mathbf{Y}_{\mathrm{E}} + (s + \mu/\eta)^{-1}\, \mathbf{Y}_{\mathrm{V}}\,, \tag{C.417}$$

where $\mathbf{Y}_{\mathrm{E}} = \mathbf{M}_1 + \mu\, \mathbf{M}_2$ is the elastic fundamental matrix and $\mathbf{Y}_{\mathrm{V}} = -(\mu^2/\eta)\, \mathbf{M}_2$. Similarly,

$$\tilde{\mathbf{Y}}^{-1} = \mathbf{Y}_{\mathrm{E}}^{-1} + s^{-1}\mathbf{Y}_{\mathrm{V}}^{-1}\,, \tag{C.418}$$

with $\mathbf{Y}_{\mathrm{E}}^{-1} = \mathbf{N}_1 + \mu^{-1}\, \mathbf{N}_2$ and $\mathbf{Y}_{\mathrm{V}}^{-1} = \eta^{-1}\mathbf{N}_2$. By using (C.417) and (C.418) into (C.390), after some simple algebra the propagator matrix can be arranged as

$$\tilde{\mathbf{\Pi}}(R_2;\, R_1) = \frac{\mathbf{A}_0 + \mathbf{A}_1 s + \mathbf{A}_2 s^2}{s(s + \mu/\eta)}\,, \tag{C.419}$$

where

$$\mathbf{A}_0 = \mathbf{Y}_{\mathrm{V}}(R_2)\mathbf{Y}_{V}^{-1}(R_1) + (\mu/\eta)\mathbf{Y}_{\mathrm{E}}(R_2)\mathbf{Y}_{V}^{-1}(R_1) \tag{C.420}$$
$$\mathbf{A}_1 = (\mu/\eta)\mathbf{Y}_{\mathrm{E}}(R_2)\mathbf{Y}_{E}^{-1}(R_1) + \mathbf{Y}_{\mathrm{E}}(R_2)\mathbf{Y}_{V}^{-1}(R_1) + \mathbf{Y}_{\mathrm{V}}(R_2)\mathbf{Y}_{E}^{-1}(R_1) \tag{C.421}$$
$$\mathbf{A}_2 = \mathbf{Y}_{\mathrm{E}}(R_2)\mathbf{Y}_{E}^{-1}(R_1) \tag{C.422}$$

are constant matrix coefficients.

We now consider a layered model in which the N_{E} outer layers correspond to the lithosphere and are perfectly elastic, while the inner N_{V} layers have a Maxwell viscoelastic rheology, with $N_{\mathrm{E}} + N_{\mathrm{V}} = N_{\mathrm{L}}$ being the total number of layers. In this case, Eq. (C.393) becomes

$$\tilde{\mathbf{\Lambda}} = \mathbf{\Lambda}_{\mathrm{E}} \left[\prod_{i=N_{\mathrm{V}}}^{1} \tilde{\mathbf{\Pi}}_i(r_i;\, r_{i-1})\,. \right] \tag{C.423}$$

where

$$\mathbf{\Lambda}_{\mathrm{E}} = \prod_{i=N_{\mathrm{L}}}^{N_{\mathrm{V}}+1} \mathbf{\Pi}_i(r_i;\, r_{i-1}) \tag{C.424}$$

is the product of elastic propagators. By substituting (C.419) in (C.423), we obtain

$$\tilde{\mathbf{\Lambda}} = s^{-N_V}\left[\prod_{i=1}^{N_V}(s+\mu_i/\eta_i)^{-1}\right]\left[\sum_{k=0}^{2N_V}\mathbf{B}_k s^k\right], \tag{C.425}$$

where the matrix polynomial in the second set of square brackets is defined by

$$\sum_{k=0}^{2N_V}\mathbf{B}_k s^k = \prod_{i=N_V}^{1}\left(\mathbf{\Lambda}_E\mathbf{A}_0^{(i)} + \mathbf{\Lambda}_E\mathbf{A}_1^{(i)}s + \mathbf{\Lambda}_E\mathbf{A}_2^{(i)}s^2\right) \tag{C.426}$$

and the $\mathbf{A}_k^{(i)}$'s are the quantities defined in (C.420–C.421) for the i-th layer. The $\tilde{\mathbf{R}}$ and $\tilde{\mathbf{Q}}$ matrices defined respectively in (C.398) and (C.401) can therefore be written in the form

$$\tilde{\mathbf{R}} = s^{-N_V}\left[\prod_{i=1}^{N_V}(s+\mu_i/\eta_i)^{-1}\right]\left[\sum_{k=0}^{2N_V}\mathbf{R}_k s^k\right] \tag{C.427}$$

and

$$\tilde{\mathbf{Q}} = s^{-N_V}\left[\prod_{i=1}^{N_V}(s+\mu_i/\eta_i)^{-1}\right]\left[\sum_{k=0}^{2N_V}\mathbf{Q}_k s^k\right], \tag{C.428}$$

with $\mathbf{R}_k = \mathbf{P}_2\mathbf{B}_k\mathbf{J}(r_0)$ and $\mathbf{Q}_k = \mathbf{P}_1\mathbf{B}_k\mathbf{J}(r_0)$. By considering that the scalar terms in (C.427–C.428) cancel out in the product $\tilde{\mathbf{Q}}\tilde{\mathbf{R}}^{-1}$, the Laplace-domain solution (C.409) takes the matrix polynomial form

$$\tilde{\mathbf{x}}(s) = \frac{\left[\sum_{k=0}^{2N_V}\mathbf{Q}_k s^k\right]\left[\sum_{k=0}^{2N_V}\mathbf{R}_k s^k\right]^{\dagger}}{N(s)}\,\mathbf{b}, \tag{C.429}$$

where

$$N(s) = \det\left(\sum_{k=0}^{2N_V}\mathbf{R}_k s^k\right) \tag{C.430}$$

is a *secular polynomial* which, by virtue of the properties of the Levi-Civita symbol ϵ_{ijk}, can be written as

$$N(s) = \sum_{i,j,k=1}^{3}\epsilon_{ijk}\sum_{\alpha,\beta,\gamma=1}^{2N_V} R_{1i}^{(\alpha)}R_{2j}^{(\beta)}R_{3k}^{(\gamma)}s^{\alpha+\beta+\gamma}, \tag{C.431}$$

where $R_{ij}^{(k)}$ are the elements of the $\mathbf{R}_k$ matrix that appears in (C.427). Recalling that, if a is a scalar and $\mathbf{A}$ a $n\times n$ matrix, we have $\det(a\mathbf{A}) = a^n\det(\mathbf{A})$ and $(a\mathbf{A})^{\dagger} = a^{n-1}\mathbf{A}^{\dagger}$, from (C.429) the elastic limit (C.407) can be written as

$$\mathbf{x}_e = \frac{\mathbf{Q}_{2N_V} \mathbf{R}^{\dagger}_{2N_V} \mathbf{b}}{\det\left(\mathbf{R}_{2N_V}\right)}, \tag{C.432}$$

while the viscoelastic residues (C.414) take the form

$$\mathbf{r}_i = \frac{\left[\sum_{k=0}^{2N_V} \mathbf{Q}_k s_i^k\right] \left[\sum_{k=0}^{2N_V} \mathbf{R}_k s_i^k\right]^{\dagger} \mathbf{b}}{N'(s_i)}. \tag{C.433}$$

From (C.431), it is evident that the degree of the secular polynomial is $6N_V$. However, numerical calculations by Spada et al. (1990, 1992b) have shown that the coefficients of the terms from s^0 to s^{2N_V} are negligible in front of the higher-order coefficients; hence, it is customarily assumed that $N(s)$ has $M = 4N_V$ non-zero roots s_k. Nevertheless, the degree of the secular polynomial can become quite large even for Earth models with a moderate number of layers, hindering the accuracy of root-finding algorithms and the numerical stability of a viscoelastic normal-mode analysis. In addition, the inclusion of more realistic rheological laws such as those accounting for transient creep (see Sect. 1.3.8), for which the functional form of $\tilde{\mu}(s)$ is more complex than for a Maxwell solid, introduces additional difficulties for a semi-analytical approach such as the one outlined above, and keeping track of the polynomial structure of the matrix products may become remarkably difficult. This motivates invoking alternative approaches to the numerical Laplace inversion, such as those described in the following sections.

C.7.3 Collocation Methods

The normal mode analysis outlined in the previous section has the remarkable advantage of providing the time-domain solution in terms of the relaxation times of the system and their associated amplitudes, which have a direct physical meaning. However, for finely-layered models, solving the secular equation $N(s) = 0$ becomes increasingly difficult and prone to numerical instabilities, since the degree of polynomial $N(s)$ scales with the number of layers. In addition, for more realistic rheologies such those accounting for transient creep, the functional form of $\tilde{\mu}(s)$ becomes more complex than that for a Maxwell solid, and as a result the analytical procedure outlined in the previous section may become unpractical.

The *collocation method*, first introduced by Schapery (1962a, b) in the context of viscoelastic stress analysis, provides a computationally simple way to obtain an approximate numerical inverse Laplace transform of (C.402) while retaining the multi-exponential form of (C.413). In the context of GIA, the collocation method has been used in the seminal works by Peltier (1974) and Wu and Peltier (1982), and its performance has been broadly discussed in Mitrovica and Peltier (1992).

We start by separating the elastic and viscous components of (C.413) as

$$\mathbf{x}_\delta(t) = \mathbf{x}^{\rm e}\delta(t) + \mathbf{x}^{\rm v}(t)\,, \tag{C.434}$$

and, for each component of the viscous term $\mathbf{x}^{\rm v}(t)$, we seek a multi-exponential approximation in the form

$$x^{\rm v*}(t) = \sum_{i=1}^{n} a_i \mathrm{e}^{-t/\alpha_i}\,, \tag{C.435}$$

where the number n of terms and the α_i constants are fixed a priori (see Mitrovica and Peltier 1992, for a discussion) while a_i are weights that shall be determined to minimise the misfit between $x^{\rm v*}(t)$ and the exact viscous component $x^{\rm v}(t)$. We remark that, in (C.435), the asterisk does not stand for complex conjugation but it indicates the approximated Laplace inverse. If we now define a quadratic error between $x^{\rm v*}$ and $x^{\rm v}(t)$ as

$$E = \int_0^\infty \left[x^{\rm v}(t) - x^{\rm v*}(t)\right]^2 \mathrm{d}t\,, \tag{C.436}$$

the values of a_i that minimise E can be found by solving the equation $\partial E/\partial a_i = 0$. We have

$$\frac{\partial E}{\partial a_i} = 2\int_0^\infty \left[x^{\rm v}(t) - x^{\rm v*}(t)\right]\mathrm{e}^{-t/\alpha_i}\mathrm{d}t \;=\; 2\left[\tilde{x}^{\rm v}(1/\alpha_i) - \tilde{x}^{\rm v*}(1/\alpha_i)\right]\,, \tag{C.437}$$

hence, the values of the weights a_i that minimise the quadratic error between $x^{\rm v}(t)$ and $x^{\rm v*}(t)$ are those for which the Laplace trasforms $\tilde{x}^{\rm v}(s)$ and $\tilde{x}^{\rm v*}(s)$ are coincident at the n points $s_1 = 1/\alpha_1, \ldots, s_n = 1/\alpha_n$. Using Table B.1, it is straightforward to write the Laplace transform of (C.435) as

$$\tilde{x}^{\rm v*}(s) = \sum_{i=1}^{n} \frac{a_i}{s + 1/\alpha_i}\,, \tag{C.438}$$

therefore, the unknown values $a_1, \ldots, a_n$ are determined by the solution of the system of n equations

$$\tilde{x}^{\rm v}(1/\alpha_j) = \sum_{i=1}^{n} \frac{a_i}{1/\alpha_i + 1/\alpha_j}\,, \qquad 1 \le j \le n\,, \tag{C.439}$$

which can be written in compact form as

$$\mathbf{v} = \chi\mathbf{a}\,, \tag{C.440}$$

where $\mathbf{a} = (a_1, \dots, a_n)^t$, $\mathbf{v} = \left[\tilde{x}^{\mathrm{v}}(1/\alpha_1), \dots, \tilde{x}^{\mathrm{v}}(1/\alpha_n)\right]^t$, and the elements of the square matrix χ are

$$\chi_{ij} = \frac{1}{1/\alpha_i + 1/\alpha_j}. \tag{C.441}$$

Therefore, once the constants α_i have been defined, the values of the Laplace transformed viscous component of the solution $\tilde{\mathbf{x}}^{\mathrm{v}}(1/\alpha_i) = \tilde{\mathbf{x}}(1/s_i) - \mathbf{x}^{\mathrm{e}}$ can be computed either by analytical or numerical methods and, for each component of the solution vector $\mathbf{x}$, the corresponding weights a_i can be obtained by solving (C.440) by standard linear algebra methods. The choice of the α_i constants in (C.435) is an important aspect of the collocation method. One possibility is to find the roots $s_1, \dots, s_M$ of the secular polynomial (C.430) and set $\alpha_i = -1/s_i$. This method, dubbed *hybrid collocation*, yields a very close approximation to the normal-mode analysis, in which the relaxation times of the system are correctly identified but the associated residuals are fitted numerically. However, keeping the physical relaxation times in (C.435) comes at the cost of not circumventing the numerical solution of the secular equation $N(s) = 0$. Another possibility is to choose a set of α_i equally spaced in logarithm across an interval of decay times encompassing the range of greatest variation in the response spectra. This method, called *pure collocation*, has been discussed in detail by Mitrovica and Peltier (1992) who suggest a spacing of three samples per decade. A trade-off between hybrid and pure collocation has been recently suggested by Brierley-Green and James (2024). According to their method, the residual viscoelastic signal that is not captured by a hybrid collocation analysis is approximated by pure collocation. In this way, only relaxation modes that are not identified by the root-finding algorithm used for the solution of the secular equation will be fitted on an equally-spaced set of inverse decay times.

In all its variants, the collocation method brings the remarkable advantage of casting the time-domain response to an impulsive load in the multi-exponential form (C.413) even if the Laplace-transformed solution can be evaluated only numerically. That would be the case, for instance, if transient rheological laws are introduced, for which $\tilde{\mu}(s)$ has a complex functional form, or if a compressible rheology is assumed, in which case the equations of motion shall be solved through numerical integration. In those cases, the availability of the impulse response in an (approximated) multi-exponential form would allow, for instance, to obtain the time-domain response to various types of forcing by analytical convolution of (C.435) with the time history of the forcing term, along the same lines of the viscoelastic normal mode method.

C.7.4 Post-Widder Laplace Inversion Formula

The Post-Widder formula (Post 1930; Widder 1934) gives an asymptotic, non-integral expression of the inverse Laplace transform of a function $f(t)$ in terms of $\tilde{f}(s)$ and its derivatives. In its original form, it reads

$$f(t) = \lim_{n\to\infty} \frac{(-1)^n}{n!} \left(\frac{n}{t}\right)^{n+1} \tilde{f}^{(n)}\left(\frac{t}{n}\right), \tag{C.442}$$

where $\tilde{f}^{(n)}(s) = d^n \tilde{f}/ds^n$ is the n-th derivative of $\tilde{f}(s)$. In general, the Post-Widder formula (C.442) is not helpful for practical applications, $\tilde{f}^{(n)}$ is usually not known in closed form and its numerical evaluation would rapidly lead to large roundoff errors. By means of a recursive discrete approximation of $\tilde{f}^{(n)}(s)$, Gaver (1966) has obtained an expression equivalent to (C.442) which reads

$$f(t) = \lim_{n\to\infty} \frac{n\log 2}{t} \binom{2n}{n} \sum_{k=0}^{n} (-1)^k \binom{n}{k} \tilde{f}\left[\frac{(n+k)\log 2}{t}\right]. \tag{C.443}$$

Equation (C.443) is more suitable for a numerical implementation since it expresses the inverse Laplace transform in terms of samples of $\tilde{f}(s)$ without invoking its derivatives. However, as discussed by Valkó and Abate (2004), the convergence of the (C.443) is logarithmically slow, and the factor $(-1)^k$ results in an alternating sum that, for large values of n, leads to *catastrophic cancellation* of significant digits, requiring the use of high-precision arithmetics. Stehfest (1970) has shown that, for numerical applications, the convergence of (C.443) can be accelerated by rewriting it in the equivalent form

$$f(t) = \lim_{M\to\infty} \frac{\log 2}{t} \sum_{k=1}^{2M} \zeta_{k,M} \tilde{f}\left(\frac{k\log 2}{t}\right), \tag{C.444}$$

where the constants $\zeta_{k,M}$ are defined as

$$\zeta_{k,M} = (-1)^{M+k} \sum_{j=\lfloor\frac{k+1}{2}\rfloor}^{\min(M,k)} \frac{j^{M+1}}{M!} \binom{M}{j} \binom{2j}{j} \binom{j}{k-j}, \tag{C.445}$$

and where $\lfloor x \rfloor$ is the largest integer not exceeding x ($\lfloor x \rfloor$ is the *floor function*).

Equation (C.444) provides a computationally suitable recipe for evaluating the inverse Laplace transform when $\tilde{f}(s)$ is only known numerically. It has been first applied to the problem of viscoelastic Love numbers in the context of GIA by Spada and Boschi (2006) and Spada (2008), and to post-seismic deformation modelling by Melini et al. (2008). If $F(t)$ is the time-history of the forcing term, by (C.405) and (C.444) the time-domain solution can be approximated as

$$\mathbf{x}_F(t) \approx \frac{\log 2}{t} \sum_{k=1}^{2M} \zeta_{k,M} \tilde{F}\left(\frac{k\log 2}{t}\right) \tilde{\mathbf{x}}\left(\frac{k\log 2}{t}\right), \tag{C.446}$$

and, for the case of a Heaviside time-history $F(t) \equiv H(t)$, for which $\tilde{F}(s) = 1/s$ (see Table B.1), it simplifies to

$$\mathbf{x}_H(t) \approx \sum_{k=1}^{2M} \frac{\zeta_{k,M}}{k} \tilde{\mathbf{x}} \left(\frac{k \log 2}{t} \right) . \tag{C.447}$$

As discussed by Spada and Boschi (2006), using $M = 8$ in (C.446) provides an excellent numerical approximation of the Laplace anti-transform. The Post-Widder Laplace inversion scheme, which has been implemented in some numerical codes (Spada 2008; Melini et al. 2022), provides an independent means of evaluating the time-domain response of a given model to arbitrary perturbations, and can be used to cross-validate other numerical approaches such as those discussed in the previous sections.

Appendix D
Rotational Transfer Functions

D.1 Centrifugal Potential Variations

In this section we the study the variations of the centrifugal potential $\Lambda(\gamma, t)$ caused by displacements of the Earth's rotation axis. We anticipate that these can be expressed as a combination of harmonic functions of degrees $n = 0$ and $n = 2$, with

$$\Lambda(\gamma, t) = \Lambda_{00} Y_{00} + \sum_{m=-2}^{2} \Lambda_{2m} Y_{2m} , \tag{D.1}$$

where the Y_{nm}'s are the 4π-normalised complex spherical harmonics and coefficients $\Lambda_{nm}(t)$ are determined in the following.

D.1.1 Rotational States

Following Milne and Mitrovica (1998), here we consider two distinct states of rotation. In the reference (zero) state, the Earth has been rotating in equilibrium until time $t_0 = 0$, with its angular velocity vector constantly aligned with the z axis,

$$\omega_i^0 = \Omega\, \delta_{i3} , \tag{D.2}$$

where Ω is a constant. Subsequently, for $t > t_0 = 0$, a new state is attained in response to the forcing imposed by some geophysical process. In this state, the rotation axis is displaced relative to the z axis, so that, according to (4.23),

$$\omega_i(t) = \Omega(m_i + \delta_{i3}) , \tag{D.3}$$

G. Spada and D. Melini, *Glacial Isostatic Adjustment*, Springer Geophysics,
https://doi.org/10.1007/978-3-031-90538-4

where the m_i's ($i = 1, 2, 3$) are the small dimensionless quantities ($m_i \ll 1$) defined in Sect. 4.2.2. We note that, in the new state, the squared modulus of the angular velocity vector is

$$\omega^2(t) \equiv \omega_1^2 + \omega_2^2 + \omega_3^2 = \Omega^2\left(1 + 2m_3 + m^2\right), \tag{D.4}$$

where $m^2(t) = m_1^2 + m_2^2 + m_3^2$. Furthermore, denoting by θ_p and λ_p the co-latitude and longitude of the instantaneous pole of rotation P, i.e., the point where the rotation axis intersects the Earth's surface (see Fig. 4.3), we have the relations:

$$\omega_1(t) = \omega \sin\theta_p \cos\lambda_p = \Omega m_1 \tag{D.5}$$

$$\omega_2(t) = \omega \sin\theta_p \sin\lambda_p = \Omega m_2 \tag{D.6}$$

$$\omega_3(t) = \omega \cos\theta_p = \Omega(1 + m_3)\,. \tag{D.7}$$

D.1.2 Centrifugal Potential

In the general rotational state, the centrifugal potential experienced by an observer A at the Earth's surface is

$$W^c(A) = \frac{1}{2}\omega^2 d^2, \tag{D.8}$$

where d is the distance between point A and the instantaneous rotation axis $\vec{\omega}$ (see Fig. 4.3). Since $d = a\sin\Theta$, where Θ is the angular separation between A and P, and a is the Earth's radius, we have

$$W^c(A) = \frac{1}{2}\omega^2 a^2\left(1 - \cos^2\Theta\right), \tag{D.9}$$

which using Table A.1 can be also expressed in terms of Legendre polynomials:

$$W^c(A) = \frac{\omega^2 a^2}{3}\left(P_0(\cos\Theta) - P_2(\cos\Theta)\right). \tag{D.10}$$

In the reference state

$$W^{c,0}(A) = \frac{\Omega^2 a^2}{3}\left(P_0(\cos\theta) - P_2(\cos\theta)\right), \tag{D.11}$$

where θ is the colatitude of A. Hence, the centrifugal potential *variation* at the same point is[35]

$$\Lambda(\gamma, t) \equiv W^c(A) - W^{c,0}(A), \tag{D.12}$$

[35] Note that in Sect. 4.2.4 of the main text we use symbol $\Delta W^c(\gamma, t)$ to denote the rotational potential variation, *in lieu* of $\Lambda(\gamma, t)$.

which can be cast in the form

$$\Lambda(\gamma, t) = \Lambda' + \Lambda'' , \tag{D.13}$$

where we have defined

$$\Lambda'(\gamma, t) \equiv \frac{a^2}{3}\left(\omega^2 - \Omega^2\right) \tag{D.14}$$

and

$$\Lambda''(\gamma, t) \equiv -\frac{a^2}{3}\left(\omega^2 P_2(\cos\Theta) - \Omega^2 P_2(\cos\theta)\right) , \tag{D.15}$$

with $\gamma = (\theta, \lambda)$ denoting the geographical coordinates of A.

D.1.3 Harmonic Degree $n = 0$

Since $\Lambda'(\gamma, t)$ in Eq. (D.14) does not actually depend on γ, $\Lambda'(t)$ is a harmonic function of degree $n = 0$, with:

$$\Lambda'(t) = \Lambda_{00} Y_{00} , \tag{D.16}$$

with $Y_{00}(\gamma) \equiv 1$ and

$$\Lambda_{00}(t) = \frac{a^2\Omega^2}{3}\left(m^2 + 2m_3\right) , \tag{D.17}$$

where we have used (D.4).

D.1.4 Harmonic Degree $n = 2$

The expansion of term $\Lambda''(\gamma, t)$ (D.15) requires some more work. First, applying the addition theorem (A.28) to the triangle with sides Θ, θ and θ_p in Fig. 4.3, we get:

$$P_2(\cos\Theta) = \frac{1}{5}\sum_{m=-2}^{2} Y^*_{2m}(\gamma_p) Y_{2m}(\gamma) . \tag{D.18}$$

This makes it possible to write $\Lambda''(\gamma, t)$ just in terms of γ and γ_p, which represent the geographical coordinates of A and P, respectively (see Fig. 4.3). Second, with the aid of Tables A.1 and A.2, we obtain the degree $n = 2$ CSHs at γ_p:

$$Y_{20}(\gamma_p) \equiv \sqrt{5}\, P_{20}(\cos\theta_p) = \frac{\sqrt{5}}{2}\left(3\cos^2\theta_p - 1\right), \tag{D.19}$$

$$Y_{21}(\gamma_p) \equiv \sqrt{\frac{5}{6}}\, P_{21}(\cos\theta_p)\, \mathrm{e}^{\mathrm{i}\lambda_p} = -3\sqrt{\frac{5}{6}} \sin\theta_p \cos\theta_p\, \mathrm{e}^{\mathrm{i}\lambda_p}, \tag{D.20}$$

$$Y_{22}(\gamma_p) \equiv \sqrt{\frac{5}{24}}\, P_{22}(\cos\theta_p)\, \mathrm{e}^{2\mathrm{i}\lambda_p} = 3\sqrt{\frac{5}{24}} \sin^2\theta_p\, \mathrm{e}^{2\mathrm{i}\lambda_p}. \tag{D.21}$$

Substitution of (D.18) into (D.15), also taking (D.19–D.21) into account, provides

$$\Lambda''(\gamma, t) = -\frac{a^2}{3}\left(\frac{\omega^2}{5}\sum_{m=-2}^{2} Y_{2m}^*(\gamma_p) Y_{2m}(\gamma) - \frac{\Omega^2}{\sqrt{5}} Y_{20}(\gamma)\right), \tag{D.22}$$

which can be decomposed as:

$$\begin{aligned}\Lambda''(\gamma, t) = &-\frac{a^2}{3}\left(\frac{\omega^2}{5} Y_{20}(\gamma_p) - \frac{\Omega^2}{\sqrt{5}}\right) Y_{20}(\gamma) + \\ &-\frac{a^2\omega^2}{15} Y_{21}^*(\gamma_p) Y_{21}(\gamma) - \frac{a^2\omega^2}{15} Y_{22}^*(\gamma_p) Y_{22}(\gamma) + [...],\end{aligned} \tag{D.23}$$

where [...] denotes terms in $Y_{2-1}(\gamma)$ and $Y_{2-2}(\gamma)$ that need not to be made explicit. Equation (D.23) has the form of a CSH expansion at degree $n = 2$, i.e.,

$$\Lambda''(\gamma, t) = \sum_{m=-2}^{2} \Lambda_{2m} Y_{2m}, \tag{D.24}$$

where the coefficients $\Lambda_{2m}(t)$ are made explicit below, and expressed in terms of the (time-dependent) direction cosines m_i. Since $\Lambda''(\gamma, t)$ is real-valued, we just need to consider the three terms Λ_{20}, Λ_{21} and Λ_{22}, being $\Lambda_{2-m} = (-1)^m \Lambda_{2m}^*$ (see A.37).

D.1.5 Term of Degree n = 2 and Order m = 0

The zonal coefficient Λ_{20} appears in front of $Y_{20}(\gamma)$ in (D.23). Using Eqs. (D.19), (D.7) and (D.4) and some algebra gives

$$\begin{aligned}\Lambda_{20}(t) &= -\frac{a^2}{3}\left(\frac{\omega^2}{5} Y_{20}(\gamma_p) - \frac{\Omega^2}{\sqrt{5}}\right) \\ &= -\frac{a^2}{6\sqrt{5}}\left(3\omega^2\cos^2\theta_p - \omega^2 - 2\Omega^2\right) \\ &= \frac{a^2\Omega^2}{6\sqrt{5}}(m_1^2 + m_2^2 - 2m_3^2 - 4m_3).\end{aligned} \tag{D.25}$$

D.1.6 Term of Degree $n = 2$ and Order $m = 1$

The tesseral coefficient Λ_{21} multiplies $Y_{21}(\gamma)$ in Eq. (D.23). Thus, using (D.20), we have

$$\begin{aligned}\Lambda_{21}(t) &= -\frac{a^2\omega^2}{15}Y_{21}^*(\gamma_p)\\ &= -\frac{a^2\omega^2}{15}\left(-3\sqrt{\frac{5}{6}}\right)\sin\theta_p\cos\theta_p\left(\cos\lambda_p - \mathrm{i}\sin\lambda_p\right)\\ &= \frac{a^2\Omega^2}{\sqrt{30}}\left(m_1(1+m_3) - \mathrm{i}\, m_2(1+m_3)\right), \end{aligned} \tag{D.26}$$

where from Eqs. (D.5) and (D.6) we have employed the useful identities $\omega_1\omega_3 = \Omega^2 m_1(1+m_3) = \omega^2\sin\theta_p\cos\theta_p\cos\lambda_p$ and $\omega_2\omega_3 = \Omega^2 m_2(1+m_3) = \omega^2\sin\theta_p \cos\theta_p\sin\lambda_p$.

D.1.7 Term of Degree $n = 2$ and Order $m = 2$

Last, the sectoral coefficient Λ_{22} is the term in front of $Y_{22}(\gamma)$ in Eq. (D.23). Thus, using (D.21) we obtain

$$\Lambda_{22}(t) = -\frac{a^2\omega^2}{15}Y_{22}^*(\gamma_p) \tag{D.27}$$

$$= -\frac{a^2\omega^2}{\sqrt{120}}\sin^2\theta_p\left(\cos 2\lambda_p - \mathrm{i}\sin 2\lambda_p\right) \tag{D.28}$$

$$= \frac{a^2\Omega^2}{2\sqrt{30}}\left(m_2^2 - m_1^2 + 2\mathrm{i}\, m_1 m_2\right), \tag{D.29}$$

where we have used the further identities $\omega_1^2 - \omega_2^2 = \omega^2\sin^2\theta_p\cos 2\lambda_p = \Omega^2(m_1^2 - m_2^2)$ and $\omega_1\omega_2 = \frac{1}{2}\omega^2\sin^2\theta_p\sin 2\lambda_p = \Omega^2 m_1 m_2$.

D.1.8 CSH Expansion of the Centrifugal Potential Variation

Following anticipation (D.1), we have found that the CSH expansion of the centrifugal potential variation only contains terms of degree $n = 0$ and $n = 2$, with:

$$\Lambda(\gamma, t) = \Lambda_{00}Y_{00} + \sum_{m=-2}^{2}\Lambda_{2m}Y_{2m}\,, \tag{D.30}$$

where the coefficients of the expansion are:

$$\Lambda_{00}(t) = \frac{a^2\Omega^2}{3}\left(m^2 + 2m_3\right) , \quad \text{(D.31)}$$

$$\Lambda_{20}(t) = \frac{a^2\Omega^2}{6\sqrt{5}}\left(m_1^2 + m^2 - 2m_3^2 - 4m_3\right) , \quad \text{(D.32)}$$

$$\Lambda_{21}(t) = \frac{a^2\Omega^2}{\sqrt{30}}\left(m_1(1+m_3) - im_2(1+m_3)\right) , \quad \text{(D.33)}$$

$$\Lambda_{22}(t) = \frac{a^2\Omega^2}{2\sqrt{30}}\left(m_2^2 - m_1^2 + 2im_1m_2\right) , \quad \text{(D.34)}$$

in agreement with the results of Milne and Mitrovica (1998) (see their Eq. 33).

Neglecting quadratic terms in $m_i \ll 1$, as it is appropriate for small excursions of the rotation axis and consistently with the linearised polar motion theory of Sect. 4.2.2.2, gives, to a high degree of approximation:

$$\Lambda_{00}(t) = \frac{2a^2\Omega^2}{3} m_3 , \quad \text{(D.35)}$$

$$\Lambda_{20}(t) = -\frac{2a^2\Omega^2}{3\sqrt{5}} m_3 , \quad \text{(D.36)}$$

$$\Lambda_{21}(t) = \frac{a^2\Omega^2}{\sqrt{30}}(m_1 - \mathrm{i}\, m_2) , \quad \text{(D.37)}$$

$$\Lambda_{22}(t) = 0 . \quad \text{(D.38)}$$

Furthermore, as pointed by Han and Wahr (1989) and Milne and Mitrovica (1998), rotational perturbations due to GIA induce m_3 variations which are several orders of magnitude smaller than either m_1 or m_2. Thus, from equations above, we conclude that the variation in the centrifugal potential is, to a good approximation, a pure degree $n = 2$ and order $m = \pm 1$ harmonic function, whose CSH expansion simply contains the coefficients $\Lambda_{2\pm1}$, where

$$\Lambda_{21}(t) = c_{21}^{\lambda} \mathbf{m}^*, \quad \text{(D.39)}$$

with

$$c_{21}^{\lambda} \equiv \frac{a^2\Omega^2}{\sqrt{30}} . \quad \text{(D.40)}$$

In closing, we note that from (D.30) and (D.39) it is possible to obtain an elegant and compact expression for the centrifugal potential variation. In fact, since $\Lambda(\gamma, t) = \Lambda_{21}Y_{21} + \Lambda_{2-1}Y_{2-1} = 2\,\mathrm{Re}\,(\Lambda_{21}Y_{21})$, we have:

$$\Lambda(\gamma, t) = \frac{a^2\Omega^2}{3}\,(m_1 \cos\lambda + m_2 \sin\lambda)\, P_{21}(\cos\theta)\,, \tag{D.41}$$

which is a pure tesseral RSH of degree $n = 2$ and order $m = 1$.

D.2 Polar Motion Transfer Function

This section is devoted to the study of the Polar Motion Transfer Functions (PMTFs), which will be obtained for both the traditional and the revised rotation theories described in Sect. 4.3.1. In addition, we shall provide an expression for the modified form of the PMTF, necessary to solve the Liouville equations for polar motion in Sect. 4.3.3.

D.2.1 Traditional Rotation Theory

Our purpose here is to show that when the *traditional rotation theory* is adopted, the Laplace-transformed PMTF defined by Eq. (4.101) takes the following form:

$$\tilde{\mathcal{A}}(s) = A^{\mathrm{e}} + \frac{A^{\mathrm{s}}}{s} + \sum_{i=1}^{M'} \frac{A_i}{s - a_i}\,, \qquad M' = M - 1\,, \tag{D.42}$$

where A^{e}, A^{s}, A^i and a_i $(i = 1, \ldots, M')$ are constants, and M is the number of viscoelastic normal modes.

First, we recall that the transformed Liouville equations (4.97) read

$$\left(1 - \frac{\tilde{k}^{\mathrm{T}}(s)}{k^{\mathrm{s}}}\right)\tilde{\mathbf{m}}(s) = \tilde{\boldsymbol{\Psi}}^{\mathrm{sur}}(s)\,, \tag{D.43}$$

where $\tilde{\mathbf{m}}(s)$ is the displacement of the pole, k^{s} is the secular Love number, $\tilde{\boldsymbol{\Psi}}^{\mathrm{sur}}(s)$ is the surface loading polar motion excitation function and

$$\tilde{k}^{\mathrm{T}}(s) = k^{\mathrm{T,e}} + \sum_{i=1}^{M} \frac{k_i^{\mathrm{T}}}{s - s_i} \tag{D.44}$$

is the Laplace-transformed 'k' TLN[36] of harmonic degree $n = 2$.

[36] To simplify notation, here we drop subscript '2', i.e., we write $k^{\mathrm{T,e}} \equiv k_2^{\mathrm{T,e}}$, $k_i^{\mathrm{T}} \equiv k_{2i}^{\mathrm{T}}$, and $s_i \equiv s_{2i}$.

Now, following e.g., Spada et al. (2011a) we make a fundamental assumption, identifying the secular Love number with the fluid TLN 'k' that characterises the Earth model adopted:

$$k^{\mathrm{s}} \equiv k^{\mathrm{T,f}}, \tag{D.45}$$

where $k^{\mathrm{T,f}} \equiv k_2^{\mathrm{T,f}}$ and where according to Eqs. (2.78) and (D.44), we have

$$k^{\mathrm{T,f}} \equiv \lim_{s\to 0} \tilde{k}^{\mathrm{T}}(s) \equiv k^{\mathrm{T,e}} - \sum_{i=1}^{M} \frac{k_i^{\mathrm{T}}}{s_i}. \tag{D.46}$$

Using Eqs. (D.45) and (D.46), the factor in front of $\tilde{\mathbf{m}}(s)$ in Eq. (D.43) can be transformed into

$$1 - \frac{\tilde{k}^{\mathrm{T}}(s)}{k^{\mathrm{s}}} = s \sum_{i=1}^{M} \frac{\gamma_i}{s - s_i}, \tag{D.47}$$

where

$$\gamma_i \equiv -\frac{1}{k^{\mathrm{T,f}}} \left(\frac{k_i^{\mathrm{T}}}{s_i} \right), \tag{D.48}$$

so that the Liouville equations become

$$\left(\sum_{i=1}^{M} \frac{\gamma_i}{s - s_i} \right) \tilde{\mathbf{m}}(s) = \frac{\tilde{\mathbf{\Psi}}^{\mathrm{sur}}(s)}{s}. \tag{D.49}$$

Comparing Eq. (D.49) with (4.100), we find that the Laplace-transformed PMTF can be expressed as

$$\tilde{\mathcal{A}}(s) = \frac{1}{s \displaystyle\sum_{i=1}^{M} \frac{\gamma_i}{s - s_i}}, \tag{D.50}$$

showing that, similar to *secular polynomial* $N(s)$, the PMTF $\tilde{\mathcal{A}}(s)$ has the property of having, in $s = s_i$, M simple zeros:

$$\tilde{\mathcal{A}}(s_i) = 0. \tag{D.51}$$

Hence, from the definition of secular polynomial (2.70), we obtain

$$\tilde{\mathcal{A}}(s) = \frac{N(s)}{s \displaystyle\sum_{i=1}^{M} \gamma_i \frac{N(s)}{s - s_i}}, \tag{D.52}$$

in which the ratio $B_i(s) \equiv N(s)/(s - s_i)$ is the degree $(M - 1)$ polynomial $B_i(s) = \sum_{k=0}^{M-1} b_{k,i} s^k$, with coefficients $b_{M-1,i} = n_M$ and $b_{k-1,i} = n_k + b_{k,i} s_i \quad (k = M - 1, M - 2, \ldots, 1)$. Thus from Eq. (D.52) we obtain

$$\tilde{\mathcal{A}}(s) = \frac{N(s)}{s\left(\sum_{i=1}^{M} \gamma_i \sum_{k=0}^{M-1} b_{k,i} s^k\right)} = \frac{N(s)}{s\left[\sum_{k=0}^{M-1}\left(\sum_{i=1}^{M} \gamma_i b_{k,i}\right) s^k\right]} = \frac{N(s)}{s \sum_{k=0}^{M-1} c_k s^k} . \tag{D.53}$$

Defining the new polynomial $C(s) = \sum_{k=0}^{M-1} c_k s^k$ with coefficients $c_k = \sum_{i=1}^{M} \gamma_i b_{k,i} \quad (k = 0, 1, \ldots M - 1)$, from Eq. (D.53) we see that the PMTF can be expressed by a rational function involving two degree M polynomials:

$$\tilde{\mathcal{A}}(s) = \frac{N(s)}{s\, C(s)} , \tag{D.54}$$

which suggests the decomposition

$$\tilde{\mathcal{A}}(s) = A^{\mathrm{e}} + \frac{A^{\mathrm{s}}}{s} + \frac{P(s)}{C(s)} , \tag{D.55}$$

where $\deg(P(s)) = M - 2$, $p_k = n_{k+1} - A^{\mathrm{e}} c_k - A^{\mathrm{s}} c_{k+1} \quad (k = 0, \ldots M - 2)$, and where we have defined the *secular rotational residue* by

$$A^{\mathrm{s}} = \lim_{s \to 0} s\tilde{\mathcal{A}}(s) = \frac{n_0}{c_0} \tag{D.56}$$

and the *elastic rotational residue* with

$$A^{\mathrm{e}} = \lim_{s \to \infty} \tilde{\mathcal{A}}(s) = \frac{n_M}{c_{M-1}} . \tag{D.57}$$

Furthermore, *assuming* that $C(s)$ has $M - 1$ distinct zeros at $s = a_i$ (i.e., $C(a_i) = 0$) we can follow the usual method for the Laplace inversion of a rational function (see e.g., Longman and Sharir 1971). Hence, defining the *rotational residues* as

$$A_i = \frac{P(a_i)}{C'(a_i)} , \quad (i = 1, \ldots M - 1) , \tag{D.58}$$

where $C'(s) \equiv dC(s)/ds$, Eq. (D.55) gives

$$\tilde{\mathcal{A}}(s) = A^{\mathrm{e}} + \frac{A^{\mathrm{s}}}{s} + \sum_{i=1}^{M'} \frac{A_i}{s - a_i} , \qquad M' = M - 1 , \tag{D.59}$$

that confirms Eqs. (D.42) and (4.105).

D.2.2 *Revised Rotation Theory*

Here we show that according to the revised rotation theory, the Laplace-transformed PMTF defined by Eq. (4.101) takes the form:

$$\tilde{\mathcal{A}}(s) = A^{\mathrm{e}} + \frac{A^{\mathrm{s}}}{s} + \sum_{i=1}^{M'} \frac{A_i}{s - a_i}, \quad M' = M, \tag{D.60}$$

where A^{e}, A^{s}, A^i and a_i ($i = 1, \ldots, M'$) are non-vanishing constants, and M is the number of VNMs of the Earth model. It is important to note that the secular rotational residue is, in this case, $A^{\mathrm{s}} = 0$.

The computations below follow quite closely those for the traditional rotation theory, but some important details differ. From Eq. (4.97), we recall that the Laplace-transformed Liouville equations read

$$\left(1 - \frac{\tilde{k}^{\mathrm{T}}(s)}{k^{\mathrm{s}}}\right) \tilde{\mathbf{m}}(s) = \tilde{\mathbf{\Psi}}^{\mathrm{sur}}(s), \tag{D.61}$$

where $\tilde{\mathbf{m}}(s)$ is position of the istantaneous pole of rotation, k^{s} is the secular Love number, $\tilde{\mathbf{\Psi}}^{\mathrm{sur}}(s)$ is the polar motion excitation function for surface loading and

$$\tilde{k}^{\mathrm{T}}(s) = k^{\mathrm{T,e}} + \sum_{i=1}^{M} \frac{k_i^{\mathrm{T}}}{s - s_i} \tag{D.62}$$

is the Laplace-transformed 'k' TLN of degree $n = 2$ where, to simplify notation, we have again dropped subscript '2', i.e., we have set $k^{\mathrm{T,e}} \equiv k_2^{\mathrm{T,e}}$, $k_i^{\mathrm{T}} \equiv k_{2i}^{\mathrm{T}}$ and $s_i \equiv s_{2i}$.

Now, following Mitrovica et al. (2005) and Mitrovica and Wahr (2011) we *do not* make the *assumption* $k^{\mathrm{s}} = k^{\mathrm{T,f}}$ that is at the basis of the traditional rotation theory (see D.45). Rather, we assume, for the moment, an arbitrary value for the secular Love number k^{s}. By substitution of Eq. (D.62) into (D.61), also using the definition of fluid TLN (2.78) and simple algebra, the factor in front of $\tilde{\mathbf{m}}(s)$ can be cast in the form

$$1 - \frac{\tilde{k}^{\mathrm{T}}(s)}{k^{\mathrm{s}}} = \varepsilon + s \sum_{i=1}^{M} \frac{\delta_i}{s - s_i}, \tag{D.63}$$

where we have defined the constant

$$\varepsilon \equiv 1 - \frac{k^{\mathrm{T,f}}}{k^{\mathrm{s}}}, \tag{D.64}$$

with $k^{\mathrm{T,f}} \equiv k_2^{\mathrm{T,f}}$ and

$$\delta_i \equiv -\frac{1}{k^{\mathrm{s}}}\left(\frac{k_i^T}{s_i}\right), \tag{D.65}$$

where we note that, for $k^{\mathrm{s}} = k^{\mathrm{T,f}}$, we have $\varepsilon = 0$ and according to Eq. (D.48) $\delta_i = \gamma_i$, so that the traditional rotation theory is re-established.

Using Eq. (D.63), the Liouville equations (D.61) become

$$\left(\varepsilon + s\sum_{i=1}^{M}\frac{\delta_i}{s - s_i}\right)\tilde{\mathbf{m}}(s) = \tilde{\boldsymbol{\Psi}}^{\mathrm{sur}}(s), \tag{D.66}$$

which compared with (4.100) gives the PMTF:

$$\tilde{\mathcal{A}}(s) = \frac{1}{\varepsilon + s\displaystyle\sum_{i=1}^{M}\frac{\delta_i}{s - s_i}}, \tag{D.67}$$

showing that, in the hypothesis that all the s_i's are distinct, the PMTF has, also in this case, M simple zeros at $s = s_i$, with:

$$\tilde{\mathcal{A}}(s_i) = 0, \tag{D.68}$$

which correspond to the zeroes of the secular polynomial $N(s)$.

Recalling from Eq. (2.70) the definition of $N(s)$, (D.67) gives

$$\begin{aligned}\tilde{\mathcal{A}}(s) &= \frac{N(s)}{\varepsilon N(s) + s\displaystyle\sum_{i=1}^{M}\delta_i\frac{N(s)}{s - s_i}} = \frac{N(s)}{\varepsilon N(s) + s\displaystyle\sum_{i=1}^{M}\delta_i B_i(s)} \\ &= \frac{N(s)}{\varepsilon N(s) + s\displaystyle\sum_{k=0}^{M-1}\left(\sum_{i=1}^{M}\delta_i b_{k,i}\right)s^k},\end{aligned} \tag{D.69}$$

where we have used (2.70) and we have defined the degree $M-1$ polynomial $B_i(s) = N(s)/(s - s_i) = \sum_{k=0}^{M-1} b_{k,i}s^k$. Hence, defining the new degree $M-1$ polynomial $D(s) = \sum_{k=0}^{M-1} d_k s^k$ with coefficients $d_k = \sum_{i=1}^{M}\delta_i b_{k,i}$ $(k = 0, 1, \ldots M-1)$, from Eq. (D.69) we see that the Laplace-transformed PMTF is expressed, as for the traditional rotation theory, as the ratio of two degree M polynomials

$$\tilde{\mathcal{A}}(s) = \frac{N(s)}{E(s)}, \tag{D.70}$$

where $E(s) \equiv \varepsilon N(s) + sD(s)$ with coefficients $e_0 = \varepsilon n_0$ and $e_k = \varepsilon n_k + d_{k-1}$, $(k = 1, \ldots M)$.

The *secular residue* of $\mathcal{A}$ is

$$A^{\mathrm{s}} = \lim_{s \to 0} s\tilde{\mathcal{A}}(s) = 0\,, \tag{D.71}$$

while the *elastic residue* is

$$A^{\mathrm{e}} = \lim_{s \to \infty} \tilde{\mathcal{A}}(s) = \frac{n_M}{e_M}\,. \tag{D.72}$$

Following the usual method of inverting the rational function (e.g., Longman and Sharir 1971), the *rotational residues* are

$$A_i = \frac{N(a_i)}{E'(a_i)}, \quad (i = 0, 1, \dots M)\,, \tag{D.73}$$

where $E'(s) = dE(s)/ds$, and a_i are the (simple) roots of algebraic equation $E(s) = 0$. Hence from (D.59) we obtain

$$\tilde{\mathcal{A}}(s) = A^{\mathrm{e}} + \frac{A^{\mathrm{s}}}{s} + \sum_{i=1}^{M'} \frac{A_i}{s - a_i}, \quad M' = M\,, \tag{D.74}$$

which along with (D.71) confirms Eqs. (4.103) and (4.106).

D.2.3 Modified Form of the PMTF

Here we show that the *modified PMTF* in the Laplace domain, defined as

$$\tilde{\mathcal{A}}'(s) \equiv \tilde{\mathcal{A}}(s)\left(1 + \tilde{k}_2^{\mathrm{L}}(s)\right) \tag{D.75}$$

can be cast in the form

$$\tilde{\mathcal{A}}'(s) = A'^{\mathrm{e}} + \frac{A'^{\mathrm{s}}}{s} + \sum_{i=1}^{M'} \frac{A'_i}{s - a_i} + \sum_{i=1}^{M} \frac{A''_i}{s - s_i}, \tag{D.76}$$

where A'^{e}, A'^{s}, A'_i and A''_i are appropriate constants (see Eq. 13 of Vermeersen et al. 1994). We start recalling, from Eq. (4.103), the general form of the PMTF

$$\tilde{\mathcal{A}}(s) = A^{\mathrm{e}} + \frac{A^{\mathrm{s}}}{s} + \sum_{i=1}^{M'} \frac{A_i}{s - a_i}, \tag{D.77}$$

where for the traditional rotation theory $M' = M - 1$ and $A^s \neq 0$, while for the revised theory $M' = M$ and $A^s = 0$. Furthermore, according to (2.68), the Laplace-transformed degree 2 LLN is

$$\tilde{k}^{\mathrm{L}}(s) = k^{\mathrm{L,e}} + \sum_{i=1}^{M} \frac{k_i^{\mathrm{L}}}{s - s_i}, \tag{D.78}$$

where, consistent with the conventions adopted in this appendix, we have dropped subscript ‘2’ to simplify notation, i.e., we have written $k^{\mathrm{L,e}} \equiv k_2^{\mathrm{L,e}}$, $k_i^{\mathrm{L}} \equiv k_{2i}^{\mathrm{L}}$, and $s_i \equiv s_{2i}$.

Direct substitution of Eqs. (D.78) and (D.77) into (D.75) gives

$$\begin{aligned}
\tilde{\mathcal{A}}'(s) &= \left(A^{\mathrm{e}} + \frac{A^{\mathrm{s}}}{s} + \sum_{i=1}^{M'} \frac{A_i}{s - a_i}\right)\left(1 + k^{\mathrm{L,e}} + \sum_{i=1}^{M} \frac{k_i^{\mathrm{L}}}{s - s_i}\right) \\
&= A^{\mathrm{e}}\left(1 + k^{\mathrm{L,e}}\right) + \sum_{i=1}^{M} \frac{A^{\mathrm{e}} k_i^{\mathrm{L}}}{s - s_i} + \frac{A^{\mathrm{s}}(1 + k^{\mathrm{L,e}})}{s} + \sum_{i=1}^{M'} \frac{(1 + k^{\mathrm{L,e}}) A_i}{s - a_i} + \\
&\quad + \sum_{i=1}^{M} \frac{A^{\mathrm{s}} k_i^{\mathrm{L}}}{s(s - s_i)} + \sum_{i'=1}^{M'} \sum_{i=1}^{M} \frac{A_{i'} k_i^{\mathrm{L}}}{(s - a_{i'})(s - s_i)},
\end{aligned} \tag{D.79}$$

where we note that since

$$\frac{1}{s(s - s_i)} \equiv \frac{1/s_i}{s - s_i} - \frac{1/s_i}{s}, \tag{D.80}$$

the fifth term is

$$\sum_{i=1}^{M} \frac{A^{\mathrm{s}} k_i^{\mathrm{L}}}{s(s - s_i)} = \sum_{i=1}^{M} \frac{A^{\mathrm{s}} k_i^{\mathrm{L}}/s_i}{s - s_i} - \frac{1}{s} \sum_{i=1}^{M} A^{\mathrm{s}} \frac{k_i^{\mathrm{L}}}{s_i}; \tag{D.81}$$

furthermore, since

$$\frac{1}{(s - a_{i'})(s - s_i)} \equiv \frac{1/(a_{i'} - s_i)}{s - a_{i'}} - \frac{1/(a_{i'} - s_i)}{s - s_i}, \tag{D.82}$$

the sixth term is

$$\sum_{i'=1}^{M'}\sum_{i=1}^{M}\frac{A_{i'}k_i^{\mathrm{L}}}{(s-a_{i'})(s-s_i)}=\sum_{i'=1}^{M'}\sum_{i=1}^{M}\frac{A_{i'}k_i^{\mathrm{L}}/(a_{i'}-s_i)}{s-a_{i'}}-\sum_{i'=1}^{M'}\sum_{i=1}^{M}\frac{A_{i'}k_i^{\mathrm{L}}/(a_{i'}-s_i)}{s-s_i}$$
$$=\sum_{i'=1}^{M'}\frac{1}{s-a_{i'}}\sum_{i=1}^{M}\frac{A_{i'}k_i^{\mathrm{L}}}{a_{i'}-s_i}-\sum_{i=1}^{M}\frac{1}{s-s_i}\sum_{i'=1}^{M'}\frac{A_{i'}k_i^{\mathrm{L}}}{a_{i'}-s_i}$$
$$=\sum_{i'=1}^{M'}\frac{F_{i'}}{s-a_{i'}}-\sum_{i=1}^{M}\frac{G_i}{s-s_i}, \tag{D.83}$$

where

$$F_{i'}\equiv\sum_{i=1}^{M}\frac{A_{i'}k_i^{\mathrm{L}}}{a_{i'}-s_i},\quad (i'=1,\ldots,M') \tag{D.84}$$

$$G_i\equiv\sum_{i'=1}^{M'}\frac{A_{i'}k_i^{\mathrm{L}}}{a_{i'}-s_i},\quad (i=1,\ldots,M). \tag{D.85}$$

Hence, substituting Eqs. (D.83) and (D.81) into (D.79) using (D.84) and (D.85), we obtain

$$\tilde{\mathcal{A}}'(s)=A^{\mathrm{e}}\left(1+k^{\mathrm{L,e}}\right)+\sum_{i=1}^{M}\frac{A^{\mathrm{e}}k_i^{\mathrm{L}}}{s-s_i}+\frac{A^{\mathrm{s}}(1+k^{\mathrm{L,e}})}{s}+\sum_{i=1}^{M'}\frac{(1+k^{\mathrm{L,e}})A_i}{s-a_i}+$$
$$+\sum_{i=1}^{M}\frac{A^{\mathrm{s}}k_i^{\mathrm{L}}/s_i}{s-s_i}+-\frac{1}{s}\sum_{i=1}^{M}A^{\mathrm{s}}\frac{k_i^{\mathrm{L}}}{s_i}+\sum_{i=1}^{M'}\frac{F_i}{s-a_i}-\sum_{i=1}^{M}\frac{G_i}{s-s_i}$$
$$=A^{\mathrm{e}}\left(1+k^{\mathrm{L,e}}\right)+\frac{A^{\mathrm{s}}}{s}\left(1+k^{\mathrm{L,f}}\right)+\sum_{i=1}^{M}\frac{k_i^{\mathrm{L}}(A^{\mathrm{e}}+A^{\mathrm{s}}/s_i)-G_i}{s-s_i}+$$
$$+\sum_{i=1}^{M'}\frac{(1+k^{\mathrm{L,e}})A_i+F_i}{s-a_i}, \tag{D.86}$$

where we have used the expression for the fluid 'k' LLN given by the first of Eq. (2.74). Thus, defining

$$A'^{\mathrm{e}}=\left(1+k^{\mathrm{L,e}}\right)A^{e}, \tag{D.87}$$
$$A'^{\mathrm{s}}=\left(1+k^{\mathrm{L,f}}\right)A^{\mathrm{s}}, \tag{D.88}$$
$$A'_i=(1+k^{\mathrm{L,e}})A_i+F_i,\quad (i=1,\ldots,M') \tag{D.89}$$
$$A''_i=k_i^{\mathrm{L}}\left(A^{\mathrm{e}}+\frac{A^{\mathrm{s}}}{s_i}\right)-G_i,\quad (i=1,\ldots,M) \tag{D.90}$$

we find that the Laplace-transformed modified PMTF can be written as

$$\tilde{\mathcal{A}}'(s) = A'^{\mathrm{e}} + \frac{A'^{\mathrm{s}}}{s} + \sum_{i=1}^{M'} \frac{A'_i}{s - a_i} + \sum_{i=1}^{M} \frac{A''_i}{s - s_i}\,, \tag{D.91}$$

which proves Eq. (D.76).

We conclude observing that a remarkable identity exists, first noted by Valentina Barletta and discussed in Spada et al. (2011a). In fact, from Eqs. (D.90) and (D.85), we obtain

$$\begin{aligned} A''_i &= k_i^{\mathrm{L}} \left(A^{\mathrm{e}} + \frac{A^{\mathrm{s}}}{s_i} \right) - G_i \\ &= k_i^{\mathrm{L}} \left(A^{\mathrm{e}} + \frac{A^{\mathrm{s}}}{s_i} \right) - \sum_{i'=1}^{M'} \frac{A_{i'} k_i^{\mathrm{L}}}{a_{i'} - s_i} \\ &= k_i^{\mathrm{L}} \left(A^{\mathrm{e}} + \frac{A^{\mathrm{s}}}{s_i} + \sum_{k=1}^{M'} \frac{A_k}{s_i - a_k} \right) \\ &= k_i^{\mathrm{L}} \mathcal{A}(s_i)\,, \quad (i = 1, \ldots, M)\,, \end{aligned} \tag{D.92}$$

where we have used the definition of PMTF given by Eq. (4.103). But from Eqs. (D.51) and (D.68), valid for the traditional and the revised rotation theories, respectively, we have

$$\tilde{\mathcal{A}}(s_i) = 0\,, \tag{D.93}$$

so that Eq. (D.92) simply gives

$$A''_i = 0\,, \quad (i = 1, \ldots, M)\,. \tag{D.94}$$

Therefore, $\tilde{\mathcal{A}}'(s)$ has in fact the same structure of $\tilde{\mathcal{A}}(s)$.

Appendix E
A Numerical Scheme for the Sea Level Equation

This appendix aims at providing discrete forms of all the terms which appear in the SLE, as they would be implemented in a numerical solution scheme such that employed in `SELEN`[4] (Spada and Melini 2019). The time discretisation, described in Sect. E.1, is accomplished on an equally spaced 1-D grid, taking advantage of the properties of the Heaviside function. The spatial discretisation, presented in Sect. E.2, is carried out by either a direct evaluation of the field variables on a spherical grid or by a spectral approach based on the CSH functions. For the computation of surface integrals, we take advantage of the properties of the equal-area, icosahedron-shaped, spherical pixelization introduced by Tegmark (1996). In Sect. E.3, the spatiotemporal discretisation scheme is applied to the SLE. The material presented here is based, with some modifications, upon the supplementary section of Spada and Melini (2019) and extends previous works by Spada and Stocchi (2006, 2007).

E.1 Temporal Discretisation

We first define the *unit step function* (or Heaviside function); then, we describe how we deal with functions characterised by a piecewise constant (PC) time-evolution, also discussing the CSH expansion of PC functions.

E.1.1 Unit Step Function

We define the *unit step function* as

$$H(t) \equiv \begin{cases} 0 \text{ if } t \leq 0 \\ 1 \text{ if } t > 0 \,, \end{cases} \tag{E.1}$$

G. Spada and D. Melini, *Glacial Isostatic Adjustment*, Springer Geophysics,
https://doi.org/10.1007/978-3-031-90538-4

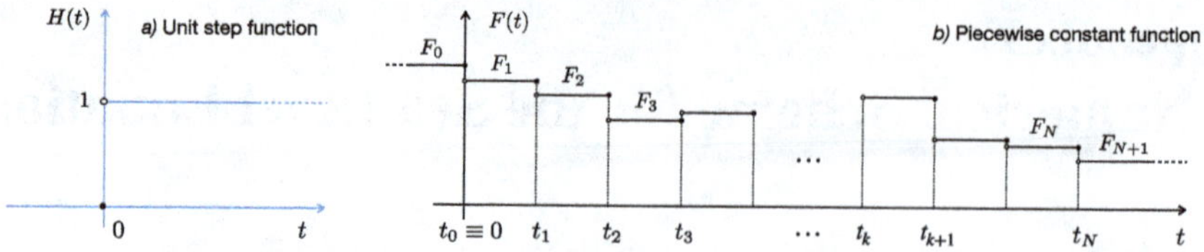

Fig. E.1 Piecewise constant function: **a** Unit step function $H(t)$ and **b** general PC function, where $F_k = F_k(\gamma)$ is the value of $F(\gamma, t)$ during the k-th time interval ($k = 1, \dots N$), $F_0 = F_0(\gamma)$ is the value of $F(\gamma, t)$ in the reference (zero) state, and $F_{N+1} = F_{N+1}(\gamma)$ is the value for $t \geq t_N$, where t_N is present time

where we have chosen $H(t)$ to be "left-continuous", i.e., $H(0) = 0$. A plot of $H(t)$ is shown in Fig. E.1a.

E.1.2 Piecewise Constant Functions

Here we denote by $F = F(\gamma, t)$ a function that, at a given point in space of coordinates $\gamma = (\theta, \lambda)$, varies with time in discrete steps as shown in Fig. E.1b. This *piecewise constant* (PC) time history is particularly suitable for thc numerical discretisation of the fields involved in the SLE.

With $F_0(\gamma)$ we denote the constant value of $F(\gamma, t)$ in the time interval $-\infty < t \leq t_0 \equiv 0$ and with $F_k(\gamma)$ the value in the time interval $t_{k-1} < t \leq t_k$ ($k = 1, \dots, N$). For $t > t_N$, the value of $F(\gamma, t)$ is $F_{N+1}(\gamma)$.

The full PC time history is:

$$F(\gamma, t) \equiv \begin{cases} F_0(\gamma), & -\infty < t \leq t_0 \equiv 0 \\ F_1(\gamma), & t_0 < t \leq t_1 \\ F_2(\gamma), & t_1 < t \leq t_2 \\ \dots & \dots \\ F_k(\gamma), & t_{k-1} < t \leq t_k \\ \dots & \dots \\ F_N(\gamma), & t_{N-1} < t \leq t_N \\ F_{N+1}(\gamma), \; t > t_N, & \end{cases} \tag{E.2}$$

where the length of the time steps is:

$$\Delta t \equiv t_k - t_{k-1} \equiv \frac{t_N}{N}, \tag{E.3}$$

although it is not necessary that all time steps have the same length. At the discrete times $t = t_k$ we have

$$F(\gamma, t_k) = F_k(\gamma), \tag{E.4}$$

because of the left-continuity of $H(t)$.

Furthermore, using definition (E.1), it is easy to see that a PC function can be expanded as:

$$F(\gamma, t) = F_0(\gamma) + \sum_{k=0}^{N} (F_{k+1}(\gamma) - F_k(\gamma))\, H(t - t_k)\,. \tag{E.5}$$

Throughout the book, we often consider the *time-variations* of a given scalar function (for which we conventionally employ a calligraphic symbol), defined as:

$$\mathcal{F}(\gamma, t) \equiv F(\gamma, t) - F_0(\gamma)\,, \tag{E.6}$$

which has a PC time history as well:

$$\begin{aligned} \mathcal{F}(\gamma, t) &= \sum_{k=0}^{N} (F_{k+1}(\gamma) - F_k(\gamma))\, H(t - t_k) \\ &= \sum_{k=0}^{N} \big((F_{k+1}(\gamma) - F_0(\gamma)) - (F_k(\gamma) - F_0(\gamma))\big) H(t - t_k) \\ &= \sum_{k=0}^{N} (\mathcal{F}_{k+1}(\gamma) - \mathcal{F}_k(\gamma))\, H(t - t_k)\,. \end{aligned} \tag{E.7}$$

E.1.3 CSH Expansion of a PC Function

Dealing with the numerical implementation of the SLE, it is often necessary to expand a PC function $\mathcal{F}(\gamma, t)$ in series of CSHs. According to the general expression (A.34), the expansion reads

$$\mathcal{F}(\gamma, t) = \sum_{nm} \mathcal{F}_{nm}(t) Y_{nm}(\gamma)\,, \tag{E.8}$$

where using Eq. (A.36) the coefficients are

$$\mathcal{F}_{nm}(t) = \frac{1}{4\pi} \int_{\gamma} \mathcal{F}(\gamma, t) Y^*_{nm}(\gamma) \mathrm{d}\gamma\,. \tag{E.9}$$

By substituting Eq. (E.7) into (E.9) we obtain

$$\mathcal{F}_{nm}(t) = \sum_{k=0}^{N} \left(\mathcal{F}_{nm,k+1} - \mathcal{F}_{nm,k}\right) H(t - t_k)\,, \tag{E.10}$$

or, equivalently

$$\mathcal{F}_{nm}(t) = \sum_{k=0}^{N} \Delta\mathcal{F}_{nm,k} H(t - t_k) , \tag{E.11}$$

where

$$\Delta\mathcal{F}_{nm,k} \equiv \left(\mathcal{F}_{nm,k+1} - \mathcal{F}_{nm,k}\right) , \quad (k = 0, \dots N) \tag{E.12}$$

and where from (E.9) we have

$$\mathcal{F}_{nm,k} \equiv \mathcal{F}_{nm}(t_k) = \frac{1}{4\pi} \int_{\gamma} \mathcal{F}_k(\gamma) Y^*_{nm}(\gamma) \mathrm{d}\gamma . \tag{E.13}$$

E.2 Spatial Discretisation

In all versions of program SELEN published so far (Spada and Stocchi 2006, 2007; Spada et al. 2012; Spada and Melini 2015, 2019), the spatial discretisation of the SLE has been accomplished with the aid of the equal-area, icosahedron-shaped, spherical pixelization introduced by Tegmark (1996) and employed in astrophysics to study the Cosmic Microwave Background radiation (e.g., Tegmark et al. 2003). The grid is suitable for the manipulation of spherical harmonics, for the numerical evaluation of surface integrals (i.e., the quadrature problem on the surface of the sphere), and for the computation of power spectra.

E.2.1 Properties of the Tegmark's Grid

The grid is characterised by a resolution parameter R that univocally determines the number of pixels (Tegmark 1996). For a given value of resolution R, the number of pixels on the whole sphere is:

$$P = 40R(R - 1) + 12 . \tag{E.14}$$

As an example, Fig. E.2 shows four Tegmark pixelizations characterised by resolutions $R = 18, 28, 44$, and 60. A Fortran code that generates the pixels coordinates is freely available on line.[37]

For a given resolution, each pixel is associated to a *cell* with area

$$A^{\mathrm{c}} = \frac{4\pi a^2}{P} , \tag{E.15}$$

[37] See https://space.mit.edu/home/tegmark/icosahedron.html [last accessed on August 13, 2024].

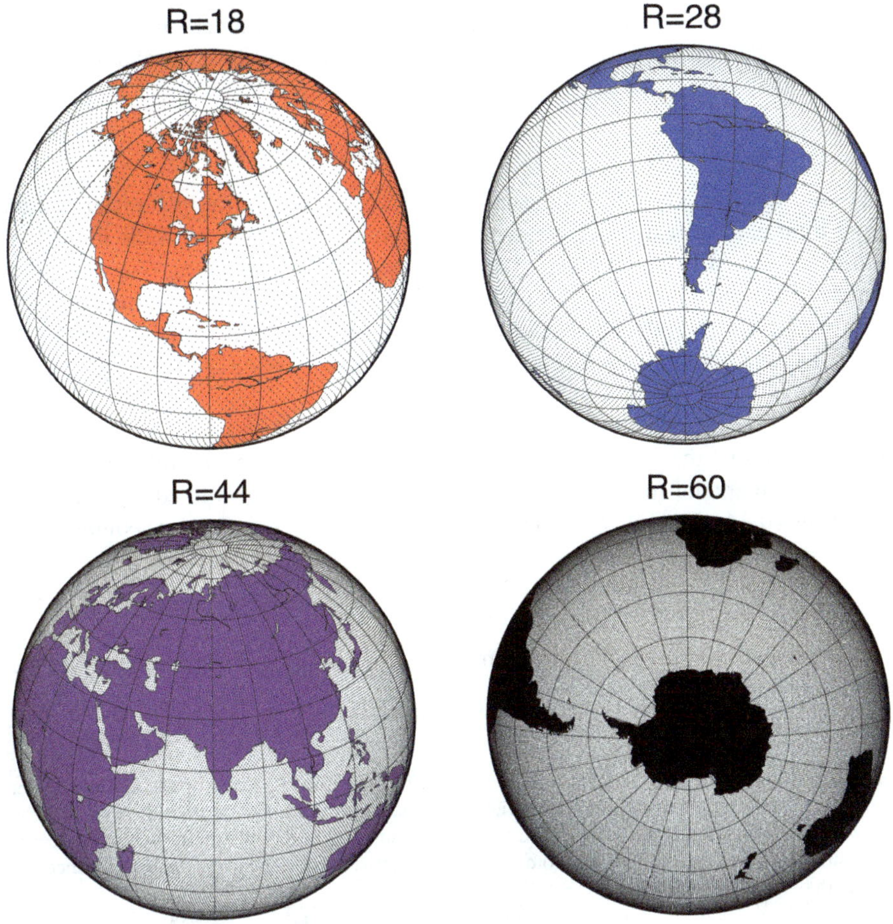

Fig. E.2 Four pixelizations of the sphere according to the icosahedron-based algorithm by Tegmark (1996), with increasing spatial resolutions $R = 18, 28, 44$ and 60. The basic geometrical properties of these and other pixelizations are presented in Table E.1

where a is the radius of the sphere. If we approximate the cell with a disc, its angular half-amplitude would be

$$\varepsilon = \left(\frac{180^\circ}{\pi}\right)\cos^{-1}\left(1 - \frac{2}{P}\right), \tag{E.16}$$

where ε is expressed in degrees. For a sufficiently large number of pixels, these can be assimilated to discs on a plane; an *approximate value* of the disc radius is

$$r_{\text{cell}} \approx \frac{2a}{\sqrt{P}}. \tag{E.17}$$

Tegmark (1996) has shown that a sufficiently accurate numerical integration of harmonic functions up to degree $n_{\max}$ demands a commensurate number of pixels, with

$$P \geq \frac{n_{\max}^2}{3}, \tag{E.18}$$

implying that, for a given P (and hence for a given resolution parameter R), an upper limit exists for the $n_{\max}$ value that can be safely employed in the numerical solution of the SLE, namely $n_{\max} \leq N_{\mathrm{M}}$. Values of $n_{\max}$ exceeding N_{M} would not guarantee that the fundamental properties of the CSHs (as their mutual orthogonality) are numerically reproduced on the grid with a sufficient precision. According to Eq. (E.18) one obtains

$$N_{\mathrm{M}} = \sqrt{3P}\,. \tag{E.19}$$

We also note that the constraint (E.18) provides a very rough *rule of thumb* giving the minimum spatial resolution R that is to be employed for a fixed $n_{\max}$. In fact, since from (E.14), $P \sim 40R^2$, (E.18) gives $40R^2 \geq n_{\max}^2/3$, hence approximately we have:

$$R \geq \frac{n_{\max}}{10}\,. \tag{E.20}$$

For some values of the resolution parameter R, the statistics of the Tegmark grid are summarised in Table E.1.

Table E.1 Statistics of the Tegmark grid. R is the grid resolution, P is the corresponding number of pixels, ε is the radius of the equivalent (i.e., with the same area) disc on the sphere, r_{cell} is the *approximate* radius of the equivalent disc on a plane, and N_{M} is the maximum harmonic degree admissible for the chosen R value. Note that a characteristic cell size of ~ 1 km is attained for $R = 2,000$

R	P	ε (°)	r_{cell} (km)	N_{M}
14	7,292	1.34	149.22	147
18	12,252	1.04	115.12	191
24	22,092	0.77	85.73	257
32	39,692	0.58	63.96	345
44	75,692	0.42	46.31	476
60	141,612	0.30	33.86	651
80	252,812	0.23	25.34	870
100	396,012	0.18	20.25	1,089
200	1,592,012	0.09	10.10	2,185
500	9,980,012	0.04	4.03	5,471
2,000	159,920,012	$\sim$0.01	$\sim$1	21,903

E.2.2 Quadrature Rule on the Tegmark Grid

In the numerical implementation of the SLE (Spada and Stocchi 2006, 2007; Spada et al. 2012; Spada and Melini 2015, 2019), the Tegmark grid is utilised intensively, especially for computing coefficients of the CSH expansions of relevant scalar fields. Of course, other integration methods are possible, such as the traditional Gauss-Legendre quadrature (e.g., Press 2007; Reeger and Fornberg 2016). However, the equal-area characteristic of the grid makes it very attractive for problems in spherical geometry, where a maximally regular pixelization is desired to optimise the algorithms and to simplify coding.

By providing a set of equal-weight integration points, the Tegmark grid allows for a straightforward, nearly optimal quadrature on the sphere, with:

$$\int_{e} G(\gamma, t)\, \mathrm{d}A \approx A^{c} \sum_{p} G(\gamma_p, t)\,, \tag{E.21}$$

where $G(\gamma, t)$ is a scalar function, the integral is over the whole surface of the sphere of radius a, $A^{c} = 4\pi a^2/P$ is the area of each cell, $\sum_p$ denotes the sum over all the pixels, we have taken advantage of the equal-area property of the grid, and $G(\gamma_p, t)$ is the value of $G(\gamma, t)$ at the centre of the cell of coordinates $\gamma_p = (\theta_p, \lambda_p)$.

Equation (E.21) constitutes the basic principle adopted in our numerical implementation of the SLE. The rule is useful, in particular, to evaluate numerically the coefficients of the CSH expansion of a general scalar field $F(\gamma, t)$, a problem very often encountered dealing with the Sea Level Equation. As we have seen in Sect. A.3, the CSH expansion of $F(\gamma, t)$ reads:

$$F(\gamma, t) = \sum_{nm} F_{nm}(t) Y_{nm}(\gamma)\,, \tag{E.22}$$

where the $F_{nm}(t)$'s are the coefficients of the expansion and $Y_{nm}(\gamma)$ is the CSH of harmonic degree n and order m. By the orthogonality property (A.26), the computation of the coefficients can be reduced to the evaluation of a surface integral over the sphere:

$$F_{nm}(t) = \frac{1}{4\pi a^2} \int_{\gamma} F(\gamma, t) Y^{*}_{nm}(\gamma) \mathrm{d}A\,, \tag{E.23}$$

where $Y^{*}_{nm}(\gamma)$ is the complex conjugate of $Y_{nm}(\gamma)$. Recalling (E.15) and using the quadrature rule (E.21) provides:

$$F_j(t) = \frac{1}{P} \sum_{p} F(\gamma_p, t) Y^{*}_{j}(\gamma_p)\,, \tag{E.24}$$

where $j = nm$. In particular, at time t_k we have:

$$F_j(t_k) = \frac{1}{P}\sum_p F(\gamma_p, t_k) Y_j^*(\gamma_p)\,, \tag{E.25}$$

so that we can define the array:

$$F_{j,k} = \frac{1}{P}\sum_p F_{p,k} Y_{j,p}^*\,, \tag{E.26}$$

where $F_{p,k} \equiv F(\gamma_p, t_j)$ and $Y_{j,p} \equiv Y_j(\gamma_p)$. Equation (E.26) provides a compact, spatiotemporally discretised form of the CSH coefficients, suitable for numerical implementation.

E.3 Discretising the Sea Level Equation

Here we provide discrete forms of all the terms in the SLE, useful for the numerical implementation in computer programs such as in SELEN[4] (Spada and Melini 2019). The time discretisation is performed on the 1-D grid introduced in Sect. E.1. The spatial discretisation on the surface of the sphere is carried out by either a direct evaluation of the field variables on the pixels of the Tegmark grid described in Sect. E.2 or by a spectral approach. A summary of the discretised variables that shall be considered in this section is given in Table E.2.

From Chap. 5 we recall that the "ocean-projected" SLE reads

$$\mathcal{Z}(\gamma, t) = \left(\mathcal{K}^{\rm i} + \mathcal{K}^{\rm s}(\mathcal{Z}) + \mathcal{K}^{\rm o}\right) + \mathcal{Z}^{\rm ave} + \mathcal{K}^{\rm rot}(\mathcal{Z})\,, \tag{E.27}$$

with the unknown

$$\mathcal{Z}(\gamma, t) = O\mathcal{S} \tag{E.28}$$

where O is the OF, $\mathcal{S}$ is sea-level change, and

$$\mathcal{Z}^{\rm ave}(t) = O\mathcal{S}^{\rm ave} = O\left(\mathcal{S}^{\rm equ} + \mathcal{S}^{\rm ofu}\right), \tag{E.29}$$

with $\mathcal{S}^{\rm equ}$ and $\mathcal{S}^{\rm ofu}$ given by Eqs. (5.17) and (5.18), respectively, and

$$\mathcal{K}^{\rm i,s,o}(\gamma, t) \equiv O\widehat{\mathcal{R}}^{\rm i,s,o} \tag{E.30}$$

$$\mathcal{K}^{\rm rot}(\gamma, t) \equiv O\widehat{\mathcal{R}}^{\rm rot}\,, \tag{E.31}$$

where

$$\widehat{\mathcal{R}}^{\rm i,s,o}(\gamma, t) = \mathcal{R}^{\rm i,s,o} - \langle \mathcal{R}^{\rm i,s,o}\rangle^{\rm o} \tag{E.32}$$

$$\widehat{\mathcal{R}}^{\rm rot}(\gamma, t) = \mathcal{R}^{\rm rot} - \langle \mathcal{R}^{\rm rot}\rangle^{\rm o}, \tag{E.33}$$

Table E.2 Scalar fields involved in the SLE (left column) and discretised forms (middle), with notes (right). The details of the discretisation are presented throughout Sect. E.3

Continuous field(s)	Discretised field(s)	Definitions, equivalences, and notes
$Y_{nm}(\gamma)$	$Y_{j,p}$	CSH of degree and order $j = nm$ at pixel p
$I(\gamma, t)$	$I_{p,n}$	Obtained by pixelizing the input ice model
$T(\gamma, t)$	$T_{p,k}$	$= T_{p,N} - (\mathcal{S}_{p,k} - \mathcal{S}_{p,N})$, with $T_{p,N}$ from e.g., ETOPO1
$O(\gamma, t)$	$O_{p,k}$, $O_{j,k}$	For $O_{p,k}$, see Eq. (E.45); $O_{j,k} = (1/P)\sum_p O_{p,k} Y^*_{j,p}$
$C(\gamma, t)$	$C_{p,k}$, $C_{j,k}$	$C_{p,k} = 1 - O_{p,k}$; $C_{j,k} = \delta_{l0}\delta_{m0} - O_{j,k}$
$\langle F\rangle^{\mathrm{o}}(t)$	$\langle F\rangle^{\mathrm{o}}_k$	$= \left(\sum_p F_{p,k} O_{p,k}\right) / \left(\sum_p O_{p,k}\right)$
$\langle F\rangle^{\mathrm{e}}(t)$	$\langle F\rangle^{\mathrm{e}}_k$	$= (1/P)\sum_p F_{p,k}$
$A^{\mathrm{o}}(t)$	A^{o}_k	$= A^{\mathrm{c}} \sum_p O_{p,k} = 4\pi a^2 O_{00,k}$
$\mu(t)$	μ_k	$= \rho^{\mathrm{i}} A^{\mathrm{c}} \sum_p \left(I_{p,k} C_{p,k} - I_{p,0} C_{p,0}\right)$
$\mathcal{S}^{\mathrm{equ}}(t)$	$\mathcal{S}^{\mathrm{equ}}_k$, $\mathcal{S}^{\mathrm{equ}}_{j,k}$	$\mathcal{S}^{\mathrm{equ}}_k = -\frac{\rho^{\mathrm{i}}}{\rho^{\mathrm{w}}}\left(\frac{\sum_p (I_{p,k}C_{p,k} - I_{p,0}C_{p,0})}{\sum_p O_{p,k}}\right)$; $\mathcal{S}^{\mathrm{equ}}_{j,k} = \mathcal{S}^{\mathrm{equ}}_k \delta_{n0}\delta_{m0}$
$\mathcal{S}^{\mathrm{ofu}}(t)$	$\mathcal{S}^{\mathrm{ofu}}_k$, $\mathcal{S}^{\mathrm{ofu}}_{j,k}$	$\mathcal{S}^{\mathrm{ofu}}_k = \left(\sum_p T_{p,0}\left(O_{p,k} - O_{p,0}\right)\right) / \left(\sum_p O_{p,k}\right)$; $\mathcal{S}^{\mathrm{ofu}}_{j,k} = \mathcal{S}^{\mathrm{ofu}}_k \delta_{n0}\delta_{m0}$
$\mathcal{S}^{\mathrm{ave}}(t)$	$\mathcal{S}^{\mathrm{ave}}_k$, $\mathcal{S}^{\mathrm{ave}}_{j,k}$	$\mathcal{S}^{\mathrm{ave}}_k = \mathcal{S}^{\mathrm{equ}}_k + \mathcal{S}^{\mathrm{ofu}}_k$; $\mathcal{S}^{\mathrm{ave}}_{j,k} = \mathcal{S}^{\mathrm{ave}}_k \delta_{n0}\delta_{m0}$
$c(t)$	c_k	$= \mathcal{S}^{\mathrm{ave}}_k - \langle\mathcal{R}^{\mathrm{i}}\rangle^{\mathrm{o}}_k - \langle\mathcal{R}^{\mathrm{s}}\rangle^{\mathrm{o}}_k - \langle\mathcal{R}^{\mathrm{o}}\rangle^{\mathrm{o}}_k - \langle\mathcal{R}^{\mathrm{rot}}\rangle^{\mathrm{o}}_k$
$\mathcal{Z}^{\mathrm{ave}}(\gamma, t)$	$\mathcal{Z}^{\mathrm{ave}}_{j,k}$	$= O_{j,k}(\mathcal{S}^{\mathrm{equ}}_k + \mathcal{S}^{\mathrm{ofu}}_k) = O_{j,k}\mathcal{S}^{\mathrm{ave}}_k$
$\mathcal{W}(\gamma, t)$	$\mathcal{W}_{j,k}$	$= (1/P)\sum_p \left(I_{p,k} - I_{p,0}\right)\left(1 - O_{p,k}\right) Y^*_{j,p}$
$\mathcal{X}(\gamma, t)$	$\mathcal{X}_{j,k}$	$= (1/P)\sum_p \mathcal{Q}_p \left(O_{p,k} - O_{p,0}\right) Y^*_{j,p}(\gamma)$
$\mathcal{L}^{\mathrm{i,s,o}}(\gamma, t)$	$\mathcal{L}^{\mathrm{i,s,o}}_{j,k}$	with $\mathcal{L}^{\mathrm{i}}_{j,k} = \rho^{\mathrm{i}}\mathcal{W}_{j,k}$, $\mathcal{L}^{\mathrm{s}}_{j,k} = \rho^{\mathrm{w}}\mathcal{Z}_{j,k}$, and $\mathcal{L}^{\mathrm{o}}_{j,k} = \rho^{\mathrm{r}}\mathcal{X}_{j,k}$
$\beta^{s,g,u}_n(t)$	$\beta^{s,g,u}_{n,k}$	$= \beta^{s,g,u}_n(k\Delta t)$
$(\mathcal{R}, \mathcal{G}, \mathcal{U})^{\mathrm{i,s,o}}(\gamma, t)$	$(\mathcal{R}, \mathcal{G}, \mathcal{U})^{\mathrm{i,s,o}}_{j,k}$	See Eqs. (E.82–E.84), (E.95–E.97), and (E.90–E.92).
$\langle\mathcal{R}^{\mathrm{i,s,o}}\rangle^{\mathrm{o}}(t)$	$\langle\mathcal{R}^{\mathrm{i,s,o}}\rangle^{\mathrm{o}}_k$	$= \left(\sum_j \mathcal{R}^{\mathrm{i,s,o}}_{j,k} O^*_{j,k}\right) / \left(O_{00,k}\right)$
$\widehat{\mathcal{R}}^{\mathrm{i,s,o}}$	$\widehat{\mathcal{R}}^{\mathrm{i,s,o}}_{j,k}$	$= \mathcal{R}^{\mathrm{i,s,o}}_{j,k} - \langle\mathcal{R}^{\mathrm{i,s,o}}\rangle^{\mathrm{o}}_k \delta_{n0}\delta_{m0}$
$\mathcal{K}^{\mathrm{i,s,o}}(\gamma, t)$	$\mathcal{K}^{\mathrm{i,s,o}}_{j,k}$	$= \sum_{(p:\ O_{p,k}=1)} \chi^{\mathrm{i,s,o}}_{p,k} Y^*_{j,p}$, with $\chi^{\mathrm{i,s,o}}_{p,k} \equiv (1/P)\sum_j \widehat{\mathcal{R}}^{\mathrm{i,s,o}}_{j,k} Y_{j,p}$
$\gamma^{s,g,u}(t)$	$\gamma^{s,g,u}_k$	$= \gamma^{s,g,u}(k\Delta t)$
$(\mathcal{R}, \mathcal{G}, \mathcal{U})^{\mathrm{rot}}(\gamma, t)$	$(\mathcal{R}, \mathcal{G}, \mathcal{U})^{\mathrm{rot}}_{j,k}$	$= \delta_{n2}\delta_{m\pm1}\left(c^{\psi}_{21} c^{\lambda}_{21}/g\right) \sum_{k'=0}^{k-1} \Delta\mathcal{L}_{j,k'} \gamma^{s,u,g}_{k-k'}$
$\langle\mathcal{R}^{\mathrm{rot}}\rangle^{\mathrm{o}}(t)$	$\langle\mathcal{R}^{\mathrm{rot}}\rangle^{\mathrm{o}}_k$	$= \left(\sum_j \mathcal{R}^{\mathrm{rot}}_{j,k} O^*_{j,k}\right) / \left(O_{00,k}\right)$
$\widehat{\mathcal{R}}^{\mathrm{rot}}(\gamma, t)$	$\widehat{\mathcal{R}}^{\mathrm{rot}}_{j,k}$	$= \mathcal{R}^{\mathrm{rot}}_{j,k} - \langle\mathcal{R}^{\mathrm{rot}}\rangle^{\mathrm{o}}_k \delta_{n0}\delta_{m0}$
$\mathcal{K}^{\mathrm{rot}}(\gamma, t)$	$\mathcal{K}^{\mathrm{rot}}_{j,k}$	$= \sum_{(p:\ O_{p,k}=1)} \chi^{\mathrm{rot}}_{p,k} Y^*_{j,p}$, with $\chi^{\mathrm{rot}}_{p,k} \equiv (1/P)\sum_j \mathcal{R}'^{\,rot}_{j,k} Y_{j,p}$
$\Psi^{\mathrm{rig}}(t)$	Ψ^{rig}_k	$= c^{\psi}_{21}\mathcal{L}^*_{21,k}$
$\mathcal{S}(\gamma, t)$	$\mathcal{S}_{j,k}$	$= \mathcal{S}^{\mathrm{ave}}_{j,k} + \widehat{\mathcal{R}}^{\mathrm{i}}_{j,k} + \widehat{R}^{s}_{j,k}(\mathcal{Z}) + \widehat{\mathcal{R}}^{\mathrm{o}}_{j,k} + \widehat{\mathcal{R}}^{\mathrm{rot}}_{j,k}(\mathcal{Z})$
$\mathcal{Z}(\gamma, t)$	$\mathcal{Z}_{j,k}$	$= \mathcal{Z}^{\mathrm{ave}}_{j,k} + \mathcal{K}^{\mathrm{i}}_{j,k} + \mathcal{K}^{\mathrm{s}}_{j,k}(\mathcal{Z}) + \mathcal{K}^{\mathrm{o}}_{j,k} + \mathcal{K}^{\mathrm{rot}}_{j,k}(\mathcal{Z})$

and where $\mathcal{R}^{i,s,o}$ denotes each of the three components of the sea-level SRF, $\mathcal{R}^{rot}$ is the sea-level RRF and $\langle\cdots\rangle^{o}$ represents the ocean-average.

The discretised spectral form of (E.27), i.e., its expansion in spherical harmonics, is:

$$\mathcal{Z}_{j,k} = \left(\mathcal{K}^{i}_{j,k} + \mathcal{K}^{s}_{j,k}(\mathcal{Z}) + \mathcal{K}^{o}_{j,k}\right) + \mathcal{Z}^{ave}_{j,k} + \mathcal{K}^{rot}_{j,k}(\mathcal{Z})\,, \tag{E.34}$$

where j stands for nm (degree and order), with $0 \le n \le n_{max}$ and subscript k denotes time t_k, with $k = 0, \ldots N+1$.[38] In the following, all terms in Eq. (E.27) and (E.29–E.33) are considered and analysed, also making available discretised forms for the terms that appear in the "unprojected" SLE:

$$\mathcal{S}_{j,k} = \left(\widehat{\mathcal{R}}^{i}_{j,k} + \widehat{\mathcal{R}}^{s}_{j,k}(\mathcal{Z}) + \widehat{\mathcal{R}}^{o}_{j,k}\right) + \mathcal{S}^{ave}_{j,k} + \widehat{\mathcal{R}}^{rot}_{j,k}(\mathcal{Z})\,, \tag{E.35}$$

which is the counterpart of Eq. (5.35).

E.3.1 Notation and Conventions

To deal with discretised fields, we shall use the following notation and conventions. With $F_{p,k}$ we indicate the scalar field $F(\gamma, t)$ evaluated on a given pixel of coordinates $\gamma_p = (\theta_p, \lambda_p)$ and at time $t = t_k$, where t_k is a point of the time grid.

Hence we define

$$F_{p,k} \equiv F(\gamma_p, t_k)\,, \tag{E.36}$$

where $p = (1, \ldots, P)$ and $k = (0, \ldots, N+1)$. The coefficient of the CSH expansion of $F(\gamma, t)$, evaluated at time $t = t_k$, shall be denoted by

$$F_{j,n} \equiv F_j(t_n)\,, \tag{E.37}$$

where $j = nm$ indicates a couple of harmonic degrees and orders. Furthermore, to denote a CSH evaluated at a given grid point of coordinates γ_p, we shall conventionally write

$$Y_{j,p} \equiv Y_j(\gamma_p)\,. \tag{E.38}$$

Finally, a spatially invariant quantity like e.g., $\mu(t)$, evaluated at time $t = t_k$, shall be simply denoted as

$$\mu_k \equiv \mu(t_k)\,. \tag{E.39}$$

[38] The range of harmonics n_{max} depends on the spatial resolution to be achieved. The computational cost of the solution, of course, increases with n_{max}. Spada and Melini (2019) have shown that the effects of increasing n_{max} in a test run. The number of time steps (i.e., parameter N) is suggested from the specific chronology of the ice sheets employed.

Note that no confusion between symbols $F_{j,k}$ (E.37) and $F_{p,k}$ (E.36) should arise, since letters j (or j') will always denote the couple composed by harmonic degree and order nm (or $n'm'$), subscript k (or k') always represents a point on the time grid, while p (or p') conventionally indicates a generic pixel on the sphere. Thus, for instance $\mathcal{X}_{j,k}$ indicates a spectral quantity (denoted by symbol j) that represents the harmonic coefficient of the CSH expansion of field $\mathcal{X}(\gamma, t)$, evaluated at time $t = t_k$, while $U_{p',k'}$ denotes scalar function $U(\gamma, t)$ in the spatial domain, evaluated at a pixel of coordinates $\gamma_{p'}$ and time $t = t_{k'}$. Similarly, $T_{p,0}$ indicates the field $T(\gamma, t)$ at pixel γ_p and time $t = t_0$. To indicate the CSHs (Sect. A.3), we shall always employ symbol Y so that, for example, $Y_{j',p}$ identifies the CSH of degree and order $j' = n'm'$, evaluated at the pixel of coordinates γ_p.

E.3.2 *Ice, Topography and the Ocean Function*

The section considers the discretized forms of the thickness of ice $I(\gamma, t)$, of topography $T(\gamma, t)$, and of the ocean function $O(\gamma, t)$.

E.3.2.1 Ice Thickness

The ice thickness distribution is the basic input of SLE, along with a description of the Earth's rheology. Function $I(\gamma, t)$, i.e., the ice thickness at any location and time, is assumed to be known based on a priori geological information, *ad hoc* hypotheses, glaciological arguments, or reconstructions published in the literature. These typically constrain the time-evolution of ice thickness distribution by the knowledge of Relative Sea Level (RSL) at specific locations since the Last Glacial Maximum (LGM, $\sim 21,000$ years ago), using predictions from the SLE. Hence, we assume that the array

$$I_{p,k} \equiv I(\gamma_p, t_k)\,, \tag{E.40}$$

where $p = (1, \ldots, P)$ and $k = (0, \ldots, N+1)$, is known a priori. The choice of the time step $\Delta t = t_{k+1} - t_k$ that characterises the time-discretisation of the SLE (see Sect. E.1) is made according to the natural time step of the ice chronology; in GIA modelling, typical values are $\Delta t = 1000$ or $\Delta t = 500$ years (see, e.g., Sect. 6.2.3). We also note that ice models currently employed in GIA studies[39] are normally defined on traditional, cartesian longitude-latitude grids. Pixelised versions of these datasets, for a given spatial resolution R, can be obtained by spherical interpolation methods.

[39] Some of the more recent models belonging to the ICE-*X* family are available from the home page of Prof. WR Peltier, see https://www.atmosp.physics.utoronto.ca/~peltier/data.php [last accessed on August 14, 2024].

E.3.2.2 Topography

Consistently with our conventions, the pixelised topography is

$$T_{p,k} \equiv T(\gamma_p, t_k)\,, \tag{E.41}$$

where $\gamma_p = (\theta_p, \lambda_p)$ denotes the pixels coordinates with $p = (1, \ldots, P)$, and time is labelled by k, with $k = (0, \ldots N + 1)$.

Contrary to $I_{p,k}$, however, array $T_{p,k}$ is not assumed to be given a priori. Indeed, according to (1.97) the past topography (or *paleo-topography*) is determined by the history of the sea-level variations, which is not known until the SLE is solved. The present-day relief serves as a "final condition" for the Earth topography, in order to constrain its evolution in the past. In this way, paleo-topography can be iteratively reconstructed using the PT equation (1.105) that in discretised form reads

$$T_{p,k} = T_{p,N} - \left(\mathcal{S}_{p,k} - \mathcal{S}_{p,N}\right), \tag{E.42}$$

where $T_{p,N}$ is topography at present time ($t = t_N$), and $\mathcal{S}_{p,k}$ is sea-level change at a given time $t = t_k$ and at pixel γ_p, as given by the SLE in Eq. (E.35). Using the definition of *true paleo-topography* PT (1.106), we also have

$$PT_{p,k} = T_{p,k} + I_{p,k}\,, \tag{E.43}$$

where $I_{p,k}$ is the input array containing the ice thickness, and $T_{p,k}$ is given by Eq. (E.42).

For practical applications, the present-day topography can be prescribed on the basis of global relief datasets such as ETOPO1 (Amante and Eakins 2009; Eakins and Sharman 2012). ETOPO1 is originally provided[40] on a traditional cartesian longitude-latitude grid with a resolution of 1 arc-minute,[41] so it needs to be interpolated on the pixels of Tegmark's grid to obtain present-day topography $T_{p,N}$.

E.3.2.3 Ocean and Continent Function

The ocean function (OF) is defined by

$$O(\gamma, t) \equiv \begin{cases} 1, \text{ if } T + \dfrac{\rho^{\mathrm{i}}}{\rho^{\mathrm{w}}} I < 0 \\ 0, \text{ if } T + \dfrac{\rho^{\mathrm{i}}}{\rho^{\mathrm{w}}} I \geq 0\,, \end{cases} \tag{E.44}$$

[40] See https://www.ngdc.noaa.gov/mgg/global/ [last accessed on February 26, 2019].

[41] One arc-minute is $(1/60)^\circ$ and corresponds to an arc of latitude of $\sim$1.85 km.

where I and T and ice thickness and topography, respectively, whereas the continent function (CF) is $C = 1 - O$. Using (E.41) and (E.40), the discretised form of the OF is:

$$O_{p,k} \equiv O(\gamma_p, t_k) \equiv \begin{cases} 1, \text{ if } T_{p,k} + \dfrac{\rho^{\mathrm{i}}}{\rho^{\mathrm{w}}} I_{p,k} < 0 \\ 0, \text{ if } T_{p,k} + \dfrac{\rho^{\mathrm{i}}}{\rho^{\mathrm{w}}} I_{p,k} \geq 0 \, , \end{cases} \tag{E.45}$$

whereas for the CF we have:

$$C_{p,k} = 1 - O_{p,k} \, . \tag{E.46}$$

According to (E.26), the harmonic coefficients of the OF are found by quadrature:

$$O_{j,k} = \frac{1}{P} \sum_p O_{p,k} Y^*_{j,p} \, , \tag{E.47}$$

with $O_{p,k}$ given by (E.45), and the CF coefficients are:

$$C_{j,k} = \delta_{n0}\delta_{m0} - O_{j,k} \, . \tag{E.48}$$

E.3.3 Spatially Invariant Terms

In the SLE, several spatially invariant terms appear. Most of them shall be analysed here, in order to provide discretised forms suitable for the numerical solution of the SLE. Some others shall be discussed in later sections devoted to quantities related with the surface and rotational response of the Earth.

E.3.3.1 Ocean Average

The ocean-average of a function $F(\gamma, t)$ is a spatially invariant term that often appears in the treatment of the SLE. Recalling (1.143), it is defined as

$$\langle F \rangle^{\mathrm{o}}(t) \equiv \frac{\displaystyle\int_{\mathrm{o}} F(\gamma, t) \mathrm{d}A}{\displaystyle\int_{\mathrm{o}} \mathrm{d}A} \, , \tag{E.49}$$

where the surface integrals are over the region where, at time t, the OF has value $O = 1$. Hence, with the aid of the OF, the ocean average can be equivalently written as

$$\langle F \rangle^{\mathrm{o}}(t) = \frac{\int_{\mathrm{e}} F(\gamma, t) O \mathrm{d}A}{\int_{\mathrm{e}} O \mathrm{d}A}, \tag{E.50}$$

which is easily discretised on the grid using the quadrature rule (E.21), obtaining

$$\langle F \rangle_k^{\mathrm{o}} \equiv \langle F \rangle^{\mathrm{o}}(t_k) = \frac{\sum_p F_{p,k}\, O_{p,k}}{\sum_p O_{p,k}}, \tag{E.51}$$

where $F_{p,k} = F(\gamma_p, t_k)$ and $O_{p,k}$ is given by Eq. (E.45).

E.3.3.2 Global Average

Recalling (1.136), the global spatial average of function $F(\gamma, t)$, defined as

$$\langle F \rangle^{\mathrm{e}}(t) \equiv \frac{\int_{\mathrm{e}} F(\gamma, t) \mathrm{d}A}{\int_{\mathrm{e}} \mathrm{d}A}, \tag{E.52}$$

is discretised on the grid according to

$$\langle F \rangle_k^{\mathrm{e}} \equiv \langle F \rangle^{\mathrm{e}}(t_k) = \frac{A^{\mathrm{c}} \sum_p F_{p,k}}{4\pi a^2} = \frac{1}{P} \sum_p F_{p,k}, \tag{E.53}$$

since the area A^{c} of each cell is given by (E.15).

E.3.3.3 Area of the Oceans

The area of the oceans is

$$A^{\mathrm{o}}(t) = \int_{\mathrm{o}} \mathrm{d}A, \tag{E.54}$$

where the integral is over the surface where the OF is, at time t, $O = 1$. Equivalently, using the OF definition and the quadrature rule (E.21), we have

$$A^{\mathrm{o}}(t) = \int_{\mathrm{e}} O \mathrm{d}A = A^{\mathrm{c}} \sum_p O(\gamma_p, t), \tag{E.55}$$

where A^{c} is the area of each of the cells and the sum is over all the pixels. Hence:

$$A_k^{\mathrm{o}} \equiv A^{\mathrm{o}}(t_k) = A^{\mathrm{c}} \sum_p O_{p,k} \,. \tag{E.56}$$

We note that A_k^{o} is proportional to the harmonic degree $n = 0$ component of the OF. In fact, since $Y_{00,p} = 1$ Eq. (E.47) provides

$$O_{00,k} = \frac{1}{P} \sum_p O_{p,k} \,, \tag{E.57}$$

which, by comparison with Eq. (E.56) and also recalling (E.15), shows that:

$$A_k^{\mathrm{o}} = 4\pi a^2 O_{00,k} \,. \tag{E.58}$$

E.3.3.4 Grounded Ice Mass Variation

According to Eq. (1.140), the mass variation of the grounded ice is

$$\mu(t) = \rho^{\mathrm{i}} \int_{\mathrm{e}} (IC - I_0 C_0)\, \mathrm{d}A \,, \tag{E.59}$$

where I is ice thickness, C is the CF, and subscript 0 indicates reference quantities. Thus, according to the quadrature rule (E.21), we have

$$\mu(t) = \rho^{\mathrm{i}} A^{\mathrm{c}} \sum_p \big(I(\gamma_p, t) C(\gamma_p, t) - I_0(\gamma_p, t_0) C_0(\gamma_p, t_0)\big) \,, \tag{E.60}$$

which at time $t = t_k$ gives

$$\mu_k \equiv \mu(t_k) = \rho^{\mathrm{i}} A^{\mathrm{c}} \sum_p \left(I_{p,k} C_{p,k} - I_{p,0} C_{p,0}\right) \,, \tag{E.61}$$

where $I_{p,k}$ and $C_{p,k}$ are given by Eqs. (E.40) and (E.46), respectively.

E.3.3.5 Term $\mathcal{S}^{\mathrm{equ}}$

This term of the SLE, defined as

$$\mathcal{S}^{\mathrm{equ}}(t) \equiv -\frac{\mu(t)}{\rho^{\mathrm{w}} A^{\mathrm{o}}(t)} \,, \tag{E.62}$$

and referred to as *equivalent sea-level change* (Sect. 5.17), describes the sea-level variation due to time-variations of the grounded ice mass $\mu(t)$, also taking the consequent changes in the area of the oceans $A^{\mathrm{o}}(t)$ into account. Taking advantage of expressions (E.56) and (E.61), the discretisation of this term is straightforward, with:

$$\mathcal{S}_k^{\mathrm{equ}} \equiv \mathcal{S}^{\mathrm{equ}}(t_k) = -\frac{\mu_k}{\rho^{\mathrm{w}} A_k^{\mathrm{o}}} = -\frac{\rho^{\mathrm{i}}}{\rho^{\mathrm{w}}} \frac{\sum_p \left(I_{p,k} C_{p,k} - I_{p,0} C_{p,0}\right)}{\sum_p O_{p,k}} . \tag{E.63}$$

Since $\mathcal{S}^{\mathrm{equ}}$ is spatially invariant, in its CSH expansion all the coefficients vanish except those of harmonic degree and order $n = m = 0$:

$$\mathcal{S}_{j,k}^{\mathrm{equ}} = \mathcal{S}_k^{\mathrm{equ}} \delta_{n0} \delta_{m0} . \tag{E.64}$$

E.3.3.6 Term $\mathcal{S}^{\mathrm{ofu}}$

This second spatially invariant term of the SLE, introduced in Sect. 1.4.4, and defined as

$$\mathcal{S}^{\mathrm{ofu}}(t) \equiv \frac{1}{A^{\mathrm{o}}} \int_{\mathrm{e}} T_0 O \mathrm{d}A , \tag{E.65}$$

accounts for time-variations of the ocean function, either associated to the horizontal migration of the shorelines or to transitions between grounded and floating ice (or *vice versa*).

Using Eq. (1.122) and applying the integration rule (E.21), we have

$$\mathcal{S}^{\mathrm{ofu}}(t) = \frac{A^{\mathrm{c}}}{A^{\mathrm{o}}(t)} \sum_p T(\gamma_p, t_0) \left(O(\gamma_p, t) - O(\gamma_p, t_0)\right) , \tag{E.66}$$

so that

$$\mathcal{S}_k^{\mathrm{ofu}} \equiv \mathcal{S}^{\mathrm{ofu}}(t_k) = \frac{\sum_p T_{p,0} \left(O_{p,k} - O_{p,0}\right)}{\sum_p O_{p,k}} , \tag{E.67}$$

where we have also used Eq. (E.56). Since this term is spatially invariant, the coefficients of its CSH expansion are

$$\mathcal{S}_{j,k}^{\mathrm{ofu}} = \mathcal{S}_k^{\mathrm{ofu}} \delta_{n0} \delta_{m0} . \tag{E.68}$$

E.3.3.7 Term $\mathcal{S}^{\text{ave}}$

According to Eq. (5.16), the spatially invariant term

$$\mathcal{S}^{\text{ave}}(t) = \mathcal{S}^{\text{equ}} + \mathcal{S}^{\text{ofu}} \tag{E.69}$$

represents the ocean-average of $\mathcal{S}(\gamma, t)$. Thus we have

$$\mathcal{S}_k^{\text{ave}} \equiv \mathcal{S}^{\text{ave}}(t_k) = \mathcal{S}_k^{\text{equ}} + \mathcal{S}_k^{\text{ofu}}\,, \tag{E.70}$$

where $\mathcal{S}_k^{\text{equ}}$ and $\mathcal{S}_k^{\text{equ}}$ are given by Eqs. (E.63) and (E.67), respectively. Furthermore, its CSH coefficients are:

$$\mathcal{S}_{j,k}^{\text{ave}} = \left(\mathcal{S}_k^{\text{equ}} + \mathcal{S}_k^{\text{ofu}}\right) \delta_{n0}\delta_{m0}\,. \tag{E.71}$$

E.3.3.8 Term $\mathcal{Z}^{\text{ave}}$

We note that the ocean-projected $\mathcal{S}^{\text{ave}}$, i.e., the term $\mathcal{Z}^{\text{ave}} = O\mathcal{S}^{\text{ave}}$ given by Eq. (E.29), is certainly *not* spatially invariant, since O is variable in space and time. However, since $\mathcal{S}^{\text{ave}}$ *is* invariant, its CSH expansion is straightforward, with

$$\mathcal{Z}_{j,k}^{\text{ave}} = O_{j,k}\mathcal{S}_k^{\text{ave}}, \tag{E.72}$$

where $\mathcal{S}_k^{\text{ave}}$ is given by Eq. (E.70).

E.3.3.9 The "*c* Constant"

The c constant, first introduced by Farrell and Clark (1976), arises from the imposition of mass conservation. According to Eq. (5.26), it is given by a combination of spatially invariant quantities

$$c(t) = \mathcal{S}^{\text{ave}} - \langle \mathcal{R}^{\text{sur}} \rangle^{\text{o}} - \langle \mathcal{R}^{\text{rot}} \rangle^{\text{o}}, \tag{E.73}$$

which can be immediately discretised as follows:

$$c_k \equiv c(t_k) = \mathcal{S}_k^{\text{ave}} - \langle \mathcal{R}^{\text{i}} \rangle_k^{\text{o}} - \langle \mathcal{R}^{\text{s}} \rangle_k^{\text{o}} - \langle \mathcal{R}^{\text{o}} \rangle_k^{\text{o}} - \langle \mathcal{R}^{\text{rot}} \rangle_k^{\text{o}}\,, \tag{E.74}$$

where terms $\langle \mathcal{R}^{\text{i,s,o}} \rangle_k^{\text{o}}$ and $\langle \mathcal{R}^{\text{rot}} \rangle_k^{\text{o}}$, are made explicit by Eqs. (E.104) and (E.119) below, respectively.

E.3.4 Surface Response

The three sea-level SRFs $\mathcal{R}^{\rm i}$, $\mathcal{R}^{\rm s}$ and $\mathcal{R}^{\rm o}$ which appear in the SLE (see Eq. E.27 and in those that immediately follow) share common structures. Below, we first discretise the surface load variation $\mathcal{L}$, then we make the CSH coefficients of the sea-level SRFs explicit; we evaluate the ocean-averages of the SRFs, and lastly we construct the "modified SRFs" defined by Eq. (E.32).

E.3.4.1 Surface Load

Based on (3.32), the surface load variation CSH coefficients are

$$\mathcal{L}_{j,k} = \mathcal{L}^{\rm i}_{j,k} + \mathcal{L}^{s}_{j,k} + \mathcal{L}^{\rm o}_{j,k}\,, \tag{E.75}$$

where

$$\mathcal{L}^{\rm i}_{j,k} \equiv \rho^{\rm i} \mathcal{W}_{j,k} \tag{E.76}$$

$$\mathcal{L}^{\rm s}_{j,k} \equiv \rho^{\rm w} \mathcal{Z}_{j,k} \tag{E.77}$$

$$\mathcal{L}^{\rm o}_{j,k} \equiv \rho^{\rm r} \mathcal{X}_{j,k}\,, \tag{E.78}$$

where the superscripts i, s, o of $\mathcal{L}_{j,k}$ stand for *ice*, *sea level* and *ocean*, respectively.

According to Eq. (1.128), $\mathcal{W} = \mathcal{I}C$, where $\mathcal{I} = I - I_0$ is the ice thickness variation and $C = 1 - O$ is the CF. The CSH coefficients $\mathcal{W}_{j,k}$ in Eq. (E.76) are obtained by direct integration of $\mathcal{W}$ on the Tegmark grid using Eq. (E.26):

$$\mathcal{W}_{j,k} = \frac{1}{P}\sum_{p} \mathcal{I}_{p,k} C_{p,k} Y^*_{j,p} = \frac{1}{P}\sum_{p} \left(I_{p,k} - I_{p,0}\right)\left(1 - O_{p,k}\right) Y^*_{j,p}\,. \tag{E.79}$$

In Eq. (E.77) $\mathcal{Z}_{j,k}$ represents the spectral discrete form of the basic unknown of the final form of the SLE (see Eq. E.27 and those that immediately follow), namely the ocean-projected sea-level change $\mathcal{Z} = O\mathcal{S}$. Within the iteration scheme of the SLE described in Chap. 6, progressively updated values of $\mathcal{Z}_{j,k}$ are used in Eq. (E.77) in order to determine $\mathcal{L}^{\rm s}_{j,k}$.

Based upon Eq. (1.130), $\mathcal{X} = \mathcal{Q}\mathcal{O}$, where variable $\mathcal{Q}$ is given by Eq. (1.131) and $\mathcal{O}$ is the OF variation. Again, integration on the spatial grid using Eq. (E.26) provides the CSH coefficients required in Eq. (E.78), namely

$$\mathcal{X}_{j,k} = \frac{1}{P}\sum_{p} \mathcal{Q}_p \mathcal{O}_{p,k} Y^*_{j,p} = \frac{1}{P}\sum_{p} \mathcal{Q}_p \left(O_{p,k} - O_{p,0}\right) Y^*_{j,p}\,, \tag{E.80}$$

where, using Eq. (1.131), we have defined

$$\mathcal{Q}_p = -\left(\frac{\rho^{\mathrm{i}}}{\rho^{\mathrm{r}}}\, I_{p,0} + \frac{\rho^{\mathrm{w}}}{\rho^{\mathrm{r}}}\, T_{p,0}\right). \tag{E.81}$$

E.3.4.2 Response Functions $\mathcal{R}^{\mathrm{i,s,o}}$, $\mathcal{U}^{\mathrm{i,s,o}}$, and $\mathcal{G}^{\mathrm{i,s,o}}$

Equations (3.36) give the CSH coefficients of the three SRFs $\mathcal{R}^{\mathrm{i}}$, $\mathcal{R}^{\mathrm{s}}$ and $\mathcal{R}^{\mathrm{o}}$, respectively. Setting $t = t_k$, where t_k is one of the time grid points, exploiting the property of the Heaviside step function, and following our conventions for notation, the SRFs harmonic coefficients are discretised as follows

$$\mathcal{R}^{\mathrm{i}}_{j,k} = \frac{3\rho^{\mathrm{i}}}{\rho^{\mathrm{e}}}\left(e^{s}_{n}\mathcal{W}_{j,k} + \sum_{k'=0}^{k-1}\Delta\mathcal{W}_{j,k'}\beta^{s}_{n,k-k'}\right) \tag{E.82}$$

$$\mathcal{R}^{\mathrm{s}}_{j,k} = \frac{3\rho^{\mathrm{w}}}{\rho^{\mathrm{e}}}\left(e^{s}_{n}\mathcal{Z}_{j,k} + \sum_{k'=0}^{k-1}\Delta\mathcal{Z}_{j,k'}\beta^{s}_{n,k-k'}\right) \tag{E.83}$$

$$\mathcal{R}^{\mathrm{o}}_{j,k} = \frac{3\rho^{\mathrm{r}}}{\rho^{\mathrm{e}}}\left(e^{s}_{n}\mathcal{X}_{j,k} + \sum_{k'=0}^{k-1}\Delta\mathcal{X}_{j,k'}\beta^{s}_{n,k-k'}\right), \tag{E.84}$$

where we have defined

$$\Delta\mathcal{W}_{j,k} = \mathcal{W}_{j,k+1} - \mathcal{W}_{j,k} \tag{E.85}$$

$$\Delta\mathcal{Z}_{j,k} = \mathcal{Z}_{j,k+1} - \mathcal{Z}_{j,k} \tag{E.86}$$

$$\Delta\mathcal{X}_{j,k} = \mathcal{X}_{j,k+1} - \mathcal{X}_{j,k}\,, \tag{E.87}$$

and

$$\beta^{s}_{n,k} \equiv \beta^{s}_{n}(k\Delta t)\,, \tag{E.88}$$

($k = 0, \ldots, N$), with $\beta^{s}_{n}(t)$ given by Eq. (3.30). We remark that since $\mathcal{W}_{j,0}$=$\mathcal{Z}_{j,0}$ =$\mathcal{X}_{j,0}$=0 and since the sums give no contributions for $k = 0$, we have $\mathcal{R}^{\mathrm{i,s,o}}_{j,0} = 0$.

We note that following the scheme above, discretised forms for the SRFs $\mathcal{U}^{\mathrm{sur}}$ (vertical displacement) and $\mathcal{G}^{\mathrm{sur}}$ (geoid height variation) are easily obtained. Indeed, from Eqs. (3.43) and (3.44) we have

$$\mathcal{G}^{\mathrm{sur}}_{j,k} = \mathcal{G}^{\mathrm{i}}_{j,k} + \mathcal{G}^{\mathrm{s}}_{j,k} + \mathcal{G}^{\mathrm{o}}_{j,k}\,, \tag{E.89}$$

with

$$\mathcal{G}^{\mathrm{i}}_{j,k} = \frac{3\rho^{\mathrm{i}}}{\rho^{\mathrm{e}}}\left(e^g_n \mathcal{W}_{j,k} + \sum_{k'=0}^{k-1} \Delta\mathcal{W}_{j,k'}\beta^g_{n,k-k'}\right) \tag{E.90}$$

$$\mathcal{G}^{\mathrm{s}}_{j,k} = \frac{3\rho^{\mathrm{w}}}{\rho^{\mathrm{e}}}\left(e^g_n \mathcal{Z}_{j,k} + \sum_{k'=0}^{k-1} \Delta\mathcal{Z}_{j,k'}\beta^g_{n,k-k'}\right) \tag{E.91}$$

$$\mathcal{G}^{\mathrm{o}}_{j,k} = \frac{3\rho^{\mathrm{r}}}{\rho^{\mathrm{e}}}\left(e^g_n \mathcal{X}_{j,k} + \sum_{k'=0}^{k-1} \Delta\mathcal{X}_{j,k'}\beta^g_{n,k-k'}\right), \tag{E.92}$$

where

$$\beta^g_{n,k} \equiv \beta^g_n(k\Delta t), \tag{E.93}$$

for any k, with $\beta^g_n(t)$ given by Eq. (3.47), and

$$\mathcal{U}^{\mathrm{sur}}_{j,k} = \mathcal{U}^{\mathrm{i}}_{j,k} + \mathcal{U}^{\mathrm{s}}_{j,k} + \mathcal{U}^{\mathrm{o}}_{j,k}, \tag{E.94}$$

with

$$\mathcal{U}^{\mathrm{i}}_{j,k} = \frac{3\rho^{\mathrm{i}}}{\rho^{\mathrm{e}}}\left(e^u_n \mathcal{W}_{j,k} + \sum_{k'=0}^{k-1} \Delta\mathcal{W}_{j,k'}\beta^u_{n,k-k'}\right) \tag{E.95}$$

$$\mathcal{U}^{\mathrm{s}}_{j,k} = \frac{3\rho^{\mathrm{w}}}{\rho^{\mathrm{e}}}\left(e^u_n \mathcal{Z}_{j,k} + \sum_{k'=0}^{k-1} \Delta\mathcal{Z}_{j,k'}\beta^u_{n,k-k'}\right) \tag{E.96}$$

$$\mathcal{U}^{\mathrm{o}}_{j,k} = \frac{3\rho^{\mathrm{r}}}{\rho^{\mathrm{e}}}\left(e^u_n \mathcal{X}_{j,k} + \sum_{k'=0}^{k-1} \Delta\mathcal{X}_{j,k'}\beta^u_{n,k-k'}\right), \tag{E.97}$$

where, for any value of index k,

$$\beta^u_{n,k} \equiv \beta^u_n(k\Delta t), \tag{E.98}$$

with $\beta^u_n(t)$ given by Eq. (3.48).

E.3.4.3 Ocean Averages of $\mathcal{R}^{\mathrm{i,s,o}}$

The (spatially invariant) ocean-averages of the three SRFs $\mathcal{R}^{\mathrm{i}}$, $\mathcal{R}^{\mathrm{s}}$, and $\mathcal{R}^{\mathrm{o}}$ are evaluated by a straightforward application of the definition of ocean-average. Hence, according to Eq. (1.143), we have

$$\langle\mathcal{R}^x\rangle^{\mathrm{o}}(t) = \frac{1}{A^{\mathrm{o}}(t)}\int_{\mathrm{o}} \mathcal{R}^x(\gamma, t)\mathrm{d}A, \qquad x = \mathrm{i, s, o} \tag{E.99}$$

then, taking advantage of the spectral decomposition given in section above for the SRF and for the OF, and since the OF is a real-valued function, we have

$$\langle \mathcal{R}^x \rangle^{\text{o}}(t) = \frac{a^2}{A^{\text{o}}(t)} \int_\gamma \mathcal{R}^x(\gamma, t) O^*(\gamma, t) \text{d}\gamma \tag{E.100}$$

$$= \frac{a^2}{A^{\text{o}}(t)} \int_\gamma \left(\sum_j \mathcal{R}_j^x(t) Y_j(\gamma) \right) \left(\sum_{j'} O_{j'}^*(t) Y_{j'}^*(\gamma) \right) \text{d}\gamma \tag{E.101}$$

$$= \frac{a^2}{A^{\text{o}}(t)} \sum_j \sum_{j'} \mathcal{R}_j^x(t) O_{j'}^*(t) \int_\gamma Y_j(\gamma) Y_{j'}^*(\gamma) \text{d}\gamma \tag{E.102}$$

$$= \frac{4\pi a^2}{A^{\text{o}}(t)} \sum_j \mathcal{R}_j^x(t)\ O_j^*(t)\,, \qquad x = \text{i, s, o} \tag{E.103}$$

where we have exploited the orthogonality of the CSHs. Hence, at a given time $t = t_k$, the ocean-averaged SRFs are

$$\langle \mathcal{R}^x \rangle_k^{\text{o}} \equiv \langle \mathcal{R}^x \rangle^{\text{o}}(t_k) = \frac{\sum_j \mathcal{R}_{j,k}\ O_{j,k}^*}{O_{00,k}}\,, \qquad x = \text{i, s, o} \tag{E.104}$$

where we have used (E.58).

E.3.4.4 Modified SRFs $\widehat{\mathcal{R}}^{\text{i,s,o}}$

According to (E.32), the modified SRFs are obtained by the SRFs simply by subtracting their ocean-averages. Hence, their discretisation is straightforward:

$$\widehat{\mathcal{R}}_{j,k}^x = \mathcal{R}_{j,k}^x - \langle \mathcal{R}^x \rangle_k^{\text{o}} \delta_{n0} \delta_{m0}\,, \qquad x = \text{i, s, o} \tag{E.105}$$

where $\langle \mathcal{R}^{\text{i,s,o}} \rangle_k^{\text{o}}$ is given by (E.104).

E.3.4.5 The Terms $\mathcal{K}^{\text{i,s,o}}$

The three terms $\mathcal{K}^{\text{i,s,o}}$ result from projecting the modified SRFs over the OF, i.e.,

$$\mathcal{K}^x(\gamma, t) \equiv O \widehat{\mathcal{R}}^x\,, \qquad x = \text{i, s, o}\,. \tag{E.106}$$

For them, we assume a CSH expansion

$$\mathcal{K}^x(\gamma, t) = \sum_j \mathcal{K}_j^x(t) Y_j(\gamma), \qquad x = \mathrm{i, s, o}, \tag{E.107}$$

where, according to (A.36) and (E.25), coefficients $\mathcal{K}_j^{\mathrm{i,s,o}}(t)$ are evaluated by a numerical quadrature on the grid:

$$\mathcal{K}_j^x(t) = \frac{1}{4\pi} \int_\gamma O(\gamma, t)\, \widehat{\mathcal{R}}^x(\gamma, t) Y_j^*(\gamma) \mathrm{d}\gamma \tag{E.108}$$

$$\approx \frac{1}{P} \sum_p O(\gamma_p, t)\, \widehat{\mathcal{R}}^x(\gamma_p, t) Y_j^*(\gamma_p) \tag{E.109}$$

$$= \frac{1}{P} \sum_p O(\gamma_p, t) \left(\sum_{j'} \widehat{\mathcal{R}}_{j'}^x(t) Y_{j'}(\gamma_p) \right) Y_j^*(\gamma_p) \tag{E.110}$$

$$= \frac{1}{P} \sum_{p:\, O(\gamma_p, t)=1} \left(\sum_{j'} \widehat{\mathcal{R}}_{j'}^x(t) Y_{j'}(\gamma_p) \right) Y_j^*(\gamma_p), \quad x = \mathrm{i, s, o} \tag{E.111}$$

hence, at time $t = t_k$, we have

$$\mathcal{K}_{j,k}^x \equiv \mathcal{K}_j^x(t_k) = \sum_{p:\, O_{p,k}=1} \chi_{p,k}^x\, Y_{j,p}^*, \quad x = \mathrm{i, s, o} \tag{E.112}$$

where we have defined the three arrays

$$\chi_{p,k}^{\mathrm{i,s,o}} \equiv \frac{1}{P} \sum_j \widehat{\mathcal{R}}_{j,k}^{\mathrm{i,s,o}} Y_{j,p}\,. \tag{E.113}$$

E.3.5 *Rotational Response*

For the RRF $\mathcal{R}^{\mathrm{rot}}$, the discretisation scheme largely follows the one adopted for the three SRFs terms $\mathcal{R}^{\mathrm{i,s,o}}$. Again, we consider separately the various terms depending on $\mathcal{R}^{\mathrm{rot}}$, which appear in the SLE.

E.3.5.1 CSH Expansion of $\mathcal{R}^{\mathrm{rot}}$

The RRF for sea level which appears in the SLE is given by Eq. (4.140). Since it is a pure harmonic function of degree $n = 2$ and order $m = \pm 1$, its discretisation is straightforward:

$$\mathcal{R}^{\text{rot}}_{j,k} = \delta_{n2}\delta_{m\pm1}\frac{c^{\psi}_{21}c^{\lambda}_{21}}{g}\sum_{k'=0}^{k-1}\Delta\mathcal{L}_{j,k'}\,\gamma^{s}_{k-k'}\,, \tag{E.114}$$

where

$$\Delta\mathcal{L}_{j,k} = \mathcal{L}_{j,k+1} - \mathcal{L}_{j,k}\,, \tag{E.115}$$

with $\mathcal{L}_{j,k}$ given by Eq. (E.75), and where for any index k, we have

$$\gamma^{s}_{k} = \gamma^{g}_{k} - \gamma^{u}_{k}\,, \tag{E.116}$$

with

$$\gamma^{g}_{k} = \gamma^{g}(k\Delta t) \tag{E.117}$$

$$\gamma^{u}_{k} = \gamma^{u}(k\Delta t)\,, \tag{E.118}$$

where $\gamma^{g}(t)$ and $\gamma^{u}(t)$ are given by (4.134) and (4.136), respectively.

E.3.5.2 Ocean-Average of $\mathcal{R}^{\text{rot}}$

The ocean-average of the RRF $\mathcal{R}^{\text{rot}}$ is evaluated following the same principles which have led to the evaluation of the ocean-averaged SRF. Indeed, using (E.104), we have immediately

$$\langle\mathcal{R}^{\text{rot}}\rangle^{\text{o}}_{k} \equiv \langle\mathcal{R}^{\text{rot}}\rangle^{\text{o}}(t_k) = \frac{\sum_j \mathcal{R}^{\text{rot}}_{j,k}\,O^{*}_{j,k}}{O_{00,k}}\,, \tag{E.119}$$

where $\mathcal{R}^{\text{rot}}_{j,k}$ is given by (E.114).

E.3.5.3 Modified RRF $\widehat{\mathcal{R}}^{\text{rot}}$

According to (E.33), the modified RRF is obtained by the corresponding RRFs by subtracting the ocean-averages; hence, its discretised form is:

$$\widehat{\mathcal{R}}^{\text{rot}}_{j,k} = \mathcal{R}^{\text{rot}}_{j,k} - \langle\mathcal{R}^{\text{rot}}\rangle_{k}\delta_{n0}\delta_{m0}, \tag{E.120}$$

where $\langle\mathcal{R}^{\text{rot}}\rangle_{k}$ is given by (E.119).

E.3.5.4 Term $\mathcal{K}^{\mathrm{rot}}$

This term of the SLE stems from the projection of the modified RRF $\widehat{\mathcal{R}}^{\mathrm{rot}}$ over the OF. In analogy with what we have done for the terms $\mathcal{K}^{\mathrm{i,s,o}}$, associated to surface loading, we write

$$\mathcal{K}^{\mathrm{rot}}_{j,k} \equiv \mathcal{K}^{\mathrm{rot}}_{j}(t_k) = \sum_{p:\, O_{p,k}=1} \chi^{\mathrm{rot}}_{p,k}\, Y^{*}_{j,p}\,, \tag{E.121}$$

where we have defined the array

$$\chi^{\mathrm{rot}}_{p,k} \equiv \frac{1}{P} \sum_{j} \widehat{\mathcal{R}}^{\mathrm{rot}}_{j,k} Y_{j,p}\,, \tag{E.122}$$

and where $\widehat{\mathcal{R}}^{\mathrm{rot}}_{j,k}$ is given by (E.120).

References

Abramowitz M, Stegun IA, Romer RH (1988) Handbook of mathematical functions with formulas, graphs, and mathematical tables. American Association of Physics Teachers

Agnew DC (2007) Earth tides. Treatise on geophysics

Alterman Z, Eyal Y, Merzer A (1974) On free oscillations of the Earth. Geophys Surv 1(4):409–428

Alterman Z, Jarosch H, Pekeris C (1959) Oscillations of the Earth. Proc R Soc Lond. Ser A. Math Phys Sci 252(1268):80–95

Amante C, Eakins B (2009) ETOPO1 arc-minute global relief model: procedures, data source and analysis. Technical Memorandum NESDIS NGDC; 24, National Oceanic and Atmospheric Administration

Atkinson K, Han W (2012) Spherical harmonics and approximations on the unit sphere: an introduction, vol 2044. Springer

Backus GE (1967) Converting vector and tensor equations to scalar equations in spherical coordinates. Geophys J Int 13(1–3):71–101

Backus G (1986) Poloidal and toroidal fields in geomagnetic field modeling. Rev Geophys 24(1):75–109

Backus G, George B, Parker RL, Parker R, Constable C (1996) Foundations of geomagnetism. Cambridge University Press

Barletta VR, Ferrari C, Diolaiuti G, Carnielli T, Sabadini R, Smiraglia C (2006) Glacier shrinkage and modeled uplift of the Alps. Geophys Res Lett 33(14)

Barrera RG, Estevez G, Giraldo J (1985) Vector spherical harmonics and their application to magnetostatics. Eur J Phys 6(4):287

Bloxham J, Jackson A (1991) Fluid flow near the surface of Earth's outer core. Rev Geophys 29(1):97–120

Bos M, Scherneck H-G (2012) Computation of Green's functions for ocean tide loading. In: Sciences of geodesy-II: innovations and future developments, pp 1–52. Springer

Brett L (1988) Methods of spherical harmonic analysis. Q J R Astron Soc 29:129

Brierley-Green C, James TS (2024) A hybrid hormal mode-collocation method for finding the response of laterally homogeneous compressible Maxwell viscoelastic Earth models. J Geophys Res Solid Earth 129(7):e2023JB027238

Cathles L (1975) The viscosity of the Earth's Mantle. Princeton University Press

Chinnery MA (1975) The static deformation of an Earth with a fluid core: a physical approach. Geophys J Int 42(2):461–475
Consorzi A, Melini D, González-Santander JL, Spada G (2024) On the love numbers of an andrade planet. Earth Space Sci 11(9):e2024EA003779
Copley L (2015) Mathematics for the physical sciences. De Gruyter Open Poland, Warsaw, Poland
Coster J, Hart HB (1991) A simple proof of the addition theorem for spherical harmonics. Am J Phys 59(4):371–373
Crossley DJ, Gubbins D (1975) Static deformation of the Earth's liquid core. Geophys Res Lett 2(1):1–4
Dahlen F (1971) Comments on Paper by D.E. Smylie and L. Mansinha: "The Elasticity Theory of Dislocations in Real Earth Models and Changes in the Rotation of the Earth". Geophys J Int 23(3):355–358
Dahlen F, Tromp J (1998) Theoretical global seismology. Princeton University Press
Dziewonski AM, Anderson DL (1981) Preliminary reference Earth model. Phys Earth Planet Inter 25(4):297–356
Eakins B, Sharman G (2012) Hypsographic curve of Earth's surface from ETOPO1. NOAA National Geophysical Data Center, Boulder CO
Farrell W (1972) Deformation of the Earth by surface loads. Rev Geophys 10(3):761–797
Farrell W, Clark J (1976) On postglacial sea level. Geophys J R Astron Soc 46(3):647–667
Ferrers NM (1877) An elementary treatise on spherical harmonics and subjects connected with them. Macmillan and Company
Fowler J, Ogle C, Bevis M (2019) An analytic method for computing the infinite sums occurring in the geoelastic disk load problem. J Geophys Res Solid Earth 124(2):2184–2201
Francioni L (2024) Tidal response and Love numbers of the moons of Uranus. Thesis, Department of Physics and Astronomy, University of Bologna
Fung Y (1965) Foundation of solid mechanics. Prentice-Hall, Englewood Cliffs, NJ
Gaver DP (1966) Observing stochastic processes, and approximate transform inversion. Oper Res 14(3):444–459
Gilbert F, Backus GE (1966) Propagator matrices in elastic wave and vibration problems. Geophysics 31(2):326–332
Gilbert F, Backus G (1968) Elastic-gravitational vibrations of a radially stratified sphere. In: Dynamics of stratified solids, pp 82–95. American Society of Mechanical Engineers, New York
Gradshteyn I, Ryzhik I, Romer RH (1988) Tables of integrals, series, and products. American Association of Physics Teachers
Han D, Wahr J (1989) Post-glacial rebound analysis for a rotating Earth. In: Cohen S, Vanicek P (eds) Slow deformations and transmission of stress in the Earth, Geophysical monograph series, vol 49, pp 1–6. American Geophysical Union
Hanyk L, Matyska C, Yuen D (1999) Secular gravitational instability of a compressible viscoelastic sphere. Geophys Res Lett 26(5):557–560
Haskell N (1935) The motion of a viscous fluid under a surface load. Physics 6(8):265–269
Heiskanen WA, Moritz H (1981) Physical geodesy. Institute of Physical Geodesy, Technical University
Hofmann-Wellenhof B, Moritz H (2006) Physical geodesy. Springer
Hurford T, Frey S, Greenberg R (2006) Numerical evaluation of Love's solution for tidal amplitude. In: Chaotic worlds: from order to disorder in gravitational n-body dynamical systems, pp 307–323. Springer
Hurford T, Greenberg R (2001) Tidal Amplitude for a self-gravitating, compressible sphere. In: AAS/division of dynamical astronomy meeting# 32, vol 32, pp 05–03
Jackson JD (1999) Classical electrodynamics. Wiley
James TS (1991) Postglacial deformation. PhD thesis, Princeton University
Jeffreys H (1952) The Earth. Cambridge University Press
Kaula W (1967) Theory of statistical analysis of data distributed over a sphere. Rev Geophys 5(1):83–107

Laden HN (1938) An historical and critical development of the theory of Legendre polynomials before 1900. PhD thesis, University of Maryland

Lamb H (1881a) On the oscillations of a viscous spheroid. Proc Lond Math Soc 1(1):51–70

Lamb H (1881b) On the vibrations of an elastic sphere. Proc Lond Math Soc 1(1):189–212

Lambeck K (1988) The Earth's variable rotation: some geophysical causes. In: Symposium-international astronomical union, vol 128, pp 1–20. Cambridge University Press

Lamé G (1854) Mémoire sur l'équilibre d'élasticité des enveloppes sphériques. Journal de Mathématiques Pures et Appliquées 51–87

Lanzano P (1982) Deformations of an elastic Earth. Academic Press, New York

Longman I (1962) A Green's function for determining the deformation of the Earth under surface mass loads: 1. Theory. J Geophys Res 67(2):845–850

Longman I (1963) A Green's function for determining the deformation of the Earth under surface mass loads. 2. Computations and numerical results. J Geophys Res 68:485–496

Longman IM (1975) On the stability of a spherical gravitating compressible liquid planet without spin. Geophys J Int 42(2):621–635

Longman I, Sharir M (1971) Laplace transform inversion of rational functions. Geophys J R Astron Soc 25(1–3):299–305

Love AEH (1911) Some problems of geodynamics: being an essay to which the Adams prize in the University of Cambridge was adjudged in 1911. CUP Archive

Lundberg KH, Miller HR, Trumper DL (2007) Initial conditions, generalized functions, and the Laplace transform troubles at the origin. IEEE Control Syst Mag 27(1):22–35

MacRobert TM (1927) Spherical harmonics: an elementary treatise on harmonic functions with applications. Dutton

Martens HR, Rivera L, Simons M (2019) Loaddef: a python-based toolkit to model elastic deformation caused by surface mass loading on spherically symmetric bodies. Earth Space Sci 6(2):311–323

Martinec Z, Klemann V, van der Wal W, Riva R, Spada G, Sun Y, Melini D, Kachuck S, Barletta V, Simon K (2018) A benchmark study of numerical implementations of the sea level equation in GIA modelling. Geophys J Int 215(1):389–414

Melini D, Cannelli V, Piersanti A, Spada G (2008) Post-seismic rebound of a spherical Earth: new insights from the application of the Post-Widder inversion formula. Geophys J Int 174(2):672–695

Melini D, Saliby C, Spada G (2022) On computing viscoelastic Love numbers for general planetary models: the ALMA^3 code. Geophys J Int 231(3):1502–1517

Milne GA, Mitrovica JX (1998) Postglacial sea-level change on a rotating Earth. Geophys J Int 133(1):1–19

Mitrovica JX, Peltier WR (1992) A comparison of methods for the inversion of viscoelastic relaxation spectra. Geophys J Int 108(2):410–414

Mitrovica JX, Wahr J (2011) Ice Age Earth rotation. Annu Rev Earth Planet Sci 39:577–616

Mitrovica JX, Wahr J, Matsuyama I, Paulson A (2005) The rotational stability of an ice-age earth. Geophys J Int 161(2):491–506

Oberhettinger F, Badii L (2012) Tables of Laplace transforms. Springer

Peltier WR (1989) Mantle convection: plate tectonics and global dynamics. CRC Press

Peltier WR (1974) The impulse response of a Maxwell Earth. Rev Geophys Space Phys 12(4):649–669

Piersanti A, Spada G, Sabadini R, Bonafede M (1995) Global post-seismic deformation. Geophys J Int 120(3):544–566

Piersanti A, Spada G, Sabadini R (1997) Global postseismic rebound of a viscoelastic Earth: theory for finite faults and application to the 1964 Alaska earthquake. J Geophys Res Solid Earth 102(B1):477–492

Plag H-P, Jüttner H-U (1995) Rayleigh-Taylor instabilities of a self-gravitating Earth. J Geodyn 20(3):267–288

Pollitz FF (1992) Postseismic relaxation theory on the spherical Earth. Bull Seismol Soc Am 82(1):422–453

Post EL (1930) Generalized differentiation. Trans Am Math Soc 32(4):723–781
Press WH (2007) Numerical recipes, 3rd edition: the art of scientific computing. Cambridge University Press
Ranalli G (1995) Rheology of the Earth. Chapman and Hall
Rayleigh L (1906) On the dilatational stability of the Earth. In: Proceedings of the Royal Society of London. Series A, containing papers of a mathematical and physical character, vol 77(519), pp 486–499
Reeger JA, Fornberg B (2016) Numerical quadrature over the surface of a sphere. Stud Appl Math 137(2):174–188
Sabadini R, Yuen DA, Boschi E (1982) Polar wandering and the forced responses of a rotating, multilayered, viscoelastic planet. J Geophys Res Solid Earth 87(B4):2885–2903
Sabadini R, Piersanti A, Spada G (1995) Toroidal/poloidal partitioning of global post-seismic deformation. Geophys Res Lett 22(8):985–988
Sabadini R, Vermeersen B, Cambiotti G (2016) Global dynamics of the Earth. Springer
Saito M (1974) Some problems of static deformation of the Earth. J Phys Earth 22(1):123–140
Schapery R (1962a) Approximate methods of transform inversion for viscoelastic stress analysis. In: 4th U.S. national congress on applied mechanics, University of California, Berkeley, United States
Schapery RA (1962b) Irreversible thermodynamics and variational principles with applications to viscoelasticity. PhD thesis, California Institute of Technology
Schey HM (2005) Div, grad, curl, and all that: an informal text on vector calculus. WW Norton New York, NY, USA
Shida T (1912) On the elasticity of the Earth and the Earth's crust. Kyoto Imperial University
Smylie D (1965) Magnetic diffusion in a spherically-symmetric conducting mantle. Geophys J Int 9(2–3):169–184
Smylie D, Mansinha L (1971a) The elasticity theory of dislocations in real Earth models and changes in the rotation of the Earth. Geophys J Int 23(3):329–354
Smylie DE, Mansinha L (1971b) Reply to comments on 'The Elasticity Theory of Dislocations in Real Earth Models and Changes in the Rotation of the Earth'. Geophys J Int 23(3):359–360
Spada G (2008) ALMA, a Fortran program for computing the viscoelastic Love numbers of a spherically symmetric planet. Comput Geosci 34(6):667–687
Spada G, Boschi L (2006) Using the Post-Widder formula to compute the Earth's viscoelastic Love numbers. Geophys J Int 166(1):309–321
Spada G, Galassi G (2015) Spectral analysis of sea level during the altimetry era, and evidence for GIA and glacial melting fingerprints. Glob Planet Change 143:34–49
Spada G, Stocchi P (2007) SELEN: a Fortran 90 program for solving the "Sea Level Equation." Comput Geosci 33(4):538–562
Spada G, Yuen D, Sabadini R, Morin P, Gasperini P (1990) A computer-aided, algebraic approach to the Post-Glacial Rebound problem. Math J 1(2):65–69
Spada G, Sabadini R, Yuen DA, Ricard Y (1992) Effects on post-glacial rebound from the hard rheology in the transition zone. Geophys J Int 109(3):683–700
Spada G, Barletta VR, Klemann V, Riva R, Martinec Z, Gasperini P, Lund B, Wolf D, Vermeersen L, King M (2011) A benchmark study for glacial isostatic adjustment codes. Geophys J Int 185(1):106–132
Spada G, Melini D (2015) SELEN: a program for solving the "Sea Level Equation" - Manual version 1.2, December 2015. Computational Infrastructure for Geodynamics (CIG). http://www.geodynamics.org/
Spada G, Melini D (2019) SELEN^4 (SELEN version 4.0): a Fortran program for solving the gravitationally and topographically self-consistent sea-level equation in glacial isostatic adjustment modeling. Geosci Model Dev 12(12):5055–5075
Spada G, Melini D, Galassi G, Colleoni F (2012) Modeling sea level changes and geodetic variations by glacial isostasy: the improved SELEN code. arXiv preprint, 1212.5061
Spada G, Stocchi P (2006) The sea level equation, theory and numerical examples. Aracne Editrice

Stehfest H (1970) Algorithm 368: numerical inversion of Laplace transforms [D5]. Commun ACM 13(1):47–49
Sternberg WJ, Smith TL (1944) The theory of potential and spherical harmonics. University of Toronto Press
Sun W, Sjöberg LE (1999) Gravitational potential changes of a spherically symmetric earth model caused by a surface load. Geophys J Int 137(2):449–468
Takeuchi H (1950) On the earth tide of the compressible earth of variable density and elasticity. Eos, Trans Am Geophys Union 31(5):651–689
Takeuchi H, Saito M, Kobayashi N (1962) Statical deformations and free oscillations of a model earth. J Geophys Res 67(3):1141–1154
Tegmark M (1996) An icosahedron-based method for pixelizing the celestial sphere. Astrophys J 470:L81
Tegmark M, de Oliveira-Costa A, Hamilton AJS (2003) High resolution foreground cleaned CMB map from WMAP. Phys Rev D 68(12)
Thomson W (1863) XXVII. On the rigidity of the Earth. Philos Trans R Soc Lond 153:573–582
Toepfer S, Narita Y, Glassmeier K-H, Heyner D, Kolhey P, Motschmann U, Langlais B (2021) The Mie representation for Mercury's magnetic field. Earth Planets Space 73(1):1–18
Valkó PP, Abate J (2004) Comparison of sequence accelerators for the Gaver method of numerical Laplace transform inversion. Comput Math Appl 48(3):629–636
Vermeersen L, Mitrovica J (2000) Gravitational stability of spherical self-gravitating relaxation models. Geophys J Int 142(2):351–360
Vermeersen L, Sabadini R (1997) A new class of stratified viscoelastic models by analytical techniques. Geophysical J Int 129(3):531–570
Vermeersen LLA, Sabadini R, Spada G, Vlaar NJ (1994) Mountain building and Earth rotation. Geophys J Int 117(3):610–624
Vermeersen LA, Sabadini R, Spada G (1996a) Analytical visco-elastic relaxation models. Geophys Res Lett 23(7):697–700
Vermeersen LLA, Sabadini R, Spada G (1996b) Compressible rotational deformation. Geophys J Int 126:735–761
Vigny C (1989) Geoide et dynamique interne de la Terre. PhD thesis, Paris 11
Widder DV (1934) The inversion of the Laplace integral and the related moment problem. Trans Am Math Soc 36(1):107–200
Wieczorek MA, Meschede M (2018) SHTools: tools for working with spherical harmonics. Geochem Geophys Geosyst 19(8):2574–2592
Winch D, Roberts P (1995) Derivatives of addition theorems for Legendre functions. Austral Math Soc Ser B 37:212–234
Wolf D (1994) Lamé's problem of gravitational viscoelasticity: the isochemical, incompressible planet. Geophys J Int 116(2):321–348
Wu P, Ni Z (1996) Some analytical solutions for the viscoelastic gravitational relaxation of a two-layer non-self-gravitating incompressible spherical Earth. Geophys J Int 126(2):413–436
Wu P, Peltier W (1982) Viscous gravitational relaxation. Geophys J Int 70(2):435–485

Index

G. Spada and D. Melini, *Glacial Isostatic Adjustment*, Springer Geophysics,
https://doi.org/10.1007/978-3-031-90538-4

R

S

T

U